Mathematics for the Technologies

Prentice-Hall Series in Technical Mathematics

Frank L. Juszli, *Editor*

Mathematics for the Technologies

Lawrence M. Clar
James A. Hart

Monroe Community College

Prentice-Hall, Inc., Englewood Cliffs, N.J. 07632

Library of Congress Cataloging in Publication Data

CLAR, LAWRENCE M   (date)
    Mathematics for the technologies.

    Includes index.
    1. Mathematics—1961–   I. Hart, James, A.
(date)   joint author. II. Title.
QA39.2.C54   510'.24'6   77-13786
ISBN 0-13-565200-6

© 1978 by Prentice-Hall, Inc., Englewood Cliffs, N.J. 07632
All rights reserved. No part of this book
may be reproduced in any form or
by any means without permission in writing
from the publisher.

Printed in the United States of America

10 9 8 7 6 5 4 3 2 1

PRENTICE-HALL INTERNATIONAL, INC., *London*
PRENTICE-HALL OF AUSTRALIA PTY. LIMITED, *Sydney*
PRENTICE-HALL OF CANADA, LTD., *Toronto*
PRENTICE-HALL OF INDIA PRIVATE LIMITED, *New Delhi*
PRENTICE-HALL OF JAPAN, INC., *Tokyo*
PRENTICE-HALL OF SOUTHEAST ASIA PTE. LTD., *Singapore*
WHITEHALL BOOKS LIMITED, *Wellington, New Zealand*

To our wives, Carol and Mary

Contents

Preface, xiii

1 Review of Numbers, Symbols, and Basic Laws, 1

1-1. The Number System, 1
1-2. Relations and Literal Symbols, 4
1-3. Absolute Value, 6
1-4. Basic Laws of Arithmetic and Algebra, 8
1-5. Operations with Zero, 11
1-6. Chapter Review, 13

2 Review of Basic Algebra, 15

2-1. Exponents and Radicals, 15
2-2. Scientific Notation, 23
2-3. Algebraic Expressions, 26
2-4. Factoring, 31
2-5. Fractions and Ratios, 36
2-6. Equations and Formulas Solved for One First Degree Variable, 45
2-7. Chapter Review, 56

3 Measurement and Dimensional Analysis, 60

3–1. The British System of Measurement, 60
3–2. The Metric System of Measurement, 64
3–3. Algebra of Dimensions, 69
3–4. Reductions and Conversions, 71
3–5. Equations Involving Dimensions, 74
3–6. Chapter Review, 76

4 Functions, Rectangular Coordinates, and Graphs, 78

4–1. Functions, 78
4–2. Functional Notation and Terminology, 82
4–3. The Rectangular Coordinate System, 86
4–4. Graphing Techniques and Graphs, 90
4–5. Graphs of Empirical Data, 97
4–6. Solving an Equation Graphically, 103
4–7. Chapter Review, 107

5 Descriptive Statistics, 112

5–1. Averages for a Group of Data, 112
5–2. Standard Deviation, 115
5–3. Frequency Distributions, 120
5–4. Frequency Curves, 124
5–5. Chapter Review, 128

6 Linear Equations with More Than One Variable, 130

6–1. Linear Equations and Their Graphs, 130
6–2. Slopes and Intercepts, 135
6–3. Systems of Linear Equations, 144
6–4. Solving Systems of Two Linear Equations in Two Unknowns Graphically, 146
6–5. Solving Systems of Two Linear Equations in Two Unknowns Algebraically, 152
6–6. Solving Systems of Two Linear Equations in Two Unknowns by Determinants, 159
6–7. Solving Systems of Three Linear Equations in Three Unknowns Algebraically, 164
6–8. Solving Systems of Three Linear Equations in Three Unknowns by Determinants, 170
6–9. Chapter Review, 177

7 Introduction to Trigonometry, 181

7–1. Angles and Their Measure, 181
7–2. Introduction to Triangles, 190
7–3. Angles in the Rectangular Coordinate System, 194
7–4. The Trigonometric Functions, 197
7–5. Functions of Acute Angles, 201
7–6. Problems Involving Right Triangles, 205
7–7. Functions of Angles That Are Not Acute, 212
7–8. Applications of Radian Measure, 219
7–9. Chapter Review, 223

8 Vectors and Oblique Triangles, 226

8–1. Introduction to Vectors, 226
8–2. Problems Involving Vectors, 236
8–3. The Law of Sines, 240
8–4. The Law of Cosines, 246
8–5. Chapter Review, 250

9 Graphs of the Trigonometric Functions, 254

9–1. Graphs of the Trigonometric Functions, 254
9–2. Amplitude, Period, and Displacement, 260
9–3. Composite Trigonometric Functions, 270
9–4. Applications of the Graphs of the Trigonometric Functions, 273
9–5. Chapter Review, 281

10 Trigonometric Identities, Equations, and Inverse Relations, 283

10–1. Additional Angle Formulas, 283
10–2. Trigonometric Identities, 289
10–3. Basic Trigonometric Equations, 294
10–4. The Inverse Trigonometric Relations, 299
10–5. The Inverse Trigonometric Functions and Their Graphs, 303
10–6. Chapter Review, 308

11 Imaginary and Complex Numbers, 311

11-1. Imaginary Numbers, 311
11-2. Complex Numbers, 313
11-3. Polar Form of a Complex Number, 315
11-4. Operations with Complex Numbers, 320
11-5. An Application to Vectors, 328
11-6. An Application to AC Circuits, 330
11-7. Chapter Review, 336

12 Logarithms, 338

12-1. Definition of a Logarithm, 338
12-2. Basic Exponential and Logarithmic Graphs, 340
12-3. Properties of Logarithms, 347
12-4. Common Logarithms, 351
12-5. Computations with Logarithms, 354
12-6. Natural Logarithms, 357
12-7. Basic Exponential and Logarithmic Equations, 360
12-8. Graphs on Logarithmic and Semilogarithmic Graph Paper, 363
12-9. Chapter Review, 377

13 Quadratic Equations, 380

13-1. The Quadratic Equation, 380
13-2. Solving Quadratic Equations by Factoring, 383
13-3. Completing the Square, 385
13-4. The Quadratic Formula, 389
13-5. The Roots of a Quadratic Equation, 392
13-6. The Graph of a Quadratic Equation, 396
13-7. Quadratic Equations Involving Trigonometric Functions, 401
13-8. Equations Solved by Quadratic Methods, 404
13-9. Chapter Review, 408

14 Systems of Equations, 410

14-1. Systems of Equations, 410
14-2. Graphical Solution of Systems of Equations, 412
14-3. Algebraic Solution of Systems of Equations, 418
14-4. Chapter Review, 422

15 Equations of Degree Greater Than Two, 424

15-1. The Remainder and Factor Theorems, 424
15-2. Synthetic Division, 427
15-3. Number and Nature of the Roots of a Polynomial Equation, 431
15-4. Rational Roots of a Polynomial Equation, 435
15-5. Irrational Roots of a Polynomial Equation, 441
15-6. Chapter Review, 444

16 Inequalities, 447

16-1. Definition and Properties of Inequalities, 447
16-2. Solution of Basic Inequalities, 449
16-3. Solution of Higher Degree Inequalities, 452
16-4. Systems of Inequalities, 457
16-5. Inequalities Involving Absolute Values, 459
16-6. Chapter Review, 463

17 Variation, 465

17-1. Definition and Examples of Variation, 465
17-2. Direct Variation, 466
17-3. Inverse Variation, 470
17-4. Joint Variation, 475
17-5. Chapter Review, 477

18 Plane Analytic Geometry, 480

18-1. Basic Definitions and Terminology, 480
18-2. The Distance Formula, 481
18-3. The Straight Line, 485
18-4. The Parabola, 492
18-5. The Circle, 499
18-6. The Ellipse, 504
18-7. The Hyperbola, 510
18-8. The General Second Degree Equation, 519
18-9. Chapter Review, 522

Appendix A: Geometric Figures and Formulas, 525

Appendix B: Operations with Hand Calculators, 531

    B–1. Exact and Approximate Numbers, 531
    B–2. Basic Operations, 535
    B–3. Scientific Notation, 537
    B–4. Powers and Roots, 539
    B–5. Trigonometric Functions, 544
    B–6. Logarithms, 547
    B–7. Combined Operations, 549
    B–8. Formulas, 551

Tables, 556

    1. Powers and Roots, 557
    2. Trigonometric Functions, 558
    3. Common Logarithms of Numbers, 563
    4. Natural Logarithms of Numbers, 565
    5. Values of e^x and e^{-x}, 566
    6. Trigonometric Formulas, 567

Answers to Odd-Numbered Problems, 569

Index, 631

Preface

The purpose of this book is to provide a mathematics text that is logical, readable, and usable for students in technical, scientific, and allied fields. The approach is nonrigorous, since students in these fields need discussion and development of a topic followed by numerous examples, not excessive theory. Several examples of each major idea are included in every section throughout the text and a large number of applied problems from the various sciences and technologies are included to show students how the various mathematical concepts may be applied. We also include a set of review problems at the end of each chapter. To stress ideas and methods, rather than specific cases, variables other than just X and Y are used throughout the text.

In Chapters 1 and 2 we give a complete review of the basic arithmetic and algebra needed for the remainder of the text. We assume that the students using this book will have some background in algebra and geometry and after a thorough study of these first chapters, the student will easily progress through the remainder of the text. An extensive description and explanation of the metric system, combined with a thorough study of dimensions and dimensional analysis, is given in Chapter 3. Both British and metric units are then used throughout the text. In Chapter 4 we give a thorough explanation of graphing techniques. Extensive work with graphs is then integrated throughout the text. A complete section is devoted to the graphs of empirical data and the writing of equations from a table of values. Also included in this chapter is a basic but thorough description of functions. A discussion of descriptive statistics (Chapter 5) is included early to aid students in their lab work. In the initial discussion of linear equations in Chapter 6, we cover the concepts of slope and intercept.

All the trigonometric information is contained in four successive chapters, 7 through 10. Thus, there is no need to review previous information before beginning a new topic. The organization of these chapters also emphasizes the continuity between the various areas of trigonometry. It is not necessary, however, to cover all of this material at once, as there are natural stopping points at the end of each chapter. Included in the study of trigonometry is a complete description of angles and triangles, and some basic geometry involving these concepts. An in-depth look at the roots, graphs, and solutions of quadratic equations, and a discussion of equations solved by quadratic methods are included in Chapter 13. We cover basic analytic geometry completely in Chapter 18, deriving the equations for each of the conic sections using the distance formula.

The reference section in the text includes a complete summary of geometric figures and formulas in Appendix A. Discussions of exact and approximate numbers and hand calculators are included in Appendix B so that they may be introduced when needed by the instructor. We also describe special iterative processes for finding square and cube roots for use with even a very basic hand calculator.

We would like to express our gratitude to everyone who assisted us in any way in the production of this book: our colleagues in the mathematics department at Monroe Community College, many of whom offered suggestions and provided needed constructive criticism; the reviewers of the original manuscript who did an excellent and conscientious job; the staff at Prentice-Hall for their faith in us and excellent work in producing and promoting this text; and most importantly, our wives, who provided us with encouragement and displayed outstanding patience throughout the project, and who typed the complete manuscript.

Rochester, New York
LAWRENCE M. CLAR
JAMES A. HART

Mathematics for the Technologies

chapter 1

Review of Numbers, Symbols, and Basic Laws

1-1. The Number System

As was pointed out in the preface, the purpose of this text is to provide the mathematical concepts and information necessary for solving basic scientific, engineering, and technical problems. Before one tries to grasp these concepts, however, it is of the utmost importance that he have a solid, working knowledge of basic arithmetic and algebraic operations. Although the material in this chapter and in Chapter 2 is not to be considered a comprehensive initial study of such ideas, but rather a review, a thorough understanding of the concepts presented will enable a person to progress smoothly through the rest of the text.

To be successful in any mathematics course, a student must have an understanding of our number system and be able to identify correctly and work with the different types of numbers encountered. Therefore, we begin in this first section by giving a brief explanation of our number system.

The first numbers used represented whole quantities, and they were used for counting. They are called *counting numbers* or *positive integers* and are represented by the symbols 1, 2, 3, 4, 5, ... These numbers can be classified as *prime* or *composite*. A prime number is divisible by 1 and itself only. If a positive integer is not prime it is composite.

Example A:
1. Numbers like 2, 3, 5, 7, 11, 13, 17, 19, 23, and 29 are prime since they are divisible only by 1 and themselves.
2. Numbers like 4, 6, 8, 9, 10, 12, 14, 15, 16, 18, and 20 are composite since they are divisible by numbers other than 1 and themselves.

NOTE: The number 1 is neither prime nor composite.

In order to be able to perform different operations with positive integers, other numbers had to be introduced. To represent answers to such problems as $(4-4)$, $(13-13)$, $(2-5)$, and $(26-31)$, *zero* and *negative integers* are used. They are represented by the symbols $0, -1, -2, -3, -4, -5, \ldots$. To represent answers to such problems as $(2 \div 3)$, $(4 \div 11)$, $(7 \div 4)$, and $(-8 \div 5)$, and to work with parts of whole quantities, *fractions* are used. They are written as $\frac{2}{3}, \frac{4}{11}, \frac{7}{4}, -\frac{8}{5}, \ldots$. Integers can also be written this way since, for example, $7 = \frac{7}{1}$ and $-9 = -\frac{9}{1}$. Therefore, we see that all integers and divisions of integers can be expressed as the quotient of two integers. Any number that can be expressed this way is called a *rational* number. If a number cannot be expressed as the ratio of two integers, but it is a real number, it is called *irrational*.

Example B:
1. Numbers such as 2 (which can be written as $\frac{2}{1}$), $\frac{7}{9}$, $\frac{14}{3}$, $-\frac{2}{5}$, $-\frac{9}{4}$, and 0 (which can be written as $\frac{0}{1}$) are rational numbers.
2. Numbers such as $\sqrt{3}, \frac{\sqrt{2}}{5}, \frac{7}{\sqrt{6}}$, and π (approximately 3.14) are irrational numbers.

NOTES:
1. The ratio of any number to 0 (e.g., $\frac{2}{0}, \frac{3}{0}, -\frac{5}{0}$) is not allowed, as will be explained in Section 1–5.
2. For a ratio of a positive integer and a negative integer, the minus sign may be placed in the numerator, in the denominator, or preceding the fraction $\left[\text{e.g.,} \frac{-1}{2} = \frac{1}{-2} = -\left(\frac{1}{2}\right)\right]$.

The rational and irrational numbers together constitute the *real* numbers. These are the numbers we will use throughout most of the text. Notable exceptions are Chapter 11 and the solutions to certain equations, such as $X^2 + 1 = 0$. Real numbers may be represented as points on a line. Although it is not necessary, a horizontal line is usually used. Some point is designated as *O* (called the *origin* or *starting point*). Then, equal intervals are marked off from this point toward the right, representing the positive integers, and toward the left, representing the negative integers. The remaining rational and irrational numbers are located between these positions. It can be proven in more advanced courses

Sec. 1–1. The Number System

that every real number can be represented by a point on this line and every point on the line corresponds to exactly one real number. For this reason, this line is usually called the *real number line* (see Figure 1–1.1).

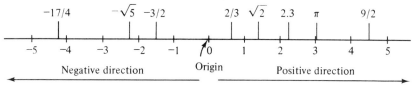

FIGURE 1-1.1

NOTES:
1. Numbers increase in value as we move toward the right on the number line.
2. When dealing with negative numbers, the larger the numerical value is, the smaller the number is (e.g., -6 is less than -2).

In addition to real numbers, we sometimes encounter *imaginary* numbers, which is the name given to even roots of negative numbers.

Example C:
1. Numbers such as 9, $\frac{2}{3}$, $-\frac{7}{4}$, $\sqrt{7}$, π, $\sqrt[3]{-8}$, and $2\sqrt{11}$ are real numbers.
2. Numbers such as $\sqrt{-1}$, $\sqrt{-4}$, $\sqrt{-7}$, $\sqrt[4]{-9}$, and $\sqrt[6]{-10}$ are imaginary numbers.

The real and imaginary numbers together make up the *complex* numbers, which is the term used to describe all numbers that we might encounter. Imaginary and complex numbers are described in detail in Chapter 11.

Figure 1–1.2 summarizes the relationships between the various types of numbers we have just mentioned.

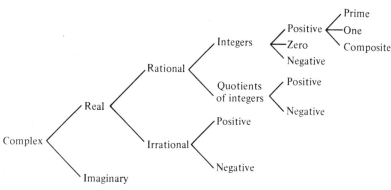

FIGURE 1-1.2

Exercises

Determine whether each of the following is prime or composite.

1. 21
2. 24
3. 37
4. 51
5. 67
6. 69
7. 87
8. 91
9. 103
10. 249
11. 323
12. 417

Determine whether each of the following is rational or irrational.

13. 6
14. $\frac{7}{5}$
15. -3
16. $\frac{\pi}{2}$
17. $\frac{-14}{9}$
18. $\sqrt{5}$
19. 4.1
20. -2.6
21. $2\sqrt{2}$
22. $5\sqrt{7}$
23. $-\sqrt{10}$
24. $\frac{-19}{17}$

Determine whether each of the following is real or imaginary.

25. $\sqrt{-2}$
26. $\frac{\pi}{3}$
27. $\frac{17}{5}$
28. $-\sqrt{2}$
29. -14
30. $\sqrt[3]{10}$
31. $\frac{-9}{11}$
32. $\sqrt{-6}$
33. $\sqrt[4]{-8}$
34. $\sqrt[3]{-9}$
35. $\frac{-\pi}{4}$
36. $-\sqrt{13}$

Plot each of the following on a real number line.

37. 8
38. -3
39. $\frac{4}{3}$
40. $\sqrt{7}$
41. $-\sqrt{2}$
42. $-\pi$
43. 2.9
44. -7.1
45. $-\sqrt{11}$
46. $\frac{5}{2}$
47. $\frac{-9}{4}$
48. $\frac{\pi}{2}$

1–2. Relations and Literal Symbols

In mathematics, relationships between quantities or expressions are explained through the use of certain symbols. In this section we mention the symbols for two basic mathematical relationships, *equality* and *inequality*.

If two quantities are equal, we express this fact by using the symbol = (equals).

Sec. 1–2. Relations and Literal Symbols

Example A: $4 = 4$, $5 = 2 + 3$, $X = 3$, $T^2 = 9$.

If two quantities are unequal, we express this fact by using the symbols $<$ (less than) or $>$ (greater than).

Example B: $4 < 7$, $0 < 3$, $-7 < -2$, $5 > 0$, $6 > -8$, $0 > -3$, $-4 > -9$, $X < 2$, $T > -1$.

Equalities and inequalities can be represented on a number line to give us a better understanding of exactly what we mean when we use any of these symbols.

Example C:
  1. If $A = 4$, then we would represent this on a number line as

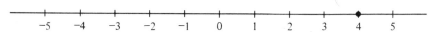

  2. If $T = -\frac{5}{2}$, then we would represent this on a number line as

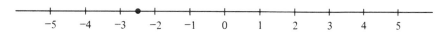

Example D:
  1. If $B < 1$, then we would represent this on a number line as

  2. If $R > -3$, then we would represent this on a number line as

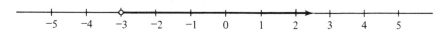

Notice that basic inequalities are *unbounded* in one direction and on a number line the arrow representing the solution extends indefinitely in that direction.

From the previous examples, we can see that sometimes we are interested not in specific numbers but in certain groups or categories of numbers. In such cases letters, called *literal symbols*, are used to represent numbers.

Example E:
  1. If we are talking about all numbers that are less than 3, we represent this as $X < 3$.
  2. If we are talking about all positive numbers, we represent this as $P > 0$.

When letters are used to represent numbers, they may be assigned a specific value and called *constants*, or they may be assigned different values and called *variables*.

Exercises

For each of the following, replace the word *and* with the appropriate symbol: $=$, $<$, or $>$.

1. 5 and -5
2. 2 and $5-3$
3. $\frac{2}{3}$ and 1.6
4. π and 2
5. -7 and -10
6. $\frac{-8}{5}$ and 0
7. 0 and $-\pi$
8. $\sqrt{5}$ and 4
9. -12 and -7
10. -3 and 3
11. $\frac{7}{9}$ and $\frac{\pi}{2}$
12. -3 and $7-10$

Represent each of the following on a number line.

13. $A = -5$
14. $T = 3\frac{1}{2}$
15. $R = -4.1$
16. $G = \frac{7}{11}$
17. $X < 6$
18. $S < 0$
19. $B < \frac{1}{2}$
20. $Y < -2$
21. $K > 3$
22. $L > \frac{4}{5}$
23. $M > 0$
24. $P > -4$

1-3. Absolute Value

An important mathematical concept associated with real numbers is *absolute value*. By the absolute value of a real number we mean the distance (disregarding direction) of that number from 0 (the origin on a number line). Therefore, since the numbers 2 and -2 are both 2 units from 0, the absolute value of both numbers is 2. The symbol used to denote absolute value is $|\ |$. Therefore, $|2| = |-2| = 2$.

Example A: $|7| = 7$, $|\frac{7}{8}| = \frac{7}{8}$, $|-5| = 5$, $|0| = 0$, $|-\sqrt{2}| = \sqrt{2}$.

Examining the definition of absolute value, we should realize that, if we are given the absolute value of a number and are asked to find the number, there will be two answers. That is, if $|A| = 2$, then $A = 2$ or $A = -2$. On a number line (Figure 1–3.1), this would be represented as two points, one at 2 and one at -2.

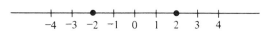

FIGURE 1–3.1

Sec. 1–3. Absolute Value

Example B:

1. If $|X| = 8$, then $X = 8$ or $X = -8$.

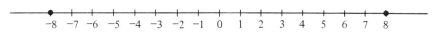

2. If $|T| = \frac{2}{3}$, then $T = \frac{2}{3}$ or $T = -\frac{2}{3}$.

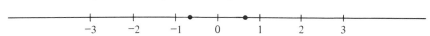

3. If $|R| = \sqrt{7}$, then $R = \sqrt{7}$ or $R = -\sqrt{7}$.

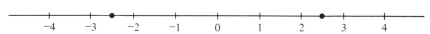

If the concept of absolute value is combined with either of the symbols $<$ or $>$, the result will no longer be a pair of numbers but a whole *set* of numbers. For example, if $|B| < 2$, we are looking for all numbers (regardless of direction) that are less than 2 units from 0. Therefore, B could equal any number between 2 and -2 and satisfy this condition. This solution is represented on a number line (Figure 1–3.2) and we clearly see that all of the numbers indicated

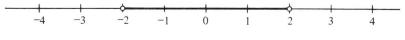

FIGURE 1–3.2

have an absolute value less than 2. If $|B| > 2$, then we are looking for all numbers (regardless of direction) that are more than 2 units from 0. Therefore, B could equal any number greater than 2 or less than -2 and satisfy this condition. This solution is represented on a number line (Figure 1–3.3) and we clearly see that all of the numbers indicated have an absolute value greater than 2.

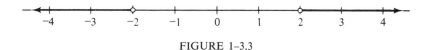

FIGURE 1–3.3

Example C:

1. If $|X| < 4$, then X is between 4 and -4.

2. If $|P| > 1$, then P is greater than 1 or P is less than -1.

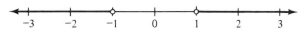

Exercises   Evaluate each of the following:

1. $|9|$
2. $|-8|$
3. $\left|\frac{4}{3}\right|$
4. $\left|\frac{-7}{9}\right|$
5. $|-2.7|$
6. $\left|\frac{\pi}{2}\right|$
7. $|-2\pi|$
8. $|\sqrt{5}|$
9. $|-3\sqrt{10}|$

Solve each of the following and represent the solution on a number line.

10. $|G| = 1$
11. $|A| = 6$
12. $|Y| = \frac{12}{19}$
13. $|P| = \sqrt{11}$
14. $|B| < 1$
15. $|X| > 3$
16. $|R| < \frac{3}{2}$
17. $|T| < \frac{14}{5}$
18. $|K| > \frac{5}{2}$
19. $|S| > 0$

1-4. Basic Laws of Arithmetic and Algebra

The general rules that apply to arithmetic of real numbers are the same as those that apply to basic algebraic operations. This is because *algebra is arithmetic* except that some numbers are not known and symbols are used in place of these numbers. These general rules are stated below and we shall accept them as valid statements without proofs.

The following table summarizes the properties that apply to the relation of equality between real numbers. A, B, and C represent real numbers.

Property	Symbolic Representation	Example
Reflexive	$A = A$	Any quantity is equal to itself.
Symmetric	If $A = B$, then $B = A$	If $X = 2$, then $2 = X$
Transitive	If $A = B$ and $B = C$, then $A = C$	If $R = S$ and $S = 3$, then $R = 3$

The following rules apply to operations with signed numbers.

1. Addition: If both numbers have the same sign, find the sum of their absolute values, and affix their common sign to the sum.

Example A:
  1. $2 + 6 = 8$
  2. $-2 + (-6) = -8$

If the numbers have different signs, find the difference between their numerical values and affix the sign of the number with the larger numerical value to the sum.

Sec. 1-4. Basic Laws of Arithmetic and Algebra

Example B:
1. $8 + (-2) = 6$
2. $-8 + 2 = -6$

2. Subtraction: Restate the problem as an addition problem by adding the opposite or negative of the number that follows the subtraction sign and proceed as in addition.

Example C:
1. $9 - 2 = 9 + (-2) = 7$
2. $-9 - 2 = -9 + (-2) = -11$
3. $9 - (-2) = 9 + 2 = 11$

3. Multiplication: If the signs of the quantities are the same, multiply the quantities and affix a positive sign to the product.

Example D:
1. $4 \times 3 = 12$
2. $(-4) \times (-3) = 12$

If the signs of the quantities are different, multiply the quantities and affix a negative sign to the product.

Example E:
1. $(-4) \times (3) = -12$
2. $(4) \times (-3) = -12$

4. Division: The rules that were used in determining the signs in multiplication apply when determining the algebraic sign of the quotient. That is, like signs yield a positive result and unlike signs yield a negative result.

Example F:
1. $\frac{8}{2} = 4$
2. $\frac{-8}{-2} = 4$
3. $\frac{-8}{2} = -4$
4. $\frac{8}{-2} = -4$

The following table summarizes the properties that apply to the operation of addition.

Property	Symbolic Representation	Examples
Identity Element	$A + 0 = A$	$2 + 0 = 2$ $4X + 0 = 4X$
Additive Inverse	$A + (-A) = 0$	$3 + (-3) = 0$ $2R + (-2R) = 0$
Commutative	$A + B = B + A$	$4 + 3 = 3 + 4$ $P + 7 = 7 + P$
Associative	$(A + B) + C = A + (B + C)$	$(2 + 3) + 8 = 2 + (3 + 8)$ $(X + 4) + 7 = X + (4 + 7)$

The following table summarizes the properties that apply to the operation of multiplication.

Property	Symbolic Representation	Examples
Identity Element	$A \times 1 = A$	$3 \times 1 = 3$ $4T \times 1 = 4T$
Reciprocal	$A \times \frac{1}{A} = 1 \; (A \neq 0)$	$2 \times \frac{1}{2} = 1$ $B \times \frac{1}{B} = 1 \quad (B \neq 0)$
Commutative	$A \times B = B \times A$	$2 \times 7 = 7 \times 2$ $3 \times (-2W) = (-2W) \times 3$
Associative	$(A \times B) \times C = A \times (B \times C)$	$(4 \times 5) \times 6 = 4 \times (5 \times 6)$ $(X \times Y) \times Z = X \times (Y \times Z)$

The distributive property of multiplication over addition is:

$$A \times (B + C) = (A \times B) + (A \times C)$$

Example G:
1. $2 \times (6 + 4) = (2 \times 6) + (2 \times 4)$
2. $-3 \times (X + 4) = -3(X) + (-3)4 = -3X - 12$
3. $-(A + B) = -A + (-B) = -A - B$

Normally, when more than one operation is involved, parentheses and brackets are used to indicate the order in which the operations are to be performed. If such order is not indicated, *multiplication and division should be performed before addition and subtraction.*

Example H:
1. Evaluate $(2 + 5) \times (3 + 8)$. Here the parentheses indicate that we should first find the two sums and then multiply the results. Therefore, $(2 + 5) \times (3 + 8) = 7 \times 11 = 77$.

Sec. 1–5. Operations with Zero   11

2. Evaluate $2 + 5 \times 3 + 8$. Here we have no grouping symbols and thus, we must first multiply and then add. Therefore, $2 + 5 \times 3 + 8 = 2 + 15 + 8 = 25$.

Exercises

What is the additive inverse of each of the following?
1. 23
2. -7
3. $\frac{2}{3}$
4. $-3\frac{1}{7}$
5. $\sqrt{3}$
6. $-\pi$

What is the reciprocal of each of the following?
7. 8
8. -12
9. $\frac{1}{7}$
10. $-\frac{2}{3}$
11. $-\sqrt{2}$
12. 0

Apply the distributive law to eliminate the parentheses in each of the following.
13. $4(M + 6)$
14. $2(5T - 3)$
15. $-3(X + 7)$
16. $-(A - 2B)$

Perform the indicated operations.
17. $17 + (-4)$
18. $(-3)13$
19. $14(-.68)$
20. $\frac{11 - 7}{-2}$
21. $\frac{2(-3) - 4(7)}{-17}$
22. $(14)(-2)(1)(-5)$
23. $\frac{4}{(2)(3)(-7)}$
24. $\frac{5 + 7(-6) - 2(-3)}{(-4)8 - 12(-2)}$
25. $5 \times 3 + 2 \times .25 - .5$
26. $6 + 6 \div 3 \times 7 - 12$

1–5. Operations with Zero

Since the basic operations with zero are frequently encountered, let us now consider each of these operations. Let A, B, and C be real numbers.

1. Addition: $A + 0 = 0 + A = A$. If you have $1,000 in the bank and you add no dollars to your account, your balance would remain the same.

Example A:
1. $2 + 0 = 2$
2. $7X + 0 = 7X$
3. $0 + 5T = 5T$
4. $0 - 3 = 0 + (-3) = -3$
5. $0 - 4B = 0 + (-4B) = -4B$

2. **Subtraction**: $A - 0 = A$. If your account has $1,000 in it and you withdraw no dollars, your balance remains the same.

Example B:
    1. $5 - 0 = 5$
    2. $9T - 0 = 9T$

3. **Multiplication**: $A \times 0 = 0 \times A = 0$. By definition, the meaning of $2A$ is $A + A$. That is, A is added two times. Therefore, $0 \times A$ means A is added zero times and the result is 0.

Example C:
    1. $6 \times 0 = 0$
    2. $7Y \times 0 = 0$
    3. $(A + B) \times 0 = 0$
    4. $0 \times 12 = 0$
    5. $0 \times (X + Y) = 0$

4. **Division**: By definition, C is the quotient of $\frac{A}{B}$ only if $A = B \times C$ where C is a unique (single) real number.

Example D: $\frac{8}{4} = 2$ only if $8 = 4 \times 2$.

1. Consider $\frac{A}{0} = C$, where $A \neq 0$. Then $A = 0 \times C$. We know from property (3) that a product must be zero if one of its multipliers is zero. But by hypothesis the product A is not equal to zero. Therefore, $\frac{A}{0}$ has no answer.

2. Consider $\frac{A}{0} = C$, where $A = 0$. Then $0 = 0 \times C$. In this case C could be any real number because by property (3) any real number times zero is equal to zero. This would not give a unique value to C, which is necessary according to our definition. Therefore, $\frac{0}{0}$ has no answer.

Consider another way of looking at division by zero. A basketball player's shooting average is calculated by dividing the number of baskets he makes by the number of times he shoots. For example, if he takes 20 shots and makes 8, his shooting average is $\frac{8}{20} = .40 = 40\%$. If he has 0 successful baskets and he has attempted 0 shots, it is impossible to give him an average. Likewise, if he has 20 successful baskets and he has attempted 0 shots, it is impossible to give

Sec. 1-6. Chapter Review 13

him an average. Note that this is different from a player taking 20 shots and making none. Thus, division by zero is undefined.

Example E: $\dfrac{0}{0}, \dfrac{2}{0}, -\dfrac{7}{0}, \dfrac{T}{0},$ and $\dfrac{X^2}{0}$ are all undefined.

Exercises Perform the indicated operations involving zero.

1. $-4 + 0$
2. $0 + 7$
3. $-9 - 0$
4. $0 - 10$
5. 0×0
6. 5×0
7. $0 \div \pi$
8. $\sqrt{2} \div 0$
9. $-17 \times 0 \times \dfrac{2}{3}$
10. $\dfrac{4 \times (5 - 5)}{5 - 5}$
11. $\dfrac{\sqrt{3} \times 13}{0 \times \pi}$
12. (a) Is $\dfrac{T}{T}$ always equal to 1? Why?
    (b) Is $\dfrac{M - 3}{M - 3}$ always equal to 1? Why?

1-6. CHAPTER REVIEW

State whether each of the following is real or imaginary. If real, state whether it is rational or irrational.

1. $\dfrac{-17}{13}$
2. $2\sqrt{3}$
3. -12.3
4. $\dfrac{0}{7}$
5. $-\sqrt{5}$
6. $\sqrt{-5}$
7. $\dfrac{\pi}{6}$
8. $\sqrt{\dfrac{19}{4}}$
9. -51
10. $\sqrt{-4}$
11. $-\sqrt[3]{-8}$
12. $\sqrt[3]{-27}$

Represent each of the following on a real number line.

13. $T = \dfrac{17}{5}$
14. $R = -\dfrac{8}{3}$
15. $X < \dfrac{7}{4}$
16. $A < -2\dfrac{1}{2}$
17. $S > 1.6$
18. $Y > -3$

Solve each of the following and represent the solution for each on a real number line.

19. $|K| = 11$
20. $|M| = \dfrac{\pi}{2}$
21. $|B| < 7$
22. $|H| < 4.3$
23. $|L| > 1.5$
24. $|P| > \dfrac{8}{5}$

For each of the following, perform any indicated operations.

25. $\dfrac{(-3)(-5) + (-2)}{8(-1) - (-4)}$

26. $\dfrac{2 + 7(-3) - 4(-2)}{-9(4) - 6(-6)}$

27. $\dfrac{-5 - 4 - 7}{4(-3) + 8(-1/2)}$

28. $\dfrac{3(-8) - 6(-4)}{9(-7) - 2(-5)}$

29. $\dfrac{(2/3)9 + 10(-1/2)}{-2(2) + (1/3)(15)}$

30. $\dfrac{12(-3) - 4(-9)}{9(2) - 3(6)}$

chapter 2

Review of Basic Algebra

2–1. Exponents and Radicals

In the preceding chapter, we discussed a number of the basic properties associated with numbers and our number system. We will now introduce two concepts that are important in the discussion of algebra and algebraic operations that occurs in the following sections. These two concepts are *exponents* and *radicals*.

An expression of the type A^N is called an *exponential expression*. A is called the *base* and is the quantity that will be used as a multiplier a certain number of times. N is called an *exponent* and indicates how many times the base A is to be used as a multiplier.

Example A:
1. 10^2 means $10 \cdot 10$.
2. 7^5 means $7 \cdot 7 \cdot 7 \cdot 7 \cdot 7$.
3. X^3 means $X \cdot X \cdot X$.
4. $(A + B)^2$ means $(A + B) \cdot (A + B)$.

When the above operation is actually performed, the result is called the *Nth power of A*.

Example B:
1. $10^2 = 10 \cdot 10 = 100$. Therefore, the second power of 10 equals 100.
2. $7^5 = 7 \cdot 7 \cdot 7 \cdot 7 \cdot 7 = 16,807$. Therefore, the fifth power of 7 equals 16,807.

While working with exponential quantities, the following basic rules must be observed.

1. Any quantity taken once obviously yields just that quantity.

$$A^1 = A$$

Example C: $10^1 = 10$, $X^1 = X$, $(A + B)^1 = A + B$

2. When multiplying, if bases are the same, add exponents.

$$(A^M) \cdot (A^N) = A^{(M+N)}$$

Example D:
1. $(10^2) \cdot (10^4) = (10 \cdot 10) \cdot (10 \cdot 10 \cdot 10 \cdot 10) = 10 \cdot 10 \cdot 10 \cdot 10 \cdot 10 \cdot 10 = 10^6$
2. $(X^3) \cdot (X^2) = (X \cdot X \cdot X) \cdot (X \cdot X) = X \cdot X \cdot X \cdot X \cdot X = X^5$

3. When dividing, if bases are the same, subtract the exponent in the denominator from the one in the numerator.

$$\frac{(A^M)}{(A^N)} = A^{(M-N)}$$

Example E:
1. $\dfrac{10^6}{10^2} = \dfrac{10 \cdot 10 \cdot 10 \cdot 10 \cdot 10 \cdot 10}{10 \cdot 10} = 10 \cdot 10 \cdot 10 \cdot 10 = 10^4$
2. $\dfrac{X^5}{X^3} = \dfrac{X \cdot X \cdot X \cdot X \cdot X}{X \cdot X \cdot X} = X \cdot X = X^2$

4. If M and N are equal, then

$$\frac{A^M}{A^M} = A^{M-M} = A^0 = 1,$$

since any quantity (other than 0) divided by itself is equal to 1.

$$A^0 = 1 \quad (A \neq 0)$$

Sec. 2–1. Exponents and Radicals

Example F:
1. $\dfrac{10^3}{10^3} = 10^0 = 1$
2. $\dfrac{X^3}{X^3} = X^0 = 1$
3. $\dfrac{(T+W)^2}{(T+W)^2} = (T+W)^0 = 1$
4. $4^0 = (-23)^0 = (156{,}000)^0 = (A+B)^0 = (P^2 R^7 S^3)^0 = 1$

5. An exponential quantity may be moved from numerator to denominator or vice versa by changing the sign of the *exponent*.

$$A^{-M} = \frac{1}{A^M}$$

Example G:
1. $\dfrac{10^2}{10^6} = \dfrac{10 \cdot 10}{10 \cdot 10 \cdot 10 \cdot 10 \cdot 10 \cdot 10} = \dfrac{1}{10 \cdot 10 \cdot 10 \cdot 10} = \dfrac{1}{10^4}$

$\dfrac{10^2}{10^6} = 10^{2-6} = 10^{-4}$

Therefore, $10^{-4} = \dfrac{1}{10^4}$

2. $\dfrac{X^3}{X^5} = \dfrac{X \cdot X \cdot X}{X \cdot X \cdot X \cdot X \cdot X} = \dfrac{1}{X \cdot X} = \dfrac{1}{X^2}$

$\dfrac{X^3}{X^5} = X^{3-5} = X^{-2}$

Therefore, $X^{-2} = \dfrac{1}{X^2}$

6. When raising an exponential quantity to some power, multiply the exponents.

$$(A^M)^N = A^{MN}$$

Example H:
1. $(10^4)^3 = 10^4 \cdot 10^4 \cdot 10^4 = 10^{12}$
2. $(X^3)^2 = X^3 \cdot X^3 = X^6$

7. When a product is raised to a power, each member of the product must be raised to the indicated power.

$$(AB)^N = A^N B^N$$

Example I:
1. $(2A)^3 = 2^3 A^3 = 8A^3$
2. $(-3X^2 Y)^4 = (-3)^4 (X^2)^4 Y^4 = 81 X^8 Y^4$

8. When a quotient is raised to a power, each member of the quotient must be raised to the indicated power.

$$\left(\frac{A}{B}\right)^N = \frac{A^N}{B^N}$$

Example J:
1. $\left(\dfrac{4}{5}\right)^2 = \dfrac{4^2}{5^2} = \dfrac{16}{25}$
2. $\left(\dfrac{3}{B}\right)^2 = \dfrac{3^2}{B^2} = \dfrac{9}{B^2}$
3. $\left(\dfrac{-2X}{Y}\right)^2 = \dfrac{(-2X)^2}{Y^2} = \dfrac{4X^2}{Y^2}$

The following examples will further illustrate and clarify the basic rules.

Example K:

$$\frac{T^5}{T^2} = T^{5-2} = T^3$$

$$\frac{T^5}{T^{-2}} = T^{5-(-2)} = T^{5+2} = T^7$$

Note that in division if a negative exponent appears in the denominator, the double negative sign yields a positive result.

Example L:

$$A^{-2} = \frac{1}{A^2}$$

$$\frac{1}{A^{-2}} = A^2$$

Note that an exponential quantity, which is a multiplier, may be moved from numerator to denominator or vice versa by changing the sign of the exponent.

Example M: If parentheses are not present, exponents apply only to the quantity immediately preceding them.

$2X^2$
$4A^0 = 4 \cdot 1 = 4$
$3B^{-2} = 3 \cdot \dfrac{1}{B^2} = \dfrac{3}{B^2}$

$(2X)^2 = 4X^2$
$(4A)^0 = 1$
$(3B)^{-2} = \dfrac{1}{(3B)^2} = \dfrac{1}{9B^2}$

Sec. 2-1. Exponents and Radicals

In some problems, we are asked the question, "What number, when used as a multiplier a certain number of times, gives a certain result?" For this reason, we define *roots* of numbers. By the *Nth root of A* (written $\sqrt[N]{A}$) is meant a number that, when used as a multiplier N times, yields a result equal to A. The sign, $\sqrt{}$, is called a *radical sign*, A is called the *radicand*, and N is called the *order* of the root.

Example N:
1. $\sqrt[2]{3}$ means the second or square root of 3.

NOTE: When working with square root, we usually do not write the 2. Therefore, $\sqrt[2]{3} = \sqrt{3}$.

2. $\sqrt[3]{7}$ means the third or cube root of 7.
3. $\sqrt[5]{12}$ means the fifth root of 12.
4. $\sqrt[3]{-10}$ means the cube root of -10.

Example O:
1. $\sqrt{4} = 2$ since $2 \cdot 2 = 4$
2. $\sqrt{25} = 5$ since $5 \cdot 5 = 25$
3. $\sqrt[3]{64} = 4$ since $4 \cdot 4 \cdot 4 = 64$
4. $\sqrt[5]{32} = 2$ since $2 \cdot 2 \cdot 2 \cdot 2 \cdot 2 = 32$
5. $\sqrt{.01} = .1$ since $.1 \cdot .1 = .01$
6. $\sqrt[3]{-8} = -2$ since $(-2) \cdot (-2) \cdot (-2) = -8$

Examining the expression $\sqrt{4}$, we see that $2 \cdot 2 = 4$ and $(-2) \cdot (-2) = 4$. Therefore, we might expect that $\sqrt{4} = 2$ or -2. To avoid confusion, we define the *principal Nth root of A* to be positive if A is positive and negative if A is negative and N is odd. This means that $\sqrt{4} = 2$, $-\sqrt{4} = -2$, and $\pm\sqrt{4} = \pm 2$. The last expression is read "plus or minus the square root of four equals plus or minus two," and indicates that both the positive and negative square roots are to be taken.

Example P:
1. $\sqrt{16} = 4$ $-\sqrt{16} = -4$ $\pm\sqrt{16} = \pm 4$
2. $\sqrt{121} = 11$ $-\sqrt{121} = -11$ $\pm\sqrt{121} = \pm 11$
3. $\sqrt[4]{16} = 2$ $-\sqrt[4]{16} = -2$ $\pm\sqrt[4]{16} = \pm 2$
4. $\sqrt[3]{27} = 3$ $-\sqrt[3]{27} = -3$ $\pm\sqrt[3]{27} = \pm 3$
5. $\sqrt[3]{-8} = -2$ $-\sqrt[3]{-8} = 2$ $\pm\sqrt[3]{-8} = \pm 2$
6. $\sqrt[3]{-27} = -3$ $-\sqrt[3]{-27} = 3$ $\pm\sqrt[3]{-27} = \pm 3$

When we actually go through the process of taking a root, we must list all possible roots.

Example Q:
1. If $A^2 = 4$, then $A = 2$ or -2 (written ± 2) since $2^2 = 4$ and $(-2)^2 = 4$.
2. If $A^2 = 25$, then $A = 5$ or -5 (± 5) since $5^2 = 25$ and $(-5)^2 = 25$.
3. If $A^4 = 81$, then $A = 3$ or -3 (± 3) since $3^4 = 81$ and $(-3)^4 = 81$.
4. If $A^3 = 27$, then $A = 3$ since only $3^3 = 27$.
5. If $A^3 = 125$, then $A = 5$ since only $5^3 = 125$.

NOTE: If the original exponent is even there will be two real roots [(1), (2), and (3) above], and if the original exponent is odd there will be one real root [(4) and (5) above].

When working with roots and radical expressions, the following basic rules must be observed.

1. If orders are the same, we may multiply radicands.

$$\sqrt[N]{A} \times \sqrt[N]{B} = \sqrt[N]{AB}$$

Example R:
1. $\sqrt{7} \times \sqrt{3} = \sqrt{21}$
2. $\sqrt{6} \times \sqrt{5} = \sqrt{30}$
3. $\sqrt[3]{10} \times \sqrt[3]{2} = \sqrt[3]{20}$
4. $\sqrt[5]{5} \times \sqrt[5]{7} = \sqrt[5]{35}$

2. If orders are the same, we may divide radicands.

$$\frac{\sqrt[N]{A}}{\sqrt[N]{B}} = \sqrt[N]{\frac{A}{B}}$$

Example S:
1. $\dfrac{\sqrt{6}}{\sqrt{2}} = \sqrt{3}$
2. $\dfrac{\sqrt{30}}{\sqrt{5}} = \sqrt{6}$
3. $\dfrac{\sqrt[3]{16}}{\sqrt[3]{4}} = \sqrt[3]{4}$

3. When finding the Nth root of a number and then raising the result to the Nth power, the final result will be the original number.

$$\sqrt[N]{A^N} = (\sqrt[N]{A})^N = A$$

Sec. 2–1. Exponents and Radicals

Example T:
1. $\sqrt{9^2} = (\sqrt{9})^2 = (3)^2 = 9$
2. $\sqrt{3^2} = (\sqrt{3})^2 = 3$
3. $\sqrt[3]{8^3} = (\sqrt[3]{8})^3 = (2)^3 = 8$
4. $\sqrt[4]{5^4} = (\sqrt[4]{5})^4 = 5$

4. When finding multiple roots, multiply the orders.

$$\sqrt[M]{\sqrt[N]{A}} = \sqrt[MN]{A}$$

Example U:
1. $\sqrt[3]{\sqrt{6}} = \sqrt[6]{6}$
2. $\sqrt[4]{\sqrt[3]{2}} = \sqrt[12]{2}$

There is a connection between exponents and radicals defined by the following two rules.

(1) $\qquad A^{1/N} = \sqrt[N]{A}$

(2) $\qquad A^{M/N} = (\sqrt[N]{A})^M$

Example V:
1. $3^{1/2} = \sqrt{3}$ $\qquad 5^{1/3} = \sqrt[3]{5}$ $\qquad 9^{1/4} = \sqrt[4]{9}$
2. $5^{3/2} = (\sqrt{5})^3$ $\qquad 6^{2/3} = (\sqrt[3]{6})^2$ $\qquad 10^{7/3} = (\sqrt[3]{10})^7$
3. $2^{-1/2} = \dfrac{1}{2^{1/2}} = \dfrac{1}{\sqrt{2}}$
4. $5^{-1/3} = \dfrac{1}{5^{1/3}} = \dfrac{1}{\sqrt[3]{5}}$
5. $8^{-3/2} = \dfrac{1}{8^{3/2}} = \dfrac{1}{(\sqrt{8})^3}$

It can be shown that these two rules satisfy all the rules previously listed for exponents.

Radical expressions may often be simplified by making use of the rules that we have mentioned.

Example W:
1. $\sqrt{8} = \sqrt{4 \times 2} = \sqrt{4}\sqrt{2} = 2\sqrt{2}$
2. $\sqrt{27} = \sqrt{9 \times 3} = \sqrt{9}\sqrt{3} = 3\sqrt{3}$
3. $\sqrt[3]{16} = \sqrt[3]{8 \times 2} = \sqrt[3]{8}\sqrt[3]{2} = 2\sqrt[3]{2}$
4. $\sqrt[3]{54} = \sqrt[3]{27 \times 2} = \sqrt[3]{27}\sqrt[3]{2} = 3\sqrt[3]{2}$

Example X:
1. $\sqrt[4]{9} = \sqrt[4]{3^2} = (\sqrt[4]{3})^2 = 3^{2/4} = 3^{1/2} = \sqrt{3}$
2. $\sqrt[6]{8} = \sqrt[6]{2^3} = (\sqrt[6]{2})^3 = 2^{3/6} = 2^{1/2} = \sqrt{2}$
3. $\sqrt[8]{25} = \sqrt[8]{5^2} = (\sqrt[8]{5})^2 = 5^{2/8} = 5^{1/4} = \sqrt[4]{5}$

Exercises Simplify each of the following expressions.

1. 4^3
2. 3^5
3. 17^0
4. $(2^4)^0$
5. $14R^0$
6. 75^0
7. $(5A)^3$
8. $(-2X)^4$
9. 5^{-2}
10. B^{-4}
11. $3X^{-2}$
12. $(4N)^{-2}$
13. $T^2 T^5$
14. $5A^3 A^4$
15. $2K(K^6)$
16. $3VT^2(T^4)$
17. $\dfrac{X^7}{X^3}$
18. $\dfrac{5B^{10}}{B^8}$
19. $\dfrac{Y^2}{Y^5}$
20. $\dfrac{2T}{T^3}$
21. $(K^3)^4$
22. $(-A^3)^4$
23. $\left(\dfrac{5}{X}\right)^3$
24. $\left(\dfrac{2B}{5K}\right)^2$
25. $AB^2(AB)^2$
26. $X^2 Y(-XY)^3$
27. $\dfrac{12R^2 S}{6RS^3}$
28. $\dfrac{K(K^2 L^3)}{L^4 M}$
29. $\sqrt{25}$
30. $-\sqrt{81}$
31. $\sqrt[3]{8}$
32. $\sqrt[3]{-64}$
33. $-\sqrt[4]{81}$
34. $-\sqrt[3]{-8}$
35. $2\sqrt{9}$
36. $5\sqrt[3]{64}$
37. $-3\sqrt{4}$
38. $-2\sqrt[4]{16}$
39. $4^{1/2}$
40. $27^{1/3}$
41. $36^{3/2}$
42. $81^{3/4}$
43. $16^{-1/2}$
44. $8^{-1/3}$
45. $9^{-3/2}$
46. $27^{-4/3}$
47. $\sqrt{X^2}$
48. $\sqrt{A^4}$
49. $\sqrt[3]{Y^6}$
50. $\sqrt[12]{T^{12}}$
51. $(T^2)^{1/2}$
52. $(B^2)^{1/4}$
53. $\sqrt{10}\sqrt{2}$
54. $\sqrt{7}\sqrt{13}$
55. $\sqrt[3]{9}\sqrt[3]{3}$
56. $\sqrt[4]{2}\sqrt[4]{7}$
57. $\dfrac{\sqrt{10}}{\sqrt{2}}$
58. $\dfrac{\sqrt{15}}{\sqrt{5}}$
59. $\sqrt[4]{\sqrt{2}}$
60. $\sqrt{\sqrt[3]{5}}$
61. $\sqrt{12}$
62. $\sqrt[3]{81}$
63. $\sqrt{32}$
64. $\sqrt{175}$
65. $\sqrt[4]{4}$
66. $\sqrt[6]{27}$

2-2. Scientific Notation

In this section we will look at an important use of exponents. When working in scientific and technical areas, we often encounter very large or very small numbers. For example, the speed of light is approximately 186,000 miles per second, the distance from the earth to the sun is about 93,000,000 miles, television signals travel at about 30,000,000,000 centimeters per second, and the mass of an electron is .000000000000000000000000000911 grams. When working with such numbers, it is convenient to use a different method of writing them, called *scientific notation*.

Writing a number in scientific notation means expressing it as a product of a number between 1 and 10 (including 1) and an integral power of 10. This can be done for any number since multiplying a number by some power of 10 is simply a case of moving the decimal point in one direction or the other.

Symbolically, the representation of a number in scientific notation is

$$P \times 10^k$$

In the equation:

1. $1 \leq P < 10$.
2. k must be an integer and $|k|$ is the number of places that the decimal point was moved in the original number.

   (a) If $k > 0$, the original number was greater than or equal to 10 and the decimal point was moved to the left.

Example A:
1. $4200 = 4.2 \times 10^3$
2. $26{,}700{,}000 = 2.67 \times 10^7$
3. $438{,}000 = 4.38 \times 10^5$
4. $87{,}300{,}000 = 8.73 \times 10^7$

   (b) If $k = 0$, the original number was greater than or equal to 1 but less than 10 and the decimal point was not moved.

Example B:
1. $6.79 = 6.79 \times 10^0$
2. $9.26 = 9.26 \times 10^0$
3. $4.05 = 4.05 \times 10^0$
4. $7.63 = 7.63 \times 10^0$

3. If $k < 0$, the original number was less than 1 and the decimal point was moved to the right.

Example C:
1. $.000128 = 1.28 \times 10^{-4}$
2. $.00359 = 3.59 \times 10^{-3}$
3. $.549 = 5.49 \times 10^{-1}$
4. $.0000672 = 6.72 \times 10^{-5}$

We should also see that if a number is written in scientific notation, it can be changed to ordinary notation by reversing the previous process. When a number is written in scientific notation ($P \times 10^k$):

1. If $k > 0$, move the decimal point $|k|$ places to the right.

Example D:
1. $2.98 \times 10^5 = 298000$
2. $7.62 \times 10^2 = 762$
3. $5.71 \times 10^6 \quad 5710000$
4. $2.69 \times 10^3 \quad 2690$

2. If $k = 0$, do not move the decimal point.

Example E:
1. $1.08 \times 10^0 = 1.08$
2. $9.403 \times 10^0 = 9.403$
3. $4.18 \times 10^0 = 4.18$
4. $6.82 \times 10^0 = 6.82$

3. If $k < 0$, move the decimal point $|k|$ places to the left.

Example F:
1. $4.92 \times 10^{-5} = .0000492$
2. $7.38 \times 10^{-1} = .738$
3. $5.37 \times 10^{-4} = .000537$
4. $8.29 \times 10^{-2} = .0829$

Sec. 2-2. Scientific Notation

Scientific notation may also be used to *estimate* answers to problems involving computations with very large or very small numbers. Such estimates are necessary when a quick approximation of an answer is needed and no calculator or other device is available for use in computing the answer. Also, slide rules can be used to obtain the correct digits for an answer but they cannot be used to place the decimal point, while some calculators do not have the capacity for dealing with extremely large or small numbers. Therefore, scientific notation allows us to work just with numbers between 1 and 10, and once the actual digits for the answer are obtained, the decimal point may be placed by using approximation and the rules for exponents.

Example G:

1. $(30,000,000) \times (420,000) \approx (3 \times 10^7) \times (4 \times 10^5) = 12 \times 10^{12}$
$= 1.2 \times 10^{13}$
2. $(48,000) \times (.0000725) \approx (5 \times 10^4) \times (7 \times 10^{-5}) = 35 \times 10^{-1}$
$= 3.5 \times 10^0$
3. $\dfrac{(92,000,000)}{(.00318)} \approx \dfrac{(9 \times 10^7)}{(3 \times 10^{-3})} = 3 \times 10^{10}$
4. $\dfrac{(.088)}{(143,000)} \approx \dfrac{(9 \times 10^{-2})}{(1.5 \times 10^5)} = 6 \times 10^{-7}$
5. $(78,000,000)^2 \approx (8 \times 10^7)^2 = 64 \times 10^{14} = 6.4 \times 10^{15}$

Exercises

Write each of the following in scientific notation.

1. 56,080
2. 93,000,000
3. 186,000
4. 30,000,000
5. 4.91
6. 7.63
7. .000000000000000000000000000911
8. .000000095
9. .251
10. .00000000000000000016

Write each of the following in ordinary notation.

11. 7.81×10^7
12. 2.19×10^{12}
13. 4.05×10^0
14. 6.002×10^4
15. 9.83×10^{-2}
16. 5.94×10^{-5}
17. 8.001×10^0
18. 8.16×10
19. 4.45×10^{-1}
20. 3.89×10^{-10}

Use scientific notation to approximate the answer for each of the following.

21. $(98,000,000)(2,700,000)$
22. $(.00083)(.0059)$
23. $(.00000723)(141,000)$
24. $\dfrac{426}{181,000,000}$
25. $\dfrac{.052}{.0000049}$
26. $\dfrac{526000}{.00017}$
27. $(194,000,000)^3$
28. $(480,000)^2$

For each of the following, change any numbers in ordinary notation to scientific notation and change any numbers in scientific notation to ordinary notation.

29. The specific gravity of hydrogen is .00008987.
30. The atomic weight of silver is 107.880.
31. In 1 atomic mass unit, there are 1.66035×10^{-27} kilograms.
32. The density of mercury at 0°C is 13595.09 kilograms per cubic meter.
33. The velocity of light in a vacuum is 2.997902×10^8 meters per second.
34. Planck's constant, used in physics, is 6.624×10^{-34} joule-second.
35. In 1 newton, there are 10^5 dynes.
36. During 1 rotation about the earth, the moon travels approximately 1,500,000 miles.
37. A photon, a "particle" of light, has a frequency of 3×10^{15} cycles per second.
38. The weight of the earth is about 6,600,000,000,000,000,000,000 tons.
39. Atmospheric pressure is 1,013,000 dynes per square centimeter.
40. Avogadro's number, the number of molecules per gram-mole of a substance, is 6.02×10^{23}.

2-3. Algebraic Expressions

In algebra, symbols are used to represent numbers. Therefore, all operations that are valid for numbers are also valid for these symbols. *Algebraic operations* include the processes of addition, subtraction, multiplication, division, raising to powers, and taking roots. When any of these operations are performed, the combination of numbers and symbols that results is called an *algebraic expression*. Algebraic expressions consist of *terms* and *factors*. Terms are separated by plus signs and minus signs. Factors are quantities that are multiplied to obtain a certain result.

Example A:
1. $2X + 3Y$ is an algebraic expression with two terms: $2X$ and $3Y$. The term $2X$ has two factors: 2 and X. The term $3Y$ has two factors: 3 and Y.
2. $5A^3 - 8\sqrt{B} + 3B\sqrt{A}$ is an algebraic expression with three terms: $5A^3$, $8\sqrt{B}$, and $3B\sqrt{A}$. The term $5A^3$ has four factors: 5, A, A, and A. The term $8\sqrt{B}$ has two factors: 8 and $\sqrt{B}$. The term $3B\sqrt{A}$ has three factors: 3, B, and $\sqrt{A}$.

Sec. 2–3. Algebraic Expressions 27

3. $8RS^2 + \dfrac{R}{S^2}$ is an algebraic expression with two terms: $8RS$ and $\dfrac{R}{S^2}$. The term $8RS^2$ has four factors: 8, R, S, and S. The term $\dfrac{R}{S^2}$ has three factors: R, $\dfrac{1}{S}$, and $\dfrac{1}{S}$.

An algebraic expression of one term is called a *monomial*. An algebraic expression containing two terms is called a *binomial*. An algebraic expression containing more than one term is called a *multinomial*. Thus, we see that a binomial could be considered a multinomial.

Example B:
1. X, $3T^2$, $5A^2B^2C$, $7X^2\sqrt{Y}$, and $\sqrt{10R^2S}$ are monomials.
2. $(A + B)$, $(2X - 3Y)$, $(8R^2 + 5S^2T)$ and $(4\sqrt{L} - 7\sqrt{K})$ are binomials.
3. $(X + Y + Z)$, $(8A^2 - 2B + 7C^3 + D^4)$, $(2R + 3\sqrt{S} - T^3)$, and $(K^3 - 2LM^2 + \sqrt{KL})$ are multinomials.

Polynomials are also very important in mathematics. A polynomial is an algebraic expression in which each term is of the type AX^N where N is either zero or a positive integer.

Example C:
1. $(7\sqrt{5X} + 2Y)$, $\left(8A^{1/3} + 2B^2 - \dfrac{7}{C}\right)$, and $(2R^{-1/2} + \sqrt{S} + 9T^2)$ are multinomials but not polynomials.
2. $(5X^2 - 3X)$, $(2R^3 + 7R^2 - R^5)$ and $(B^5 - 3B^4 + 9B^2)$ are multinomials and also polynomials.

In a polynomial, the exponent associated with each term gives the *degree* of that term and the degree of the highest power term is the degree of the polynomial.

Example D: In the expression $(2R^5 - 5R^3 + 8R)$, the term $2R^5$ is of degree 5, the term $5R^3$ is of degree 3, and the term $8R$ is of degree 1; thus, the polynomial is of degree 5.

In the terms of an algebraic expression, the numbers and literal symbols that multiply any given factor are called the *coefficients* of that factor.

Example E:
1. In the term $3X$, 3 is the coefficient of X and X is the coefficient of 3.

2. In the term $8A^2\sqrt{B}$, 8 is the coefficient of $A^2\sqrt{B}$, A^2 is the coefficient of $8\sqrt{B}$, and $\sqrt{B}$ is the coefficient of $8A^2$.

Terms that differ only in their coefficients are called *similar* or *like* terms.

Example F:
1. $7X$ and $9X$ are similar terms since only the numerical coefficients are different.
2. $2R$ and $2T$ are similar terms since only the literal coefficients are different.

When working with algebraic expressions the following rules must be observed.

1. Addition and Subtraction: Combine similar terms by adding or subtracting their coefficients.

Example G:
1. $2T + 7T - 5T = 4T$
2. $3X^2 + 4Y - 2X^2 = X^2 + 4Y$
3. $(5A + 3B) + (6A - 9B) = 5A + 6A + 3B - 9B = 11A - 6B$
4. $(4R - 3S) - (R - 2S + T) = 4R - R - 3S + 2S - T$
$= 3R - S - T$
5. $8K^3 - 5(K^3 + L^2) = 8K^3 - 5K^3 - 5L^2 = 3K^3 - 5L^2$
6. $5\sqrt{T} + 7\sqrt{T} - 3\sqrt{T} = 9\sqrt{T}$
7. $2A\sqrt{B} + 7\sqrt{B} - A\sqrt{B} = (A + 7)\sqrt{B}$
8. $[XY^2 - 2(XY + Y^2)] - [3Y^2 + 4(XY - XY^2)]$
$= (XY^2 - 2XY - 2Y^2) - (3Y^2 + 4XY - 4XY^2)$
$= XY^2 - 2XY - 2Y^2 - 3Y^2 - 4XY + 4XY^2$
$= 5XY^2 - 6XY - 5Y^2$
9. $3\sqrt{2} + 4\sqrt[3]{2} - \sqrt{2} + \sqrt[3]{2} = 2\sqrt{2} + 5\sqrt[3]{2}$

2. Multiplication:
(a) Multiply two monomials by using the laws of multiplying signed numbers and the laws of exponents.

Example H:
1. $(7T^3)(4T^2) = 28T^5$
2. $(5AB^2)(8A^2B) = 40A^3B^3$
3. $(2X\sqrt{Y})(3X^2) = 6X^3\sqrt{Y}$

(b) Multiply a monomial and a multinomial by using the distributive law and observing the laws of multiplication of signed numbers and the laws of exponents.

Sec. 2–3. Algebraic Expressions

Example I:
1. $2A(AB^2 + C) = 2A^2B^2 + 2AC$
2. $-3R(RS - R^2S + 2ST^2) = -3R^2S + 3R^3S - 6RST^2$
3. $2\sqrt{X}(Y^2 - 2Z) = 2Y^2\sqrt{X} - 4Z\sqrt{X}$

(c) Multiply two multinomials by multiplying each term in one expression by each term in the other and observing the laws of multiplication of signed numbers and the laws of exponents.

Example J:
1. $(2A + 3B)(5A - 4B) = 2A(5A - 4B) + 3B(5A - 4B)$
$= 10A^2 - 8AB + 15AB - 12B^2$
$= 10A^2 + 7AB - 12B^2$
2. $(\sqrt{X} - \sqrt{Y})(2\sqrt{X} + \sqrt{Y}) = \sqrt{X}(2\sqrt{X} + \sqrt{Y}) - \sqrt{Y}(2\sqrt{X} + \sqrt{Y})$
$= 2X + \sqrt{XY} - 2\sqrt{XY} - Y$
$= 2X - \sqrt{XY} - Y$
3. $(R - 2S)(R^2 - 3RS^2 + ST)$
$= R(R^2 - 3RS^2 + ST) - 2S(R^2 - 3RS^2 + ST)$
$= R^3 - 3R^2S^2 + SRT - 2R^2S + 6RS^3 - 2S^2T$
4. $(K + L)^2 = (K + L)(K + L) = K(K + L) + L(K + L)$
$= K^2 + KL + KL + L^2$
$= K^2 + 2KL + L^2$
5. $3A(A - B)(2B + C) = [3A(A - B)] \times (2B + C)$
$= (3A^2 - 3AB)(2B + C)$
$= 3A^2(2B + C) - 3AB(2B + C)$
$= 6A^2B + 3A^2C - 6AB^2 - 3ABC$

3. Division:
(a) Divide two monomials by using the laws of dividing signed numbers and the laws of exponents.

Example K:
1. $\dfrac{8A^4}{2A} = 4A^3$
2. $\dfrac{9B^2}{7B^5} = \dfrac{9}{7B^3}$ or $\dfrac{9B^{-3}}{7}$
3. $\dfrac{10\sqrt{X}}{2\sqrt{Y}} = 5\sqrt{\dfrac{X}{Y}}$
4. $\dfrac{6AX}{3AX^{-2}} = 2X^3$

(b) Divide a multinomial by a monomial by rewriting as sums or differences of monomials divided by monomials.

Example L:

1. $\dfrac{4A^5 - 2B^2}{AB} = \dfrac{4A^5}{AB} - \dfrac{2B^2}{AB} = \dfrac{4A^4}{B} - \dfrac{2B}{A}$

2. $\dfrac{7RS^2 - 3RS + 5R^2S^2}{RS} = \dfrac{7RS^2}{RS} - \dfrac{3RS}{RS} + \dfrac{5R^2S^2}{RS} = 7S - 3 + 5RS$

3. $\dfrac{5T^2\sqrt{X} - 8T\sqrt{Y}}{\sqrt{X}} = \dfrac{5T^2\sqrt{X}}{\sqrt{X}} - \dfrac{8T\sqrt{Y}}{\sqrt{X}} = 5T^2 - 8T\sqrt{\dfrac{Y}{X}}$

(c) The different cases of division of multinomials will be taken up as they arise throughout the text.

Exercises Perform the indicated operations in each of the following.

1. $2A - 5A - (-3A)$
2. $4B^2 - 12B^2 + 7B$
3. $(R - S) - (5R - S)$
4. $(5T^2 - 3V^2) - (8T^2 + V)$
5. $3X^3 - 2(X^3 - 5Y^3)$
6. $K^2L^2 - K(KL^2 - 2K)$
7. $-[-(X - Y)]$
8. $-2[-3(A - B) + 2A]$
9. $-5[2(3A - B) - 4(A - B)] + B$
10. $2\sqrt{X} - 3\sqrt{Y} + 5\sqrt{X}$
11. $(4\sqrt{T}) - 5\sqrt{V}) + (2\sqrt{T} + 3\sqrt{V})$
12. $(3\sqrt{K} + \sqrt{L}) - 2(\sqrt{K} - 2\sqrt{L})$
13. $5\sqrt{8} + 2\sqrt{32} - 4\sqrt{50}$
14. $3\sqrt{12} + \sqrt{108} - 2\sqrt{75}$
15. $(7A^3)(5A^5)$
16. $(2B^2C)(-6B^3C^4)$
17. $(3X^{-2}Y)(5X^4Y^{-1})$
18. $(2RS^{-1})(-4R^2S^{-3})$
19. $(\sqrt{2}X^2)(\sqrt{3}X^5Y)$
20. $(\sqrt{3}T^3S)(-\sqrt{3}T)$
21. $(2A - 3B)(A - 5B)$
22. $(5V + 4W)(V - 3W)$
23. $4R^2(2RS - 3S^2 + R^2S)$
24. $\sqrt{2}(\sqrt{3}A - \sqrt{5}B^2 + \sqrt{7}AB)$
25. $(2X - Y)(X^2 + 2XY - Y^2)$
26. $(3K - 2L)(KL^2 - KL + 2K^2L^2)$
27. $(A + 3B)^2$
28. $(2 - 4X)^3$
29. $(R - S)^2(R + S)$
30. $(T + 3)(T - 4)(2T + 5)$
31. $(\sqrt{5} + X)(X^2 - \sqrt{5})$
32. $(A - \sqrt{3})(2A + \sqrt{3})$
33. $(\sqrt{7} - 2\sqrt{3})(2\sqrt{7} + 5\sqrt{3})$
34. $(3\sqrt{2} - \sqrt{5})(\sqrt{2} + 4\sqrt{5})$
35. $\dfrac{15A^3X}{5AX}$
36. $\dfrac{27B^2Y^5}{9B^3Y^3}$

Sec. 2-4. Factoring

37. $\dfrac{14R^2S}{7RS^{-3}}$

38. $\dfrac{12T^2 - 8T^3 + 6T}{2T}$

39. $\dfrac{35K^3 - 5K + KL^2}{5KL}$

40. $\dfrac{10A\sqrt{X} - 16B^2\sqrt{X} + 2}{4\sqrt{X}}$

2-4. Factoring

Factoring is an operation that is frequently used in mathematics. It is used to simplify fractions, perform operations with fractions, and solve equations. In this section we will review some techniques that will allow us to factor an algebraic expression.

Factoring is the reverse process of multiplication. Since a number of factors are multiplied to obtain a product, the reverse process is to determine the factors when a product is given.

It was pointed out in Section 2-3 that a factor is a quantity that is used as a multiplier to obtain a certain result.

Example A:
1. Since $3 \times 4 = 12$, 3 and 4 are factors of 12.
2. Since $5 \times 3 \times X = 15X$, 5, 3, and X are factors of $15X$.
3. Since $4(A + B) = 4A + 4B$, 4 and $(A + B)$ are factors of $4A + 4B$.

A *prime factor* is a factor that is divisible only by itself and 1. *Divisible* means the quotient is an integer or a prime algebraic expression.

Example B:
1. $3T(4 - T) = 12T - 3T^2$ 3, T, and $(4 - T)$ are prime factors.
2. $5M(2M + 6N) = 10M^2 + 30MN$ 5 and M are prime factors, $2M + 6N$ is not prime because it is divisible by 2.

It is possible to check to see if you have the factors of an expression by multiplying together the quantities you believe to be the factors. If the product of these quantities is equal to the original expression, then you have the correct quantities as factors.

Example C:
1. Are 50 and 2 factors of 100? Since $50 \times 2 = 100$, we are certain that 50 and 2 are factors of 100.
2. Are $-3X$ and $(2X - 5)$ factors of $15X - 6X^2$? Since $-3X(2X - 5) = 15X - 6X^2$, $-3X$ and $(2X - 5)$ are factors of $15X - 6X^2$.

The first type of factoring we should be familiar with is recognition of a *common monomial factor* in an algebraic expression.

Example D: Note that the expression $10X + 15Y$ also can be written as $(5 \cdot 2X) + (5 \cdot 3Y)$ and that 5 is a common monomial factor in each term. Therefore, by the distributive property, this expression can be written as $5(2X + 3Y)$. To check this factored form, perform the multiplication and the product of $5(2X + 3Y)$ must equal the original expression $10X + 15Y$.

Example E:

Algebraic Expression	Prime Factors of Each Term	Factored Form
$12A^2 - 6AB$	$(3 \cdot 2 \cdot 2 \cdot A \cdot A) - (3 \cdot 2 \cdot A \cdot B)$	$3A(4A - 2B)$
$4MN - 20M^2N + 28MN^2$	$(2 \cdot 2MN) - (2 \cdot 2 \cdot 5M \cdot M \cdot N) + (2 \cdot 2 \cdot 7MNN)$	$4MN(1 - 5M + 7N)$
$6RJ + 10RS$	$(2 \cdot 3 \cdot R \cdot J) + (2 \cdot 5 \cdot R \cdot S)$	$2R(3J + 5S)$
$-3TS - 6S^2T + 9T^2$	$(-1 \cdot 3 \cdot T \cdot S) + (-1 \cdot 3 \cdot 2 \cdot S \cdot S \cdot T) - (-1 \cdot 3 \cdot 3 \cdot T \cdot T)$	$-3T(S + 2S^2 - 3T)$
		$3T(-S - 2S^2 + 3T)$

Note that 1 and -1 are common factors in every algebraic expression. At times it is useful to factor -1 from the expression when performing operations with algebraic fractions.

$$Y - X = [(-1)(-1)] \cdot Y + [-1 \cdot 1] \cdot X = -1 \cdot (-Y + X) = -(X - Y)$$

In general,

$$(B - A) = -(A - B) \text{ for any } A \text{ and } B.$$

At times a binomial will appear as a common factor.

Example F:
1. $2(A + B) + 3(A + B) = (2 + 3)(A + B) = 5(A + B)$
2. $X(T + D) - Y^2(T + D) = (X - Y^2)(T + D)$

When multiplying two binomials, where one binomial is the sum of two quantities and the other binomial is the difference of the same two quantities, the product is always the difference between two perfect squares. Since factoring is the reverse process of multiplication, the difference between two perfect squares can be factored into a product of a binomial that is the sum of the square roots of the two terms and a binomial that is the difference of the square roots of the same two terms. This idea is illustrated by an equation:

$$A^2 - B^2 = (A + B)(A - B)$$

Example G: The following examples will demonstrate this type of factoring:
1. $T^2 - 49 = (T - 7)(T + 7)$

Sec. 2-4. Factoring

2. $9S^2 - 25 = (3S + 5)(3S - 5)$
3. $16X^2 - 81Y^2 = (4X - 9Y)(4X + 9Y)$
4. $V^6 - 4W^6 = (V^3 - 2W^3)(V^3 + 2W^3)$

The remaining types of factoring that we will consider apply to *trinomials*; that is, algebraic expressions with three terms.

Whenever a binomial is *squared* the product will be one of the following:

$$(A + B)^2 = (A + B)(A + B) = A^2 + 2AB + B^2$$
$$(A - B)^2 = (A - B)(A - B) = A^2 - 2AB + B^2$$

In each case the first and last terms of the product are perfect squares and the middle term is twice the product of the terms of the binomial.

Therefore, to factor this type of expression the procedure should be reversed. That is, examine the given expression to see whether the first and last terms are positive and perfect squares, and whether the middle term is plus or minus two times the product of the square roots of the first and last terms.

Example H:

Algebraic Expression	Inspection of Terms	Factored Form
$9T^2 - 12ST + 4S$	$(3T)^2 - 2(3T)(2S) + (2S)^2$	$(3T - 2S)^2$
$A^2 + 10AB + 25B^2$	$A^2 + 2(A)(5B) + (5B)^2$	$(A + 5B)^2$
$25X^2 - 70X + 49$	$(5X)^2 - 2(5X)(7) + (7)^2$	$(5X - 7)^2$

Another type of product that is more general than the preceding types is represented by the following: If A and B are real numbers, then

$$(X + A)(X + B) = X^2 + BX + AX + AB = X^2 + (B + A)X + AB$$

Inspecting this product, we see that A and B are multiplied to obtain the last term and added to obtain the coefficient of the middle term. Therefore, to reverse the process, we inspect the last term to see if there are two factors whose product is the last term and whose sum is the coefficient of the middle term.

Example I:

Algebraic Expression	Inspection of Terms	Factored Form
$X^2 + 10X + 21$	$X^2 + (3 + 7)X + (3 \cdot 7)$	$(X + 7)(X + 3)$
$T^2 - 6T + 8$	$T^2 + (-2 - 4)X + (-2)(-4)$	$(T - 2)(T - 4)$
$V^2 - 4V - 32$	$V^2 + (-8 + 4)V + (-8)(4)$	$(V - 8)(V + 4)$
$R^2 + 4R - 12$	$R^2 + (6 - 2)R + 6(-2)$	$(R + 6)(R - 2)$

The last type of factoring that we shall consider will be the factoring of a trinomial where the coefficient of the second degree term is not equal to 1. Our technique for factoring an expression of this type will be by inspection and trial and error.

In order to factor $3X^2 + 7X + 2$ we will consider the factors of the second degree term and the constant term, $3X^2$ and 2 respectively, and put these factors into the binomials where the factors of $3X^2$ are the first terms and the factors of 2 are the second terms of the binomials. Then multiply the binomials together to see if the selected combination of terms will yield the correct middle term in the original expression.

TRIAL: $(X + 1)(3X + 2) = 3X^2 + 5X + 2$; since the middle term is $5X$ this selection is *incorrect*.

TRIAL: $(3X + 1)(X + 2) = 3X^2 + 7X + 2$; since the middle term is $7X$ this selection of factors is *correct*.

Example J: Factor $5P^2 + 37P + 14$ and $4R^2 + 17R - 15$.

Algebraic Expression	Possible Factors	Products
1. $5P^2 + 37P + 14$	$(5P + 14)(P + 1)$	$5P^2 + 19P + 14$
	$(P + 14)(5P + 1)$	$5P^2 + 71P + 14$
	$(5P + 7)(P + 2)$	$5P^2 + 17P + 14$
	*$(P + 7)(5P + 2)$	$5P^2 + 37P + 14$

Since $(P + 7)(5P + 2)$ yields the original expression to be factored, it is the correct factored form of $5P^2 + 37P + 14$.

Algebraic Expression	Possible Factors	Products
2. $4R^2 + 17R - 15$	$(2R - 15)(2R + 1)$	$4R^2 - 28R - 15$
	$(2R + 15)(2R - 1)$	$4R^2 + 28R - 15$
	$(4R - 15)(R + 1)$	$4R^2 - 11R - 15$
	$(4R + 15)(R - 1)$	$4R^2 + 11R - 15$
	*$(4R - 3)(R + 5)$	$4R^2 + 17R - 15$
	$(4R + 3)(R - 5)$	$4R^2 - 17R - 15$
	$(R - 15)(4R + 1)$	$4R^2 - 59R - 15$
	$(R + 15)(4R - 1)$	$4R^2 + 59R - 15$
	$(R - 3)(4R + 5)$	$4R^2 - 7R - 15$
	$(R + 3)(4R - 5)$	$4R^2 + 7R - 15$

Since $(4R - 3)(R + 5)$ yields the original expression, it is the only correct factored form.

As the number of factors in the coefficient of the first term and the constant term increase, the more *possible* binomial factors there will be. From the previous

Sec. 2-4. Factoring

example you can see that a number of these factors can be eliminated because they differ only in the algebraic sign of the middle term.

Sometimes more than one type of factoring can be applied to the same problem. A general rule to follow is first to look for a common monomial factor and then to inspect for other types of factoring.

Example K:

1. $16A^4 - 81B^4 = (4A^2 + 9B^2)(4A^2 - 9B^2)$
$= (4A^2 + 9B^2)(2A + 3B)(2A - 3B)$
2. $242AZ^2 - 450AV^2 = 2A(121Z^2 - 225V^2)$
$= 2A(11Z - 15V)(11Z + 15V)$
3. $2R^2 - 12RS + 18S^2 = 2(R^2 - 6RS + 9S^2)$
$= 2(R - 3S)(R - 3S) = 2(R - 3S)^2$
4. $KD^2 - 3KD - 10K = K(D^2 - 3D - 10)$
$= K(D - 5)(D + 2)$
5. $2X^3 + 11X^2 + 5X = X(2X^2 + 11X + 5)$
$= X(2X + 1)(X + 5)$

Exercises

Find a common factor for each of the following.

1. $7M - 28N$
2. $18P - 12Q$
3. $21Z^2 - 7Z$
4. $X^3Y + Y$
5. $18A + 3B - 15C$
6. $49R^2 - 14RS + 21RS^2$
7. $(X - Y)A - (X - Y)B$
8. $2X^MY - 4X^M$
9. $X^{3A+2} + X^{3A}$
10. $D^{3K} + 3D^K$

Each of the following involves the difference between two perfect squares. Write each in its complete factored form.

11. $X^2 - 4$
12. $T^2 - 1$
13. $I^2 - 16$
14. $16S^2 - 1$
15. $4V^2 - 25$
16. $3M^2 - 27$
17. $121 - Z^2$
18. $V^6 - 49Z^6$
19. $(A + B)^2 - Y^2$
20. $16B^4 - 81C^4$

Each of the following involves a perfect trinomial square. Write each in its complete factored form.

21. $A^2 + 4A + 4$
22. $Y^2 - 6Y + 9$
23. $4V^2 - 12V + 9$
24. $16T^2 + 24T + 9$
25. $4M^2 + 20MN + 25N^2$
26. $121Z^2 + 22ZT + T^2$
27. $180R^2 - 300R + 125$
28. $3X^2 + 12X + 12$
29. $4I^2 + 12I + 9$
30. $.04S^2 - 2S + 25$

Each of the following is a general expression. Factor each expression completely.

31. $A^2 + 6A + 5$
32. $Y^2 - 4Y + 5$
33. $U^2 + 6U + 8$
34. $Q^2 + Q - 2$
35. $B^2 + 4B - 12$
36. $I^2 - 7I + 6$
37. $V^2 - 6V - 16$
38. $C^2 + 9CD + 8D^2$
39. $3X^2 + 6XY - 144Y^2$
40. $24 - 14X + 2X^2$
41. $4R^2 + 12RS + 5S^2$
42. $3F^2 + 11F + 6$
43. $6Y^2 + 11Y + 3$
44. $2A^2M - 23AM - 39M$
45. $6Z^2 - 13ZW - 6W^2$
46. $3S^3B + 27S^2B + 54SB$
47. $15T^2 - 21T - 18$
48. $15A^2 + 5A - 10$
49. $9S^2 - 12S + 4$
50. $8A^2 - 24A - 1440$

2-5. Fractions and Ratios

One type of expression that anyone studying mathematics must be able to work with correctly and efficiently is a fraction. In this section we will take a detailed look at fractional quantities.

A *fraction* is an expression of the type $\frac{A}{B}$, where A is called the *numerator* and B is called the *denominator*. The denominator of a fraction is its "identification tag" that tells us what type of "things" we are dealing with. The numerator of a fraction tells us how many of these things we are talking about. So, just as "2 pounds" tells us we are talking about pounds and have "two" of them, "$\frac{2}{3}$" tells us we are talking about "thirds" and have "two" of them.

Example A:
1. $\frac{7}{9}$ is a fraction with a numerator of 7 and a denominator of 9.
2. $\frac{-12}{5}$ is fraction with a numerator of -12 and a denominator of 5.

NOTE: Recall from Section 1-1 that $\frac{-12}{5} = \frac{12}{-5} = -\frac{12}{5}$.

3. $\frac{X + 3}{Y - 4}$ is a fraction with a numerator of $(X + 3)$ and a denominator of $(Y - 4)$.
4. $\frac{T^2 - 4V}{5\sqrt{T} + 3}$ is a fraction with a numerator of $(T^2 - 4V)$ and a denominator of $(5\sqrt{T} + 3)$.

Two fractions, $\frac{A}{B}$ and $\frac{C}{D}$, are said to be equivalent if there is a nonzero

Sec. 2-5. Fractions and Ratios

number K such that $C = KA$ and $D = KB$. In other words, *the numerator and denominator of a fraction may be multiplied or divided by the same nonzero number without changing the value of the fraction.* This is called the *basic principle of fractions.*

Example B:

1. $\dfrac{1}{2} = \dfrac{1 \cdot 4}{2 \cdot 4} = \dfrac{4}{8}$

2. $\dfrac{-7}{-9} = \dfrac{-7 \cdot (-1)}{-9 \cdot (-1)} = \dfrac{7}{9}$

3. $\dfrac{3}{9} = \dfrac{3 \div 3}{9 \div 3} = \dfrac{1}{3}$

4. $\dfrac{24}{60} = \dfrac{24 \div 12}{60 \div 12} = \dfrac{2}{5}$

5. $\dfrac{X}{Y^2} = \dfrac{X \cdot T}{Y^2 \cdot T} = \dfrac{XT}{Y^2 T}$

6. $\dfrac{A + B}{A - B} = \dfrac{(A + B) \cdot 3}{(A - B) \cdot 3} = \dfrac{3A + 3B}{3A - 3B}$

7. $\dfrac{2}{\sqrt{K}} = \dfrac{2 \cdot \sqrt{K}}{\sqrt{K} \cdot \sqrt{K}} = \dfrac{2\sqrt{K}}{K}$

8. $\dfrac{X^2}{X^5} = \dfrac{X^2 \div X^2}{X^5 \div X^2} = \dfrac{1}{X^3}$

This basic principle is used to reduce fractions to lowest terms. A fraction is in *lowest terms* when the numerator and denominator have no common factor except 1 or -1. *To reduce a fraction to lowest terms, divide numerator and denominator by their greatest common factor.*

Example C:

1. $\dfrac{8}{12} = \dfrac{2 \cdot 4}{3 \cdot 4} = \dfrac{2}{3}$ The greatest common factor is 4.

2. $\dfrac{-18}{81} = \dfrac{-2(9)}{9(9)} = \dfrac{-2}{9}$ The greatest common factor is 9.

3. $\dfrac{12A^2 B}{18AB} = \dfrac{2A(6AB)}{3(6AB)} = \dfrac{2A}{3}$ The greatest common factor is $6AB$.

4. $\dfrac{25T^2 \sqrt{K}}{10T} = \dfrac{5T(5T)\sqrt{K}}{2(5T)} = \dfrac{5T\sqrt{K}}{2}$ The greatest common factor is $5T$.

5. $\dfrac{R^2 - S^2}{R + S} = \dfrac{(R - S)(R + S)}{(R + S)} = R - S$ The greatest common factor is $(R + S)$.

6. $\dfrac{K^2 - 5K + 6}{K^2 - 9} = \dfrac{(K - 2)(K - 3)}{(K + 3)(K - 3)} = \dfrac{K - 2}{K + 3}$ The greatest common factor is $(K - 3)$.

One important point to be mentioned here is that we may only divide numerator and denominator by common factors, *not* common terms.

Example D: The fraction $\dfrac{X^2 - 3}{X^2}$ may not be reduced by dividing numerator and denominator by X^2, since X^2 is a term in the numerator and not a factor. This fraction cannot be reduced at all since $(X^2 - 3)$ and X^2 have no common factors.

When adding or subtracting fractions, we must observe the general rule of adding or subtracting similar quantities. Since the denominator of a fraction identifies a fractional quantity, *we may only add or subtract fractions that have the same denominator.* So, just as 2 pounds + 3 pounds = 5 pounds and $2X + 3X = 5X$, $\frac{2}{7} + \frac{3}{7} = \frac{5}{7}$.

Example E:
1. $\frac{4}{9} + \frac{2}{9} = \frac{6}{9} = \frac{2}{3}$
2. $\frac{11}{12} - \frac{9}{12} = \frac{2}{12} = \frac{1}{6}$
3. $\frac{7}{8} - \frac{2}{8} + \frac{5}{8} = \frac{10}{8} = \frac{5}{4}$
4. $\dfrac{2}{X} + \dfrac{7}{X} = \dfrac{9}{X}$
5. $\dfrac{5}{T^2} - \dfrac{9}{T^2} + \dfrac{A}{T^2} = \dfrac{A - 4}{T^2}$
6. $\dfrac{B}{K+2} - \dfrac{7}{K+2} = \dfrac{B-7}{K+2}$

NOTE: All answers should be reduced to lowest terms.

If fractions do not have the same denominators, we must make use of the *basic principle* to make all denominators the same. That is, we must have a *common denominator*. Symbolically, addition of fractions with unlike denominators is represented by

$$\frac{A}{B} + \frac{C}{D} = \frac{AD + BC}{BD}$$

Although any common denominator will work, the one that is most convenient is the *lowest common denominator*. This is the simplest expression into which each of the given denominators will divide evenly. For example, if we wish to add $\frac{2}{3}$ and $\frac{3}{4}$, we can certainly see that the lowest common denominator would be 12, since it is the simplest expression into which 3 and 4 (the given denominators) divide evenly. If the lowest common denominator cannot be found by inspection, it can be determined by factoring each of the given denominators completely. Adding the fractions

$$\frac{3}{X+1} \text{ and } \frac{9}{X^2 - 1},$$

we write the denominators $(X + 1)$ and $(X - 1)(X + 1)$ respectively. We can see that the lowest common denominator is $(X + 1) \cdot (X - 1)$ since it is the

Sec. 2–5. Fractions and Ratios

simplest expression into which $(X+1)$ and $(X-1)(X+1)$ divide evenly. Once the lowest common denominator is obtained, we use the basic principle to change each of the given fractions to an equivalent fraction having the lowest common denominator. Therefore, in the first example

$$\frac{2}{3} = \frac{2 \cdot 4}{3 \cdot 4} = \frac{8}{12}$$

and

$$\frac{3}{4} = \frac{3 \cdot 3}{4 \cdot 3} = \frac{9}{12}$$

and thus,

$$\tfrac{2}{3} + \tfrac{3}{4} = \tfrac{8}{12} + \tfrac{9}{12} = \tfrac{17}{12}.$$

In the second example,

$$\frac{3}{X+1} = \frac{3(X-1)}{(X+1)(X-1)}$$

and

$$\frac{9}{(X+1)(X-1)}$$

stays the same. Therefore

$$\frac{3}{X+1} + \frac{9}{X^2-1} = \frac{3(X-1)}{(X+1)(X-1)} + \frac{9}{(X+1)(X-1)}$$
$$= \frac{3X-3+9}{(X+1)(X-1)} = \frac{3X+6}{(X+1)(X-1)}$$

Example F:

1. $\dfrac{7}{5} + \dfrac{2}{3} - \dfrac{5}{6} = \dfrac{7(6)}{30} + \dfrac{2(10)}{30} - \dfrac{5(5)}{30} = \dfrac{42+20-25}{30} = \dfrac{37}{30}$

2. $\dfrac{4}{R^2} - \dfrac{2}{S} + \dfrac{3}{RS^2} = \dfrac{4(S^2)}{R^2S^2} - \dfrac{2(R^2S)}{R^2S^2} + \dfrac{3(R)}{R^2S^2} = \dfrac{4S^2 - 2R^2S + 3R}{R^2S^2}$

3. $\dfrac{A}{(B+1)} - \dfrac{7}{(B-1)} = \dfrac{A(B-1)}{(B+1)(B-1)} - \dfrac{7(B+1)}{(B+1)(B-1)}$
$= \dfrac{AB-A-7B-7}{(B+1)(B-1)}$

NOTE: If a minus sign precedes an expression, the signs of all terms in that expression must be changed.

4. $\dfrac{5T}{K^2} + \dfrac{6}{K-1} - \dfrac{3T}{K^2-K} = \dfrac{5T}{K^2} + \dfrac{6}{(K-1)} - \dfrac{3T}{K(K-1)}$
$= \dfrac{5T(K-1)}{K^2(K-1)} + \dfrac{6(K^2)}{K^2(K-1)} - \dfrac{3T(K)}{K^2(K-1)}$
$= \dfrac{5TK - 5T + 6K^2 - 3TK}{K^2(K-1)}$
$= \dfrac{2TK - 5T + 6K^2}{K^2(K-1)}$

5. $\dfrac{X}{4-X} + \dfrac{3X}{X^2 - 16} = \dfrac{X}{-(X-4)} + \dfrac{3X}{(X+4)(X-4)}$

$= \dfrac{-(X+4)X}{(X+4)(X-4)} + \dfrac{3X}{(X+4)(X-4)}$

$= \dfrac{-X^2 - 4X + 3X}{(X+4)(X-4)} = \dfrac{-X^2 - X}{(X+4)(X-4)}$

NOTE: In this problem, $(4 - X) = -(X - 4)$.

6. $\dfrac{2T}{T^2 - 9} - \dfrac{5T - 7}{T^2 + T - 6} = \dfrac{2T}{(T+3)(T-3)} - \dfrac{5T - 7}{(T+3)(T-2)}$

$= \dfrac{2T(T-2)}{(T+3)(T-3)(T-2)}$

$\quad - \dfrac{(5T-7)(T-3)}{(T+3)(T-3)(T-2)}$

$= \dfrac{2T^2 - 4T - (5T^2 - 15T - 7T + 21)}{(T+3)(T-3)(T-2)}$

$= \dfrac{2T^2 - 4T - 5T^2 + 15T + 7T - 21}{(T+3)(T-3)(T-2)}$

$= \dfrac{-3T^2 + 18T - 21}{(T+3)(T-3)(T-2)}$

When multiplying fractions, the product will be a fraction whose numerator is the product of the given numerators and whose denominator is the product of the given denominators. Symbolically, this would be represented as:

$$\dfrac{A}{B} \times \dfrac{C}{D} = \dfrac{AC}{BD}$$

Since all answers should be in simplest form, and since all factors in the numerators and all factors in the denominators are to be multiplied, it is usually easier to factor each expression, indicate the multiplication, and simplify before multiplying. This is called *canceling* common factors.

Example G:

1. $\dfrac{7}{9} \cdot \dfrac{3}{14} \cdot \dfrac{6}{11} = \dfrac{(7)(3)(3)(2)}{(3)(3)(7)(2)(11)} = \dfrac{1}{11}$

2. $\dfrac{4}{5} \cdot \dfrac{-10}{13} \cdot \dfrac{5}{8} = \dfrac{(2)(2)(-5)(2)(5)}{(5)(13)(2)(2)(2)} = \dfrac{-5}{13}$

3. $\dfrac{5}{X} \cdot \dfrac{X^2}{Y} \cdot \dfrac{3}{XY} = \dfrac{(5)(X)(X)(3)}{(X)(Y)(X)(Y)} = \dfrac{15}{Y^2}$

4. $\dfrac{K+3}{K} \cdot \dfrac{K^2}{K^2 - 9} = \dfrac{(K+3)(K)(K)}{(K)(K+3)(K-3)} = \dfrac{K}{K-3}$

5. $\dfrac{T^2 - 4}{T + 1} \cdot \dfrac{T^2 + T}{T - 2} = \dfrac{(T+2)(T-2)(T)(T+1)}{(T+1)(T-2)} = T(T+2)$

6. $\dfrac{A - 3}{5} \cdot \dfrac{2A}{9 - A^2} \cdot \dfrac{5A + 10}{3A^2} = \dfrac{(A-3)(2)(A)(5)(A+2)}{(5)(3 - A)(3 + A)(3)(A)(A)}$

$= \dfrac{2(A+2)}{-3A(3+A)}$

Sec. 2-5. Fractions and Ratios

NOTE: $(A - 3)$ and $(3 - A)$ differ only by a factor of (-1). Therefore, they cancel each other and the (-1) appears in the answer since $(-3) = (-1) \cdot 3$.

Division of fractions is represented as

$$\frac{A}{B} \div \frac{C}{D} = \frac{AD}{BC}$$

By rewriting the indicated division as

$$\frac{\frac{A}{B}}{\frac{C}{D}}$$

and using the basic principle, we can verify this rule.

$$\frac{\frac{A}{B} \cdot \frac{D}{C}}{\frac{C}{D} \cdot \frac{D}{C}} = \frac{\frac{AD}{BC}}{1} = \frac{AD}{BC}$$

Examining this rule closely, we can see that the procedure for dividing one fraction by another is to take the reciprocal of the divisor and change the division to multiplication.

Example H:

1. $\dfrac{2}{3} \div \dfrac{7}{9} = \dfrac{2}{3} \cdot \dfrac{9}{7} = \dfrac{(2)(3)(3)}{(3)(7)} = \dfrac{6}{7}$

2. $\dfrac{6T}{V} \div \dfrac{8T}{V^2} = \dfrac{6T}{V} \cdot \dfrac{V^2}{8T} = \dfrac{(2)(3)(T)(V)(V)}{(V)(2)(2)(2)(T)} = \dfrac{3V}{4}$

3. $\dfrac{X + 2}{X^2} \div \dfrac{(X^2 - 4)}{7} = \dfrac{X + 2}{X^2} \cdot \dfrac{7}{X^2 - 4} = \dfrac{(X + 2)(7)}{(X)(X)(X + 2)(X - 2)}$
$= \dfrac{7}{X^2(X - 2)}$

4. $\dfrac{\frac{9A^2}{B}}{\frac{3A}{B^3}} = \dfrac{9A^2}{B} \cdot \dfrac{B^3}{3A} = \dfrac{(3)(3)(A)(A)(B)(B)(B)}{(B)(3)(A)} = 3AB^2$

5. $\dfrac{\frac{K^2 - 16}{K^3}}{\frac{K^2 - 5K + 4}{K}} = \dfrac{K^2 - 16}{K^3} \cdot \dfrac{K}{K^2 - 5K + 4}$
$= \dfrac{(K + 4)(K - 4)(K)}{(K)(K)(K)(K - 4)(K - 1)} = \dfrac{K + 4}{K^2(K - 1)}$

Another way of describing fractions is to say that they can be used to represent *ratios*. This was first pointed out in Section 1–1. A ratio is a way of indicating a comparison. Thus, $\frac{A}{B}$ can be interpreted as the ratio of A to B. Ratios can also be represented in the form $A:B$. Therefore,

$$\frac{A}{B} = A:B = \text{the ratio of } A \text{ to } B.$$

It is important to point out that only like or similar quantities may be compared. Thus, we may form ratios of money to money, feet to feet, and pounds to pounds. From this last statement, we should realize that *not every fraction is a ratio*.

Example I:
1. The length of a certain field is 150 feet and the width is 30 feet. Therefore, the ratio of length to width is

$$\frac{150 \text{ ft}}{30 \text{ ft}} = \frac{5}{1}$$

2. The specific gravity of silver is 10.5 and the specific gravity of zinc is 7.0. Therefore, the ratio of the specific gravity of silver to that of zinc is

$$\frac{10.5}{7.0} = \frac{1.5}{1} = 1.5$$

Example J: The fraction

$$\frac{1000 \text{ mi}}{8 \text{ hr}} = 125 \text{ mi/hr}$$

is not a ratio but merely a division problem since the terms *miles* and *hours* are not similar (and cannot be made similar). We are really not comparing quantities here.

Measurements can also be considered as ratios. For example, when we say that a person is 6 feet tall, we are saying that the ratio of that person's height to the accepted unit of height measurement, the foot, is 6 to 1. If a force acting on an object is measured to be 120 pounds, we mean that the ratio of the magnitude of this force to the accepted unit for measuring force, the pound, is 120 to 1.

When a given quantity is to be divided into parts that must satisfy a certain ratio, we must remember that the stated ratio must be satisfied *after* the parts are selected and therefore may not be used directly to find the parts.

Example K: Divide a 150 pound force into two separate forces that are in the ratio of $2:3$. We can see that if the ratio is used in the form $\frac{2}{3}$, $\frac{2}{3}$ of 150 is 100.

Sec. 2–5. Fractions and Ratios 43

Thus, we would get forces of 50 and 100 pounds, which are certainly not in the ratio of 2:3. What we must do is add 2 and 3 to get 5. Then we take $\frac{2}{5}$ of 150 to get 60 and $\frac{3}{5}$ of 150 to get 90. Thus, the two required forces are 60 pounds and 90 pounds, which are obviously in the ratio 2:3.

Example L:
1. Divide a board of length 16 feet into two pieces that are in the ratio 3:5.

$$3 + 5 = 8$$
$$\tfrac{3}{8} \times \tfrac{16}{1} = \tfrac{48}{8} = 6$$
$$\tfrac{5}{8} \times \tfrac{16}{1} = \tfrac{80}{8} = 10$$

The pieces must be 6 feet and 10 feet.

2. Divide a 200-pound force into three separate forces that are in the ratio 2:3:5.

$$2 + 3 + 5 = 10$$
$$\frac{2}{10} \times \frac{200}{1} = \frac{400}{10} = 40, \quad \frac{3}{10} \times \frac{200}{1} = \frac{600}{10} = 60,$$
$$\frac{5}{10} \times \frac{200}{1} = \frac{1000}{10} = 100$$

The forces must be 40 pounds, 60 pounds, and 100 pounds.

Exercises Reduce each of the following fractions to simplest form.

1. $\dfrac{36}{132}$
2. $\dfrac{42}{78}$
3. $\dfrac{5A^3X}{10AX^4}$
4. $\dfrac{2M + 6}{M^2 - 9}$
5. $\dfrac{T^2 - 25}{T^2 + 10T + 25}$
6. $\dfrac{2K^2 - 12K - 14}{K^3 - K}$

For each of the following, perform the indicated operation(s) and express each answer in simplest form.

7. $\dfrac{5}{6} - \dfrac{3}{4} + \dfrac{9}{8}$
8. $\dfrac{7}{9} + \dfrac{11}{12} - \dfrac{7}{4}$
9. $\dfrac{2}{A^2} - \dfrac{7}{A} + \dfrac{5}{A^3}$
10. $\dfrac{X}{Y} + \dfrac{3X}{Y^2} - \dfrac{6}{5Y}$
11. $\dfrac{T}{V + 1} - \dfrac{T^2}{V^2 + V}$
12. $\dfrac{M + 3}{M^2} + \dfrac{5M}{M^3 - M^2}$
13. $\dfrac{K + 1}{K^2 - 4} - \dfrac{K}{K + 2} + \dfrac{2K - 3}{K^2 - 2K}$

14. $\dfrac{R}{R-3} + \dfrac{R+1}{R^2 - 4R + 3} - \dfrac{R-4}{2R^2 - R - 1}$

15. $\dfrac{6}{7} \cdot \dfrac{5}{12} \cdot \dfrac{3}{10}$

16. $\dfrac{5}{9} \cdot \dfrac{6}{25} \cdot \dfrac{27}{4}$

17. $\dfrac{A^2}{B} \cdot \dfrac{A}{3B^4} \cdot \dfrac{B^2}{5A}$

18. $\dfrac{M^4}{N} \cdot \dfrac{N^2}{2M} \cdot \dfrac{4}{MN}$

19. $\dfrac{R^2 - 1}{R^3} \cdot \dfrac{R}{R^2 - 5R + 4}$

20. $\dfrac{V^3 + 2V^2}{V - 5} \cdot \dfrac{2V^2 - 9V - 5}{V^2 - 4}$

21. $\dfrac{T^3 - 4T^2 + 4T}{T^4} \cdot \dfrac{T^2}{4T + 8} \cdot \dfrac{4T}{3T^2 - 7T + 2}$

22. $\dfrac{K - 6}{2K} \cdot \dfrac{K^3}{2K^2 - 10K - 12} \cdot \dfrac{K + 1}{K^2}$

23. $\dfrac{5}{8} \div \dfrac{15}{32}$

24. $\dfrac{16}{21} \div \dfrac{28}{33}$

25. $\dfrac{9A^2}{2B} \div \dfrac{12A^3}{6B^4}$

26. $\dfrac{6XY}{Z} \div \dfrac{2X}{YZ^2}$

27. $\dfrac{M - 4}{M^5} \div \dfrac{2M^2 - 32}{4M^2}$

28. $\dfrac{R^2 - 2R - 3}{2R + 2} \div \dfrac{3R^2 - 27}{5R^2 + 17R + 6}$

29. $\dfrac{5}{T^2} - \left(\dfrac{6T}{T + 1} \cdot \dfrac{T^2 - 1}{8T} \right)$

30. $\dfrac{6}{X - 2} + \left(\dfrac{X + 3}{X} \div \dfrac{X^2 - 9}{2X - 6} \right)$

31. $\dfrac{Y + 1}{Y^3} \div \left(\dfrac{Y - 2}{Y} \cdot \dfrac{Y^2 - 1}{2Y^2 - 4Y} \right)$

32. $\dfrac{K^2}{K - 5} \cdot \left(\dfrac{K - 1}{K} \div \dfrac{K^2 - 9K + 8}{K^2 - 13K + 40} \right)$

33. $\dfrac{\dfrac{2}{X^2} - \dfrac{5}{X}}{\dfrac{7}{X} + \dfrac{6}{X^3}}$

34. $\dfrac{\dfrac{A^2}{3B} + \dfrac{2A}{5B}}{\dfrac{A}{2B^2} - \dfrac{3A}{B^4}}$

For each of the following, express the ratios in simplest form.

35. 32 feet to 8 feet
36. 6 kilograms to 24 kilograms
37. 12 amps to 9 amps
38. 186 meters to 78 meters
39. 3 hours to 25 minutes
40. 6 pounds to 8 ounces
41. Divide a force of 240 pounds into two forces that are in the ratio of $3:5$.
42. Divide a field of area 600 square feet into two smaller fields whose areas are in the ratio of $7:5$.

Sec. 2–6. Equations and Formulas Solved for One First Degree Variable

43. Divide a warehouse of area 4500 square feet into three sections whose areas are in the ratio of $3:5:7$.

44. Divide a distance of 70 miles into four shorter distances that are in the ratio of $2:3:4:5$.

2–6. Equations and Formulas Solved for One First Degree Variable

The fields of engineering and technology deal with physical problems that can be solved if a mathematical equation can be written expressing the relationship between the quantities involved in each problem.

This section will deal with solving equations that involve one first degree variable and writing such equations from a verbal statement.

An *equation* is a statement of equality between two quantities. For example, the statements $(5 + 3 = 8)$ and $(X + 4 = 6)$ are equations. The second statement is a *conditional* equation. That is, it is a true statement only if 2 is substituted in place of the variable X. If any other real number is substituted in place of X it will be a false statement. An equation such as $X^2 - Y^2 = (X + Y)(X - Y)$ will be true for any real numbers substituted in place of X and Y. An equation of this type is an *identity*. More specialized statements of equality are formulas that express a relationship between a number of quantities. An example of an equation of this type is the formula for the circumference of a circle, $C = \pi D$, where C represents the circumference, D represents the diameter of the circle, and π is a constant. Two equations are *equivalent* if they have the same solution. The equations $X + 4 = 6$ and $X = 2$ are equivalent because if 2 is substituted for X in each of the two equations, the resulting statements are true.

To solve an equation for a first degree variable it is necessary to write an equivalent equation in which the variable for which we are solving is isolated on one side of the equation and all other terms and factors are on the other side of the equation.

There are two basic properties of equations that will result in equivalent equations. These properties are:

1. Any real number may be added to both members of the equation.
2. Each member of the equation may be multiplied by the same nonzero real number.

The following table shows these properties applied to four basic equations and it also shows a transformation that gives the same result as the basic properties.

Equation	Equivalent Equation	Transformation
1. $T + 4 = 12$	$T + 4 + (-4) = 12 + (-4)$ $T = 8$	$T = 12 - 4$ $T = 8$
2. $V - 7 = 14$	$V - 7 + 7 = 14 + 7$ $V = 21$	$V = 14 + 7$ $V = 21$
3. $2S = 10$	$\frac{1}{2}(2S) = 10(\frac{1}{2})$ $S = 5$	$S = 10(\frac{1}{2})$ $S = 5$
4. $\frac{5}{4}B = 30$	$(\frac{4}{5})\frac{5}{4}B = 30(\frac{4}{5})$ $B = 24$	$B = 30(\frac{4}{5})$ $B = 24$

Example A: Solve the equation:

$$\frac{5X}{6} + 7 = 47$$

Add -7 to both members.

$$\frac{5X}{6} = 40$$

Multiply both members by $\frac{6}{5}$.

$$X = 48$$

Check: $\frac{5}{6}(48) + 7 = 47$
$40 + 7 = 47$
$47 = 47$

Since 48 makes the original equation true, it is the solution to the equation.
NOTE: It is not always *necessary* to check the solution to an equation. The basic properties of an equation guarantee that if one of these properties is applied to an equation the resulting equation is equivalent to the original one. The only time a solution must be checked is when an operation, other than addition or subtraction, involving the variable has been performed. The only purpose the check serves in other cases is to assure you that the computations are correct.

Example B: Solve the equation:

$$\tfrac{3}{4} - 5X = -2X + 3$$

Add $2X$ to both members.

$$\tfrac{3}{4} - 3X = 3$$

Add $-\tfrac{3}{4}$ to both members.

$$-3X = \tfrac{12}{4} - \tfrac{3}{4} = \tfrac{9}{4} \quad (3 = \tfrac{12}{4})$$

Multiply both members by $(-\tfrac{1}{3})$.

Sec. 2–6. Equations and Formulas Solved for One First Degree Variable   47

$$X = -\tfrac{3}{4}$$

Check: $\tfrac{3}{4} - 5(-\tfrac{3}{4}) = -2(-\tfrac{3}{4}) + 3$
$\tfrac{3}{4} + \tfrac{15}{4} = \tfrac{6}{4} + \tfrac{12}{4}$
$\tfrac{18}{4} = \tfrac{18}{4}$

Example C: Solve the equation:

$$4 - (2R + 3) = 39$$

Use the distributive property to eliminate the parentheses.

$$4 - 2R - 3 = 39$$

Simplify the left member.

$$-2R + 1 = 39$$

Add -1 to both members.

$$-2R = 38$$

Multiply both members by $(-\tfrac{1}{2})$.

$$R = -19$$

Check: $4 - [2(-19) + 3] = 39$
$4 - (-38 + 3) = 39$
$4 - (-35) = 39$
$39 = 39$

Example D: Solve the equation:

$$\frac{G}{5} + 3 = \frac{14}{3}$$

Multiply both members by 15 (the LCD of 5 and 3).

$$15 \cdot \left(\frac{G}{5} + 3\right) = \frac{14}{3} \cdot 15$$

Simplify both members.

$$3G + 45 = 70$$

Add -45 to both members.

$$3G = 25$$

Multiply both members by $\tfrac{1}{3}$.

$$G = \frac{25}{3}$$

Check: $\dfrac{\frac{25}{3}}{5} + 3 = \dfrac{14}{3}$

$$\dfrac{25}{15} + \dfrac{45}{15} = \dfrac{14}{3}$$

$$\dfrac{70}{15} = \dfrac{14}{3}$$

$$\dfrac{14}{3} = \dfrac{14}{3}$$

Example E: Solve the equation:

$$\dfrac{A+2}{3} - \dfrac{2A-7}{2} = \dfrac{3}{4}$$

Multiply both members by 12 (the LCD of 3, 2, and 4).

$$12\left(\dfrac{A+2}{3} - \dfrac{2A-7}{2}\right) = \dfrac{3}{4} \cdot 12$$

$$4(A+2) - 6(2A-7) = 9$$

Eliminate the parentheses.

$$4A + 8 - 12A + 42 = 9$$

Simplify the left member.

$$-8A + 50 = 9$$

Add -50 to both members.

$$-8A = -41$$

Multiply both members by $-\tfrac{1}{8}$.

$$A = \dfrac{41}{8}$$

Check: $\dfrac{\frac{41}{8}+2}{3} - \dfrac{2(\frac{41}{8})-7}{2} = \dfrac{3}{4}$

$$\dfrac{\frac{41}{8}+\frac{16}{8}}{3} - \dfrac{\frac{82}{8}-\frac{56}{8}}{2} = \dfrac{3}{4}$$

$$\dfrac{\frac{57}{8}}{3} - \dfrac{\frac{26}{8}}{2} = \dfrac{3}{4}$$

$$\dfrac{19}{8} - \dfrac{13}{8} = \dfrac{3}{4}$$

$$\dfrac{6}{8} = \dfrac{3}{4}$$

$$\dfrac{3}{4} = \dfrac{3}{4}$$

Sec. 2–6. Equations and Formulas Solved for One First Degree Variable

Example F: Solve the equation:

$$\frac{2}{M-1} = \frac{3}{M}$$

Multiply both members by $M(M-1)$ (the LCD of M and $M-1$).

$$M(M-1)\frac{2}{M-1} = \frac{3}{M}M(M-1)$$

$$2M = 3M - 3$$

Add $-3M$ to both members.

$$-M = -3$$

Multiply both members by -1.

$$M = 3$$

Check: Since the LCD contains the variable M it is *necessary* to check the solution in the original equation.

$$\frac{2}{3-1} = \frac{3}{3}$$

$$1 = 1$$

Example G: Solve the equation:

$$\frac{4}{V^2 - V} - \frac{4}{V} = \frac{4}{V-1}$$

Multiply both members by $V(V-1)$ [the LCD of V, $V-1$, and $V(V-1)$].

$$V(V-1)\left[\frac{4}{V(V-1)} - \frac{4}{V}\right] = \frac{4}{V-1}V(V-1)$$

$$4 - 4V + 4 = 4V$$

Simplify the left member.

$$8 - 4V = 4V$$

Add $4V$ to both members.

$$8 = 8V$$

Multiply both members by $\frac{1}{8}$.

$$1 = V$$

Check: Since the LCD contains the variable V it is *necessary* to check the solution in the original equation.

$$\frac{4}{1-1} - \frac{4}{1} = \frac{4}{1-1}$$

$$\frac{4}{0} - \frac{4}{1} = \frac{4}{0}$$

Since the denominator contains zero the fraction is undefined and $V \neq 1$. Thus, there is no solution to this equation.

A *radical equation* is an equation, containing a radical expression ($\sqrt[N]{}$), in which the radicand contains the variable. The procedure for solving an equation of this type is to isolate the radical and then raise both members of the equation to the Nth power. Since this procedure is not a basic property of an equation it is possible to have extraneous roots (extra roots of the equation). Therefore, *it is necessary to perform the check* in order to find the solution to the equation. In this section we will restrict ourselves to radical equations that will result in a first degree variable after squaring both members of the equation. In doing so, we make use of the rule $\sqrt[N]{A^N} = A$, with $N = 2$.

Example H: Solve the following equations:
1. $\sqrt{X+1} = 2$
   Square both members.

   $$(\sqrt{X+1})^2 = 2^2$$
   $$X + 1 = 4$$

   Add -1 to both members.

   $$X = 3$$

   Check: $\sqrt{3+1} = 2$
   $\sqrt{4} = 2$ (Recall that $\sqrt{4}$ means the "principal square root of 4".)
   $2 = 2$

2. $3\sqrt{X} + 8 = 14$
   Add -8 to both members.

   $$3\sqrt{X} = 6$$

   Square both members.

   $$(3\sqrt{X})^2 = 6^2$$
   $$9X = 36$$

   Multiply both members by $\frac{1}{9}$.

   $$X = 4$$

   Check: $3\sqrt{4} + 8 = 14$
   $(3)(2) + 8 = 14$
   $14 = 14$

3. $5 - \sqrt{X} = 7$

Add -5 to both members.

$$-\sqrt{X} = 2$$

Square both members.

$$(-\sqrt{X})^2 = 2^2$$
$$X = 4$$

Check: $5 - \sqrt{4} = 7$
$5 - 2 = 7$
$3 \neq 7$

Since the equation is not true when $X = 4$, there is no solution to this equation.

A number of *technical problems* involve *formulas* that must be solved for a single variable, regardless of the number of variables appearing in the formula. It is possible to isolate any one of the variables on one side of the equation, using the basic properties of an equation.

Example I: Solve for D:

$$C = \pi D$$

Recall that the formula for the circumference of a circle is $C = \pi D$. Multiply both members by $\dfrac{1}{\pi}$.

$$\frac{C}{\pi} = D$$

Example J: Solve for C:

$$F = \tfrac{9}{5}C + 32$$

This formula states the relationship between Fahrenheit and Celsius temperatures. Add -32 to both members.

$$F - 32 = \frac{9C}{5}$$

Multiply both members by $\tfrac{5}{9}$.

$$\tfrac{5}{9}(F - 32) = C$$

Example K: Solve for n:

$$L = a + (n - 1)d$$

This formula is used to find the last term of an arithmetic progression. Add $-a$ to both members.

$$L - a = (n - 1)d$$

Multiply both members by $\frac{1}{d}$.

$$\frac{L-a}{d} = n-1$$

Add 1 to both members.

$$\frac{L-a}{d} + 1 = n$$

Example L: An equation encountered when dealing with gases in chemistry is $PV = RT$. Solve this equation for V.
Basic equation:

$$PV = RT$$

Multiply both members by $\frac{1}{P}$.

$$V = \frac{RT}{P}$$

Example M: The kinetic energy of an object is given by $KE = \frac{1}{2}mv^2$. Solve this equation for m. Basic equation:

$$KE = \frac{1}{2}mv^2$$

Multiply both members by 2.

$$2KE = mv^2$$

Multiply both members by $\frac{1}{v^2}$.

$$\frac{2KE}{v^2} = m$$

An equation involving absolute value can be solved only after the absolute value symbol has been removed. By definition, the absolute value of a number is the distance (disregarding direction) of a number from some reference point (usually zero). Thus, $|X| = 8$ means all numbers that are eight units from zero. Therefore $X = 8$ and -8. (Refer to Section 1–3.)

In general, an absolute value equation can be written without the absolute value symbol in the following manner.

$$|X| = A \text{ is equivalent to } X = A \text{ or } X = -A$$

Example N: Solve the equation:

$$|X| = 4$$
$$X = 4 \text{ or } X = -4$$

Sec. 2–6. Equations and Formulas Solved for One First Degree Variable

Check: $|4| = 4$ and $|-4| = 4$

Example O: Solve the equation:
$$|2R - 3| = 9$$

$2R - 3 = 9$	or	$2R - 3 = -9$
$2R = 12$	Add 3 to both members	$2R = -6$
$R = 6$	Multiply both members by $\frac{1}{2}$	$R = -3$
Check: $|12 - 3| = 9$	and	$|-6 - 3| = 9$

Example P: Solve the equation:
$$|A + 2| = 6$$

$A + 2 = 6$	or	$A + 2 = -6$
$A = 4$	Add -2 to both members	$A = -8$
Check: $|4 + 2| = 6$	and	$|(-8) + 2| = 6$

Before an equation can be solved it is often necessary to write the equation from a verbal statement. A careful reading of the statement will usually make it possible to translate the verbal statement into mathematical symbols.

Example Q: The current in one electrical component is $1\frac{1}{2}$ times the current in another component. If the sum of the currents is 7.5 amperes, what is the current in each component?

Let $I =$ the current (in amperes) in one component. $1.5I =$ the current (in amperes) in the other component.

$$1.5I + I = 7.5$$
$$2.5I = 7.5$$
$$I = \frac{7.5}{2.5}$$
$$I = 3$$

The current in one component is 3 amperes. The current in the other component is 4.5 amperes.

Check: Since $1\frac{1}{2}(3) + (3) = 7\frac{1}{2}$, and since 3 amperes fits the description of the verbal statement, then it satisfies the problem.

Example R: A certain type of spring increases $\frac{1}{2}$ foot in length for each pound it supports. If the spring is 10 feet long when 8 pounds are attached to it, what was the original length of the spring?

Let $L =$ the original length of the spring in feet.

$$L + .5(8) = 10$$
$$L + 4 = 10$$
$$L = 6$$

Check: If the original length of the spring is 6 feet, and it increases $\frac{1}{2}$ foot for each pound added to it and there are 8 pounds added, then the spring would be stretched an additional 4 feet. Since the sum of 6 feet and 4 feet is equal to the overall length of 10 feet, the solution is correct.

Example S: A sewage-treatment plant has a water aerator system featuring the use of a circular tank 100 feet in diameter. What is the surface area of the water exposed to the air?

This problem can be solved by finding the area of a circle. The formula for the area of a circle is $A = \pi R^2$. Since the formula calls for the radius as a substitution, we use 50 feet (half of the diameter of 100 feet).

$$A = \pi(50)^2$$
$$A = \pi(2500 \text{ sq ft})$$
$$A = (3.14)(2500 \text{ sq ft})$$
$$A = 7850 \text{ sq ft}$$

Example T: A certain earth satellite completes a circular orbit in 2 hours and 15 minutes. If it is 180 miles above the surface of the earth and the radius of the earth is 4000 miles, how fast does the satellite travel?

In this problem we must find the total distance travelled by the satellite and then divide this distance by the time elapsed. The distance traveled will be the circumference of a circle.

$$C = 2\pi R$$
$$= 6.28(4180 \text{ mi})$$
$$= 26250.4 \text{ mi}$$

$$\text{Speed} = \frac{26250.4 \text{ mi}}{2.25 \text{ hr}} = 11666.8 \text{ mph}$$

Exercises

Solve each of the equations for the variable involved.

1. $2T - 5 = 15$
2. $-3B + 6 = 30$
3. $3R - 8 = -R$
4. $-15 - 6M = 7M$
5. $5(N + 3) = N$
6. $5P - (P - 5) = 4 + 2P$

Sec. 2–6. Equations and Formulas Solved for One First Degree Variable

7. $-(5+V) = 3(2V-3) - 3$
8. $\dfrac{3X}{4} + 6 = 9$
9. $\dfrac{X}{10} + 2 = \dfrac{3+X}{2}$
10. $\dfrac{3A}{6} - \dfrac{3}{2} = \dfrac{A}{3}$
11. $\dfrac{1}{2B} - \dfrac{1}{2} = 4$
12. $\dfrac{1}{3P} - \dfrac{1}{3} = \dfrac{2}{P}$
13. $\dfrac{1}{Z^2 - Z} - \dfrac{1}{Z} = \dfrac{1}{Z-1}$
14. $\dfrac{4}{T} = \dfrac{6}{T-1}$
15. $\dfrac{5}{2V+4} + \dfrac{3}{V+2} = 2$
16. $\sqrt{2X-1} = 5$
17. $\sqrt{4P} - 6 = 0$
18. $\sqrt{4M+1} = 2$
19. $\sqrt{Z-12} + 7 = 0$
20. $2 - \sqrt{X} = 6$
21. $\sqrt{3X} + 5 = 2$
22. $|A| = 7$
23. $|P| = 12$
24. $|B| = 2$
25. $|T-2| = 4$
26. $|2M+1| = 11$
27. $|4N| = 16$
28. $|-2Z| = 4$
29. $|-2S-4| = -1$
30. $|3R-1| = 5$

For each of the following, solve for the variable indicated after each equation.

31. $V = IR$. Solve for R.
32. $S = \tfrac{1}{2}gt^2$. Solve for t.
33. $\dfrac{F}{W} = \dfrac{L}{H}$. Solve for W.
34. $F = \dfrac{MV^2}{R}$. Solve for R.
35. $M = RX - P(X - A)$. Solve for A.
36. The formula for the volume of a rectangular solid is $V = LWH$. Solve for H.
37. A cubic foot contains 7.48 gallons. Calculate the number of gallons in a cyclindrical tank 10 feet high with a radius of 6 feet.
38. An acre is 43,560 square feet. How many acres are there in a rectangular plot of ground 925 feet long and 480 feet wide?
39. The resistance in a certain resistor is four times that in another resistor. If the total resistance in the two resistors is 150 ohms, what is the resistance in each resistor?
40. One motor develops 3 horsepower more than another motor. A third motor develops horsepower equal to the sum of the other two motors. If the three motors develop 12 horsepower, what is the horsepower of each motor?
41. The life support pack of an astronaut allows him to walk 12.57 miles on the moon before his supplies are exhausted. If he walks in a square path with the lunar base as one corner, how far can he walk in any one direction?

42. If the astronaut in the previous problem walks in a circle, what is the maximum distance he will be from the lunar base?

2-7. CHAPTER REVIEW

Simplify each of the following.

1. $-\sqrt{9}$
2. $\sqrt[3]{8}$
3. $-\sqrt[4]{16}$
4. $-\sqrt[3]{-27}$
5. $36^{1/2}$
6. $64^{1/3}$
7. $8^{5/3}$
8. $16^{7/2}$
9. $8^{-(1/3)}$
10. $49^{-3/2}$
11. $\sqrt{16X^2Y^4}$
12. $\sqrt[3]{27A^6B^{12}}$

For each of the following, perform any indicated operations and express all answers in simplest form.

13. $3K - (2K - 5M) + (6K - 7M)$
14. $(2\sqrt{7} - \sqrt{5}) - 4\sqrt{7} - 3(-\sqrt{7} - 2\sqrt{5})$
15. $X^2 - 3Y^2 - X(2X - Y) + 5XY$
16. $2R^3 - S^2 + S(5R - 6S) - 7RS$
17. $2T - [T + V - 3(T - 4V) - 5V] + V$
18. $L^2 - [-L(L + P) + 2LP - 4L^2]$
19. $(5X + 2Y)(3X - 7Y)$
20. $(A^2 - 3B^2)(4A - 5B^2)$
21. $(2\sqrt{3} - 4\sqrt{7})(\sqrt{3} - 2\sqrt{7})$
22. $(3\sqrt{X} + \sqrt{Y})(2\sqrt{X} - 4\sqrt{Y})$
23. $(R^2 + 3S^2)^2$
24. $(5K - 2M)^3$
25. $(2A^3)^2 - (3A^0)^3$
26. $(\sqrt{3}X^4)^2 - (2X^0Y)^0$
27. $(T + 5)(T - 1)(2T - 3)$
28. $2V^2(3V - 1)(V^2 - V + 4)$
29. $\dfrac{12A^3X^5}{4A^5X}$
30. $\dfrac{18R^4S}{4RS^{-2}}$
31. $\dfrac{20M^3N - 10MN - M^2N^3}{15M^2N^2}$
32. $\dfrac{18LP^4 - 21LP + 9L^3P^3}{3LP}$

Factor each of the following expressions completely.

33. $2M^2N^2 - 8M^2N + 12MN$
34. $49A^2 - 64B^2$
35. $X^2 - 14X + 49$
36. $T^2 + 7T + 12$
37. $2L^2 + 7L - 30$
38. $6Y^2 + 17Y + 5$
39. $24T^2 - 42T - 45$
40. $5T^3 + 28T^2 - 12T$

Sec. 2–7. Chapter Review

For each of the following, perform the indicated operations and express each answer in simplest form.

41. $\dfrac{2K^2 + 17K + 8}{4K^2 + 2K}$

42. $\dfrac{2M - 10}{M^2 + 4M - 45}$

43. $\dfrac{3X}{4Y^2} - \dfrac{5X}{8Y} + \dfrac{6X}{2Y^3}$

44. $\dfrac{A^3}{2B} - \dfrac{2A}{5} - \dfrac{4A^2}{B}$

45. $\dfrac{3V}{T-4} + \dfrac{V^2}{4-T} - \dfrac{5V}{T}$

46. $\dfrac{S+2}{S^2-49} + \dfrac{35}{S^2-8S+7} - \dfrac{S-1}{S+7}$

47. $\dfrac{9A^2 - 4B^2}{12A - 8B} \times \dfrac{12B^2}{3AB + 2B^2}$

48. $\dfrac{2K^2 - 12K - 54}{4K^3} \times \dfrac{2K^3 + 18K^2}{K^2 - 81}$

49. $\dfrac{8R^2 + 46R - 12}{2R + 12} \div \dfrac{12R - 3}{10R^2}$

50. $\dfrac{12M^2 - 27N^2}{6M + 4N} \div \dfrac{4M^2 - 12MN + 9N^2}{12M - 18N}$

51. $\dfrac{X}{2Y} + \left[\dfrac{5X}{3Y^2} \div \dfrac{4X^2}{7Y}\right]$

52. $\left[\dfrac{T-2}{V} \times \dfrac{T+1}{2V - TV}\right] - \dfrac{5T^2}{V^3}$

Solve each of the following for T.

53. $5T + 8 = 23$

54. $8T - 3 = 9$

55. $3(T - 2) = 4(2T - 1)$

56. $6(2T + 5) = 2(3T - 8)$

57. $\dfrac{3T - 7}{2} = T$

58. $\dfrac{2T + 1}{5} = 4T + 3$

59. $\dfrac{T}{24} = \dfrac{36}{54}$

60. $\dfrac{25}{T} = 15$

61. $3T + 4S = T + 5$

62. $AT + 3BT = 7$

63. $R^2T + 7 = RST + 12$

64. $\dfrac{XT}{Y} - 4 = 3T + 6$

65. $T^2 + 7 = 23$

66. $2T^2 - 5 = 13$

67. $\sqrt{T + 10} + 12 = 17$

68. $\sqrt{8 - T} - 7 = -5$

69. $4 = \sqrt{2T - 6} + 2$

70. $-5 = \sqrt{12 - T} - 6$

71. $|2T - 1| = 5$

72. $|4 - 3T| = 8$

73. An expression used to find the combined resistance of three resistances (R_1, R_2, R_3) connected in parallel in an electric circuit is

$$\frac{1}{R_1} + \frac{1}{R_2} + \frac{1}{R_3}$$

Combine these into a single fraction.

74. The expression

$$E + \frac{P}{RG} + \frac{V}{2G}$$

is found in hydrokinetics. Combine into a single fraction.

75. The volume of a right circular cone can be found by using the formula $V = \frac{1}{3}\pi R^2 H$. Solve for H.

76. A formula found in the study of gravitation is

$$F = \frac{GM_1 M_2}{R^2}$$

Solve for G.

77. A formula for the effective height of an antenna is

$$H = \frac{ED}{.2WI}$$

Solve for D.

78. The formula

$$P = P_0 \left[\frac{T_0 - \lambda H^x}{T_0} \right]$$

is encountered in the study of atmospheric phenomena. Solve for λ.

79. A formula for finding the equivalent resistance of two resistances connected in parallel is

$$R = \frac{R_1 R_2}{R_1 + R_2}$$

Solve for R_1.

80. A formula for determining the exit speed of water flowing from a tank is $v = \sqrt{2gd}$. Solve for d.

81. A construction company rents two pieces of heavy equipment each month for a total of $2,400. If the rental for one piece is three times as much as for the other, what is the monthly rent for each piece?

82. One train leaves a railroad station and travels due east at a rate of 50 miles per hour. An hour later, a second train leaves the same station and travels due west at a rate of 60 miles per hour. How long will it take after the second train leaves the station for the trains to be 380 miles apart?

Sec. 2–7. Chapter Review

83. The *spring rate* is the amount of force needed to extend or compress a spring each inch. If a certain spring has a rate of 8 pounds per inch, what is its unloaded length if it is 22 inches long with a 34-pound weight attached?

84. A person made a 320 mile trip between two cities by car in 6 hours and 45 minutes. The return trip by bus took 21 minutes less. How fast did the bus travel?

chapter 3

Measurement and Dimensional Analysis

3-1. The British System of Measurement

When numbers are used in scientific and technical problems, they usually involve some type of measurement. Therefore, in order for the calculations and answers for problems involving these numbers to be accurate and meaningful, we must know the units of measurement associated with these numbers. There are two basic systems of units which are used in problems involving measurement. They are the *British* system and the *metric* system. In this section, we will discuss the British system.

Regardless of which system we are working with, there are certain *basic* or *fundamental* dimensions, and all other dimensions are expressed in terms of them. These basic dimensions are *length*, *time*, *force* or *mass* (since they are proportional, either may be used), *temperature*, and *electric charge*. All other dimensions can be derived from or expressed in terms of these five.

Example A:
1. Area is the product of two lengths.
2. Velocity is the quotient of a length and a time.
3. Work is the product of a length and a force.
4. Voltage is a combination of a force, a length, and a charge.
5. Power is a combination of a length, a force, and a time.

Sec. 3-1. The British System of Measurement

The following table lists the most common dimensions, the symbols used to represent them, and the basic unit of measurement associated with each in the British system.

DIMENSION	SYMBOL	UNIT OF MEASUREMENT
Basic Dimensions:		
length	l or s	foot (ft)
time	t	second (sec)
force (weight)	$F(w)$	pound (lb)
mass	m	slug
temperature	T	degrees Fahrenheit (°F)
electric charge	q	coulomb (coul)
Derived Dimensions:		
area	A	square feet (ft^2)
volume	V	cubic feet (ft^3)
velocity	v	feet per second (ft/sec)
acceleration	a	feet per second per second (ft/sec^2)
density	d	pounds per cubic foot (lb/ft^3)
work (energy)	$W(E)$	foot pound (ft lb or ft-lb)
heat		British thermal unit (Btu)
power	P	horsepower (hp)
electric potential	V or E	volt
electric current	I	ampere
resistance	R	ohm
capacitance	C	farad
inductance	L	henry

Following is a list of the *basic reduction factors* for the most commonly used dimensions in the British system.

Length: 1 foot = 12 inches
   1 yard = 3 feet = 36 inches
   1 rod = 5.5 yards = 16.5 feet
   1 mile = 320 rods = 1,760 yards = 5,280 feet
   1 nautical mile = 6,080 feet

Time: 1 minute = 60 seconds
   1 hour = 60 minutes
   1 day = 24 hours
   1 week = 7 days = 168 hours
   1 month = $4\frac{1}{3}$ weeks
   1 year = 12 months = 52 weeks = 365 days (366 days in a leap year)

Force: 1 pound = 16 ounces
   1 ton = 2,000 pounds

Area: 1 square foot = 144 square inches
   1 square yard = 9 square feet
   1 acre = 4,840 square yards = 43,560 square feet

Volume: 1 cubic foot = 1,728 cubic inches
1 cubic yard = 27 cubic feet
1 cup = 8 ounces = 14.43 cubic inches
1 pint = 2 cups = 16 ounces = 28.86 cubic inches
1 quart = 2 pints = 4 cups = 32 ounces = 57.71 cubic inches
1 gallon = 4 quarts = 8 pints = 16 cups = 128 ounces
= 230.84 cubic inches

Heat: 1 Btu = 778 foot pounds

Power: 1 horsepower = 550 foot pounds per second (ft-lb/sec)

Example B: How many feet are contained in 6.5 yards? Since there are 3 feet in 1 yard, 3 × 6.5 = 19.5 feet.

$$6.5 \text{ yd} = 19.5 \text{ ft}$$

Example C: How many yards are contained in 7.2 rods? Since there are 5.5 yards in a rod, 5.5 × 7.2 = 39.6 yards.

$$7.2 \text{ rd} = 39.6 \text{ yd}$$

Example D: How many square feet are contained in 5.2 square yards? Since there are 9 square feet in 1 square yard, 9 × 5.2 = 46.8 square feet.

$$5.2 \text{ sq yd} = 46.8 \text{ sq ft}$$

Example E: How many hours are contained in 3 days? Since there are 24 hours in 1 day, 24 × 3 = 72 hours.

$$3 \text{ days} = 72 \text{ hr}$$

Example F: How many square inches are contained in 3 square yards? This problem involves two steps:
1. Since there are 9 square feet in 1 square yard, 9 × 3 = 27 square feet.
2. Since there are 144 square inches in 1 square foot, 144 × 27 = 3,888 square inches.

$$3 \text{ sq yd} = 3,888 \text{ sq in.}$$

Example G: 51.3 square feet would equal how many square yards? Since 9 square feet equal 1 square yard, 51.3 ÷ 9 = 5.7 square yards.

$$51.3 \text{ sq ft} = 5.7 \text{ sq yd}$$

Example H: 48,300 feet would equal how many miles? Since there are 5,280 feet in 1 mile, 48,300 ÷ 5,280 = 9.15 miles.

$$48,300 \text{ ft} = 9.15 \text{ mi}$$

Example I: 473.2 cubic inches would equal how many quarts? Since there are 57.71 cubic inches in 1 quart, 473.2 ÷ 57.71 = 8.2 quarts.

$$473.2 \text{ cu in.} = 8.2 \text{ qt}$$

Sec. 3-1. The British System of Measurement

Example J: 56,000 pounds would equal how many tons? Since there are 2,000 pounds in 1 ton, 56,000 ÷ 2,000 = 28 tons.

$$56{,}000 \text{ lb} = 28 \text{ tn}$$

Example K: 50,000 gallons would equal how many cubic feet? This problem involves two steps:
1. Since there are 230.84 cubic inches in 1 gallon, 50,000 × 230.84 = 11,542,000 cubic inches.
2. Since there are 1,728 cubic inches in 1 cubic foot, 11,542,000 ÷ 1,728 = 6,679.4 cubic feet.

$$50{,}000 \text{ gal} = 6{,}679.4 \text{ cu ft}$$

We should note from these examples that when changing from a larger unit to a smaller unit (e.g., feet to inches) we multiply by the appropriate factor. When changing from a smaller unit to a larger unit (e.g., minutes to hours) we divide by the appropriate factor.

Exercises

Solve each of the following reduction problems.
1. How many inches are contained in 17.3 feet?
2. How many feet are contained in 12 rods?
3. How many square feet are contained in 3.5 acres?
4. How many seconds are contained in $4\frac{1}{2}$ hours?
5. How many pounds are contained in 2.4 tons?
6. How many inches are contained in 5.2 yards?
7. How many yards are contained in 10.7 miles?
8. How many foot pounds per second are contained in 120 horsepower?
9. How many quarts are contained in 112 ounces?
10. How many cubic inches are contained in 14.1 pints?
11. How many yards would equal 520 miles?
12. How many square yards would equal 96 square feet?
13. How many acres would equal 72,000 square yards?
14. How many days would equal 82 hours?
15. How many miles would equal 600 rods?
16. How many tons would equal 142,000 ounces?
17. How many horsepower would equal 7100 foot pounds per second?
18. How many cubic yards would equal 110 cubic feet?
19. How many rods would equal 1500 feet?
20. How many cubic feet would equal 24,000 gallons?

3-2. The Metric System of Measurement

Whereas the British system is used in the United States and a few other countries for both general purposes and technical applications, the metric system is used for general purposes in the rest of the world and in scientific areas in all parts of the world. There is also a strong possibility that even some (if not all) of those countries presently using the British system will be adopting the metric system as their standard in the not-so-distant future. Therefore, any person who works or will be working in a scientific or technical area should have a basic knowledge of the metric system.

The metric system actually is broken into two branches, the *cgs* (centimeter-gram-second) *system* and the *mks* (meter-kilogram-second) *system*. The cgs system is more appropriate for measuring smaller dimensions, whereas the mks system is more appropriate for measuring larger dimensions.

The following table lists the most common dimensions, the symbols used to represent each, and the basic unit of measurement associated with each in the metric system.

DIMENSION	SYMBOL	UNIT OF MEASUREMENT	
		cgs	mks
Basic Dimensions:			
length	l or s	centimeter (cm)	meter (m)
time	t	second (sec)	second (sec)
force (weight)	$F(w)$	dyne	newton (n)
mass	m	gram (gm)	kilogram (kg)
temperature	T	degrees Celsius (°C)	degrees Celsius (°C)
electric charge	q	abcoulomb (abcoul) statcoulomb (statcoul)	coulomb (coul)
Derived Dimensions:			
area	A	square centimeters (cm^2)	square meters (m^2)
volume	V	cubic centimeters (cm^3)	cubic meters (m^3)
velocity	v	centimeters per second (cm/sec)	meters per second (m/sec)
acceleration	a	centimeters per second per second (cm/sec^2)	meters per second per second (m/sec^2)
density	d	grams per cubic centimeter (gm/cm^3)	kilograms per cubic meter (kg/m^3)
work (energy)	$W(E)$	ergs (dyne–cm)	joule (newton-meter)
heat		calories (cal)	calories (cal)
power	P	ergs per second (erg/sec)	watts (joules/sec)
electric potential	V or E	abvolt or statvolt	volt
electric current	I	abampere or statampere	ampere
resistance	R	abohm or statohm	ohm
capacitance	C	abfarad or statfarad	farad
inductance	L	abhenry or stathenry	henry

Sec. 3-2. The Metric System of Measurement

Essentially, the metric system is a system of measurement based on different powers of ten. Therefore, basic reduction problems involve multiplication or division by some power of ten, which simply means that the decimal point is moved one way or the other. The appropriate power of ten is indicated by the prefix that is adjoined to the basic unit involved. The following table lists the most common prefixes used in the metric system and the power of ten represented by each one.

Prefix	Change in Basic Unit	Power of Ten
micro	.000001	10^{-6}
milli	.001	10^{-3}
centi	.01	10^{-2}
deci	.1	10^{-1}
deka	10	10^{1}
hecto	100	10^{2}
kilo	1,000	10^{3}
mega	1,000,000	10^{6}

Example A:
1. 1 milliliter = .001 liters.
2. 1 kilometer = 1,000 meters.
3. 1 centigram = .01 grams.

Following is a list of the basic reduction factors for the most commonly used dimensions in the metric system.

Length: 1 micrometer = .000001 meters   1 dekameter = 10 meters
         1 millimeter = .001 meters   1 hectometer = 100 meters
         1 centimeter = .01 meters   1 kilometer = 1,000 meters
         1 decimeter = .1 meters   1 megameter = 1,000,000 meters

Time: Exactly the same as for the British system.

Mass: 1 microgram = .000001 grams   1 dekagram = 10 grams
        1 milligram = .001 grams   1 hectogram = 100 grams
        1 centigram = .01 grams   1 kilogram = 1,000 grams
        1 decigram = .1 grams   1 megagram = 1,000,000 grams

Area: 1 square micrometer = .000000000001 square meters
       1 square millimeter = .000001 square meters
       1 square centimeter = .0001 square meters
       1 square decimeter = .01 square meters
       1 square dekameter = 100 square meters
       1 square hectometer = 10,000 square meters
       1 square kilometer = 1,000,000 square meters
       1 square megameter = 1,000,000,000,000 square meters

Volume: 1 liter = 1,000 cubic centimeters
1 microliter = .000001 liters   1 dekaliter = 10 liters
1 milliliter = .001 liters   1 hectoliter = 100 liters
1 centiliter = .01 liters   1 kiloliter = 1,000 liters
1 deciliter = .1 liters   1 megaliter = 1,000,000 liters

Heat: 1 calorie = 4.18 joules
1 joule = 10,000,000 ergs

Power: 1 watt (joule per second) = 10,000,000 ergs per second

Example B: How many meters are contained in 12.4 kilometers? Since there are 1,000 meters in 1 kilometer, 1,000 × 12.4 = 12,400 meters.

$$12.4 \text{ km} = 12{,}400 \text{ m}$$

Example C: How many milligrams are contained in 126 grams? Since there are 1,000 milligrams in 1 gram, 1,000 × 126 = 126,000 milligrams.

$$126 \text{ g} = 126{,}000 \text{ mg}$$

Example D: How many centilitiers are contained in 3.8 liters? Since there are 100 centiliters in a liter, 100 × 3.8 = 380 centiliters.

$$3.8 \text{ l} = 380 \text{ cl}$$

Example E: How many square centimeters are contained in 5.1 square meters? Since there are 10,000 square centimeters in 1 square meter, 10,000 × 5.1 = 51,000 square centimeters.

$$5.1 \text{ m}^2 = 51{,}000 \text{ cm}^2$$

Example F: How many cubic centimeters are contained in 3.8 dekaliters? This problem involves two steps:
1. Since there are 10 liters in 1 dekaliter, 10 × 3.8 = 38 liters.
2. Since there are 1,000 cubic centimeters in 1 liter, 1,000 × 38 = 38,000 cubic centimeters.

$$3.8 \text{ dl} = 38{,}000 \text{ cm}^3$$

Example G: 183 centimeters would equal how many meters? Since there are 100 centimeters in 1 meter, 183 ÷ 100 = 1.83 meters.

$$183 \text{ cm} = 1.83 \text{ m}$$

Example H: 24,300 grams would equal how many kilograms? Since there are 1,000 grams in 1 kilogram, 24,300 ÷ 1,000 = 24.3 kilograms.

$$24{,}300 \text{ g} = 24.3 \text{ kg}$$

Example I: 56,000 square millimeters would equal how many square centimeters? Since there are 100 square millimeters in 1 square centimeter, $56,000 \div 100 = 560$ square centimeters.

$$56,000 \text{ mm}^2 = 560 \text{ cm}^2$$

Example J: 483.7 cubic centimeters would equal how many liters? Since there are 1,000 cubic centimeters in 1 liter, $483.7 \div 1,000 = .4837$ liters.

$$483.7 \text{ cm}^3 = .4837 \text{ l}$$

Example K: 88.5 centigrams would equal how many kilograms? This problem involves two steps:
1. Since there are 100 centigrams in a gram, $88.5 \div 100 = .885$ grams.
2. Since there are 1,000 grams in 1 kilogram, $.885 \div 1,000 = .000885$ kilograms.

$$88.5 \text{ cg} = .000885 \text{ kg}$$

Once again we can see that changing from larger units to smaller units (e.g., meters to centimeters) we multiply by the appropriate factor. In changing from smaller units to larger units (e.g., liters to hectoliters) we divide by the appropriate factor.

Besides changing to units of different sizes within the same system (a *reduction*), we can also change from one system to another (a *conversion*). A list of conversion factors for the most commonly used dimensions follows. Although a few brief examples of conversion problems will be given here, the majority of these problems involve some dimensional algebra and will be discussed fully in Sections 3–3 and 3–4.

Length: 1 inch = 2.54 centimeters
1 foot = 30.48 centimeters = .3048 meters
1 yard = 91.44 centimeters = .9144 meters
1 meter = 39.37 inches = 3.28 feet = 1.09 yards
1 mile = 1.61 kilometers or 1 kilometer = .621 miles

Force or *Mass:* 1 pound = 453.6 grams = .4536 kilograms
or 1 kilogram = 2.2 pounds

Temperature: $°F = \frac{9}{5}C° + 32°$
$°C = \frac{5}{9}(F° - 32°)$

Volume: 1 teaspoon = 5 milliliters
1 tablespoon = 15 milliliters
1 cup = 250 milliliters
1 quart = .946 liters or 1 liter = 1.057 quarts
1 gallon = 3.785 liters
1 cubic foot = 28.32 liters

Work: 1 joule = .739 foot pounds or 1 foot pound = 1.35 joules.

Heat: 1 Btu = 252 calories or 1 calorie = .004 Btu

Power: 1 horsepower = 744.66 watts or 1 watt = .00134 horsepower.

Example L: How many kilometers are contained in 55 miles? Since there are 1.61 kilometers in 1 mile, 1.61 × 55 = 88.55 kilometers.

$$55 \text{ mi} = 88.55 \text{ km}$$

Example M: How many grams are contained in 32 pounds? Since there are 453.6 grams in 1 pound, 453.6 × 32 = 14,515.2 grams.

$$32 \text{ lb} = 14,515.2 \text{ g}$$

Example N: How many quarts are contained in 12 liters? Since there are 1.057 quarts in 1 liter, 1.057 × 12 = 12.684 quarts.

$$12 \text{ l} = 12.684 \text{ qt}$$

Example O: How many British thermal units are contained in 2,400 calories? Since there are .004 Btus in 1 calorie, .004 × 2,400 = 9.6 Btu.

$$2,400 \text{ cal} = 9.6 \text{ Btu}$$

Exercises

Solve each of the following reduction problems.
1. How many micrometers are contained in 31.6 meters?
2. How many grams are contained in 27.2 hectograms?
3. How many square meters are contained in 16.8 square kilometers?
4. How many joules are contained in 220 calories?
5. How many liters are contained in 3.7 megaliters?
6. How many centimeters are contained in 26 kilometers?
7. How many calories would equal 125.4 joules?
8. How many square meters would equal 95.8 square centimeters?
9. How many grams would equal 143.6 centigrams?
10. How many liters would equal 42,600 cubic centimeters?
11. How many kiloliters would equal 1,940 deciliters?
12. How many dekameters would equal 9,200 millimeters?

Sec. 3–3. Algebra of Dimensions   69

Solve each of the following conversion problems.

13. How many centimeters are contained in 13.4 inches?

14. How many kilograms are contained in 175 pounds?

15. How many degrees Celsius are contained in 86° Fahrenheit?

16. How many joules are contained in 96 foot pounds?

17. How many miles are contained in 480 kilometers?

18. How many pounds are contained in 3,300 grams?

19. How many quarts are contained in 27.8 liters?

20. How many British thermal units are contained in 4,100 calories?

3–3. Algebra of Dimensions

Applied mathematics is a problem-solving situation in which the answer consists of two parts: the *numerical value* and the *unit of dimension*. The answer will be correct only if both of these parts are correct. In this section the emphasis will be on the mathematics of dimensional units.

A *dimension* is any property of an object that can be measured. According to this definition an object may have a number of dimensions. For example, consider a brick at rest. Some of the properties that can be measured would be its length, width, height, total surface area, volume, mass, density, and temperature. If the brick is thrown through the air, its velocity could be measured. When the brick hits a wall, its power could be measured.

Recall that dimensions are classified in two categories, the basic dimensions (length, force or mass, time, temperature, and electric charge) and derived dimensions such as area, volume, density, force, velocity, and any other dimensions that are derived from or expressed in terms of the basic ones.

A *dimensional quantity* is a quantity expressed as a combination of a real number and a unit of dimension.

Example A: Two feet, 40 feet per second, 3.5 square feet, and $\frac{1}{2}$ foot pound per second are dimensional quantities.

There are four basic operations that can be performed with dimensional quantities.

1. Addition: The sum of two dimensional quantities in which the units of dimension are the same is defined as the sum of their real parts and the common unit of dimension. If the units of dimension are not the same, addition is not defined.

Example B: $\qquad$ 2 cm + 4 cm = 6 cm

$\qquad\qquad\qquad\qquad$ 12 m² + 6 m² = 18 m²

The equation, 2 ft + 4 sec, is undefined.

2. Subtraction: The difference of two dimensional quantities in which the units of dimension are the same is defined as the difference of their real parts and the common unit of dimension. If the units of dimension are not the same, subtraction is not defined.

Example C: $\qquad$ 240 cm − 40 cm = 200 cm

$\qquad\qquad\qquad\qquad$ 40 ft/sec − 60 ft/sec = −20 ft/sec

3. Multiplication:
   (a) The product of a dimensional quantity and a real number is the product of their real parts and the same unit of dimension.

Example D: $\qquad$ 7(13 ft) = 91 ft

$\qquad\qquad\qquad\qquad$ 2(6 cm³) = 12 cm³

   (b) The product of two dimensional quantities is the product of their real parts and the product of their units of dimension. The laws of exponents apply to the multiplication of units of dimension.

Example E: $\qquad$ (4 m)(6 m) = 24 m²

$\qquad\qquad\qquad\qquad$ (5 lb)(10 ft/sec) = 50 ft-lb/sec

$\qquad\qquad\qquad\qquad$ (3 ft²)(2 ft) = 6 ft³

4. Division: The quotient of two dimensional quantities is the quotient of their real parts and the quotient of their units of dimension. The laws of exponents apply to the division of units of dimension.

Example F:

1. $\dfrac{8 \text{ ft}^3}{2} = 4 \text{ ft}^3$

2. $\dfrac{21 \text{ ft}^2}{7 \text{ ft}} = 3 \text{ ft}$

3. $\dfrac{81 \text{ sec}^2}{9 \text{ sec}^2} = 9 \text{ sec}^0 = 9$

4. $1 \text{ ft}^{-1} = \dfrac{1}{\text{ft}}$

NOTE: ft^{-1} is the reciprocal dimension of foot.

Sec. 3-4. Reductions and Conversions

5. $\dfrac{4}{8 \text{ ft}^2} = \dfrac{1}{2 \text{ ft}^2} = \dfrac{1}{2} \text{ ft}^{-2}$

6. $\dfrac{12 \text{ ft/sec}}{3 \text{ ft/sec}^2} = \dfrac{12 \text{ ft}}{1 \text{ sec}} \times \dfrac{1 \text{ sec}^2}{3 \text{ ft}} = 4 \text{ sec}$

Exercises For each of the following, perform the indicated operations and express the answer in simplest form.

1. 4,000 cal − 1,000 cal
2. $\frac{3}{4}$ tn − $\frac{1}{2}$ tn
3. 5 Btu + 4 Btu
4. 15 sec + $2\frac{1}{2}$ min
5. 17 ft − 12 ft + 9 ft
6. 27 kg + 12 kg − 18 kg
7. 48 m² − 109 m² + 217 m²
8. 143 cm³ − 74 cm³ − (58 cm³ − 41 cm³)
9. 130 ft/sec − 96 ft/sec − (48 ft/sec − 16 ft/sec)
10. 627 yd² − 514 m² + 126 yd²
11. 3 (4 ft)
12. 7 (83 kg)
13. (43 ft)(126 ft)
14. (92 m)(260 m²)
15. (16 in.²)(3 in.)
16. (4 ft/sec²)(2 sec)
17. (75 kg/m³)(40 m²)
18. (120 ft)(28 lb/ft³) − (70 lb/ft²)
19. (46 ft/sec) ÷ 2
20. (81 m³) ÷ (9 m)
21. (98 ft²) ÷ (14 ft)
22. (862 ft-lb) ÷ (158 ft)
23. (73 m/min ÷ 40 m/min²) + 12 min
24. (1,500 lb/ft² × 90 ft³) + 180 ft-lb
25. (100 ft/sec) ÷ (47 m/sec)
26. (405 lb/ft²) ÷ (34 kg/ft³)

3-4. Reductions and Conversions

When *reducing* a dimensional quantity within a system of measurement or when *converting* a dimensional quantity from one system of measurement to another system, the same basic principle of mathematics is involved: multiplying the dimensional quantity by one. That is, if B is any dimensional quantity, then $B \times 1 = B$.

Example A: Reduce 56 inches to feet.

$$\frac{56 \text{ in.}}{1} \times \frac{1 \text{ ft}}{12 \text{ in.}} = 4\tfrac{2}{3} \text{ ft}$$

Since 1 ft = 12 in., multiplying by

$$\frac{1 \text{ ft}}{12 \text{ in.}}$$

is equivalent to multiplying by 1.

NOTE: Since 1 ft = 12 in. is an equation, we can divide both members of the equation by either 1 ft or 12 in. and have an equivalent equation $\frac{1 \text{ ft}}{12 \text{ in.}} = 1$ and $1 = \frac{12 \text{ in.}}{1 \text{ ft}}$. In order to eliminate the dimensional quantity of inch and have a final answer with a dimensional quantity of foot, the appropriate reduction factor must be in the form $\frac{1 \text{ ft}}{12 \text{ in.}}$. This procedure must be used in all reduction and conversion problems. That is, use the appropriate reduction or cenversion factor in the form that will eliminate the unwanted dimension and give the required dimension for the answer.

Example B: Reduce $\frac{2}{3}$ square foot to square inches.

$$\frac{2 \text{ ft}^2}{3} \times \frac{144 \text{ in.}^2}{1 \text{ ft}^2} = 96 \text{ in.}^2$$

Since 1 ft² = 144 in.², multiplying by

$$\frac{144 \text{ in.}^2}{1 \text{ ft}^2}$$

is equivalent to multiplying by 1.

Example C: Convert 6 inches to centimeters.

$$\frac{6 \text{ in.}}{1} \times \frac{2.54 \text{ cm}}{1 \text{ in.}} = 15.24 \text{ cm}$$

Since 1 in. = 2.54 cm, multiplying by

$$\frac{2.54 \text{ cm}}{1 \text{ in.}}$$

is equivalent to multiplying by 1.

Example D: Convert 18 centimeters to inches.

$$\frac{18 \text{ cm}}{1} \times \frac{1 \text{ in.}}{2.54 \text{ cm}} = 7.1 \text{ in.}$$

Sec. 3-4. Reductions and Conversions

Since 1 in. = 2.54 cm, multiplying by

$$\frac{1 \text{ in.}}{2.54 \text{ cm}}$$

is equivalent to multiplying by 1.

Example E: Reduce 45 miles per hour to feet per second.

$$\frac{45 \cancel{\text{mi}}}{1 \cancel{\text{hr}}} \times \frac{5{,}280 \text{ ft}}{1 \cancel{\text{mi}}} \times \frac{1 \cancel{\text{hr}}}{3{,}600 \text{ sec}} = \frac{66 \text{ ft}}{\text{sec}}$$

Since 1 mi = 5,280 ft, multiplying by

$$\frac{5{,}280 \text{ ft}}{1 \text{ mi}}$$

is equivalent to multiplying by 1. A second step in this problem is to use the factor of 1 hr = 3,600 sec. Multiplying by

$$\frac{1 \text{ hr}}{3{,}600 \text{ sec}}$$

is also equivalent to multiplying by 1.

Exercises Use the tables in Sections 3-1 and 3-2 and the algebra of dimensional analysis to perform the required operations in each of the following:

1. Reduce 3.2 miles to feet.
2. Reduce 100,000 square feet to acres.
3. Convert 16 miles to kilometers.
4. Convert 12 liters to quarts.
5. Convert 150 pounds to kilograms.
6. Reduce 20 rods to inches.
7. Reduce 48 miles per hour to feet per second.
8. Reduce 4 centimeters to kilometers.
9. Convert 1200 grams to ounces.
10. Reduce 38 kilograms to centigrams.
11. Convert 1200 yards to kilometers.
12. Convert 62 pounds per square foot to kilograms per square meter.

13. Reduce 30 calories to ergs.
14. Convert 60 miles per hour to kilometers per hour.
15. Reduce 4200 cubic centimeters to liters.
16. Reduce 45 Btu per minute to horsepower.
17. Convert 15 miles per gallon to kilometers per liter.
18. Convert 20,000 ft-lb to calories.
19. Convert 264 liters to cubic inches.
20. Convert 250 horsepower to kg-cm per sec.

3–5. Equations Involving Dimensions

When solving an equation or formula involving dimensional quantities, it is necessary to substitute units of dimension as well as numerical values into the equation or formula in order to obtain an appropriate result. In doing so, we must observe the laws of dimensional algebra as stated in Section 3–3.

Example A: $P = H \times W$ is a formula in which P represents pressure, H represents height, and W represents density. What is the pressure when $H = 100$ feet and $W = 62.4$ pounds per cubic foot?

$$P = \frac{100 \text{ ft}}{1} \times \frac{62.4 \text{ lb}}{\frac{\text{ft}^3}{\text{ft}^2}} = 6{,}240 \text{ lb/ft}^2$$

NOTE: The answer is expressed in the proper units of dimension for pressure.

Example B: Suppose in the above formula $H = 4$ meters and $W = 2$ pounds per cubic foot. If we substitute these values,

$$P = \frac{4 \text{ m}}{1} \times \frac{2 \text{ lb}}{\text{ft}^3} = \frac{8 \text{ m–lb}}{\text{ft}^3}$$

We can certainly see that this is an incorrect dimension for pressure and thus we have substituted the wrong units. We must convert meters to feet.

$$\frac{4 \text{ m}}{1} \times \frac{1 \text{ ft}}{.3048 \text{ m}} = 13.123 \text{ ft}$$

Then,

$$P = \frac{13.123 \text{ ft}}{1} \times \frac{2 \text{ lb}}{\frac{\text{ft}^3}{\text{ft}^2}} = 26.246 \text{ lb/ft}^2$$

Sec. 3–5. Equations Involving Dimensions

From this last example we should realize that in any problem involving dimensions, the units must be consistent. First of all, we must use all British or all metric units, but not both. Secondly, within each system we must use similar units in any given problem (all feet, all inches, all grams, or all liters).

Example C: If $E = Z + \dfrac{P}{W} + \dfrac{V^2}{2G}$, find E when $Z = 2$ feet, $W = 4$ pounds per cubic foot, $V = 2$ feet per second, $G = 4$ feet per second squared, and $P = 2$ pounds per square foot.

$$E = 2\text{ ft} + \frac{2\text{ lb/ft}^2}{4\text{ lb/ft}^3} + \frac{(2\text{ ft/sec})^2}{2(4\text{ ft/sec}^2)}$$

$$= 2\text{ ft} + \frac{2\cancel{\text{lb}}}{\cancel{\text{ft}^2}} \times \frac{\overset{\text{ft}}{\cancel{\text{ft}^3}}}{4\cancel{\text{lb}}} + \frac{4\cancel{\text{ft}^2}}{\cancel{\text{sec}^2}} \times \frac{\overset{\text{ft}}{\cancel{\text{sec}^2}}}{8\cancel{\text{ft}}}$$

$$= 2\text{ ft} + \tfrac{1}{2}\text{ ft} + \tfrac{1}{2}\text{ ft}$$

$$= 3\text{ ft}$$

Example D: If $t = \dfrac{1}{2\pi}\sqrt{\dfrac{l}{g}}$, find l if $t = 3$ seconds, and $g = 32$ feet per second per second.

$$t^2 = \frac{1}{4\pi^2}\frac{l}{g}$$

$$t^2(4\pi^2)g = l \quad \text{or} \quad l = t^2(4\pi^2)g$$

Therefore,
$$l = \frac{9\text{ sec}^2}{1} \times \frac{39.48}{1} \times \frac{32\text{ ft}}{\text{sec}^2}$$

$$= 11{,}370.24\text{ ft}$$

Exercises

1. The formula for the distance that an object falls, starting from rest, is $s = \tfrac{1}{2}gt^2$. Find s if $g = 32$ ft/sec^2 and $t = 3$ sec.
2. If $s = v_o t + \tfrac{1}{2}at^2$, find a when $s = 169.4$ ft, $v_o = 1.75$ ft/sec, and $t = 3.2$ sec.
3. If the radius of a circle is 2.1 cm, find the area.
4. The volume of a right circular cone is given by the formula

$$V = \frac{\pi R^2 H}{3}$$

Find the volume if the radius is 4 meters and the height is 12 yards.

5. The volume of a rectangular solid is given by the formula $V = LWH$. Find H if $V = 11{,}594$ ft³, $L = 22$ ft, and $W = 31$ ft.

6. The rate at which water flows from a tank is given by the formula $v = \sqrt{2gd}$. Find v when $g = 32$ ft/sec² and $d = 25$ ft.

7. The stress in a certain type of spring is given by the formula $T = \dfrac{4P}{\pi d^2}$. Find T if $P = 1{,}500$ lb and $d = 2.8$ cm.

8. When a car accelerates at a constant rate, this acceleration can be found by using the formula

$$a = \frac{v_f - v_i}{t_f - t_i}$$

where $v_f =$ final speed, $v_i =$ initial speed, $t_f =$ final time, and $t_i =$ initial time. If a certain car changes its speed from 10 m/sec to 28 m/sec in 6 seconds, what is the acceleration?

9. The useful output power of a pump is expressed as $P = WQ$ where W is the useful work done on each pound of fluid and is measured in foot pounds per pound. If P is measured in foot pounds per second, what is the unit of dimension of Q?

10. The density of an object is defined as the mass of the object divided by its volume. Find the density of the earth if its mass is 5.98×10^{24} kilograms and its radius is 6.38×10^6 meters. (Assume the earth to be a sphere with volume given by $V = \tfrac{4}{3}\pi R^3$.)

3–6. CHAPTER REVIEW

Use the appropriate reduction or conversion factors and dimensional algebra to solve each of the following.

1. Change 17.8 rods to feet.
2. Change 248 ounces to pounds.
3. Change 72,000 cubic inches to cubic yards.
4. Change 162 inches to yards.
5. Change 12 weeks to hours.
6. Change 148 cubic feet to gallons.
7. Change 73.8 centimeters to meters.
8. Change 52 grams to milligrams.
9. Change 27.4 liters to cubic centimeters.
10. Change 14.72 joules to calories.

Sec. 3–6. Chapter Review

11. Change 6,300 square meters to square kilometers.
12. Change 52,300 centigrams to hectograms.
13. Convert 78 kilometers to miles.
14. Convert 195 pounds to kilograms.
15. Convert 261 liters to quarts.
16. Convert 30 degrees Celsius to degrees Fahrenheit.
17. Convert 120 pounds per square inch to kilograms per square centimeter.
18. Convert 87.6 feet to centimeters.
19. Convert 900 milliliters to quarts.
20. Convert 5,000 foot pounds to joules.
21. Convert 420 square yards to square meters.
22. Convert 54,000 milliliters to cubic feet.
23. Convert 45 miles per hour to feet per second.
24. Convert 120 kilometers to yards.
25. Convert 62 kilometers per hour to meters per second.
26. Convert 196 square centimeters to square feet.
27. Solve the equation $V = \pi R^2 H$ for H if $V = 170$ ft^3 and $R = 3$ ft.
28. Solve the equation $A = \frac{1}{2}(a + b)H$ for b if $A = 52$ sq m, $a = 12$ m, and $H = 4$ m.
29. Solve the equation $v = v_o - gt$ for g if $v = 80$ ft/sec, $v_o = 128$ ft/sec, and $t = 1.5$ sec.
30. Solve the equation $s = s_o + v_o t - \frac{1}{2}at^2$ for v_o if $s = 120$ ft, $s_o = 220$ ft, $t = 5$ sec, and $a = 32$ ft/sec^2.
31. A formula for finding centrifugal force is $F = \dfrac{4\pi^2 R N^2 W}{g}$. Find F if $R = 2.5$ ft, $N = 300$/minute, $W = 3$ lb, and $g = 32$ ft/sec^2.
32. A formula for determining the deflection of a certain beam is $d = \dfrac{wl^3}{PI}$. Find d if $w = 6,000$ lb, $l = 2$ ft, $P = 150,000$ lb/sq in., and $I = .9$ in.^{4}

chapter 4

Functions, Rectangular Coordinates, and Graphs

4–1. Functions

A great deal of technology deals with both the study of sets of data that are collected from physical phenomena and the determination and interpretation of any existing relationships between the sets of data.

In this chapter we will discuss *relationships* between quantities and sets of data. Mathematically, there are two types of relationships that are of special interest to us, *relations* and *functions*. In this first section, we will describe both relations and functions, and point out the difference between them. The remainder of the chapter will be devoted primarily to the study of functions.

A relationship between quantities may be described in a number of different ways. An equation or formula, a list or table of corresponding values, or even a verbal statement may be used.

Example A:
 1. The formula $C = 2\pi R$ describes a relationship between the radius and circumference of a circle.
 2. The equation $R = S^2$ describes a relationship between the variables R and S.

Sec. 4–1. Functions

Example B:
1. This table describes a relationship between the variables X and Y.

X	0	1	2	3	4	5
Y	1	3	5	7	9	11

2. This table describes a relationship between voltage and current in a circuit.

Voltage	20	40	60	80	100
Current	160	183	211	268	305

Example C:
1. "The interest earned on a savings account depends on the rate of interest paid by the bank" is a statement that describes a relationship between interest earned and the rate at which interest accumulates.
2. "The cost of drilling a well depends on the depth of the well" is a statement that describes a relationship between the cost of drilling and the depth of a well.

If we look closely at the formula $C = 2\pi R$, we should note that C is given in terms of R. For any particular value of R there will be exactly one value for C. Such a relationship is called a *function*. On the other hand, looking closely at the equation $T = \pm\sqrt{V}$, we note that T is given in terms of V. For any particular value of V there will be two values for T. Such a relationship is called a *relation*. In general, whenever A is given in terms of B, the relationship is a function if no value substituted for B yields more than one value for A. If any value substituted for B yields more than one value for A, then we have a relation. We should recall here that the sign preceding the radical indicates the root to be taken. For example, $\sqrt{4} = 2$, $-\sqrt{4} = -2$, and $\pm\sqrt{4} = 2$ and -2.

Example D:
1. The equation $A = B^2$ is a function since A is given in terms of B and any selected value for B will yield exactly one value for A.
2. The equation $Y = \frac{1}{X}$ is a function since Y is given in terms of X and any selected value for X will yield exactly one value for Y or none at all. (If $X = 0$, we get no value for Y but this fact does not contradict our description of a function).

Example E:
1. $P = \pm\sqrt{1 - Q^2}$ is a relation since P is given in terms of Q and any value substituted for Q will yield two values for P. For example, if $Q = 0$, then $P = 1$ or -1.
2. $K^2 = 2L^2 + 7$ is a relation since K is given in terms of L and any value substituted for L will yield two values for K. For example, if $L = 3$, then $K = 5$ or -5.

When a relationship is given, it is important to know which quantity is given in terms of the other in order to determine whether we have a function or relation.

Example F: If $A = B + 3$, A is given in terms of B and this relationship is a function. If we rearrange this relationship to read $B = A - 3$, then B is given in terms of A and we still have a function.

Example G: If $Y = X^2$, Y is given in terms of X and this relationship is a function. If we rearrange this relationship to read $X = \pm\sqrt{Y}$, then X is given in terms of Y and we no longer have a function but a relation.

The same equation could determine either a relation or a function, or both, depending on which variable is given in terms of the other.

If A is given in terms of B and the relationship is a function, then we say that *A is a function of B*. If B is given in terms of A and the relationship is a function then we say that *B is a function of A*.

Example H:
1. $Y = 3X - 2$ is a function since Y is given in terms of X and any value substituted for X yields exactly one value for Y. Therefore, Y is a function of X.
2. $A = \pi R^2$ is a function since A is given in terms of R and any value substituted for R yields exactly one value for A. Therefore, A is a function of R.

Most of the geometric formulas with which we are familiar can be interpreted as functions since they satisfy our basic description.

Example I:
1. $C = 2\pi R$ expresses the circumference of a circle as a function of its radius.
2. $P = 4S$ expresses the perimeter of a square as a function of its side.
3. $A = 6E^2$ expresses the area of a cube as a function of its edge.

Exercises Determine whether each of the following is a function or a relation.

1. $Y = 5X - 7$
2. $A = 3B^2 + 1$
3. $R^2 = 4 - S^2$
4. $A^2 = 6B^2 + 9$
5. $K = 3L^2 - 2L + 5$
6. $X = \dfrac{1}{Y + 3}$

Sec. 4–1. Functions

7. $M = \pm\sqrt{N + 3}$
8. $Y = \sqrt{X - 5}$
9. $P = \dfrac{Q + 1}{Q - 1}$
10. $T^3 = V^2 - 1$
11. Given the table

X	0	1	2	3	4	5
Y	−5	−2	1	4	7	10

    Is Y a function of X? Is X a function of Y?

12. Given the table

S	1	2	3	−1	−2	−3
R	2	5	10	2	5	10

    Is R a function of S? Is S a function of R?

13. Given the table

B	2	12	30	0	6	20
A	1	3	5	−1	−3	−5

    Is A a function of B? Is B a function of A?

14. Given the table

V	2	5	−1	7	2	4
T	1	3	1	−2	−6	−5

    Is T a function of V? Is V a function of T?

15. If $Y = 4X + 1$, is Y a function of X? Rewrite this equation to solve for X in terms of Y. Is X a function of Y?
16. If $R = 6 - 5S$, is R a function of S? Rewrite this equation to solve for S in terms of R. Is S a function of R?
17. If $T = \pm\sqrt{V + 1}$, is T a function of V? Rewrite this equation to solve for V in terms of T. Is V a function of T?
18. If $K = 2 - 3L^2$, is K a function of L? Rewrite this equation to solve for L in terms of K. Is L a function of K?
19. Express the area of a square as a function of its side.
20. Express the volume of a cube as a function of its edge.
21. Express the perimeter of a rectangle of width 8 as a function of its length.

22. Express the area of a triangle of height 12 as a function of its base.

23. Express the area of a circle as a function of its circumference.

24. Express the area of a square as a function of its perimeter.

4–2. Functional Notation and Terminology

In this section we will discuss the terminology associated with the relationships mentioned in the last section, and a special notation associated with functions.

In a mathematical relationship it is important to determine the independent and dependent variables. The *independent variable* is the variable for which values are arbitrarily substituted. The *dependent variable* is the variable that is determined after a value has been substituted in place of the independent variable. Thus, the value of the dependent variable *depends* on the value of the independent variable.

Example A: Consider the formula $C = 2\pi R$. If the radius of a circle is 2 centimeters, then the circumference is approximately 12.56 centimeters.

$$2 \cdot \pi \cdot 2 \text{ cm} = 12.56 \text{ cm}$$

If the radius of a circle is 3 feet, then the circumference is approximately 18.84 feet.

$$2 \cdot \pi \cdot 3 \text{ ft} = 18.84 \text{ ft}$$

For each positive real number R, the formula determines a corresponding real number C. Thus R is the independent variable and C is the dependent variable.

Example B: Consider the equation $Y = 2X - 4$.

$$\text{If } X = -2, \text{ then } Y = -8.$$
$$\text{If } X = 0, \text{ then } Y = -4.$$

For each real number X, the equation determines a corresponding real number Y. Thus X is the independent variable and Y is the dependent variable.

Example C: Consider the equation $S = 2T^2$.

$$\text{If } T = 4, \text{ then } S = 32.$$
$$\text{If } T = -4, \text{ then } S = 32.$$

For each real number T, the equation determines a corresponding real number S. Thus T is the independent variable and S is the dependent variable.

Sec. 4–2. Functional Notation and Terminology

Example D: Consider the equation $4A + 5B = 20$.
1. If the equation is solved for A, that is, $A = 5 - \frac{5}{4}B$, the equation is in a form where the variable B can be replaced arbitrarily with real numbers and the corresponding values of A determined. Thus B would be the independent variable and A would be the dependent variable.
2. If the equation is solved for B, that is, $B = 4 - \frac{4}{5}A$, the equation is in a form where the variable A can be arbitrarily replaced with real numbers and the corresponding values of B determined. Thus A is the independent variable and B is the dependent variable.

The complete set of numbers that may be used as replacements for the independent variable is called the *domain*. The complete set of numbers that may represent the dependent variable is called the *range*.

Example E: Determine the domain and range for the equation $A = \pi R^2$. Since this is the formula for finding the area of a circle, the radius must be a positive real number and the area must also be a positive real number. Thus the domain is all positive real numbers, $R > 0$, and the range is all positive real numbers, $A > 0$.

Example F: Determine the domain and range for the equation $Y = 2X - 4$. Since any real number can be substituted for X and yield a corresponding real number for Y, then both the domain and range would be the set of real numbers.

Example G: Determine the domain and range for the equation $T = \sqrt{1 - V}$. In this equation, in order for T to be a real number it is necessary for the radicand $(1 - V)$ to be equal to 0 or greater than 0. In order for this radicand to be equal to or greater than 0, we should see that V must be less than or equal to 1. Thus the domain is all real numbers less than or equal to 1 ($V \leq 1$).

NOTE: Whenever a square root expression is contained in an equation the radicand must be 0 or greater than 0. Setting the radicand equal to or greater than 0 and solving will determine the maximum and minimum allowable values for the independent variable.

From the original equation we note that T must be 0 or a positive real number. Thus the range is the set of positive real numbers or 0 ($T \geq 0$).

Whenever time is a variable in a table or an equation it is the independent variable and the domain is the replacement values for time. The remaining variable would be the dependent variable and thus, the range would be the replacement set for that variable.

Example H: The following table defines a relationship between time and temperature.

Time	6 A.M.	7	8	9	10	11	12
Temp (°F)	34°	36°	37°	40°	42°	44°	45°

Time is the independent variable and the domain is the replacements for time (6, 7, 8, ..., 12). Temperature is the dependent variable and the range is the replacements for temperature (34°, 36°, 37°, 40°, 42°, 44°, 45°).

We now formally define a *function* as a mathematical relationship such that, for a given value of the independent variable, there is not more than one corresponding value for the dependent variable.

Example I:
1. The equation $Y = 2X - 1$ is a function because for any value of X there is only one corresponding value of Y.
2. The equation $A = \pm\sqrt{B}$ is not a function since for any value of B (other than 0) there are two corresponding values of A. In other words, A is not unique for all values of B.
3. The equation $R = \dfrac{1}{S}$ is a function because for any value of S there is not more than one value for R. (If $S = 0$, we get no value for R, but this fact does not contradict our definition.)

When referring to a function, we often use the notation $f(x)$, read "f of x". This symbolism does not mean that x is multiplied by f. It is a convenient way of referring to a function at a given value of the independent variable. When this notation is used, other letters may be used in place of f. For example, $g(x)$ denotes a function of x and $g(R)$ denotes a function of R. The letter in parentheses always represents the independent variable, while $f(x)$, $g(x)$, or $g(R)$ are the corresponding values for the dependent variable.

Example J: The equation $Y = 4X^2 - 2X + 4$ could also be written as

$$f(X) = 4X^2 - 2X + 4.$$
$$f(2) = 4(2)^2 - 2(2) + 4 = 16.$$

That is, if 2 is substituted for the independent variable, the corresponding value for the dependent variable is $f(2)$ or 16.

$$f(-1) = 4(-1)^2 - 2(-1) + 4 = 10.$$

That is, if -1 is substituted for the independent variable, the corresponding value for the dependent variable is $f(-1)$ or 10.

Sec. 4-2. Functional Notation and Terminology

Example K: If $G(T) = \dfrac{2T-1}{T+1}$, find $G(-2)$ and $G(-1)$.

$$G(-2) = \frac{2(-2)-1}{-2+1} = \frac{-5}{-1} = 5$$

$$G(-1) = \frac{2(-1)-1}{-1+1} = \frac{-3}{0} \text{ which is undefined.}$$

Example L: If $q(S) = S^2 - 3$, find $q(\tfrac{3}{2})$ and $q(2A)$.

$$q(\tfrac{3}{2}) = (\tfrac{3}{2})^2 - 3 = \tfrac{9}{4} - \tfrac{12}{4} = -\tfrac{3}{4}$$

$$q(2A) = (2A)^2 - 3 = 4A^2 - 3$$

It should be pointed out that it is the equation, not the letters used, that determines the function.

Example M:
1. $f(X) = X^3 - 8$ and $g(X) = X^3 - 8$ are equivalent functions.
2. $f(X) = X^3 - 8$ and $f(T) = T^3 - 8$ are the same function.
3. $f(X) = X^3 - 8$ and $g(X) = X^2 - 2X + 1$ are different functions.

Exercises

In each of the following equations:
  (a) Identify the independent and dependent variable.
  (b) Find the domain and range.
  (c) Determine which equations are functions.

1. $Y = 4X^2 - 3$
2. $s = 120 - 32t$ ($s =$ distance, $t =$ time)
3. $M = \sqrt{X}$
4. $V = \dfrac{34}{T^2}$
5. $Z = \pm\sqrt{25 - W^2}$
6. $G = \dfrac{S+2}{S-2}$
7. $A = \dfrac{1}{B}$
8. $C = D^2$
9. $3X = 6Y - 4$
10. $Y = \sqrt{X^2 - 1}$

For each of the following, evaluate the functions as indicated.
11. If $f(X) = 4 - 2X + X^3$, find $f(0)$ and $f(2)$.
12. If $g(S) = 4\sqrt{S} - 1$, find $g(9)$ and $g(4)$.
13. If $k(T) = \dfrac{3T^2 - T}{2T}$, find $k(-3)$ and $k(-1)$.

14. If $f(R) = \sqrt{2R^3 + 4}$, find $f(-1)$ and $f(0)$.
15. If $f(X) = \frac{1}{X^2}$, find $f(0)$ and $f(\frac{2}{3})$.
16. If $g(B) = \frac{B+1}{B-1}$, find $g(1)$ and $g(-1)$.
17. If $k(V) = V^2 - V + 3$, find $k(2V)$ and $k(V+1)$.
18. If $f(X) = X^3 - 2X + 1$, find $f(-X)$ and $f(X+2)$.

4–3. The Rectangular Coordinate System

One of the best ways of analyzing functions and relations is to construct their *graphs*. These graphs are really "pictures" of the functions or relations, and as such help us to determine certain information and characteristics about different relations and functions.

We should recall from Chapter 1 that sets of real numbers can be represented on a real number line. Since functions and relations involve two sets of real numbers (the domain and the range), we must use two real number lines in constructing their graphs. Although it is not absolutely necessary, it is most convenient to place the two number lines perpendicular to each other (i.e., one, a horizontal line and the other, a vertical line). Each of these number lines is called an *axis*. The horizontal axis is called the *independent axis* since values for the independent variable are placed on this axis. The vertical axis is called the *dependent axis* since values for the dependent variable are placed on this axis. The point of intersection of the two axes is called the *origin*. Horizontally, positive values are to the right of the origin and negative values to the left. Vertically, positive values are above the origin and negative values below. The two axes divide the plane into four parts called *quadrants*, which are numbered I, II, III, and IV starting at the upper right and moving counterclockwise.

Any point in the plane is designated by a pair of numbers (X, Y), where X (the first number of the pair) is always the value of the independent variable and Y (the second number of the pair) is always the value of the dependent variable. The pair (X, Y) is called an *ordered pair* since the order in which the numbers are written is extremely important. That is, the pair (X, Y) is not the same as the pair (Y, X). The first number of the ordered pair is called the *abscissa* and denotes the distance of the point from the vertical axis. The second number of the ordered pair is called the *ordinate* and denotes the distance of the point from the horizontal axis. We should also note that the abscissa will be positive in the first and fourth quadrants and negative in the second and third quadrants, while the ordinate is positive in the first and second quadrants and negative in the third and fourth quadrants. The abscissa and ordinate together form the *coordinates* of a point. From our discussion of real numbers and real number lines in Chapter 1, it follows that this representation allows for one point for any

Sec. 4–3. The Rectangular Coordinate System 87

ordered pair (X, Y). This system we have just described is called the *rectangular coordinate system* and is pictured in Figure 4–3.1.

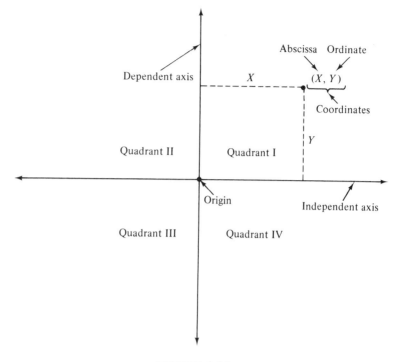

FIGURE 4–3.1

Example A: Locate the points $(2, 3)$, $(3, 2)$, $(-4, 1)$, $(-3, -4)$, $(5, -2)$, $(4, 0)$ and $(0, 1)$ in the rectangular coordinate system.

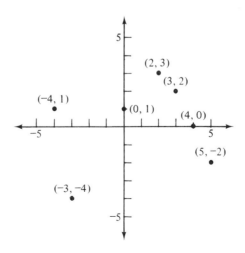

Example B:
1. Where are all the points whose abscissas are 3? This means we are looking for all points 3 units to the right of the vertical axis. Therefore, all such points would be on a line 3 units to the right of the vertical axis. Whenever a condition is put on the abscissa, but no corresponding condition is put on the ordinate, we obtain a vertical line.

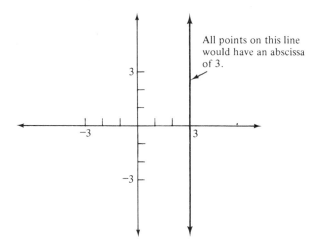

2. Where are all the points whose ordinates are −1? This means we are looking for all points 1 unit below the horizontal axis. Therefore, all such points would be on a line 1 unit below the horizontal axis. Whenever a condition is put on the ordinate, but no corresponding condition is put on the abscissa, we obtain a horizontal line.

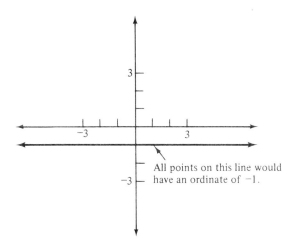

Sec. 4–3. The Rectangular Coordinate System

Exercises

1. What are the coordinates of points *A*, *B*, *C*, and *D*?

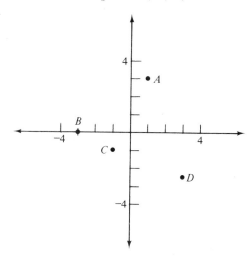

2. What are the coordinates of points *A*, *B*, *C*, and *D*?

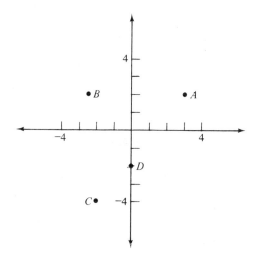

3. Plot the points $(3, -5)$, $(-4, 0)$, $(-2, -\frac{1}{2})$, and $(-1, \frac{5}{2})$ in the rectangular coordinate system.

4. Plot the points $(2, -1)$, $(0, \frac{3}{2})$, $(-4, -2)$, and $(\frac{1}{2}, -\frac{7}{2})$ in the rectangular coordinate system.

5. Where are all the points with an abscissa of 5?

6. Where are all the points with an abscissa of -2?

7. Where are all the points with an ordinate of 2?
8. Where are all the points with an ordinate of $-3\frac{1}{2}$?
9. If the abscissa of a point is positive and the ordinate of the same point is negative, in which quadrant is the point located?
10. If the abscissa of a point is negative and the ordinate of the same point is negative, in which quadrant is the point located?
11. What is the abscissa of all points on the dependent axis?
12. What is the ordinate of all points on the independent axis?
13. In which quadrant(s) is the ratio of abscissa to ordinate positive?
14. In which quadrant(s) is the ratio of ordinate to abscissa negative?
15. Where are all the points whose abscissas equal their ordinates?
16. Where are all the points whose abscissas are greater than their ordinates?
17. Do the points $(-3, -2)$, $(-1, 1)$ and $(1, 4)$ lie on a straight line?
18. Three vertices of a parallelogram are $(-2, -1)$, $(-1, 2)$, and $(4, 2)$. What are the coordinates of the fourth vertex?
19. In constructing the graph of $Y = X^2 - 2X + 1$, which variable would be plotted horizontally? Why?
20. In constructing the graph of $A = 2B - 3$, which variable would be plotted vertically? Why?

4–4. Graphing Techniques and Graphs

Having introduced the rectangular coordinate system, we are now able to construct graphs of different functions and relations in order to get a "picture" of the relationships that these functions and relations represent.

The *graph of a function or relation* is the set of all points whose coordinates satisfy the given relationship.

Example A:
1. If $Y = X^2$, this is a function and its graph would be the set of points (X, Y) that satisfy the relationship $Y = X^2$. A few of these points would be $(0, 0)$, $(2, 4)$, $(-3, 9)$, $(\frac{1}{2}, \frac{1}{4})$, and $(5, 25)$.
2. If $A = 3B - 1$, this is a function and its graph would be the set of points (B, A) that satisfy the relationship $A = 3B - 1$. A few of these points would be $(0, -1)$, $(1, 2)$, $(-2, -7)$, and $(\frac{1}{3}, 0)$.

Sec. 4–4. Graphing Techniques and Graphs   91

3. If $R = \pm\sqrt{S}$, this is a relation and its graph would be the set of points (S, R) that satisfy the relationship $R = \pm\sqrt{S}$. A few of these values would be $(0, 0)$, $(4, 2)$, $(4, -2)$, $(9, 3)$, and $(9, -3)$.

To construct an accurate graph, the following procedure should be followed.

1. *Determine the independent variable.* This variable will be plotted horizontally and the dependent variable will be plotted vertically.

2. *Determine any restrictions on the independent variable.* Since we are only graphing real numbers, we allow no value for the independent variable that would lead to division by zero or the square root of a negative number.

3. *Make a table of values.* This is done by substituting values for the independent variable and determining corresponding values for the dependent variable. In doing so, we should substitute values for the independent variable that are relatively easy to work with, such as 0, 1, −1, 2, and −2. The number of values to be substituted will vary from one graph to another, but we should usually start with five or six values and then use more if they are needed.

4. *Determine an appropriate scale and plot the points corresponding to the ordered pairs listed in the table of values.* While it is desirable where possible, it is not absolutely necessary to use identical scales both horizontally and vertically.

5. *Connect the points that have been plotted with a smooth curve.* Points are usually connected from left to right taking into account any restrictions that have been determined. Also, in most cases, the curve does not end at the last points we have plotted, but continues. We should examine the given equation to determine how the graph should continue past these points and then indicate the continuation.

Once we have constructed a graph, certain parts of it are of particular importance. *Intercepts* are points that a graph has in common with either axis. A horizontal intercept would have coordinates $(X, 0)$ where X could be any real number. A vertical intercept would have coordinates $(0, Y)$ where Y could be any real number. *Asymptotes* are lines that a curve approaches but never actually meets. The only asymptotes we will consider at this time will be those that correspond to restrictions on the independent and dependent variable.

Example B: Construct the graph of $Y = 2X - 1$. X is the independent variable and will be plotted horizontally. There is no real number that X cannot equal.

X	0	1	−1	2	−2
Y	−1	1	−3	3	−5

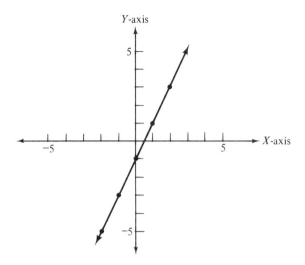

Analyzing the equation for both larger positive and larger negative values of X, we can see that the graph continues in the directions indicated. From the graph we should observe that there is a horizontal intercept at $(\frac{1}{2}, 0)$ and a vertical intercept at $(0, -1)$.

Example C: Construct the graph of $A = B^2 - 4B + 3$. B is the independent variable and will be plotted horizontally. There is no real number that B cannot equal.

B	0	1	−1	2	−2	3	4	5
A	3	0	8	−1	15	0	3	8

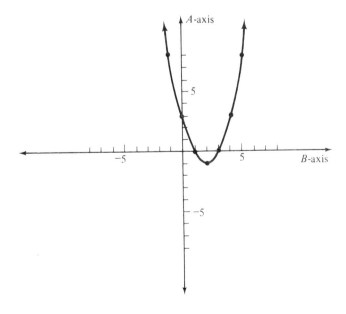

Sec. 4–4. Graphing Techniques and Graphs 93

Analyzing the equation for both larger positive and larger negative values of B, we can see that the graph continues in the directions indicated. Note that additional pairs were needed in our table of values in order to obtain an accurate graph. From the graph we should observe that there are horizontal intercepts at $(1, 0)$ and $(3, 0)$ and a vertical intercept at $(0, 3)$.

Example D: Construct the graph of $R = 2 + S - S^2$. S is the independent variable and will be plotted horizontally. There is no real number that S cannot equal.

S	0	1	−1	2	−2	3	$\frac{1}{2}$
R	2	2	0	0	−4	−4	$2\frac{1}{4}$

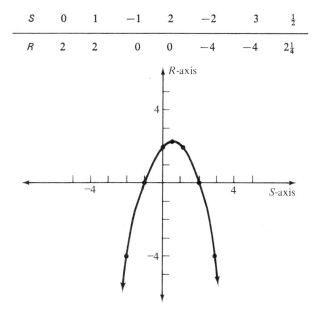

Analyzing the equation for both larger positive and negative values of S, we can see that the graph continues in the directions indicated. Note that obtaining the same value for R for both $S = 0$ and $S = 1$ led to the substitution of the value $\frac{1}{2}$ for S and the inclusion of the corresponding point on the graph. From the graph we should observe that there are horizontal intercepts at $(-1, 0)$ and $(2, 0)$ and a vertical intercept at $(0, 2)$.

Example E: Construct the graph of $T = V^3 - V^2 - 4V + 4$. V is the independent variable and will be plotted horizontally. There is no real number that V cannot equal.

V	0	1	−1	2	−2	3	−3	$1\frac{1}{2}$
T	4	0	6	0	0	10	−20	$-\frac{7}{8}$

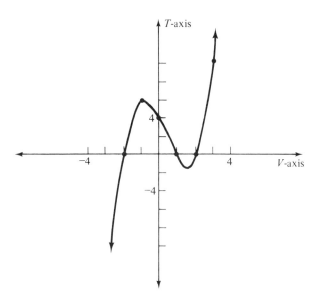

Analyzing the equation for larger positive and negative values of V, we can see that the graph continues in the directions indicated. Note that it is convenient to use different scales horizontally and vertically. Also note the use of the value $1\frac{1}{2}$ for V after obtaining the same value for T when $V = 1$ and $V = 2$. From the graph we should observe that there are horizontal intercepts at $(-2, 0)$, $(1, 0)$, and $(2, 0)$ and a vertical intercept at $(0, 4)$.

Example F: Construct the graph of $Y = \dfrac{1}{X-1}$. X is the independent variable and will be plotted horizontally. X cannot equal 1 since this would lead to division by 0. Therefore, there can be no point of the graph corresponding to $X = 1$.

X	0	1	−1	2	−2	3	−3	4	$\frac{1}{2}$	$1\frac{1}{2}$
Y	−1	?	$-\frac{1}{2}$	1	$-\frac{1}{3}$	$\frac{1}{2}$	$-\frac{1}{4}$	$\frac{1}{3}$	−2	2

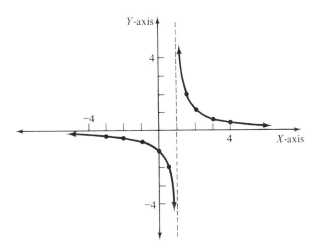

Analyzing the equation for larger positive values of X, larger negative values of X, and values close to $X = 1$, we can see that the graph continues in the directions indicated. Note that the values $\frac{1}{2}$ and $1\frac{1}{2}$ are substituted for X in order to get the proper shape of the graph. As mentioned above, we can see that there is no point of the graph corresponding to $X = 1$. Thus we have a vertical asymptote, the vertical line $X = 1$. We also have a horizontal asymptote of $Y = 0$, the horizontal axis. Recall that earlier in this section we stated that an asymptote is a line that the graph approaches but never actually touches. From the graph we should observe that there are no horizontal intercepts and there is a vertical intercept at $(0, -1)$.

Example G: Construct the graph of $P = \sqrt{Q + 1}$. Q is the independent variable and will be plotted horizontally. Q cannot equal any number less than -1 since this would lead to square roots of negative numbers.

Q	0	1	−1	2	3	4	5
P	1	1.42	0	1.73	2	2.24	2.45

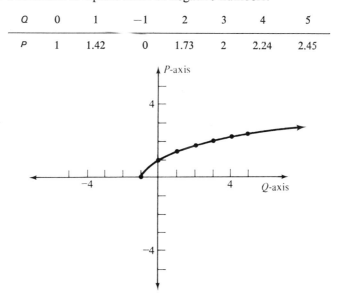

Analyzing the equation for larger positive values of Q we can see that the graph continues in the direction indicated. Note that the graph stops at the point $(-1, 0)$ since Q can equal no number less than -1. From the graph we should observe that there is a horizontal intercept at $(-1, 0)$ and a vertical intercept at $(0, 1)$.

Example H: Construct the graph of $K = \pm\sqrt{L - 2}$. L is the independent variable and will be plotted horizontally. L cannot equal any number less than 2 since this would lead to square roots of negative numbers.

L	2	3	4	5	6	7
K	0	±1	±1.42	±1.73	±2	±2.24

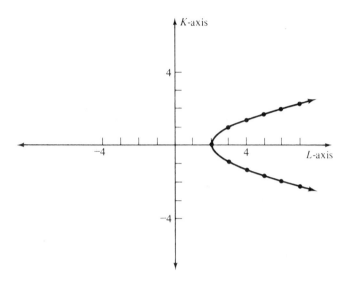

Analyzing the equation for larger positive values of L we can see that the graph continues in the directions indicated. Note that with the $\pm$ preceding the radical, we must list both square roots and plot both corresponding points. From the graph we should observe that there is a horizontal intercept at $(2, 0)$ and there are no vertical intercepts.

Example 1: Construct the graph of $Y = 2^X$. X is the independent variable and will be plotted horizontally.

X	0	1	−1	2	−2	3	−3
Y	1	2	$\frac{1}{2}$	4	$\frac{1}{4}$	8	$\frac{1}{8}$

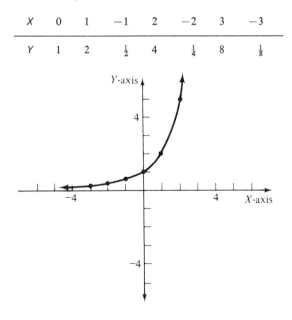

Analyzing the equation for larger positive and negative values of X we can see that the graph continues in the directions indicated. Since 2 raised to any power

Sec. 4–5. Graphs of Empirical Data

will yield a positive result, Y will never be 0 or a negative number. Thus, the left-hand portion of the graph approaches the independent axis but never actually touches it. Hence, the independent axis is a horizontal asymptote. From the graph we should observe that there are no hortizontal intercepts and there is a vertical intercept at (0, 1).

The graph in Example H represents a relation since any value substituted for the independent variable yields more than one value (in this case two values) for the dependent variable. The graphs in the other examples (Examples B–G and Example I) represent functions since any value substituted for the independent variable yields no more than one value for the dependent variable. Graphically, then, we can distinguish a function from a relation since *no vertical line would intersect a function in more than one point, whereas some vertical line would intersect a relation in at least two points.*

Exercises

For each of the following, construct the graph, list any intercepts and/or asymptotes, and state whether the graph represents a function or a relation.

1. $Y = 3X - 2$
2. $A = 5B + 1$
3. $R = 4 - 2S$
4. $T = 3 + 4V$
5. $K = L^2 + L - 2$
6. $P = Q^2 - 9$
7. $Y = 2X - X^2$
8. $R = 8 + 2S - S^2$
9. $B = A^3 + 2A - 1$
10. $V = T^3 - 8$
11. $L = K^3 + 2K^2 - K - 2$
12. $X = Y^3 - 7Y - 6$
13. $S = \dfrac{2}{R}$
14. $Y = \dfrac{1}{X + 2}$
15. $A = \dfrac{B}{B - 1}$
16. $T = \dfrac{V - 1}{V + 1}$
17. $Y = \sqrt{4 - 2X}$
18. $R = \sqrt{S^2 + 2}$
19. $K = \pm\sqrt{3L - 1}$
20. $Q = \pm\sqrt{4 - P^2}$
21. $A = 3^B$
22. $Y = 4^X$

4–5. Graphs of Empirical Data

Experimental situations involve collecting data from an experiment in order to study certain physical phenomena. Information collected in this manner is called *empirical data*. The procedure usually followed is to collect the data in a table, and then graph this information in order to get a clear idea of the relationship between the sets of data.

The following technique should be used for graphing empirical data:

1. *Determine the independent and dependent variables.* Recall that whenever time is involved it is always the independent variable.

2. *Set up an appropriate scale (determined from the range of values in the table) on each axis.* When graphing empirical data, all four quadrants usually are not needed.

3. *Plot the points corresponding to the ordered pairs of data in the table of values.* Circle each point on the graph.

4. *Join the points with a smooth curve or with straight line segments.* A smooth curve would indicate a definite relationship between the sets of data, whereas straight line segments would indicate that there is no definite relationship between the sets of data.

Example A: A certain experiment consisted of measuring the amplitude (in centimeters) of a pendulum at the end of each unit of time (in seconds). The following data was collected. Plot the graph of amplitude as a function of time.

Time (sec)	0	1	2	3	4	5
Amplitude (cm)	15	10.4	7.12	5.34	4.57	3.82

Since amplitude is a function of time, time is the independent variable. Therefore, time will be plotted on the horizontal axis and amplitude will be plotted on the vertical axis. The range of values for time is 0–5. Therefore, the independent axis should only include the values 0–5. The range of the amplitude is 3.82–15.0, and therefore a meaningful scale on the dependent axis would include a range of the values 3–15.

Plot the points, circle each point, and connect the points with a smooth curve. It is now possible to make a conjecture based on the data collected from the graph.

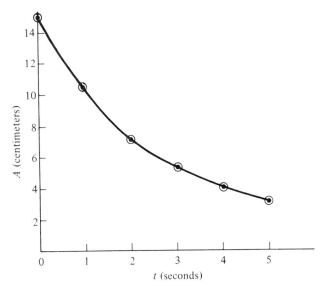

Sec. 4–5. Graphs of Empirical Data

Example B: The hourly temperature (in degrees Celsius) for the hours 8 A.M. to 4 P.M. were recorded as follows:

Time	8	9	10	11	12	1	2	3	4
Temperature (°C)	−5	−2	0	3	7	15	15	13	10

Time is the independent variable and temperature is the dependent variable.

The range of values for time should only include the time from 8 A.M. to 4 P.M. and the meaningful range for temperature would only include the temperatures from −5°C to 15°C.

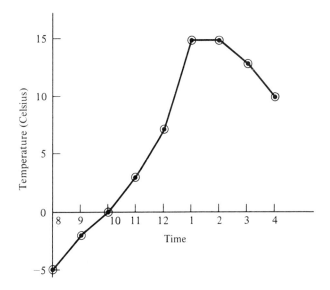

Since there is no definite relationship between temperature and time, a series of straight lines is used to join the points. It is not possible to predict the temperature at any other time during the day because of the lack of a relationship between time and temperature.

Once the empirical data has been collected and put into tabular form, and a graph has been constructed, it is useful to write a precise description of the relationship that exists between the sets of values, if this is possible. An equation that gives an accurate description of the relationship of one set of observed values to the other is called an *empirical equation*. An empirical equation is useful not only for giving a description, but also for finding a precise value of one of the variables when one is given a value of the other variable that is not included in the table.

Example C: The following values represent the displacement (in meters) of a particle moving in a straight line in relation to time (in seconds).

t (sec)	0	.5	1.5	2	2.5	3
s (m)	0	1	3	4	5	6

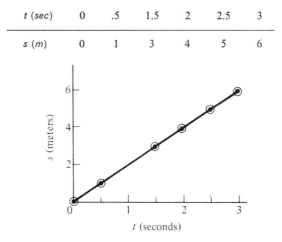

Write an equation that would give an accurate description of the relationship that exists between time and displacement. From the table a comparison between consecutive values of time and displacement shows an increase of 1 meter for each corresponding increase of $\frac{1}{2}$ second. This means there is a $2:1$ ratio between the corresponding values of displacement and time. This relationship can be expressed in an empirical equation where t is the independent variable and s is the dependent variable:

$$s = 2t.$$

This equation gives a description of the relationship between displacement and time under the conditions that prevail in this particular situation.

The empirical equation is also useful for finding the precise value of displacement for a particular value of time that is not given. For example, the displacement when $t = 1.75$ sec can be determined by substituting 1.75 for t in the empirical equation: $s = 2(1.75) = 3.5$ m.

This same information could be estimated from a graph, but the information from the graph, in most cases, would only be an approximation.

Example D: A particle was observed moving along a path with uniform acceleration. The following table of values was obtained. Write an empirical equation from these values.

t (sec)	0	1	2	3	4	5
a (ft/sec²)	0	1	4	9	16	25

Sec. 4–5. Graphs of Empirical Data

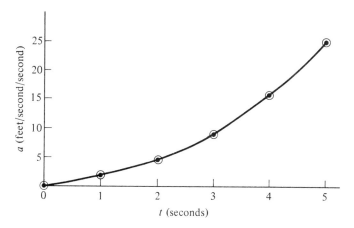

In this case there is not a constant ratio between consecutive pairs of values of time and acceleration. Therefore, a different relationship between the sets of data must be sought. It should be observed that the acceleration is the square of each corresponding value of time:

$$a = t^2$$

This equation gives a description of the relationship between acceleration and time.

Exercises

1. Plot the graph of atmospheric pressure as a function of altitude above sea level.

Altitude (ft)	0	5000	10000	15000	20000
Pressure (lb/in.^{2})	14.7	12.2	10.1	8.3	6.8

2. The density of water was measured at various temperatures. Plot the graph showing the density of water as a function of temperature.

Temperature (°C)	0	4	8	12	16	20	24
Density (gm/cm^3)	.99988	1.0	.99993	.99958	.99894	.99823	.99710

3. The boiling point of water was recorded in degrees Fahrenheit at various atmospheric pressures. Plot the graph showing the effect of pressure on the boiling point of water.

Pressure (lb/in.^{2})	.0880	.1	.2	.8	1.2	2.2	4.0	6.0	8.8	13.6
Temperature (°F)	30°	50°	70°	90°	110°	130°	150°	170°	190°	210°

4. The distance that an object is above ground level was recorded as a function of time. Plot the distance as a function of time.

Time (sec)	0	2	4	6	8
Distance (m)	0	192	256	192	0

5. The saturated vapor pressure of carbon disulfide was recorded as a function of temperature. Plot the graph of this relationship.

Temperature (°F)	0	20	40	60	80	100
Pressure (lb/in.²)	1	2	3.1	5.2	7.5	11

6. Plot a graph of the following data on the solubility of sodium sulfate as a function of temperature.

Temperature (°C)	0	10	20	30	40	50	60	70	80	90
S (gm/100 gm of H_2O)	5	9.0	19.4	40.8	48.8	46.7	45.3	44.4	43.7	43.1

7. The formula for the velocity of a pulse traveling in a string is $v = 100\sqrt{T}$ where T represents the tension on the string (in pounds) and v represents the velocity of the pulse (in feet per second). Plot the graph of this function and answer the following questions.
   (a) What is the velocity when the tension is 20 pounds?
   (b) What is the velocity when the tension is 50 pounds?
   (c) What is the tension when the velocity is 400 feet per second?

8. The formula for the surface area of a sphere is $A = 4\pi R^2$. Plot the graph of the surface area as a function of its radius.

9. The range of a television station is a function of the height of the antenna. The range is given by the formula $D = 3.56\sqrt{h}$ where D is the distance in kilometers and h is the height of the antenna in meters.
   (a) Plot the range as a function of the antenna height.
   (b) Determine from the graph the range for an antenna height of 40 meters.

10. According to Ohm's Law the current (I) is directly proportional to the voltage (V) and inversely proportional to the resistance (R). The formula $I = \dfrac{V}{R}$ expresses this relationship.
    (a) If the resistance remains constant at 25 ohms, plot the current (in amperes) as a function of voltage.
    (b) At what value of voltage will the current be 4 amperes?
    (c) At what value of current will the voltage be 50 volts?

Sec. 4–6. Solving an Equation Graphically

11. The following table gives the breaking point of a rod (in kilograms) vs. the diameter (in centimeters) of the rod.

D (cm)	1	$1\frac{1}{2}$	2	$2\frac{1}{2}$	3	$3\frac{1}{2}$	4
F (kg)	4	$5\frac{1}{2}$	7	$8\frac{1}{2}$	10	$11\frac{1}{2}$	13

Graph the relationship and write an empirical equation for the relationship between F and D.

12. The length and weight of 7 wooden spools with the same diameter were measured and recorded.

L (m)	2.5	5.1	7.5	10	12.5	15.0
W (g)	5.5	13.3	20.5	28.0	35.5	43.0

Graph the relationship and write an empirical equation for the relationship between W and L.

4–6. Solving an Equation Graphically

To find the solution for an equation means to find a value(s) for the variable(s) involved that will make the equation a true statement.

Example A:
1. If $2X + 1 = 7$, then $X = 3$ would be the solution since this value, when substituted for X, makes the relationship valid.
2. If $A^2 = 16$, then $A = 4$ or -4 since these values, when substituted for A, make the relationship valid.
3. If $P + Q = 8$, then $P = 2$ and $Q = 6$ would be one of the unlimited number of solutions for this equation. Any replacements for P and Q that add up to 8 would constitute a valid solution.

NOTE: Any time we have a single equation with more than one variable, there will always be more than one solution. Therefore, we will only consider in this section the solution of single equations with one variable.

To solve such equations graphically we must:

1. Rewrite the equation (using the basic principles discussed in Chapter 1) so that zero is one of the members.
2. Set the nonzero member equal to some variable.
3. Construct the graph of the resulting equation.

4. From the graph, read the values of the independent variable that yield a value of zero for the dependent variable. These values of the independent variable will occur where the graph crosses the horizontal axis (if it does) and they are called *zeros* of the function or relation that was graphed. These zeros correspond to the solution of the original equation.

Example B: Find any zeros of the relationship $P = Q^2 + Q - 2$. Q is the independent variable and will be plotted horizontally. There is no real number that Q cannot equal.

Q	0	1	-1	2	-2	3	-3	$-\frac{1}{2}$
P	-2	0	-2	4	0	10	4	$-\frac{9}{4}$

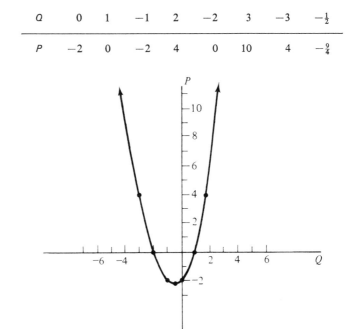

We can see that the graph crosses the horizontal axis at the points $(1, 0)$ and $(-2, 0)$. Therefore $Q = 1$ and $Q = -2$ are the zeros of this relationship.

Example C: Solve the equation $2X + 1 = 7$ graphically. First, we rewrite this equation as $2X - 6 = 0$. Now, we let $Y = 2X - 6$. X is the independent variable and will be plotted horizontally. There is no real number that X cannot equal.

X	0	1	-1	2	-2	3	4
Y	-6	-4	-8	-2	-10	0	2

Sec. 4–6. Solving an Equation Graphically

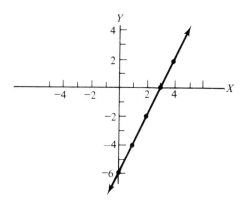

From the graph, we can see that when $X = 3$, $Y = 0$. Therefore, $X = 3$ is the solution to the original equation.

Example D: Solve the equation $A^2 - 2 = A$ graphically. First, we rewrite this equation as $A^2 - A - 2 = 0$. Now, we let $B = A^2 - A - 2$. A is the independent variable and will be plotted horizontally. There is no real number that A cannot equal.

A	0	1	-1	2	-2	3	-3	$\frac{1}{2}$
B	-2	-2	0	0	4	4	10	$-\frac{9}{4}$

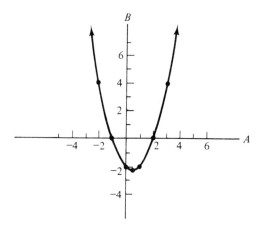

From the graph, we can see that when $A = -1$ or 2, $B = 0$. Therefore, $A = -1$ and $A = 2$ are the solutions to the original equation.

Example E: Solve the equation $T^2 = -1$ graphically. First, we rewrite this equation as $T^2 + 1 = 0$. Now, we let $V = T^2 + 1$.

T	0	1	-1	2	-2
V	1	2	2	5	5

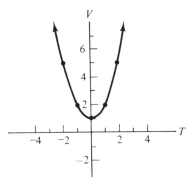

From the graph, we can see that no value for T will yield a value of 0 for V. Therefore, there is no real solution to the original equation.

Example F: A projectile is fired into the air with an initial velocity of 160 feet per second. If its distance above the ground is given by $s = 160t - 16t^2$, where s is the distance in feet and t is the time in seconds, in how many seconds after it is fired will the projectile be back on the ground? After how many seconds will it be 100 feet above the ground? Solve graphically. t is the independent variable and will be plotted horizontally. Since t represents time, it can only be positive.

t	0	1	2	3	4	5	6	7	8	9	10
s	0	144	256	336	384	400	384	336	256	144	0

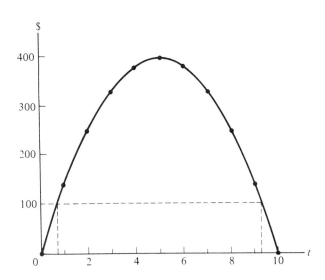

(a) We can see from the graph that when $t = 0$ or 10, $s = 0$. Therefore, 10 seconds after it is fired, the projectile hits the ground.

(b) We can see from the graph that when $t = .7$ or 9.3, $s = 100$. Therefore, .7 seconds after it is fired and again 9.3 seconds after it is fired, the projectile is 100 feet above the ground.

Exercises

For each of the following relationships, find any real zeros.

1. $B = 5 - 3A$
2. $Y = 2X + 6$
3. $V = 6T^2 + T - 2$
4. $P = Q^2 + Q + 1$
5. $R = S^3 + 4S^2$
6. $K = 6 - L^3$

Solve each of the following equations graphically.

7. $5A = 8$
8. $2T - 7 = 0$
9. $5 = 2X^2 + 4X$
10. $R^3 = -4R$
11. $M^2 - M - 7 = 0$
12. $Y^3 - Y = -4$
13. $V^3 + 1 = -8$
14. $-S^2 - 5S = 7$
15. $M^2 + 5 = -3M$
16. $\frac{1}{X} = 3$
17. $\sqrt{5 - T} = 4$
18. $\sqrt{2B + 1} = 3$

Solve each of the following graphically.

19. The equation for a square law detector used in a microwave receiver is $2V^2 = 5 - 3V$. Find the solution(s) for this equation.

20. The formula for the volume of a right circular cone is $V = \frac{1}{3}\pi R^2 H$. If a certain cone has a height of 6 meters, what radius would result in a volume of 15.7 cubic meters?

21. A projectile is fired into the air with an initial velocity of 64 feet per second. If its distance above the ground is given by $s = 64t - 16t^2$, where s is the distance in feet and t is the time in seconds, in how many seconds after it is fired will the projectile hit the ground? After how many seconds will it be 50 feet above the ground?

22. The force needed to counterbalance a weight by means of a lever is given by the equation $F = 7.7d$, where F is the force in newtons and d is the distance of the force from the fulcrum in meters. How far from the fulcrum should a force of 20 newtons be placed in order for the weight to be counterbalanced?

4-7. CHAPTER REVIEW

For each of the following, determine whether the equation defines a relation or a function, and then find the domain and range.

1. $A = 5B + 6$
2. $Y = 3X^2 - 7X$
3. $R = S^3 - 8$
4. $T = \dfrac{4}{V}$
5. $Y = \dfrac{X+1}{X-1}$
6. $K = \sqrt{4 - L^2}$
7. $P = \pm\sqrt{Q^2 + 1}$
8. $A = \dfrac{B-1}{2}$
9. If $f(X) = 4X - 7$, find $f(2)$ and $f(-3)$.
10. If $f(T) = T^2 - 5T$, find $f(-3)$ and $f(\tfrac{1}{2})$.
11. If $f(R) = 3 - 2R^2$, find $f(-2)$ and $f(\tfrac{1}{3})$.
12. If $f(V) = V^3 - 3V^2 + 6V$, find $f(-1)$ and $f(\tfrac{2}{3})$.
13. If $f(Y) = \dfrac{Y+1}{Y}$, find $f(0)$ and $f(-3)$.
14. If $f(S) = \dfrac{S-2}{S+1}$, find $f(2)$ and $f(-2)$.
15. If $g(X) = X^2 - 3X - 4$, find $g(4)$ and $g(2X)$.
16. If $k(T) = T^3 + 2T - 4$, find $k(-\tfrac{1}{2})$ and $k(3T)$.

For each of the following, construct the graph and list any intercepts, asymptotes, and zeros.

17. $A = 2B - 3$
18. $R = 5S - 1$
19. $Y = 2X^2 - 5X - 3$
20. $T = 6V^2 + V - 2$
21. $P = Q^3 - 2Q^2 - Q + 2$
22. $K = L^4 - 4L^2$
23. $Y = \sqrt{X^2 - 4}$
24. $A = \sqrt{9 - B^2}$
25. $P = \dfrac{3}{Q}$
26. $T = \dfrac{V-1}{V}$
27. $R = \pm\sqrt{S^2 - 1}$
28. $Y = \pm\sqrt{5 - X^2}$

Solve each of the following equations graphically.

29. $2X + 7 = 0$
30. $5 - 3B = 0$
31. $6R^2 + 11R = 10$
32. $8V^2 - 3 = -2V$
33. $T^3 = 3T^2$
34. $X^3 - 6X^2 = 6 - 11X$
35. $\dfrac{K}{K+1} = 0$
36. $\dfrac{1}{Y-3} = 0$
37. $\sqrt{9 - A^2} = 0$
38. $\sqrt{2S + 3} = 0$
39. The formula for the circumference of a circle is $C = 2\pi R$. Construct the graph of this relationship.

Sec. 4–7. Chapter Review   109

40. The formula for the surface area of a cube is $A = 6E^2$. Construct the graph of this relationship.

41. The number of central air-conditioning units sold each year for the past ten years at a certain heating and air-conditioning distributor is shown in the following table (-1 = last year).

Year	-10	-9	-8	-7	-6	-5	-4	-3	-2	-1
Units	52	69	65	90	98	101	133	148	165	197

Represent this information graphically.

42. The net sales for each month of the past year at a certain department store are given by the following table.

Month	Sales	Month	Sales
Jan.	$305,000	July	$225,000
Feb.	$240,000	Aug.	$298,000
March	$203,000	Sept.	$311,000
April	$165,000	Oct.	$287,000
May	$280,000	Nov.	$346,000
June	$262,000	Dec.	$394,000

Represent this information graphically.

43. The temperatures for a 12-hour period for a certain city are given by the following table.

Time	6 A.M.	7	8	9	10	11	12	1	2	3	4	5 P.M.
Temperature (°F)	26	28	31	31	35	36	40	42	43	45	45	42

Represent this information graphically.

44. The voltage and current for a certain electric circuit are given by the following table.

Voltage (volts)	10	20	30	40	50	60	70	80	90	100
Current (milliamps)	142	160	175	183	196	211	232	268	290	305

Represent this information graphically.

45. The rainfall for the past 8 years in a certain city is given by the following table (-1 = last year).

110 Ch. 4. Functions, Rectangular Coordinates, and Graphs

Year	−8	−7	−6	−5	−4	−3	−2	−1
Rainfall (in.)	36.4	35.2	37.8	41.3	40.8	38.9	37.2	41.6

Represent this information graphically.

46. The current in relation to time for a certain electric circuit is given by the following table.

Time (sec)	0	$\frac{1}{360}$	$\frac{1}{180}(\frac{2}{360})$	$\frac{1}{120}(\frac{3}{360})$	$\frac{1}{90}(\frac{4}{360})$	$\frac{1}{72}(\frac{5}{360})$	$\frac{1}{60}(\frac{6}{360})$
Current (amps)	0	2.6	4.1	6	4.1	2.6	0

Represent this information graphically.

47. The formula for the volume of a sphere is $V = \frac{4}{3}\pi R^3$. What radius (R) will result in a volume of 905 cubic feet? Solve graphically.

48. The formula for the volume of a right circular cone is $V = \frac{1}{3}\pi R^2 H$. If a certain cone has a radius of 3 feet, how high should the cone be in order to have a volume of 75.4 cubic feet? Solve graphically.

49. The distance above the ground of an object with an initial vertical velocity v_o is given by $s = v_o t - \frac{1}{2}at^2$. If $a = 32$ and $v_o = 88$, when will an object be 120 feet above the ground? Solve graphically.

50. The kinetic energy of an object is given by $E = \frac{1}{2}mv^2$, where m is the mass (in grams) and v is the velocity (in centimeters per second) of the object. If an object weighs 30 grams, what velocity will result in a value of 135 for the kinetic energy? Solve graphically.

51. Construct the graph and determine an equation relating X and Y.

X	0	1	2	3	−1	−2	−3
Y	3	5	7	9	1	−1	−3

52. Construct the graph and determine an equation relating R and S.

R	0	1	2	3	−1	−2	−3
S	0	1	8	27	1	8	27

53. The visibility at sea (in miles) from certain heights (in feet) above sea level is given by the following table.

Height	10	20	30	40	50	60
Visibility	3.5	6	8.5	11	13.5	16

Construct the graph and determine an equation for this relationship.

54. The mileage of a certain car (in miles per gallon) at different speeds (in miles per hour) is given by the following table.

Speed	30	35	40	45	50	55
Mileage	17.5	18.0	18.5	19.0	19.5	20

Construct the graph and determine an equation for this relationship.

chapter 5

Descriptive Statistics

5-1. Averages for a Group of Data

One task that technicians and engineers often must perform is the tabulation and analysis of statistical information obtained from research and experiments. When there is a great deal of statistical information to interpret, it is most useful and efficient to use a representative value to analyze the data involved. Such representative values are called *averages*. In this section, we will look at the three most common types of averages.

The *mean* is the arithmetic average of a group of data. It is calculated by finding the sum of all the values and then dividing this sum by the number of values. The *median* of a group of data is the middle item, that item for which there are as many values above as below when the values are arranged in numerical order. If there is an even number of values, the median is halfway between the two numbers nearest the middle. The *mode* of a group of data is the value that appears most frequently. If each value occurs exactly once, there is no mode. It is also possible to have more than one mode.

Example A: The fifteen employees of a certain heating corporation have annual salaries of $10,000, $14,200, $16,200, $12,800, $14,200, $15,000, $11,400,

Sec. 5–1. Averages for a Group of Data

$16,000, $17,300, $22,000, $19,600, $10,700, $14,200, $16,000, and $16,600. Find the mean, median, and mode for these salaries.

1. If we add all of the salaries together, we obtain a sum of $226,200. Dividing $226,200 by 15, we obtain $15,080. Therefore, the mean salary is $15,080.
2. Arranging these salaries in numerical order, we obtain the following list:

$10,000 $12,800 $14,200 $16,000 $17,300
$10,700 $14,200 $15,000 $16,200 $19,600
$11,400 $14,200 $16,000 $16,600 $22,000

The median then will be $15,000 since there are seven salaries higher than this figure and seven salaries lower than this figure.
3. The mode of these salaries is $14,200 since it occurs in the list three times and no other value occurs that often.

Example B: The temperatures (in degrees Celsius) for a certain 12-hour period are given by the following table:

6 A.M.	7	8	9	10	11	12 P.M.	1	2	3	4	5
20°	21°	23°	24°	25°	24°	26°	26°	29°	30°	28°	24°

Find the mean, median, and mode of these temperatures.
1. If we add all of the temperatures, we obtain a sum of 300°. Dividing 300° by 12, we obtain 25°. Therefore, the mean temperature is 25°.
2. Arranging these temperatures in numerical order, we obtain the following list:

20° 23° 24° 25° 26° 29°
21° 24° 24° 26° 28° 30°

The median will be 24.5° since we have an even number of values, and this value is halfway between the two numbers nearest the middle (24° and 25°).
3. The mode of these temperatures is 24° since it occurs more often than any other temperature.

It should be realized that the more evenly dispersed the individual values are over a certain range, the more meaningful will be our averages. On the other hand, our averages become less representative the more our data is dispersed over a large range of values. Also, one or more of the averages may be completely inaccurate if one or two values are a large distance from the remaining values.

Example C: Given the numbers 24, 64, 68, 70, and 71, find the mean, median, and mode.

1. Mean $= \dfrac{24 + 64 + 68 + 70 + 71}{5} = \dfrac{297}{5} = 59.4$

2. Median $= 68$

3. There is no mode.

We should see here that, because of the inclusion of the number 24 in the list, the mean is not very representative of the data.

Example D: The salaries of four people at a consulting firm are $12,000, $15,000, $42,000, and $47,000. Find the mean, median, and mode.

1. Mean $= \dfrac{\$12000 + \$15000 + \$42000 + \$47000}{4} = \dfrac{\$116000}{4} = \$29,000$

2. Median $= \$28,500$.

3. There is no mode.

We should see here that neither the mean nor the median is very representative of the data.

We must be careful in our judgement, however, since the same set of values may represent a large range in one problem but a small range in another problem. For example, if the difference between the largest and smallest values in a given problem is 45, this would be a large range if the values represent daily temperatures and a very small range if the values represent annual salaries. Whenever we calculate these averages, we should check the given data to see just how representative the averages are before attaching any importance to them.

Exercises

Find the mean, median, and mode for each of the following lists of numbers.

1. 12, 14, 8, 9, 16, 14, 10, 12, 11, 17, 15, 14, 17.
2. 163, 210, 184, 202, 196, 198, 170.
3. 47, 81, 56, 59, 72, 56, 64, 60, 50, 69, 72, 70.
4. 26, 28, 22, 40, 36, 39, 31, 27, 32, 28.
5. The ages of the eleven employees of a certain consulting firm are 30, 32, 37, 29, 40, 38, 32, 38, 30, 36, and 32. Find the mean, median, and mode of these ages.

Sec. 5-2. Standard Deviation   115

6. The monthly salaries of the twelve employees of a certain tool and die corporation are $650, $590, $550, $620, $780, $630, $620, $590, $610, $540, $490, and $500. Find the mean, median, and mode of these salaries.

7. The temperatures (in degrees Fahrenheit) for a 15-hour period on a certain day are listed in the following table:

6 A.M.	7	8	9	10	11	12 P.M.	1	2	3	4	5	6	7	8
43°	44°	47°	49°	53°	55°	56°	58°	63°	63°	58°	55°	54°	50°	47°

Find the mean, median, and mode of these temperatures.

8. During an experiment, the following values for electric current (in milliamperes) were recorded: 180, 197, 206, 188, 201, 194, 212, 185, 215, and 199. Find the mean, median, and mode for these readings.

9. The rainfall (in inches) in a certain city for the past 10 years was recorded as follows: 32.1, 35.6, 40.3, 38.5, 37.4, 39.2, 41.7, 37.8, 36.9, and 38.5. Find the mean, median, and mode for these measurements.

10. The daily rents received by a computer-leasing firm for a two-week period were $3,220, $3,500, $3,360, $3,500, $3,220, $3,920, $4,060, $3,640, $3,220, $3,780, $4,480, $4,340, $4,200, and $4,620. Find the mean, median, and mode of these figures.

5-2. Standard Deviation

As we pointed out in the preceding section, the range of a given group of values has a definite effect on our interpretation of the various averages. The variance, or deviation, of each value from the averages (especially the mean) is also important if one is to analyze correctly the given group of data. Numbers that measure such deviations provide invaluable assistance to us in our interpretation of data and their averages. The measure of deviation that we will discuss here is called the *standard deviation*, which measures the average distance of each value from the mean. The smaller the standard deviation the closer the values are to the mean and the more representative the mean will be.

To find the standard deviation for a given group of data:

1. Find the mean of the values using the method described in the previous section.
2. Subtract the mean from each individual value.
3. Square these differences.
4. Find the mean of these squares.
5. Find the square root of this last mean.

Example A:
1. Find the standard deviation for the numbers 74, 76, 70, 68, and 72.

$$\text{Mean} = \frac{74 + 76 + 70 + 68 + 72}{5} = \frac{360}{5} = 72$$

Values	Value − Mean	(Value − Mean)²
74	74 − 72 = 2	2² = 4
76	76 − 72 = 4	4² = 16
70	70 − 72 = −2	(−2)² = 4
68	68 − 72 = −4	(−4)² = 16
72	72 − 72 = 0	0² = 0
		40

$$40 \div 5 = 8$$
$$\sqrt{8} = 2.8$$

The standard deviation is 2.8.

2. Find the standard deviation for the numbers 70, 90, 64, 82, and 54.

$$\text{Mean} = \frac{70 + 90 + 64 + 82 + 54}{5} = \frac{360}{5} = 72$$

Value	(Value − Mean)	(Value − Mean)²
70	70 − 72 = −2	(−2)² = 4
90	90 − 72 = 18	18² = 324
64	64 − 72 = −8	(−8)² = 64
82	82 − 72 = 10	10² = 100
54	54 − 72 = −18	(−18)² = 324
		816

$$816 \div 5 = 163.2$$
$$\sqrt{163.2} = 12.8$$

The standard deviation is 12.8. Note that even though the mean is 72 in each of these examples, the standard deviation is much larger when the data is spread out over a larger range.

Example B: The grades received by the 20 students in a technical mathematics course on their most recent test were as follows:

```
78  70  90  50  83
68  98  48  48  50
85  88  50  90  68
85  78  70  73  70
```

Sec. 5–2. Standard Deviation

Find the standard deviation for these grades. Adding these grades together, we obtain a sum of 1440.

$$1440 \div 20 = 72$$

Therefore, the mean is 72.

Grade	(Grade − Mean)	(Grade − Mean)^2
78	78 − 72 = 6	$6^2 = 36$
68	68 − 72 = −4	$(-4)^2 = 16$
85	85 − 72 = 13	$(13)^2 = 169$
85	85 − 72 = 13	$(13)^2 = 169$
70	70 − 72 = −2	$(-2)^2 = 4$
98	98 − 72 = 26	$26^2 = 676$
88	88 − 72 = 16	$16^2 = 256$
78	78 − 72 = 6	$6^2 = 36$
90	90 − 72 = 18	$18^2 = 324$
48	48 − 72 = −24	$(-24)^2 = 576$
50	50 − 72 = −22	$(-22)^2 = 484$
70	70 − 72 = −2	$(-2)^2 = 4$
50	50 − 72 = −22	$(-22)^2 = 484$
48	48 − 72 = −24	$(-24)^2 = 576$
90	90 − 72 = 18	$18^2 = 324$
73	73 − 72 = 1	$1^2 = 1$
83	83 − 72 = 11	$11^2 = 121$
50	50 − 72 = −22	$(-22)^2 = 484$
68	68 − 72 = −4	$(-4)^2 = 16$
70	70 − 72 = −2	$(-2)^2 = 4$
		4760

$$4760 \div 20 = 238$$
$$\sqrt{238} = 15.4$$

Therefore, the standard deviation for these grades is 15.4.

Example C: The muzzle velocities (in feet per second) of bullets fired from a certain rifle were as follows:

```
2750  2830  2800  2850
2810  2760  2810  2840
2790  2810  2770  2760
2780  2770  2860  2810
```

Find the standard deviation for these velocities. Adding these velocities together we obtain a sum of 44,800.

$$44{,}800 \div 16 = 2800$$

Therefore, the mean is 2800.

Velocity	Velocity − Mean	(Velocity − Mean)^2
2750	$2750 - 2800 = -50$	$(-50)^2 = 2500$
2810	$2810 - 2800 = 10$	$10^2 = 100$
2790	$2790 - 2800 = -10$	$(-10)^2 = 100$
2780	$2780 - 2800 = -20$	$(-20)^2 = 400$
2830	$2830 - 2800 = 30$	$30^2 = 900$
2760	$2760 - 2800 = -40$	$(-40)^2 = 1600$
2810	$2810 - 2800 = 10$	$10^2 = 100$
2770	$2770 - 2800 = -30$	$(-30)^2 = 900$
2800	$2800 - 2800 = 0$	$0^2 = 0$
2810	$2810 - 2800 = 10$	$10^2 = 100$
2770	$2770 - 2800 = -30$	$(-30)^2 = 900$
2860	$2860 - 2800 = 60$	$60^2 = 3600$
2850	$2850 - 2800 = 50$	$50^2 = 2500$
2840	$2840 - 2800 = 40$	$40^2 = 1600$
2760	$2760 - 2800 = -40$	$(-40)^2 = 1600$
2810	$2810 - 2800 = 10$	$(10)^2 = 100$
		17,000

$$17{,}000 \div 16 = 1062.5$$
$$\sqrt{1062.5} = 32.6$$

Therefore, the standard deviation for these velocities is 32.6.

Besides yielding additional information about the data we are analyzing, the standard deviation is also significant since in most practical problems approximately 68% of the values lie within one standard deviation (in either direction) of the mean.

Example D: In Example B, the mean is 72 and the standard deviation is 15.4. Therefore, the range, 56.6–87.4, would include one standard deviation in each direction from the mean. Examining the given grades, we can see that 11 out of 20, or 55%, of the grades fall within this range.

Example E: In Example C, the mean is 2800 and the standard deviation is 32.6. Therefore, the range, 2767.4–2832.6, would include one standard deviation in each direction from the mean. Examining the given velocities, we can see that 10 out of 16, or $62\frac{1}{2}$%, of the velocities fall within this range.

This property of the standard deviation helps in the derivation of what is called the *normal distribution curve*. This curve represents the theoretical distribution of values about the mean for a given group of data. This theoretical curve is pictured in Figure 5–2.1.

Sec. 5–2. Standard Deviation

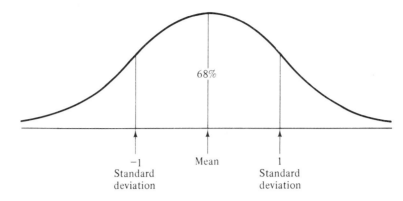

FIGURE 5–2.1

Exercises

1. Find the standard deviation for the following numbers.

   48 60 41 38 52
   58 39 50 49 45

2. Find the standard deviation for the following numbers.

   205 210 184 202 224 195 230
   198 217 191 190 186 217 193

3. The ages of 15 randomly selected students were recorded as follows:

   18 19 19 18 18
   20 17 19 17 20
   18 21 23 19 19

   Find the standard deviation for these ages.

4. The number of years of experience for 12 salesmen is given as follows:

   7   4   9   8
   9 12 10 15
   10   8   5 11

   Find the standard deviation for these figures.

5. The temperatures (in degrees Fahrenheit) for 24 consecutive hours beginning at midnight of a certain day were recorded as follows:

   35° 30° 34° 45° 53° 46°
   34° 30° 38° 46° 53° 45°
   34° 31° 39° 50° 52° 42°
   32° 31° 43° 52° 50° 39°

   Find the standard deviation for these temperatures. What percentage of values lie within one standard deviation of the mean?

6. In a certain electrical experiment the following voltage readings were taken.

109	116	113	103	104	106
117	124	115	123	111	103
121	106	113	118	110	110
102	125	119	114	104	125
115	113	108	115	110	118

Find the standard deviation for these voltages. What percentage of values lie within one standard deviation of the mean?

5-3. Frequency Distributions

Groups of statistical data often consist of a large number of values. As the number of values gets larger, there is a need to present the data in a condensed form that reveals the nature of the data quickly and easily.

A *frequency distribution* is a tabular arrangement of data by intervals, with the corresponding number of values in each interval.

Example A: The following table is a frequency distribution of the weights of 100 students at a certain university.

Weight (lb)	Number of Students
100–109	1
110–119	4
120–129	0
130–139	8
140–149	3
150–159	20
160–169	29
170–179	20
180–189	15

When data is summarized by means of a frequency distribution it is called *grouped data*. Although grouping the data results in losing some of the original detail and accuracy, an advantage is gained by having a concise summary of the original set of data.

To make a frequency distribution from a set of values, a table is made up with three headings: *interval, tabulation* or *tally*, and *frequency*. The number of intervals that you use for a frequency distribution is a matter of judgement. Too few intervals will not give a true summary of the data and too many intervals will defeat the purpose of constructing a frequency distribution. The size of the intervals (if not specified) are determined by dividing the range of the given

Sec. 5-3. Frequency Distributions

values by the number of intervals that you want for the frequency distribution. The *range* is the difference between the largest and the smallest value from the set of given values.

The tabulation, or tally, column is used to signify, by a tally mark, that a value has been channeled into a class interval from the set of given values. The frequency column is simply the sum of the marks, in the tally column for each interval.

Example B: The following numbers represent the life span in hours of electronic relay switches at a temperature of $-15°C$.

$$
\begin{array}{ccccccc}
31 & 51 & 58 & 46 & 51 & 50 & 55 \\
41 & 43 & 42 & 56 & 69 & 41 & 56 \\
50 & 55 & 52 & 63 & 43 & 65 & 47 \\
46 & 35 & 61 & 46 & 50 & 59 & 47
\end{array}
$$

Make a frequency distribution consisting of 10 class intervals. Range $= 69 - 31 = 38$. Size of intervals $= \frac{38}{10} = 3.8$. Therefore, the intervals will each be of size 4.

Class Interval	Tally	Frequency
66–69	|	1
62–65	||	2
58–61	|||	3
54–57	||||	4
50–53	⋌⋋⋌⋋ |	6
46–49	⋌⋋⋌⋋	5
42–45	|||	3
38–41	||	2
34–37	|	1
30–33	|	1

Frequency distributions can be used to determine the mean, median, and the modal class for grouped data.

By definition, the mean is the arithmetic average. In order to calculate the mean of grouped data it is necessary to calculate the sum of the products of the frequency times the midpoint of the interval for each of the intervals in the frequency distribution. This sum is then divided by the total number of values. The *midpoint* of each interval is obtained by adding the lower and upper limits of the interval and dividing by two. All values within a given interval are assumed to coincide with the midpoint.

The *median* of grouped data is determined by locating the value (in one of the intervals) that divides the grouped data into two groups of equal size.

The *modal class* is the interval that contains the greatest number of values.

Example C: Determine the mean, median, and modal class from the following frequency distribution (refer to Example B for the original information).

Interval	Frequency	Midpoint	Frequency × Midpoint
66–69	1	67.5	67.5
62–65	2	63.5	127.0
58–61	3	59.5	178.5
54–57	4	55.5	222.0
50–53	6	51.5	309.0
46–49	5	47.5	237.5
42–45	3	43.5	130.5
38–41	2	39.5	79.0
34–37	1	35.5	35.5
30–33	1	31.5	31.5
	28		1418.0

Midpoints: $\frac{30+33}{2} = 31.5$, $\frac{34+37}{2} = 35.5$, ...

Mean: $\frac{1418}{28} = 50.6429 = 50.6$ hr

Median: 1. $\frac{\text{Total number of values}}{2} = \frac{28}{2} = 14$

2. Locate the fourteenth value: $1 + 1 + 2 + 3 + 5 = 12$

$14 - 12 = 2$ the number of scores needed from the interval 50–53

$\frac{2}{6} = \frac{N}{4}$ (4 = size of the interval)

$N = \frac{2}{6}(4) = \frac{1}{3}(4) = \frac{4}{3} = 1\frac{1}{3}$

Add $1\frac{1}{3}$ onto lower limit of the interval 50–53.

$$50 + 1\frac{1}{3} = 51\frac{1}{3}$$

Therefore, the median is $51\frac{1}{3}$ hours.

The modal class is the interval 50–53 since this interval contains the greatest number of values.

The true mean for this group of values is 50.3 (calculated by the method described in Section 5–1). When the mean of the grouped data is compared to this true mean there is a difference of .3 or a relative error of $\frac{.3}{50.3} = .596\%$.

If the number of class intervals were decreased there would be more error in the mean and if the number of class intervals were increased there would be less error in the mean.

Sec. 5-3. Frequency Distributions

The true median is 50. When the median of the grouped data is compared to this true median there is a difference of $1\frac{1}{3}$ or a relative error of $\frac{1\frac{1}{3}}{50} = 2\frac{2}{3}\%$.

There are two modes, 46 and 50. One of these is included in the modal class determined from the grouped data.

Example D: The number of Australian pine trees that germinated from 40 flats, each containing 100 seeds, is listed below:

```
70  80  62  65  45  61  49  55  59  73
76  57  75  58  58  68  59  86  60  66
83  52  41  60  71  69  63  80  44  67
74  56  62  68  65  57  65  67  78  69
```

Starting with the smallest number and using intervals of size 5, construct a frequency distribution. Then find the mean, median, and modal class for these numbers from the frequency distribution.

Interval	Tally	Frequency	Midpoint	Frequency × Midpoint
86–90	|	1	88	88
81–85	|	1	83	83
76–80	||||	4	78	312
71–75	||||	4	73	292
66–70	⊬⊬ |||	8	68	544
61–65	⊬⊬ ||	7	63	441
56–60	⊬⊬ ||||	9	58	522
51–55	||	2	53	106
46–50	|	1	48	48
41–45	|||	3	43	129
		40		2565

Mean: $\frac{2565}{40} = 64.1$

Median: $\frac{\text{Total number of values}}{2} = \frac{40}{2} = 20$

Locate the twentieth value: $3 + 1 + 2 + 9 = 15$

$20 - 15 = 5$ (needed from the interval $61 - 65$)

$\frac{5}{7} = \frac{N}{5}$ ($5 =$ size of the interval)

$N = \frac{25}{7} = 3.6$

$61 + 3.6 = 64.6$

Modal class = 56–60 since this interval contains the greatest number of values.

Exercises

1. The following values represent kilograms of force applied to 40 different cables before the cables broke.

138	164	150	132	144	125	149	157
146	158	140	147	136	148	152	144
168	126	138	176	163	119	154	165
146	173	142	147	135	153	140	135
161	145	135	142	150	156	145	128

Start with the smallest value, and using intervals of size 9 kilograms, construct a frequency distribution. Then find the mean, median, and modal class for these forces from the frequency distribution.

2. The following values represent the diameter (in inches) of a sample of 36 ball bearings manufactured by a certain company.

.741	.741	.746	.743	.735	.740	.736	.734	.738
.738	.732	.730	.746	.731	.730	.724	.735	.734
.754	.730	.732	.742	.740	.727	.728	.741	.732
.733	.741	.741	.732	.735	.746	.739	.733	.735

Start with the smallest value, and using intervals of size .003 inch, construct a frequency distribution. Then find the mean, median, and modal class for these values from the frequency distribution.

3. In testing a certain gasoline mileage ingredient, an oil company recorded the following distances (in miles) traveled by 30 different cars, each starting with exactly 10 gallons of gasoline.

183	176	181	186	180	172	185	181	179	175
180	177	188	178	173	184	188	177	188	189
171	182	175	177	186	179	187	176	180	184

Start with the smallest distance, and using 5 intervals, construct a frequency distribution. Then find the mean, median, and modal class for these distances from the frequency distribution.

4. The high water level of a certain lake (in feet) was measured for 32 consecutive weeks and recorded as follows:

242.6	244.9	246.3	246.7	246.6	245.4	244.3	242.1
243.1	244.3	246.8	247.2	247.2	246.5	243.6	242.7
244.1	245.2	247.1	247.9	246.3	245.9	243.4	243.4
244.5	245.7	246.5	247.0	245.9	244.8	242.9	244.0

Start with the lowest number, and using 6 intervals, construct a frequency distribution. Then find the mean, median, and modal class for these measurements from the frequency distribution.

5–4. Frequency Curves

At times it is helpful to look at a graphical representation of a frequency distribution.

Sec. 5–4. Frequency Curves

One such graphical representation is a histogram. A *histogram* consists of a set of rectangles having bases on the horizontal axis with the centers of the bases corresponding to the interval midpoints, the widths of the rectangles corresponding to the size of the intervals, and the heights of the rectangles corresponding to the frequency of the particular intervals.

Example A: Construct a histogram for the frequency distribution of Example C in Section 5–3. Frequency distribution:

Interval	Frequency	Midpoint	Frequency × Midpoint
66–69	1	67.5	67.5
62–65	2	63.5	127.0
58–61	3	59.5	178.5
54–57	4	55.5	222.0
50–53	6	51.5	309.0
46–49	5	47.5	237.5
42–45	3	43.5	130.5
38–41	2	39.5	79.0
34–37	1	35.5	35.5
30–33	1	31.5	31.5

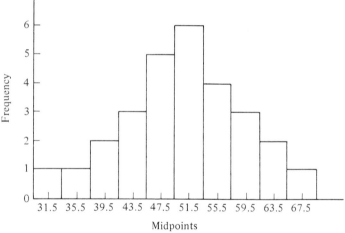

If collected data belong to a sample drawn from a large population and if the collected sample is representative of this large population, the frequencies corresponding to the midpoints of the widths of the rectangles that make up the histogram can be joined with a smooth curve. A curve constructed in this manner is a *frequency curve*. Theoretically, this curve is a representation of the entire population. As the size of the sample population increases, and if the sample is a random sample, then the frequency curve will be an accurate representation of the entire population.

Example B: Construct a frequency curve from the histogram in Example A.

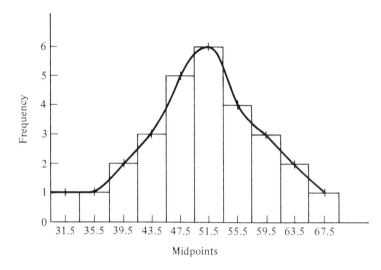

Example C: Construct a histogram and a frequency curve for the frequency distribution obtained in Example D of Section 5–3. Frequency distribution:

Interval	Tally	Frequency	Midpoint	Frequency × Midpoint
86–90	|	1	88	88
81–85	|	1	83	83
76–80	||||	4	78	312
71–75	||||	4	73	292
66–70	⨰ |||	8	68	544
61–65	⨰ ||	7	63	441
56–60	⨰ ||||	9	58	522
51–55	||	2	53	106
46–50	|	1	48	48
41–45	|||	3	43	129

Histogram:

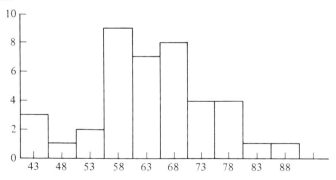

Sec. 5-4. Frequency Curves

Frequency curve:

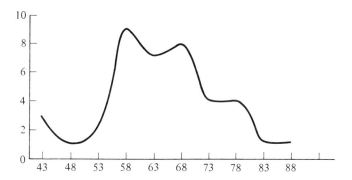

Frequency curves take on many different shapes. One of these shapes which is very important is the *bell-shaped*, or *normal distribution curve* (mentioned in Section 5-2). It is characterized by the fact that it is symmetric with respect to its central maximum point (the mean). That is, the frequencies of values equidistant from the central maximum point (regardless of direction) are the same. This curve is a theoretical curve and is pictured in Figure 5-4.1 with some important observations pertaining to it.

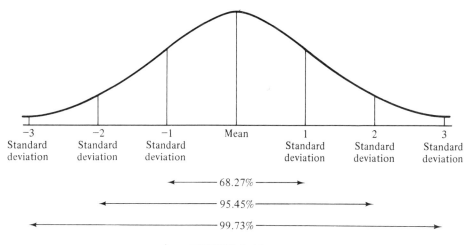

FIGURE 5-4.1

1. Theoretically, 68.27% of the given values will be within ± one standard deviation.

2. Theoretically, 95.45% of the given values will be within ± two standard deviations.

3. Theoretically, 99.73% of the given values will be within ± three standard deviations.

Exercises   Refer to the exercises at the end of Section 5–3.
1. Construct a histogram for the data in problem (1).
2. Construct a histogram for the data in problem (2).
3. Construct a histogram for the data in problem (3).
4. Construct a histogram for the data in problem (4).
5. Construct a frequency curve for the data in problem (1).
6. Construct a frequency curve for the data in problem (2).
7. Construct a frequency curve for the data in problem (3).
8. Construct a frequency curve for the data in problem (4).

5–5. CHAPTER REVIEW

Find the mean, median, mode, and standard deviation for the following groups of numbers.

1. 114, 172, 125, 183, 164, 176, 172, 149, 157, 172, 181, 164, 122, 168, 136.
2. 2430, 2240, 2520, 2480, 2590, 2510, 2360, 2520, 2360, 2440, 2470, 2520, 2230, 2310, 2590, 2470.
3. The systolic blood pressures (in millimeters) of twenty different people were recorded as follows:

    132 123 114 130 130 129 126 130 160 128
    115 123 132 125 107 98 154 128 134 136

    Find the mean, median, mode, and standard deviation for these blood pressures. What percentage of values lie within one standard deviation of the mean?

4. The mileage lives of a certain group of automobile tires being tested were as follows:

    39200 37500 42300 43500 41600 37300
    36200 39800 41600 38400 39100 41600
    38400 44100 37200 39000 41200 39800
    41600 44700 38800 40900 42000 41400

    Find the mean, median, mode, and standard deviation for these mileages. What percentage of values lie within one standard deviation of the mean?

5. In the testing of a new reinforced safety glass tubing, the following readings, representing the weight (in pounds) at which each of 30 randomly selected sections of this tubing broke, were obtained.

    1430 1380 1056 1180 1112 1488
    1450 1560 1350 1231 1420 1445
    1396 1435 1210 1240 1500 1468
    1432 1024 1445 1414 1450 1350
    1247 1200 1214 1565 1105 1315

Sec. 5–5. Chapter Review

Find the mean, median, mode, and standard deviation for these weights. What percentage of values lie within one standard deviation of the mean?

6. Following is a list of lumber production figures (in millions of board feet) for the United States for a 30-year period.

3074	1291	2096	2745	2742	3030
2171	1628	2411	2344	3242	3154
1377	2030	2801	2843	3126	3216
902	2166	3028	2950	3122	2851
1225	1804	2857	3064	3062	2798

Find the mean, median, mode, and standard deviation for these production figures. What percentage of values lie within one standard deviation of the mean?

7. The total net weekly sales of the branches of a certain corporation are as follows:

$23600	$24700	$34300	$27000	$31400	$35300	$24900
$28500	$37400	$26800	$30100	$39600	$25900	$33600
$36800	$31100	$29300	$39400	$36200	$27700	$35800
$34100	$24800	$39700	$28600	$27700	$34000	$32200
$24800	$36400	$34500	$28300	$22000	$35700	$25000

Prepare a frequency distribution, starting with the smallest sales figure and using $2000 intervals, and then find the mean, median, and modal class for this data.

8. The lifetimes (in hours) of 40 radio tubes that were tested are listed as follows:

420	300	630	1012	791	941	587	655	591	779
643	520	783	383	490	562	627	890	1034	538
826	612	418	618	513	964	720	454	668	754
312	805	572	732	699	476	352	671	983	853

Prepare a frequency distribution starting with the shortest lifetime and using 100-hour intervals, and then find the mean, median, and modal class for this data.

9. Construct a histogram for the data in problem (7).
10. Construct a histogram for the data in problem (8).
11. Construct a frequency curve for the data in problem (7).
12. Construct a frequency curve for the data in problem (8).

chapter 6

Linear Equations with More than One Variable

6-1. Linear Equations and Their Graphs

Probably the most important procedure for anyone who must make extensive use of mathematics is the solution of equations. In this chapter, we will take a detailed look at the most basic type of equation, the linear equation.

A *linear equation* is an equation in which each term either contains one variable raised to the first power or is a constant.

Example A:
1. $2X + 3Y = 7$ is a linear equation.
   $W = 5T - 3$ is a linear equation.
   $2P + 4Q - 5R = 0$ is a linear equation.
2. $3X^2 - 2X = 7$ is not a linear equation due to the presence of X^2.
   $4RS - 3 = 0$ is not a linear equation due to the presence of RS in the same term.
   $5T - \frac{6}{V} + 7 = 0$ is not a linear equation due to the presence of V in the denominator.

A *solution* of a linear equation (or any equation) is a value (s) that, when sub-

Sec. 6-1. Linear Equations and Their Graphs

stituted for the variable (s) contained in the equation, makes the equation a valid statement.

Example B:
1. If $X + 3 = 8$, then $X = 5$ is a solution for this equation.
2. If $R - 2S = 1$, then $R = 5$ and $S = 2$ is a solution for this equation.
3. If $4K - L + 3T = 0$, then $K = 2$, $L = -1$, and $T = -3$ is a solution for this equation.

The *complete solution* of a linear equation (or of any equation) is the set of all such individual solutions.

Example C:
1. If $X + 3 = 8$, then $X = 5$ is the complete solution of this equation since this is the only value of X that makes the equation valid.
2. If $R - 2S = 1$, the complete solution would be infinite since any number of values could be substituted for R and S to make this equation valid. A few of these substitutions would be $R = 1$ and $S = 0$, $R = 7$ and $S = 3$, and $R = -1$ and $S = -1$.

From this last example, we should realize that any linear equation (or any equation) containing more than one variable will have more than one solution.

Any linear equation containing one variable can be written in the form $X = K$ (X could be replaced by any variable). The graph of such an equation will be a vertical (if X is the independent variable) or horizontal (if X is the dependent variable) line.

Example D: Construct the graph of $2X = 5$, considering X as the independent variable. First, rewrite this equation to obtain $X = \frac{5}{2}$.

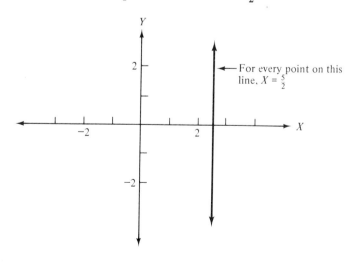

Example E: Construct the graph of $3Y + 5 = 0$, considering Y as the dependent variable. First, rewrite this equation to obtain $Y = -\frac{5}{3}$

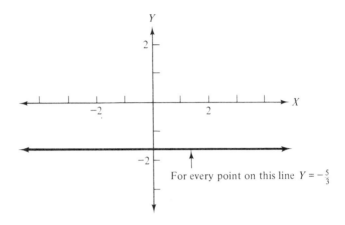

Any linear equation containing two variables can be written in the form $Y = MX + B$, where M and B are constants (X and Y could be replaced by any variables). Looking closely at this general equation, we should realize that *equal changes in the variable X will result in equal changes in the variable Y*. For example: if $X = 1$, then $Y = M + B$; if $X = 2$, then $Y = 2M + B$; and if $X = 3$, then $Y = 3M + B$. Therefore, the graph of an equation of this type will be a straight line.

Example F: Construct the graph of $2X + 3Y = 6$.

1. If X is the independent variable, $Y = \dfrac{6 - 2X}{3}$ or $Y = -\dfrac{2}{3}X + 2$.

X	0	1	−1	2	−2	3
Y	2	$1\frac{1}{3}$	$2\frac{2}{3}$	$\frac{2}{3}$	$3\frac{1}{3}$	0

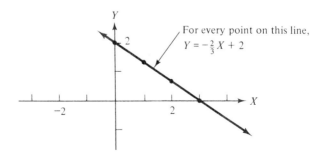

2. If Y is the independent variable, $X = \dfrac{6 - 3Y}{2}$ or $X = -\dfrac{3}{2}Y + 3$.

Sec. 6-1. Linear Equations and Their Graphs

Y	0	1	−1	2	−2
X	3	1½	4½	0	6

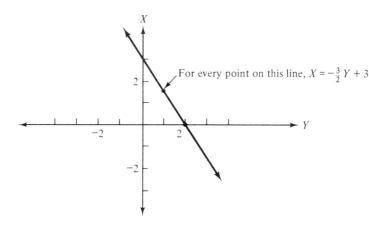

For every point on this line, $X = -\frac{3}{2}Y + 3$

NOTE: Although the graphs are different depending on which variable is independent, the corresponding values of X and Y are the same.

Example G: Construct the graph of $R - 2S - 3 = 0$.
  1. If S is the independent variable, $R = 2S + 3$.

S	0	1	−1	2	−2
R	3	5	1	7	−1

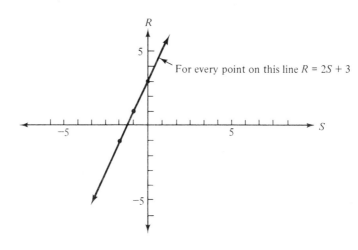

For every point on this line $R = 2S + 3$

  2. If R is the independent variable, then $-2S = -R + 3$ or $S = \frac{1}{2}R - \frac{3}{2}$.

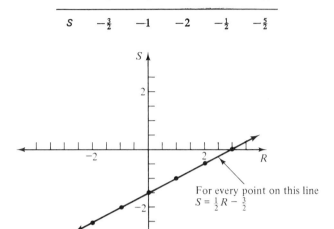

NOTE: Although the graphs are different depending on which variable is independent, the corresponding values of R and S are the same.

We should see, then, that the graph of a linear equation containing either one or two variables will always be a straight line, and each point on the straight line will be a solution for the linear equation which the straight line represents.

In Chapter 3, we pointed out that variables represent different sets of numbers and a number line is used to represent a given set of numbers. To represent accurately the relationship between sets of numbers, we need one number line for each set of numbers (i.e., each variable). Thus, if three variables are involved we need three number lines, if four variables are involved we need four number lines, etc. Therefore, we will not construct the graph of a linear equation (or any equation) containing more than two variables, since the procedure would be extremely difficult, if not impossible.

Exercises State whether or not each of the following is a linear equation.

1. $2X - 3Y = 8$
2. $5R - 2S = R + 1$
3. $2VT - 1 = 0$
4. $2K = L$
5. $3P - 4Q = 2R$
6. $M^2 - 3M = 6$
7. $A = \sqrt{B + 3}$
8. $W - 2X + 3Y - Z = 0$
9. $\dfrac{1}{Y} = 2$
10. $X + 4Y = \sqrt{7}$

Sec. 6–2. Slopes and Intercepts

For each of the following, state whether the indicated values are solutions for the given equations.

11. $2B - 3 = 0$ $\quad (B = \frac{3}{2})$
12. $4A + 1 = 2$ $\quad (A = \frac{1}{4})$
13. $5X + 4 = -1$ $\quad (X = 1)$
14. $2Y - 1 = 0$ $\quad (Y = -\frac{1}{2})$
15. $3R + S = 2$ $\quad (R = 1, S = -1)$
16. $5K - 2L = 1$ $\quad (K = 2, L = 4)$
17. $-2T + 4V = 6$ $\quad (T = \frac{3}{2}, V = \frac{4}{3})$
18. $W + 4Z = 0$ $\quad (W = -3, Z = \frac{3}{4})$
19. $P - 2Q + 3R = 0$ $\quad (P = 2, Q = 1, R = 0)$
20. $3L + M - 4N = 0$ $\quad (L = 1, M = 0, N = 1)$

Construct the graph of each of the following.

21. $2X + 3 = 0$ $\quad$ (Consider X the independent variable.)
22. $5Y + 2 = 0$ $\quad$ (Consider Y the dependent variable.)
23. $3R = 7$ $\quad$ (Consider R the dependent variable.)
24. $2S = -5$ $\quad$ (Consider S the independent variable.)
25. $5A - B = 10$ $\quad$ (Consider A the independent variable.)
26. $2K + 3L = 9$ $\quad$ (Consider L the independent variable.)
27. $-3M + N = 4$ $\quad$ (Consider M the independent variable.)
28. $4P - 2Q = 8$ $\quad$ (Consider Q the independent variable.)
29. $W + 3 = -2Z$ $\quad$ (Consider Z the independent variable.)
30. $2T + 1 = 3V + 2$ $\quad$ (Consider T the independent variable.)

6–2. Slopes and Intercepts

An aid in constructing the graph of a linear equation is to determine the slope of a straight line, and the vertical and horizontal intercepts of the straight line.

The *slope* of a straight line is the ratio of the vertical change between two points to the horizontal change between the same two points. A formula for determining the slope is

$$M = \frac{\text{vertical change}}{\text{horizontal change}}$$

where M is a symbol that represents the numerical value of the slope.

Example A:
1.

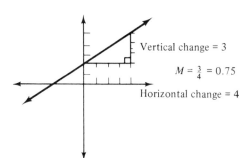

2.

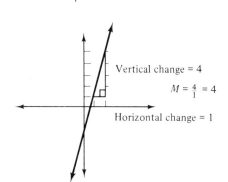

3.

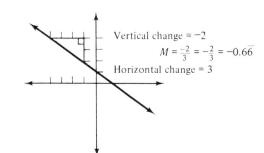

4.

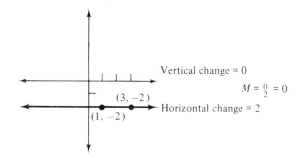

Therefore, a horizontal line has a slope of zero.

Sec. 6-2. Slopes and Intercepts 137

5.

(3, 3) Vertical change = 3

$M = \frac{3}{0}$ (undefined)

Horizontal change = 0
(3, 0)

Therefore, a vertical line has no slope.

From the previous examples we should realize that a line going upward to the right has a positive slope while a line going downward to the right has a negative slope.

The *intercepts* are the points where the graph crosses the vertical and horizontal axes (see Section 4-4).

Example B:

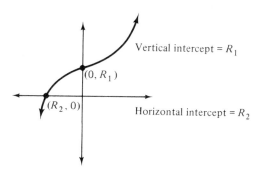

Vertical intercept = R_1

$(0, R_1)$

$(R_2, 0)$

Horizontal intercept = R_2

A procedure for finding the *vertical intercept* is to substitute zero in the place of the independent variable and solve the resulting equation for the dependent variable.

A procedure for finding the *horizontal intercept* is to substitute zero in place of the dependent variable and solve the resulting equation for the independent variable. Since two points will determine a straight line, the graph of a linear equation in two variables can be constructed by drawing a straight line through the intercepts.

Example C: Find the intercepts of $A = 4B - 4$.

1. Find the A-intercept by substituting 0 for B and solve the resulting equation for A.

$$A = 4B - 4$$
$$A = 4(0) - 4$$
$$A = -4$$

2. Find the *B*-intercept by substituting 0 for *A* and solve the resulting equation for *B*.

$$A = 4B - 4$$
$$0 = 4B - 4$$
$$4 = 4B$$
$$1 = B$$

The intercepts are $(0, -4)$ and $(1, 0)$ since *A* is the dependent variable and *B* is the independent variable. These are points on the *A* and *B* axes through which the straight line will pass.

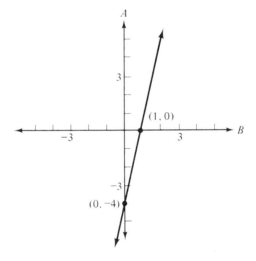

Example D: Determine the vertical and horizontal intercepts of $Y = \frac{2}{3}X + 1$ and plot the graph of this equation.

1. The *Y*-intercept is determined by substituting 0 for *X*.

$$Y = \tfrac{2}{3}X + 1$$
$$Y = \tfrac{2}{3}(0) + 1$$
$$Y = 1$$

2. The *X*-intercept is determined by substituting 0 for *Y*.

$$Y = \tfrac{2}{3}X + 1$$
$$0 = \tfrac{2}{3}X + 1$$

Sec. 6–2. Slopes and Intercepts

$$-1 = \tfrac{2}{3}X$$
$$-\tfrac{3}{2} = X$$

3. Plotting these points and constructing a straight line through these points yields the graph of $Y = \tfrac{2}{3}X + 1$.

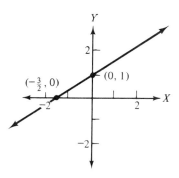

Example E: Determine the vertical and horizontal intercepts of $S + 2T = -1$ and plot the graph of this equation. Consider T the independent variable.

1. The S-intercept is determined by substituting 0 for T.

$$S + 2T = -1$$
$$S + 2(0) = -1$$
$$S = -1$$

2. The T-intercept is determined by substituting 0 for S.

$$S + 2T = -1$$
$$0 + 2T = -1$$
$$T = -\tfrac{1}{2}$$

3. Plotting these points and constructing a straight line through these points yields the graph of $S + 2T = -1$.

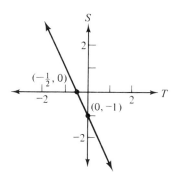

If a linear equation is expressed with the dependent variable isolated on one side of the equation, the constant term in the equation is the vertical intercept and the coefficient of the independent variable represents the slope of the straight line.

Example F: Determine the vertical intercept and slope of $Y = 3X + 2$.
1. Determine vertical intercept:

$$Y = 3(0) + 2$$
$$Y = 2$$

NOTE: The vertical intercept and the constant term in the original equation are equal.

2. Determine a second point:

$$0 = 3X + 2$$
$$-2 = 3X$$
$$-\tfrac{2}{3} = X$$

3. Construct the graph.

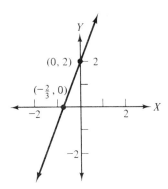

4. Determine the slope. The slope of this line can be determined from the formula

$$M = \frac{\text{vertical change}}{\text{horizontal change}} = \frac{2}{\tfrac{2}{3}} = 3$$

NOTE: The slope of 3 is the coefficient of the independent variable.

In general, when a linear equation is expressed in the form $Y = MX + B$, the slope-intercept form of a linear equation, the value of B is the *vertical intercept* and the value of M is the *slope* of the line. This form of an equation is

Sec. 6–2. Slopes and Intercepts 141

very useful for constructing the graph of a linear equation in one or two variables.

Example G: Graph the straight line representing $Y = \frac{2}{3}X - 3$.
1. The vertical intercept is -3.
2. The slope of the line is $\frac{2}{3}$.
3. Plot two points as follows:
   (a) The vertical intercept first.
   (b) Use the slope to determine another point.

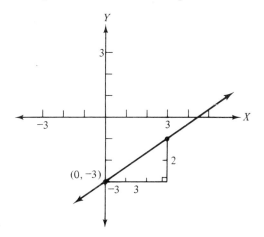

Example H: Graph the straight line representing $3A - 6B + 12 = 0$. Let A be the dependent variable.
1. Solve for A: $A = 2B - 4$.
2. The vertical intercept is -4.
3. The slope is 2.

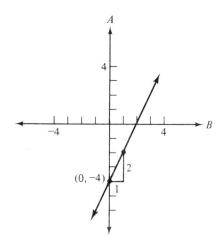

Example I: Graph the straight line representing $M = -\frac{2}{5}N + 5$.
  1. The vertical intercept is 5.
  2. The slope is $\frac{-2}{5}$ or $\frac{2}{-5}$.

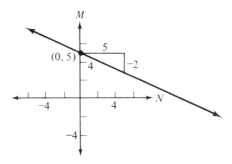

Example J: Graph the straight line representing $C = \frac{1}{2}D$.
  1. The vertical intercept is 0.
  2. The slope is $\frac{1}{2}$.

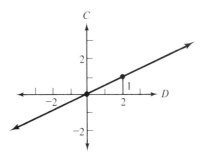

In summary then, the graph of a linear equation in one or two variables will always be a straight line, and the graph may be constructed very quickly by using either the two intercepts or the slope-intercept method.

Exercises   Determine the slope and vertical intercept of each line:

1. 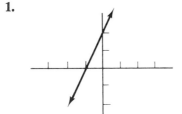   2.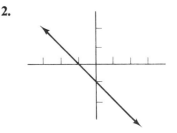

Sec. 6–2. Slopes and Intercepts   143

3. 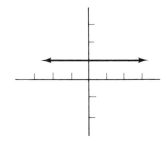   4.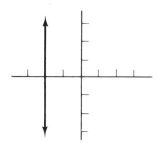

5. 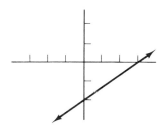   6.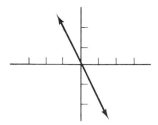

For each of the following, determine the vertical and horizontal intercepts and use them to construct the graph.

7. $3Y - 4X = 12$
8. $7S = 14T + 21$
9. $A = 3B - 4$
10. $4V = \frac{2}{3}T - 8$
11. $R = -5T$
12. $3D + 4E = -12$

Write each equation in the slope intercept form.

13. $4S - 5T - 40 = 0$ S is the dependent variable.
14. $-6Y + 4X = 12$ Y is the dependent variable.
15. $4C = 5D - 1$ C is the dependent variable.
16. $16S + 14T - 32 = 0$ S is the dependent variable.
17. $25V - 75T = 100$ V is the dependent variable.
18. $3A - 7B = 12$ B is the dependent variable.

Determine the graph of each equation by using the vertical intercept and slope from the equation.

19. $A = 3B - 1$
20. $R = -2S + 4$
21. $M = \frac{2}{3}N - \frac{1}{2}$
22. $T = \frac{3}{2}S$
23. $Y = -\frac{1}{3}X - 2$
24. $Z = 4$
25. The speed of a ball increases uniformly in relation to time as it rolls down

an inclined plane. If a ball is given an initial velocity of 60 centimeters per second down an inclined plane and 2 seconds later the velocity of the ball is 90 centimeters per second, determine an equation relating the velocity of the ball as a function of time. Put the equation into the slope-intercept form and sketch the graph of the function.

26. The function for converting degrees Celsius (C) to Fahrenheit (F) is $F = \frac{9}{5}C + 32$. Write an equivalent equation expressing C as a function of F. What is the slope and vertical intercept of this new equation? Sketch the graph of this new function.

27. The general equation relating velocity with time is $v = v_o + at$ where v_o is the initial velocity and a is the acceleration.

   (a) Discuss the vertical intercept and the range of values it may assume.
   (b) Discuss the slope and the range of values it may assume.
   (c) Sketch a general graph of velocity versus time.

6–3. Systems of Linear Equations

In Section 6–1, we pointed out that the complete solution for a single linear equation containing more than one variable will always be infinite. Therefore, if we wish to solve such an equation, we would list a few individual solutions and indicate that these are only part of the complete infinite solution. In the case of a linear equation in two variables, we could construct the straight line graph and state that each point on the line is a solution for the equation. However, if we want or need a finite solution when working with linear equations, as is the case in most applied problems, then whenever more than one variable is involved we must have more than one equation. In the next five sections of this chapter we will deal with problems involving the same number of equations as variables. A system that is an array of N linear equations in the same N variables is called a system of *simultaneous linear equations*. In applied situations, if different conditions in a problem and different relationships between or among the involved variables can be represented as such a system, appropriate solutions may be obtained.

Example A:

1. $2X - 3Y = 8$ This is a system of two simultaneous linear equations
   $X + 4Y = 7$ in the variables X and Y.
2. $3A - B + 2C = 1$ This is a system of three simultaneous linear
   $A + 4B - C = 3$ equations in the variables A, B, and C.
   $-2A - 3B + 5C = 0$

The solution for a system of simultaneous linear equations is a set of values that, when substituted for the variables involved, satisfies each equation

Sec. 6–3. Systems of Linear Equations

in the system *simultaneously*. The *complete solution* for such a system of equations would be the set of all individual solutions.

Example B: Consider the system of equations: $2X - Y = 3$
$3X + 2Y = 8$
1. $X = 4$ and $Y = 5$ is a solution of $2X - Y = 3$, but not of $3X + 2Y = 8$. Therefore, $X = 4$ and $Y = 5$ is not a solution of the system.
2. $X = -2$ and $Y = 7$ is a solution of $3X + 2Y = 8$, but not of $2X - Y = 3$. Therefore, $X = -2$ and $Y = 7$ is not a solution of the system.
3. $X = 2$ and $Y = 1$ is a solution for both $2X - Y = 3$ and $3X + 2Y = 8$. Therefore, $X = 2$ and $Y = 1$ is a solution of the system.

Example C: Consider the system of equations: $R - 2S = 6$
$3R + S = 4$
1. $R = 6$ and $S = 0$ is a solution of $R - 2S = 6$, but not of $3R + S = 4$. Therefore, $R = 6$ and $S = 0$ is not a solution of the system.
2. $R = 1$ and $S = 1$ is a solution of $3R + S = 4$, but not of $R - 2S = 6$. Therefore, $R = 1$ and $S = 1$ is not a solution of the system.
3. $R = 2$ and $S = -2$ is a solution for both $R - 2S = 6$ and $3R + S = 4$. Therefore, $R = 2$ and $S = -2$ is a solution of the system.

Example D: Consider the system of equations: $A + 2B - C = -5$
$3A + 3B + 4C = 5$
$2A - B + 2C = 8$
1. $A = 2$, $B = 1$, and $C = 9$ is a solution of $A + 2B - C = -5$, but not of either of the other two equations. Therefore, these values would not be a solution of the system.
2. $A = 1$, $B = 0$, and $C = \frac{1}{2}$ is a solution of $3A + 3B + 4C = 5$, but not of either of the other two equations. Therefore, these values would not be a solution of the system.
3. $A = \frac{1}{2}$, $B = 1$, and $C = 4$ is a solution of $2A - B + 2C = 8$, but not of either of the other two equations. Therefore, these values would not be a solution of the system.
4. $A = 1$, $B = -2$, and $C = 2$ is a solution for all three equations. Therefore, these values would be a solution of the system.

Exercises

For each of the following, determine whether or not the indicated values are solutions for the system of equations.

1. $A - 3B = 0 \quad A = 3$ and $B = 1$
   $2A + B = 5$
2. $-3T + 4V = 10 \quad T = -2$ and $V = 1$
   $T - 5V = -7$

3. $4K - L = 3$ $K = 1$ and $L = 1$
   $2K + 3L = 5$

4. $3P - 4Q = 6$ $P = 2$ and $Q = 0$
   $-P + Q = 8$

5. $5W - 2Z = 10$ $W = 1$ and $Z = \frac{1}{2}$
   $2W + 7Z = 4$

6. $-2X + 4Y = 8$ $X = \frac{1}{2}$ and $Y = \frac{9}{4}$
   $3X - 2Y = -3$

7. $7A + B + 3C = 52$ $A = 7, B = 3, C = 0$
   $A + 15B - 9C = 52$
   $4A - 5B + 6C = 13$

8. $2K + 3L + 4M = -7$ $K = 2, L = 1, M = \frac{1}{2}$
   $-3K + 4L + 5M = -3$
   $K + 2L - 3M = 7$

9. $3P + 4Q + 6R = 3$ $P = 0, Q = 0, R = \frac{1}{2}$
   $12P + 18Q - 24R = -5$
   $5P - 6Q - 9R = 5$

10. $2X - Y - 4Z = 3$ $X = 2, Y = -3, Z = 1$
    $3X + 2Y - 2Z = -2$
    $-X + 3Y + Z = -10$

6–4. Solving Systems of Two Linear Equations in Two Unknowns Graphically

A linear equation in two variables will always be represented by a straight line. A system of two linear equations in two variables will be represented by two straight lines and the point of intersection of these lines (if a point of intersection exists) will be the solution to the system of equations. The point of intersection will be an ordered pair of numbers that will satisfy both equations.

Example A: Solve the following system of equations graphically.

$$X - 2Y = 4$$
$$-2X + Y = 1$$

1. Put each equation into the slope-intercept form.

$$Y = \tfrac{1}{2}X - 2$$
$$Y = 2X + 1$$

2. Use the vertical intercept and slope to graph the straight line representing each equation.

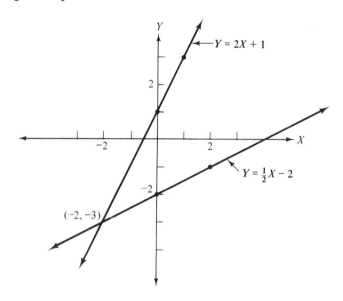

3. Read the coordinates of the point of intersection. The coordinates of the point of intersection in this case are $(-2, -3)$.
4. As a check, the ordered pair must satisfy each equation.

$$X - 2Y = 4 \qquad\qquad -2X + Y = 1$$
$$(-2) - 2(-3) = 4 \qquad\qquad -2(-2) + (-3) = 1$$
$$-2 + 6 = 4 \qquad\qquad 4 + (-3) = 1$$
$$4 = 4 \qquad\qquad 1 = 1$$

Example B: Solve the following system of equations graphically.

$$-A + 3B = -5$$
$$-3A + B = -7$$

1. Put each equation into the slope-intercept form.

$$B = \frac{A}{3} - \frac{5}{3}$$
$$B = 3A - 7$$

2. Use the vertical intercept and slope to graph the straight line representing each equation.

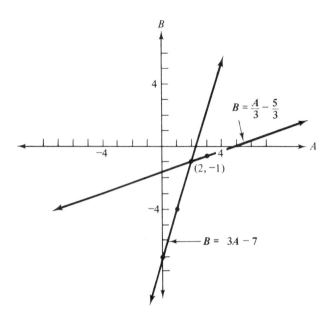

3. Read the coordinates of the point of intersection. The coordinates of the point of intersection are $(2, -1)$.
4. As a check, the ordered pair must satisfy each equation.

$$-A + 3B = -5 \qquad\qquad -3A + B = -7$$
$$-(2) + 3(-1) = -5 \qquad\qquad -3(2) + (-1) = -7$$
$$-2 - 3 = -5 \qquad\qquad -6 + (-1) = -7$$
$$-5 = -5 \qquad\qquad -7 = -7$$

If A is considered to be the dependent variable and thus each equation is solved for A instead of B as was done in the first part of this example, the same solution would be found.

1. Put each equation into the slope-intercept form.

$$A = 3B + 5$$
$$A = \frac{B}{3} + \frac{7}{3}$$

2. Use the vertical intercept and slope to graph the straight line representing each equation.

Sec. 6–4. Solving Two Linear Equations in Two Unknowns Graphically   149

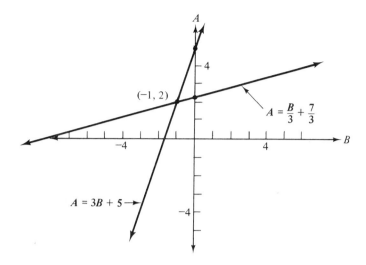

3. Read the coordinates of the point of intersection. In this case they are are $(-1, 2)$.
4. Check

$$-A + 3B = -5 \qquad\qquad -3A + B = -7$$
$$-2 + 3(-1) = -5 \qquad\qquad -3(2) + (-1) = -7$$
$$-2 - 3 = -5 \qquad\qquad -6 - 1 = -7$$
$$-5 = -5 \qquad\qquad -7 = -7$$

NOTE: A graphical solution will not always yield the exact answer. A careful reading from a graph should yield a reading correct to the nearest tenth. The accuracy of an answer depends on the degree of care taken when constructing the graph.

Example C: Solve the following system of equations graphically.

$$2V + T = -1$$
$$2V + T = 2$$

1. Put each equation into the slope-intercept form.

$$T = -2V - 1$$
$$T = -2V + 2$$

2. Use the vertical intercept and slope to graph the straight line representing each equation.

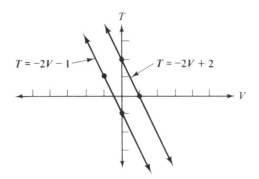

3. These lines are parallel and do not have a common solution. Therefore, there is no solution to this system of equations. This system of equations is called *inconsistent*.

Example D: Solve the following system of equations graphically.

$$R - 3T = -1$$
$$2R = 6T - 2$$

1. Put each equation into the slope-intercept form.

$$R = 3T - 1$$
$$R = 3T - 1$$

NOTE: Since both equations have the same slope and vertical intercept, the lines representing the equations would coincide and thus the solution to the system of equations is infinite. This system of equations is called *dependent*.

Example E: A computer can do two operations in 20 seconds. One operation takes 6 seconds longer than the other. How much time is required for each operation? Let A be the time required for one operation. Let B be the time required for the other operation. Analyzing the verbal statements we obtain the following equations:

(1) $$A + B = 20$$
(2) $$A = B + 6$$

Use the intercepts to construct the graphs.

(1)

A	0	20
B	20	0

(2)

A	0	6
B	−6	0

Sec. 6–4. Solving Two Linear Equations in Two Unknowns Graphically

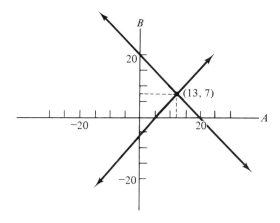

Therefore, one operation takes 13 seconds and the other 7 seconds.

The graphical solution to a system of linear equations is not always accurate enough, but even though an exact solution may be obtained by other methods, the graphical method is still quick and useful in many types of problems.

Exercises Find the solution to the following systems of equations by the graphical method.

1. $X + Y = 4$
    $X - Y = 8$

2. $3A + 2B = 3$
    $5A + 4B = 7$

3. $S = \frac{1}{2}T$
    $S + 2T - 1 = 0$

4. $R_1 + .5R_2 + 11.5 = 0$
    $3R_2 = -6R_1 - 69$

5. $Y = 2X - 1$
    $Y + X = 2$

6. $V - 3T = 5$
    $2V + T = 3$

7. $.05V_1 + .05V_2 = .1$
    $.5V_1 - .5V_2 = 6$

8. $V - H = -1$
    $2.5V - 6H = 4.5$

9. $S + 2T = \frac{1}{3}$
    $3S = 6T + 1$

10. $R = .3T - 1$
    $R = .3T + 1$

For each of the following, set up an appropriate system of equations and solve graphically.

11. A boat travels at a constant rate in still water. It travels downstream 24 miles in 2 hours while the return trip requires 3 hours. Find the rate of the boat in still water and the rate of the current.

12. A sum of $10,000 is invested, one part at 5% and the other part at 7%.

The total annual interest earned is $560. What is the amount invested at each rate?

13. A man purchased 80 acres of land for $29,000. He paid $300 per acre for part of the land and $400 per acre for the rest. How many acres did he buy at each price?

6–5. Solving Systems of Two Linear Equations in Two Unknowns Algebraically

The graphical solution of a system of two linear equations in two unknowns that we have just looked at is a relatively easy method for solving such problems and also gives us a "picture" of the solution. However, as is the case with any graphical procedure, this method is only approximate. Therefore, if exact solutions are required, we must use other methods. In this section, we will look at the algebraic solution of two linear equations in two unknowns.

Algebraically, systems of two linear equations in two unknowns may be solved by:

1. *Substitution:*
    (a) Rewrite one of the equations so that one variable is expressed in terms of the other.
    (b) Substitute this expression into the other equation. This will result in one equation with one variable.
    (c) Solve this equation for the involved variable and then substitute this solution into one of the original equations to obtain the solution for the other variable.
2. *Elimination:*
    (a) Take appropriate multiples of one or both equation(s) so that the coefficients of one of the variables are numerically equal but of opposite sign.
    (b) Add the two equations. This will result in one equation with one variable.
    (c) Solve this equation for the involved variable and then substitute this solution into one of the original equations to obtain the value for the other variable.

Example A: Use the method of substitution to solve the following system of equations:

$$X + 3Y = 6$$
$$2X + 3Y = 3$$

Rewrite the first equation to solve for X.

Sec. 6–5. Solving Two Linear Equations in Two Unknowns Algebraically

$$X = 6 - 3Y$$

NOTE: The choice of which equation to rewrite and which variable to express in terms of the other depends only on the ease of manipulation.
Substitute this expression into the second equation.

$$2(6 - 3Y) + 3Y = 3$$

Solve this equation for Y.

$$12 - 6Y + 3Y = 3$$
$$12 - 3Y = 3$$
$$-3Y = -9$$
$$Y = 3$$

Let $Y = 3$ in the first of the original equations.

$$X + 3(3) = 6$$
$$X + 9 = 6$$
$$X = -3.$$

Therefore, the solution is $X = -3$, $Y = 3$.

Example B: Use the method of substitution to solve the following system of equations:

$$3R + 3S = -1$$
$$-5R - 6S = 1$$

Rewrite the first equation to solve for S.

$$S = \frac{-1 - 3R}{3}$$

Substitute this expression into the second equation.

$$-5R - 6\left(\frac{-1 - 3R}{3}\right) = 1$$
$$-5R + 2 + 6R = 1$$
$$R + 2 = 1$$
$$R = -1$$

Let $R = -1$ in the first of the original equations.

$$3(-1) + 3S = -1$$
$$-3 + 3S = -1$$
$$3S = 2$$
$$S = \tfrac{2}{3}$$

Therefore, the solution is $R = -1$, $S = \tfrac{2}{3}$.

Example C: Use the method of elimination to solve the following system of equations:

$$X + 3Y = 6$$
$$2X + 3Y = 3$$

Multiply the first equation by -1.

$$-X - 3Y = -6$$

Add the second equation to this result.

$$-X - 3Y = -6$$
$$2X + 3Y = 3$$
$$\overline{}$$
$$X = -3$$

Let $X = -3$ in the first of the original equations.

$$-3 + 3Y = 6$$
$$3Y = 9$$
$$Y = 3$$

Therefore, the solution is $X = -3$, $Y = 3$.

Example D: Use the method of elimination to solve the following system of equations:

$$3R + 3S = -1$$
$$-5R - 6S = 1$$

Multiply the first equation by 2.

$$6R + 6S = -2$$

Add the second equation to this result.

$$6R + 6S = -2$$
$$-5R - 6S = 1$$
$$\overline{}$$
$$R = -1$$

Sec. 6–5. Solving Two Linear Equations in Two Unknowns Algebraically

Let $R = -1$ in the first of the original equations.

$$3(-1) + 3S = -1$$
$$-3 + 3S = -1$$
$$3S = 2$$
$$S = \tfrac{2}{3}$$

Therefore, the solution is $R = -1$, $S = \tfrac{2}{3}$.

Example E: Use the method of elimination to solve the following system of equations:

$$2M - 3N = 4$$
$$3M - 2N = -2$$

1. Eliminate M.
   Multiply the first equation by 3.

   $$6M - 9N = 12$$

   Multiply the second equation by -2.

   $$-6M + 4N = 4$$

   Add these results.

   $$\begin{aligned} 6M - 9N &= 12 \\ -6M + 4N &= 4 \\ \hline -5N &= 16 \end{aligned}$$

   $$N = -\tfrac{16}{5}$$

   Let $N = -\tfrac{16}{5}$ in the first of the original equations.

   $$2M - 3\left(-\tfrac{16}{5}\right) = 4$$
   $$2M + \tfrac{48}{5} = \tfrac{20}{5}$$
   $$2M = \tfrac{-28}{5}$$
   $$M = -\tfrac{14}{5}$$

   Therefore, the solution is $M = -\tfrac{14}{5}$, $N = -\tfrac{16}{5}$.

2. Eliminate N.
   Multiply the first equation by 2.

   $$4M - 6N = 8$$

Multiply the second equation by -3.
$$-9M + 6N = 6$$
Add these results.
$$\begin{array}{r} 4M - 6N = 8 \\ -9M + 6N = 6 \\ \hline -5M = 14 \end{array}$$
$$M = -\frac{14}{5}$$

Let $M = -\frac{14}{5}$ in the first of the original equations.
$$2\left(\frac{-14}{5}\right) - 3N = 4$$
$$\frac{-28}{5} - 3N = \frac{20}{5}$$
$$-3N = \frac{48}{5}$$
$$N = -\frac{16}{5}$$

Therefore, the solution is $M = -\frac{14}{5}$, $N = -\frac{16}{5}$.

We should see from this last example that it makes no difference which variable is eliminated first in the solution procedure. The only determining factor should be the ease of manipulation.

Example F: Use the method of elimination to solve the following system of equations:
$$3X - 2Y = 5$$
$$6X - 4Y = 2$$

Multiply the first equation by -2.
$$-6X + 4Y = -10$$

Add the second equation to this result.
$$\begin{array}{r} -6X + 4Y = -10 \\ 6X - 4Y = 2 \\ \hline 0 = -8 \end{array}$$

We should certainly realize that $0 \neq -8$. Therefore, there is no solution.

Sec. 6–5. Solving Two Linear Equations in Two Unknowns Algebraically

When solving algebraically (by substitution or elimination), if we obtain a result of $0 = K$ ($K \neq 0$), the system of equations is inconsistent and there is no solution. This would correspond to obtaining parallel lines when solving graphically. If we obtain a result of $0 = 0$, the system of equations is dependent and there is an *infinite solution*. This would correspond to obtaining coincident lines when solving graphically.

Example G: A certain circuit contains two resistances, R_1 and R_2. The sum of R_1 and R_2 is 12 ohms. Also, R_2 is 3 ohms less than twice R_1. Find R_1 and R_2. Analyzing the verbal statements, we obtain the following equations:

$$R_1 + R_2 = 12$$
$$R_2 = 2R_1 - 3$$

Writing these equations in standard form yields the system:

$$R_1 + R_2 = 12$$
$$-2R_1 + R_2 = -3$$

Multiply the first equation by -1.

$$-R_1 - R_2 = -12$$

Add the second equation to this result.

$$-R_1 - R_2 = -12$$
$$-2R_1 + R_2 = -3$$
$$\overline{-3R_1 \qquad\; = -15}$$
$$R_1 \qquad\quad = 5$$

Let $R_1 = 5$ in the first of the original equations.

$$5 + R_2 = 12$$
$$R_2 = 7$$

Therefore, $R_1 = 5$ ohms and $R_2 = 7$ ohms.

Exercises

Solve each of the following systems of equations by algebraic substitution.

1. $P + Q = 0$
   $9P - 5Q = -42$

2. $2R + S = 5$
   $R + 2S = 19$

3. $3T - 2V = 10$
   $T - V = 1$

4. $5W + 6Z = 17$
   $3W - 2Z = 27$

5. $2X - 3Y = -5$
   $3X + 2Y = 12$

6. $4M - 6N = 2$
   $30M - 42N = 17$

7. $5A + 7B = 1$
   $2A - 3B = 12$

8. $4K + 5L = 30$
   $6K + L = 19$

9. $3X + 5Y + 1 = 0$
   $2Y = 4X + 10$

10. $2R = 28 - 5S$
    $5R - 29 = 8S$

Solve each of the following systems of equations by algebraic elimination.

11. $X + Y = 8$
    $X - Y = 2$

12. $2A - B = 3$
    $2A - 3B = 11$

13. $3R - S = 4$
    $2R - 6S = -8$

14. $3K + 5L = 9$
    $7K - 10L = 8$

15. $4M - 3N = 10$
    $6M - 8N = -20$

16. $4W - 3Z = 24$
    $W - 2Z = 28$

17. $3P + 5Q = 9$
    $-7P + 10Q = -8$

18. $2R + 4S = 28$
    $5R - 8S = 25$

19. $X - 2Y + 6 = 0$
    $5X + 4Y - 8 = 0$

20. $7U = 5 - 3V$
    $3U + 20 = -2V$

21. Two cars start at the same place and the same time but travel in opposite directions. After 6 hours, they are 540 miles apart. If one car travels 10 miles per hour slower than the other, what is the speed of each (in miles per hour)?

22. An airplane took 4 hours to fly 1800 miles with a tail wind, and $4\frac{1}{2}$ hours for the return flight with a head wind. Find the wind velocity and the average air speed of the airplane (both in miles per hour).

23. A druggist wishes to prepare 100 gallons of 45% alcohol. He has two kinds of alcohol solution in stock. One is 55% pure and the other is 30% pure. How many gallons of each kind must be used for the mixture?

24. One tank contains a mixture of 20 gallons of water and 4 gallons of insect spray. Another tank contains a mixture of 12 gallons of water and 4 gallons of insect spray. How many gallons must be drawn from each tank to have 6 gallons of a mixture that is 20% insect spray?

25. The perimeter of a rectangle is 40 meters. The length is 4 meters less than twice the width. What are the dimensions of the rectangle?

26. Four cameras and five radios cost $250. Three of the same camera and seven of the same radio cost $285. Find the cost of a single camera and a single radio.

6–6. Solving Systems of Two Linear Equations in Two Unknowns by Determinants

Let us consider the general case of two linear equations in two unknowns:

(1) $$A_1 X + B_1 Y = C_1$$
(2) $$A_2 X + B_2 Y = C_2$$

In these equations, the A's and B's represent general numerical coefficients of X and Y and the C's are general constants.

1. We will solve algebraically for X. If we multiply equation (1) by B_2 we obtain:

$$A_1 B_2 X + B_1 B_2 Y = C_1 B_2$$

If we multiply equation (2) by $-B_1$ we obtain:

$$-A_2 B_1 X - B_1 B_2 Y = -C_2 B_1$$

If we add these two results we obtain:

$$A_1 B_2 X - A_2 B_1 X = C_1 B_2 - C_2 B_1$$

Factoring on the left we obtain:

$$(A_1 B_2 - A_2 B_1) X = C_1 B_2 - C_2 B_1$$

Dividing both members by $A_1 B_2 - A_2 B_1$ we obtain:

$$X = \frac{C_1 B_2 - C_2 B_1}{A_1 B_2 - A_2 B_1}$$

2. We will solve algebraically for Y. If we multiply equation (1) by A_2 we obtain:

$$A_1 A_2 X + A_2 B_1 Y = A_2 C_1$$

If we multiply equation (2) by $-A_1$ we obtain:

$$-A_1 A_2 X - A_1 B_2 Y = -A_1 C_2$$

If we add these two results we obtain:

$$A_2 B_1 Y - A_1 B_2 Y = A_2 C_1 - A_1 C_2$$

Factoring on the left we obtain:

$$(A_2B_1 - A_1B_2)Y = A_2C_1 - A_1C_2$$

Multiplying both members by -1 we obtain:

$$(A_1B_2 - A_2B_1)Y = A_1C_2 - A_2C_1$$

Dividing both members by $A_1B_2 - A_2B_1$ we obtain:

$$Y = \frac{A_1C_2 - A_2C_1}{A_1B_2 - A_2B_1}$$

The expressions that appear in the numerator and denominator of the solutions for X and Y ($A_1B_2 - A_2B_1$, $C_1B_2 - C_2B_1$, $A_1C_2 - A_2C_1$) are examples of special types of expressions which are called *determinants of the second order*. When considered as a determinant, $A_1B_2 - A_2B_1$ would be represented symbolically as $\begin{vmatrix} A_1 & B_1 \\ A_2 & B_2 \end{vmatrix}$. A_1 and B_1 are elements of the first row of the determinant, A_2 and B_2 are elements of the second row, A_1 and A_2 are elements of the first column, and B_1 and B_2 are elements of the second column of the determinant. A_1 and B_2 are elements of the principal diagonal, whereas A_2 and B_1 are elements of the secondary diagonal. Thus, we see that the value of a second order determinant can be found by finding the product of the elements on the principal diagonal and subtracting the product of the elements on the secondary diagonal.

$$\begin{vmatrix} A_1 & B_1 \\ A_2 & B_2 \end{vmatrix} = A_1B_2 - A_2B_1$$

Example A:

1. $\begin{vmatrix} 2 & 4 \\ 5 & 3 \end{vmatrix} = 2(3) - 5(4) = 6 - 20 = -14$

2. $\begin{vmatrix} -3 & 1 \\ -1 & 4 \end{vmatrix} = -3(4) - (-1)1 = -12 + 1 = -11$

3. $\begin{vmatrix} 0 & -3 \\ -2 & -1 \end{vmatrix} = 0(-1) - (-2)(-3) = 0 - 6 = -6$

Our solutions for X and Y can now be written in determinant form.

$$X = \frac{\begin{vmatrix} C_1 & B_1 \\ C_2 & B_2 \end{vmatrix}}{\begin{vmatrix} A_1 & B_1 \\ A_2 & B_2 \end{vmatrix}} \qquad Y = \frac{\begin{vmatrix} A_1 & C_1 \\ A_2 & C_2 \end{vmatrix}}{\begin{vmatrix} A_1 & B_1 \\ A_2 & B_2 \end{vmatrix}}$$

Sec. 6-6. Solving Two Linear Equations in Two Unknowns by Determinants

Examining these expressions we should note that the denominators are the same and consist of the coefficients of X and Y from our original system. In the numerator of the solution for X, the X cofficients (A's) have been replaced by the C's. In the numerator of the solution for Y, the Y coefficients (B's) have been replaced by the C's. Knowing this scheme makes it very easy to solve systems of two equations in two unknowns by determinants.

Example B: Solve the following system of equations by using determinants:

$$2X - 3Y = 8$$
$$4X + Y = 2$$

The denominator of the solution for both variables will consist of the coefficients of X and Y, respectively:

$$\begin{vmatrix} 2 & -3 \\ 4 & 1 \end{vmatrix}$$

In the numerator of the solution for X, the X coefficients are replaced by 8 and 2:

$$\begin{vmatrix} 8 & -3 \\ 2 & 1 \end{vmatrix}$$

In the numerator of the solution for Y, the Y coefficients are replaced by 8 and 2:

$$\begin{vmatrix} 2 & 8 \\ 4 & 2 \end{vmatrix}$$

Therefore:

$$X = \frac{\begin{vmatrix} 8 & -3 \\ 2 & 1 \end{vmatrix}}{\begin{vmatrix} 2 & -3 \\ 4 & 1 \end{vmatrix}} = \frac{8(1) - 2(-3)}{2(1) - 4(-3)} = \frac{8 + 6}{2 + 12} = \frac{14}{14} = 1$$

$$Y = \frac{\begin{vmatrix} 2 & 8 \\ 4 & 2 \end{vmatrix}}{\begin{vmatrix} 2 & -3 \\ 4 & 1 \end{vmatrix}} = \frac{2(2) - 4(8)}{2(1) - 4(-3)} = \frac{4 - 32}{2 + 12} = \frac{-28}{14} = -2$$

Example C: Solve the following system of equations by using determinants:

$$3R - 5S = 2$$
$$2R + 6S = -7$$

$$R = \frac{\begin{vmatrix} 2 & -5 \\ -7 & 6 \end{vmatrix}}{\begin{vmatrix} 3 & -5 \\ 2 & 6 \end{vmatrix}} = \frac{2(6) - (-7)(-5)}{3(6) - 2(-5)} = \frac{12 - 35}{18 + 10} = \frac{-23}{28}$$

$$S = \frac{\begin{vmatrix} 3 & 2 \\ 2 & -7 \end{vmatrix}}{\begin{vmatrix} 3 & -5 \\ 2 & 6 \end{vmatrix}} = \frac{3(-7) - 2(2)}{3(6) - 2(-5)} = \frac{-21 - 4}{18 + 10} = \frac{-25}{28}$$

This last example illustrates the advantages of using determinants. We get the exact answer as opposed to the approximate answer determined graphically, and if the answers are fractions, since each variable is solved separately, there is no need to substitute these fractions into an equation.

Example D: Solve the following system of equations by using determinants:

$$P = 3Q - 1$$
$$2Q - 5 = 4P$$

We must first put the equations in standard form.

$$P - 3Q = -1$$
$$-4P + 2Q = 5$$

$$P = \frac{\begin{vmatrix} -1 & -3 \\ 5 & 2 \end{vmatrix}}{\begin{vmatrix} 1 & -3 \\ -4 & 2 \end{vmatrix}} = \frac{-1(2) - 5(-3)}{1(2) - (-4)(-3)} = \frac{-2 + 15}{2 - 12} = \frac{13}{-10} = -\frac{13}{10}$$

$$Q = \frac{\begin{vmatrix} 1 & -1 \\ -4 & 5 \end{vmatrix}}{\begin{vmatrix} 1 & -3 \\ -4 & 2 \end{vmatrix}} = \frac{1(5) - (-4)(-1)}{1(2) - (-4)(-3)} = \frac{5 - 4}{2 - 12} = \frac{1}{-10} = -\frac{1}{10}$$

Example E: A man invests $\frac{1}{4}$ of his money at 6% interest per year and the remainder at 5% interest per year. His total yearly interest from the two investments is $210. Use determinants to find the amount invested at each rate. Let A represent the amount invested at 6% and B represent the amount invested at 5%.

The interest earned on the 6% investment is .06A. The interest earned on the 5% investment is .05B. Since the total interest earned is $210, then .06$A$ + .05B = 210. If $\frac{1}{4}$ of the money is invested at 6%, then $\frac{3}{4}$ must be invested at 5%. Therefore, $B = 3A$. Our system then is:

Sec. 6–6. Solving Two Linear Equations in Two Unknowns by Determinants

$$.06A + .05B = 210$$
$$-3A + B = 0$$

$$A = \frac{\begin{vmatrix} 210 & .05 \\ 0 & 1 \end{vmatrix}}{\begin{vmatrix} .06 & .05 \\ -3 & 1 \end{vmatrix}} = \frac{210(1) - 0(.05)}{.06(1) - (-3)(.05)} = \frac{210 - 0}{.06 + .15} = \frac{210}{.21} = 1000$$

$$B = \frac{\begin{vmatrix} .06 & 210 \\ -3 & 0 \end{vmatrix}}{\begin{vmatrix} .06 & .05 \\ -3 & 1 \end{vmatrix}} = \frac{.06(0) - (-3)(210)}{.06(1) - (-3)(.05)} = \frac{0 + 630}{.06 + .15} = \frac{630}{.21} = 3000$$

Therefore, \$1000 is invested at 6% and \$3000 is invested at 5%.

Some general points regarding determinants should be known. The equations must be in standard form (see Example D) before setting up the determinants. If either variable is missing in one of the equations, its coefficient is taken as zero, and zero is used in the appropriate position(s) in the determinants when finding the solution. If the determinant of the denominator is zero and that of the numerator is not zero, the system is inconsistent and there is no solution. This situation corresponds to obtaining parallel lines when solving graphically or $0 = K$ ($K \neq 0$) when solving algebraically. If the determinant of the denominator is zero and that of the numerator is also zero, the system is dependent and there is an infinite solution. This situation corresponds to obtaining coincident lines when solving graphically or $0 = 0$ when solving algebraically.

Exercises Evaluate each of the following determinants.

1. $\begin{vmatrix} 2 & 3 \\ 4 & 1 \end{vmatrix}$
2. $\begin{vmatrix} -1 & 3 \\ 6 & 2 \end{vmatrix}$
3. $\begin{vmatrix} -2 & -8 \\ 7 & -11 \end{vmatrix}$
4. $\begin{vmatrix} 4 & -5 \\ 2 & -7 \end{vmatrix}$
5. $\begin{vmatrix} 12 & -9 \\ -11 & 10 \end{vmatrix}$
6. $\begin{vmatrix} 8 & -5 \\ -2 & 6 \end{vmatrix}$
7. $\begin{vmatrix} 6 & 8 \\ 3 & 4 \end{vmatrix}$
8. $\begin{vmatrix} 5 & 2 \\ -10 & 4 \end{vmatrix}$
9. $\begin{vmatrix} -10 & 6 \\ -5 & 3 \end{vmatrix}$
10. $\begin{vmatrix} -9 & -3 \\ -15 & -5 \end{vmatrix}$

Solve each of the following systems of equations, using determinants.

11. $2X - 3Y = 8$
    $4X + Y = 5$

12. $A + B = -4$
    $-2A - 2B = 8$

13. $-4W + Z = 0$
    $7W + 9Z = 10$

14. $3R + 5S = 2$
    $10R - 6S = 4$

15. $-3P + 5Q = -6$
    $2Q = 7$

16. $2M = 9$
    $-M + 3N = 6$

17. $-3A + 4B = -7$
    $-A + 6B = 1$

18. $2K - 3L = 0$
    $-4K + 6L = 1$

19. $6X - 2Y = 3$
    $9X - 3Y = -5$

20. $5P - Q = 8$
    $-4P + 2Q = 3$

21. $W - 5Z = 2$
    $6W + 7Z = 9$

22. $7M - 2N = -3$
    $-3M + 10N = -6$

In each of the following, set up an appropriate system of equations and solve, using determinants.

23. A boat travels at a constant rate in still water. It travels downstream 30 miles in 2 hours while the return trip requires 3 hours and 20 minutes. Find the rate of the boat in still water and the rate of the current.

24. The perimeter of an isosceles triangle (a triangle with two equal sides) is 30 feet. Each of the equal sides is twice as long as the third side. Find the three sides of the triangle.

25. A piece of lumber is 16 feet long. It is to be cut into two pieces with one piece being three times as long as the other. Find the lengths of the two pieces.

26. In assembling a certain structure, 112 rivets are used. There are two different kinds of rivets and 18 more of one type are used than the other. How many of each type are used?

6–7. Solving Systems of Three Linear Equations in Three Unknowns Algebraically

The procedure for solving algebraically a system of three equations with three variables is to eliminate one of the variables from any two of the three equations, thus generating a new equation with two variables. Then eliminate the same variable from a different pair of equations in the original system. This will generate another equation with two variables that are the same variables as those in the first equation that was generated. The procedure for solving the two new equations that are generated is the same as the procedure used in Section 6–5.

Example A: Solve the following system of equations algebraically.

(1) $\quad X + 2Y + 3Z = 8$
(2) $\quad -X - 4Y + Z = -1$
(3) $\quad 2X - 2Y + Z = 3$

Sec. 6–7. Solving Three Linear Equations in Three U

1. Eliminate Z from equations (1) and (2).
   Multiply equation (2) by -3.
   $$3X + 12Y - 3$$

   Add equations (1) and (2).

   (1) $X + 2Y + 3Z$
   (2) $3X + 12Y - 3Z$
   (4) $4X + 14Y$

   A new equation, (4), is generated.

2. Eliminate Z from equations (2) and (3).
   Multiply equation (3) by -1.
   $$-2X + 2Y - Z = -3$$

   Add equations (2) and (3).

   (2) $-X - 4Y + Z = -1$
   (3) $-2X + 2Y - Z = -3$
   (5) $-3X - 2Y = -4$

   A new equation, (5), is generated.

3. Solve the system of two equations with two variables just generated using the techniques described in Section 6–5. Multiply equation (5) by 7.
   $$-21X - 14Y = -28$$

   Add equations (4) and (5) in order to eliminate Y.

   (4) $4X + 14Y = 11$
   (5) $-21X - 14Y = -28$
                    $-17X = -17$
                    $X = 1$

4. Let $X = 1$ in equation (5).
   $$-3X - 2Y = -4$$
   $$-3(1) - 2Y = -4$$
   $$-2Y = -1$$
   $$Y = \tfrac{1}{2}$$

5. Let $X = 1$ and $Y = \tfrac{1}{2}$ in equation (1).
   $$X + 2Y + 3Z = 8$$
   $$1 + 2(\tfrac{1}{2}) + 3Z = 8$$
   $$3Z = 6$$
   $$Z = 2$$

Therefore, the solution is $X = 1$, $Y = \frac{1}{2}$, and $Z = 2$. We may write this as an ordered triple $(1, \frac{1}{2}, 2)$ with the order of the values corresponding to the order in which the variables appeared in the original system of equations.

NOTE: To find Y, either equation (4) or equation (5) could have been used and to find Z, equations (1), (2), or (3) could have been used.

6. Check the solution in the original system.

(1) $$X + 2Y + 3Z = 8$$
$$(1) + 2(\tfrac{1}{2}) + 3(2) = 8$$
$$1 + 1 + 6 = 8$$
$$8 = 8$$

(2) $$-X - 4Y + Z = -1$$
$$-(1) - 4(\tfrac{1}{2}) + (2) = -1$$
$$-1 = -1$$

(3) $$2X - 2Y + Z = 3$$
$$2(1) - 2(\tfrac{1}{2}) + (2) = 3$$
$$3 = 3$$

Example B: Solve the system of equations.

(1) $$2A + 3B + 4C = 2$$
(2) $$3A - B + 6C = -8$$
(3) $$B + C = 2$$

1. Since the third equation only has the variables B and C, the variable A should be eliminated from equations (1) and (2). This procedure will cut down on the amount of work that must be done in order to solve the system of equations. Multiply equation (1) by 3.

$$6A + 9B + 12C = 6$$

Multiply equation (2) by -2.

$$-6A + 2B - 12C = 16$$

Add these two equations.

$$6A + 9B + 12C = 6$$
$$-6A + 2B - 12C = 16$$

(4) $$11B = 22$$
$$B = 2$$

Sec. 6–7. Solving Three Linear Equations in Three Unknowns Algebraically

2. Letting $B = 2$ in equation (3) will give a value of $C = 0$.

(3)
$$B + C = 2$$
$$(2) + C = 2$$
$$C = 0$$

3. Letting $B = 2$ and $C = 0$ in equation (1).

(1)
$$2A + 3B + 4C = 2$$
$$2A + 3(2) + 4(0) = 2$$
$$2A = -4$$
$$A = -2$$

Thus, the solution to this system of equations is $(-2, 2, 0)$.

4. Check this result in the original system of equations.

(1)
$$2A + 3B + 4C = 2$$
$$2(-2) + 3(2) + 4(0) = 2$$
$$-4 + 6 + 0 = 2$$
$$2 = 2$$

(2)
$$3A - B + 6C = -8$$
$$3(-2) - (2) + 6(0) = -8$$
$$-6 - 2 + 0 = -8$$
$$-8 = -8$$

(3)
$$B + C = 2$$
$$(2) + (0) = 2$$
$$2 = 2$$

NOTE: Your choice of the variable to be eliminated depends only on the ease of manipulation. Thus, the system of equations should be examined to see whether any of the variables are missing from one of the equations or whether any of the variables have the same coefficient in any two of the equations.

Example C: Solve the system of equations.

(1) $\quad\quad\quad\quad\quad\quad .5R + 3S - 6T = 7$
(2) $\quad\quad\quad\quad\quad\quad 3R - S + 2T = 1$
(3) $\quad\quad\quad\quad\quad\quad 1.5R - .5S + T = .5$

1. Eliminate S from equations (1) and (2).
   Multiply equation (2) by 3.

$$9R - 3S + 6T = 3$$

Add equations (1) and (2).

$$.5R + 3S - 6T = 7$$
$$9R - 3S + 6T = 3$$

(4)
$$9.5R = 10$$
$$R = \frac{10}{9.5} = \frac{100}{95} = \frac{20}{19}$$

2. Eliminate S from equations (2) and (3).
Multiply equation (3) by 2.

$$3R - S + 2T = 1$$

NOTE: This makes equation (3) the same as equation (2). When these two equations are added we obtain $0 = 0$. Therefore, this system of equations is dependent and it has an infinite solution.

Example D: Find the three angles of a triangle if it is known that angle C is one third the measure of angle A and angle B is 40° larger than angle A.

1. This problem can be solved by setting up a system of three equations with three variables, where the variables represent the number of degrees in each of the three angles of a triangle.

(1) $A + B + C = 180°$ The sum of the angles of a triangle equal 180°.

(2) $-\frac{1}{3}A + C = 0°$ C is $\frac{1}{3}$ the measure of A.

(3) $-A + B = 40°$ B is 40° larger than A.

2. Multiply equation (2) by -1.

(4) $\qquad\qquad\qquad \frac{1}{3}A - C = 0°$

Add equations (1) and (4).

(1) $\qquad A + B + C = 180°$
(4) $\qquad \frac{1}{3}A \quad\;\; - C = 0°$
(5) $\qquad 1\frac{1}{3}A + B = 180°$

3. Multiply equation (3) by -1.

(5) $\qquad\qquad\qquad A - B = -40°$

Add equations (3) and (5).

Sec. 6-7. Solving Three Linear Equations in Three Unknowns Algebraically

$$1\tfrac{1}{3}A + B = 180°$$
$$A - B = 40°$$
$$\overline{2\tfrac{1}{3}A = 140°}$$
$$A = \frac{140°}{1} \cdot \frac{3}{7} = 60°$$

4. Let $A = 60°$ in equation (3).

$$-A + B = 40°$$
$$-60° + B = 40°$$
$$B = 100°$$

5. Let $A = 60°$ and $B = 100°$ in equation (1).

$$A + B + C = 180°$$
$$60° + 100° + C = 180°$$
$$160° + C = 180°$$
$$C = 20°$$

Exercises

Solve each of the following systems of equations algebraically.

1. $3A - 2B + 4C = -2$
    $5A + 7B = -1$
    $2A - 3C = 23$

2. $6X - 2Y + 3Z = 24$
    $X - 2Y - Z = -16$
    $2X + 2Z = 20$

3. $2L + M - 4N = 6$
    $-L + 3M = 18$
    $ 2M - N = 8$

4. $I - 2J + 3K = 2$
    $2I - 3K = 3$
    $I + J + K = 6$

5. $U + V - W = 4$
    $-U + V + 2W = 12$
    $2U - V + 4W = 8$

6. $P + Q - R = 5$
    $P + Q + 2R = 4$
    $-P + 2Q + 3R = 1$

7. $X - 3Y + Z = 14$
    $4.5X - .5Y + 3Z = .75$
    $9X - Y + 6Z = 1.5$

8. $2A - B + 3C = 12$
    $A + 2B - 3C = -10$
    $A + B - C = -3$

9. $P - R = 14$
    $ Q + R = 21$
    $P - Q + R = -10$

10. $2T - 3U + V = -2$
    $T - 6U + 3V = -2$
    $3T + 3U - 2V = 2$

11. $3L + M - 4N = -1$
    $-2L - M - 2N = -5$
    $L - M - 8N = -5$

12. $I - J + K = 2$
    $2I - J + 3K = 6$
    $3I + 3J + 3K = 18$

For each of the following, set up an appropriate system of equations and solve algebraically.

13. Stainless steel is an alloy consisting of steel, nickel, and chromium. The percentage of chromium is 2% more than twice the percentage of nickel, while the percentage of steel is 2% more than nine times the percentage of nickel. What is the percentage of each metal in this alloy?

14. The sum of the specific gravities of platinum, gold, and lead is 52. The specific gravity of platinum is 2.1 more than that of gold, while the specific gravity of gold is 8 more than that of lead. Find the specific gravities of each of these elements.

6–8. Solving Systems of Three Linear Equations in Three Unknowns by Determinants

Let us now consider the general case of three linear equations in three unknowns:

$$A_1 X + B_1 Y + C_1 Z = D_1$$
$$A_2 X + B_2 Y + C_2 Z = D_2$$
$$A_3 X + B_3 Y + C_3 Z = D_3$$

In these equations, the A's, B's, C's, and D's represent general constants. This arrangement is called the *standard form* for a system of three linear equations in three unknowns.

When these equations are solved for X, Y, and Z by using the algebraic procedure described in Section 6-7, the following expressions are obtained:

$$X = \frac{D_1 B_2 C_3 + D_3 B_1 C_2 + D_2 B_3 C_1 - D_3 B_2 C_1 - D_1 B_3 C_2 - D_2 B_1 C_3}{A_1 B_2 C_3 + A_3 B_1 C_2 + A_2 B_3 C_1 - A_3 B_2 C_1 - A_1 B_3 C_2 - A_2 B_1 C_3}$$

$$Y = \frac{A_1 D_2 C_3 + A_3 D_1 C_2 + A_2 D_3 C_1 - A_3 D_2 C_1 - A_1 D_3 C_2 - A_2 D_1 C_3}{\text{same}}$$

$$Z = \frac{A_1 B_2 D_3 + A_3 B_1 D_2 + A_2 B_3 D_1 - A_3 B_2 D_1 - A_1 B_3 D_2 - A_2 B_1 D_3}{\text{same}}$$

The expressions that appear in the numerator and denominator of these solutions for X, Y, and Z are examples of special types of expressions that are called *determinants of the third order*. When considered as determinants, these expressions would be represented as follows:

1. $A_1 B_2 C_3 + A_3 B_1 C_2 + A_2 B_3 C_1 - A_3 B_2 C_1 - A_1 B_3 C_2 - A_2 B_1 C_3$

$$= A_1(B_2 C_3 - B_3 C_2) - A_2(B_1 C_3 - B_3 C_1) + A_3(B_1 C_2 - B_2 C_1)$$

$$= A_1 \begin{vmatrix} B_2 & C_2 \\ B_3 & C_3 \end{vmatrix} - A_2 \begin{vmatrix} B_1 & C_1 \\ B_3 & C_3 \end{vmatrix} + A_3 \begin{vmatrix} B_1 & C_1 \\ B_2 & C_2 \end{vmatrix}$$

$$= \begin{vmatrix} A_1 & B_1 & C_1 \\ A_2 & B_2 & C_2 \\ A_3 & B_3 & C_3 \end{vmatrix}$$

Sec. 6–8. Solving Three Linear Equations in Three Unknowns by Determinants

2. $D_1B_2C_3 + D_3B_1C_2 + D_2B_3C_1 - D_3B_2C_1 - D_1B_3C_2 - D_2B_1C_3$
$= D_1(B_2C_3 - B_3C_2) - D_2(B_1C_3 - B_3C_1) + D_3(B_1C_2 - B_2C_1)$
$= D_1 \begin{vmatrix} B_2 & C_2 \\ B_3 & C_3 \end{vmatrix} - D_2 \begin{vmatrix} B_1 & C_1 \\ B_3 & C_3 \end{vmatrix} + D_3 \begin{vmatrix} B_1 & C_1 \\ B_2 & C_2 \end{vmatrix}$
$= \begin{vmatrix} D_1 & B_1 & C_1 \\ D_2 & B_2 & C_2 \\ D_3 & B_3 & C_3 \end{vmatrix}$

3. $A_1D_2C_3 + A_3D_1C_2 + A_2D_3C_1 - A_3D_2C_1 - A_1D_3C_2 - A_2D_1C_3$
$= A_1(D_2C_3 - D_3C_2) - A_2(D_1C_3 - D_3C_1) + A_3(D_1C_2 - D_2C_1)$
$= A_1 \begin{vmatrix} D_2 & C_2 \\ D_3 & C_3 \end{vmatrix} - A_2 \begin{vmatrix} D_1 & C_1 \\ D_3 & C_3 \end{vmatrix} + A_3 \begin{vmatrix} D_1 & C_1 \\ D_2 & C_2 \end{vmatrix}$
$= \begin{vmatrix} A_1 & D_1 & C_1 \\ A_2 & D_2 & C_2 \\ A_3 & D_3 & C_3 \end{vmatrix}$

4. $A_1B_2D_3 + A_3B_1D_2 + A_2B_3D_1 - A_3B_2D_1 - A_1B_3D_2 - A_2B_1D_3$
$= A_1(B_2D_3 - B_3D_2) - A_2(B_1D_3 - B_3D_1) + A_3(B_1D_2 - B_2D_1)$
$= A_1 \begin{vmatrix} B_2 & D_2 \\ B_3 & D_3 \end{vmatrix} - A_2 \begin{vmatrix} B_1 & D_1 \\ B_3 & D_3 \end{vmatrix} + A_3 \begin{vmatrix} B_1 & D_1 \\ B_2 & D_2 \end{vmatrix}$
$= \begin{vmatrix} A_1 & B_1 & D_1 \\ A_2 & B_2 & D_2 \\ A_3 & B_3 & D_3 \end{vmatrix}$

Each of these determinants has three rows, three columns, a principal diagonal, and a secondary diagonal. For example, in the determinant

$$\begin{vmatrix} A_1 & B_1 & C_1 \\ A_2 & B_2 & C_2 \\ A_3 & B_3 & C_3 \end{vmatrix}$$

A_1, B_1, and C_1 are elements of the first row; A_1, A_2, and A_3 are elements of the first column (similarly for the second and third rows and columns); A_1, B_2, and C_3 are elements of the principal diagonal; and A_3, B_2, and C_1 are elements of the secondary diagonal. From these four expressions, we should see that a third order determinant may be evaluated as follows:

1. Multiply the element in the first row and first column times the second order determinant that remains after the first row and first column are eliminated.
2. Subtract the product of the element in the second row and first column times the second order determinant that remains after the second row and first column are eliminated.

3. Add the product of the element in the third row and first column times the second order determinant that remains after the third row and first column are eliminated.

Example A:

1. $\begin{vmatrix} 1 & 3 & 5 \\ 2 & -4 & 6 \\ 7 & 4 & -2 \end{vmatrix} = 1\begin{vmatrix} -4 & 6 \\ 4 & -2 \end{vmatrix} - 2\begin{vmatrix} 3 & 5 \\ 4 & -2 \end{vmatrix} + 7\begin{vmatrix} 3 & 5 \\ -4 & 6 \end{vmatrix}$

$= 1(8 - 24) - 2(-6 - 20) + 7(18 + 20)$
$= 1(-16) - 2(-26) + 7(38)$
$= -16 + 52 + 266$
$= 302$

2. $\begin{vmatrix} -2 & 5 & 2 \\ 4 & -1 & -4 \\ 1 & -3 & 0 \end{vmatrix} = -2\begin{vmatrix} -1 & -4 \\ -3 & 0 \end{vmatrix} - 4\begin{vmatrix} 5 & 2 \\ -3 & 0 \end{vmatrix} + 1\begin{vmatrix} 5 & 2 \\ -1 & -4 \end{vmatrix}$

$= -2(0 - 12) - 4(0 + 6) + 1(-20 + 2)$
$= -2(-12) - 4(6) + 1(-18)$
$= 24 - 24 - 18$
$= -18$

3. $\begin{vmatrix} 3 & 2 & 0 \\ -4 & -2 & -3 \\ -1 & 5 & 1 \end{vmatrix} = 3\begin{vmatrix} -2 & -3 \\ 5 & 1 \end{vmatrix} - (-4)\begin{vmatrix} 2 & 0 \\ 5 & 1 \end{vmatrix} + (-1)\begin{vmatrix} 2 & 0 \\ -2 & -3 \end{vmatrix}$

$= 3(-2 + 15) + 4(2 - 0) - 1(-6 - 0)$
$= 3(13) + 4(2) - 1(-6)$
$= 39 + 8 + 6$
$= 53$

Our solutions for X, Y, and Z can now be written in determinant form:

$$X = \frac{\begin{vmatrix} D_1 & B_1 & C_1 \\ D_2 & B_2 & C_2 \\ D_3 & B_3 & C_3 \end{vmatrix}}{\begin{vmatrix} A_1 & B_1 & C_1 \\ A_2 & B_2 & C_2 \\ A_3 & B_3 & C_3 \end{vmatrix}} \qquad Y = \frac{\begin{vmatrix} A_1 & D_1 & C_1 \\ A_2 & D_2 & C_2 \\ A_3 & D_3 & C_3 \end{vmatrix}}{\begin{vmatrix} A_1 & B_1 & C_1 \\ A_2 & B_2 & C_2 \\ A_3 & B_3 & C_3 \end{vmatrix}} \qquad Z = \frac{\begin{vmatrix} A_1 & B_1 & D_1 \\ A_2 & B_2 & D_2 \\ A_3 & B_3 & D_3 \end{vmatrix}}{\begin{vmatrix} A_1 & B_1 & C_1 \\ A_2 & B_2 & C_2 \\ A_3 & B_3 & C_3 \end{vmatrix}}$$

Sec. 6-8. Solving Three Linear Equations in Three Unknowns by Determinants

Examining these expressions, we should note once again that the denominators are the same and consist of the coefficients of X, Y, and Z from our original system. In the numerator of the solution for X, the X coefficients (A's) have been replaced by the D's. In the numerator of the solution for Y, the Y coefficients (B's) have been replaced by the D's. In the numerator of the solution for Z, the Z coefficients (C's) have been replaced by the D's.

Example B: Solve the following system of equations by using determinants:

$$2X - Y + Z = 3$$
$$X + 2Y + 2Z = 1$$
$$4X + Y + 2Z = 0$$

$$X = \frac{\begin{vmatrix} 3 & -1 & 1 \\ 1 & 2 & 2 \\ 0 & 1 & 2 \end{vmatrix}}{\begin{vmatrix} 2 & -1 & 1 \\ 1 & 2 & 2 \\ 4 & 1 & 2 \end{vmatrix}} = \frac{3\begin{vmatrix} 2 & 2 \\ 1 & 2 \end{vmatrix} - 1\begin{vmatrix} -1 & 1 \\ 1 & 2 \end{vmatrix} + 0\begin{vmatrix} -1 & 1 \\ 2 & 2 \end{vmatrix}}{2\begin{vmatrix} 2 & 2 \\ 1 & 2 \end{vmatrix} - 1\begin{vmatrix} -1 & 1 \\ 1 & 2 \end{vmatrix} + 4\begin{vmatrix} -1 & 1 \\ 2 & 2 \end{vmatrix}}$$

$$= \frac{3(4-2) - 1(-2-1) + 0(-2-2)}{2(4-2) - 1(-2-1) + 4(-2-2)}$$

$$= \frac{3(2) - 1(-3) + 0(-4)}{2(2) - 1(-3) + 4(-4)} = \frac{6+3+0}{4+3-16} = \frac{9}{-9} = -1$$

$$Y = \frac{\begin{vmatrix} 2 & 3 & 1 \\ 1 & 1 & 2 \\ 4 & 0 & 2 \end{vmatrix}}{-9} = \frac{2\begin{vmatrix} 1 & 2 \\ 0 & 2 \end{vmatrix} - 1\begin{vmatrix} 3 & 1 \\ 0 & 2 \end{vmatrix} + 4\begin{vmatrix} 3 & 1 \\ 1 & 2 \end{vmatrix}}{-9}$$

$$= \frac{2(2-0) - 1(6-0) + 4(6-1)}{-9}$$

$$= \frac{2(2) - 1(6) + 4(5)}{-9} = \frac{4-6+20}{-9} = \frac{18}{-9} = -2$$

$$Z = \frac{\begin{vmatrix} 2 & -1 & 3 \\ 1 & 2 & 1 \\ 4 & 1 & 0 \end{vmatrix}}{-9} = \frac{2\begin{vmatrix} 2 & 1 \\ 1 & 0 \end{vmatrix} - 1\begin{vmatrix} -1 & 3 \\ 1 & 0 \end{vmatrix} + 4\begin{vmatrix} -1 & 3 \\ 2 & 1 \end{vmatrix}}{-9}$$

$$= \frac{2(0-1) - 1(0-3) + 4(-1-6)}{-9}$$

$$= \frac{2(-1) - 1(-3) + 4(-7)}{-9} = \frac{-2+3-28}{-9} = \frac{-27}{-9} = 3$$

Example C: Solve the following system of equations by using determinants:

$$2A + B - C = 4$$
$$4A - 3B - 2C = -2$$
$$8A - 2B - 3C = 3$$

$$A = \frac{\begin{vmatrix} 4 & 1 & -1 \\ -2 & -3 & -2 \\ 3 & -2 & -3 \end{vmatrix}}{\begin{vmatrix} 2 & 1 & -1 \\ 4 & -3 & -2 \\ 8 & -2 & -3 \end{vmatrix}} = \frac{4\begin{vmatrix} -3 & -2 \\ -2 & -3 \end{vmatrix} - (-2)\begin{vmatrix} 1 & -1 \\ -2 & -3 \end{vmatrix} + 3\begin{vmatrix} 1 & -1 \\ -3 & -2 \end{vmatrix}}{2\begin{vmatrix} -3 & -2 \\ -2 & -3 \end{vmatrix} - 4\begin{vmatrix} 1 & -1 \\ -2 & -3 \end{vmatrix} + 8\begin{vmatrix} 1 & -1 \\ -3 & -2 \end{vmatrix}}$$

$$= \frac{4(9-4) + 2(-3-2) + 3(-2-3)}{2(9-4) - 4(-3-2) + 8(-2-3)}$$

$$= \frac{4(5) + 2(-5) + 3(-5)}{2(5) - 4(-5) + 8(-5)} = \frac{20 - 10 - 15}{10 + 20 - 40} = \frac{-5}{-10} = \frac{1}{2}$$

$$B = \frac{\begin{vmatrix} 2 & 4 & -1 \\ 4 & -2 & -2 \\ 8 & 3 & -3 \end{vmatrix}}{-10} = \frac{2\begin{vmatrix} -2 & -2 \\ 3 & -3 \end{vmatrix} - 4\begin{vmatrix} 4 & -1 \\ 3 & -3 \end{vmatrix} + 8\begin{vmatrix} 4 & -1 \\ -2 & -2 \end{vmatrix}}{-10}$$

$$= \frac{2(6+6) - 4(-12+3) + 8(-8-2)}{-10}$$

$$= \frac{2(12) - 4(-9) + 8(-10)}{-10} = \frac{24 + 36 - 80}{-10} = \frac{-20}{-10} = 2$$

$$C = \frac{\begin{vmatrix} 2 & 1 & 4 \\ 4 & -3 & -2 \\ 8 & -2 & 3 \end{vmatrix}}{-10} = \frac{2\begin{vmatrix} -3 & -2 \\ -2 & 3 \end{vmatrix} - 4\begin{vmatrix} 1 & 4 \\ -2 & 3 \end{vmatrix} + 8\begin{vmatrix} 1 & 4 \\ -3 & -2 \end{vmatrix}}{-10}$$

$$= \frac{2(-9-4) - 4(3+8) + 8(-2+12)}{-10}$$

$$= \frac{2(-13) - 4(11) + 8(10)}{-10} = \frac{-26 - 44 + 80}{-10} = \frac{10}{-10} = -1$$

Example D: In a certain triangle, angle B is 6° less than twice angle A, and angle C is 6° more than three times angle A. Use determinants to find angles A, B, and C.

Since there are 180° in a triangle, $A + B + C = 180°$. Since angle B is 6° less than twice angle A, $B = 2A - 6°$. Since angle C is 6° more than three times angle A, $C = 3A + 6°$. Therefore,

Sec. 6-8. Solving Three Linear Equations in Three Unknowns by Determinants

$$A + B + C = 180°$$
$$-2A + B = -6°$$
$$-3A + C = 6°$$

$$A = \frac{\begin{vmatrix} 180 & 1 & 1 \\ -6 & 1 & 0 \\ 6 & 0 & 1 \end{vmatrix}}{\begin{vmatrix} 1 & 1 & 1 \\ -2 & 1 & 0 \\ -3 & 0 & 1 \end{vmatrix}} = \frac{180\begin{vmatrix} 1 & 0 \\ 0 & 1 \end{vmatrix} - (-6)\begin{vmatrix} 1 & 1 \\ 0 & 1 \end{vmatrix} + 6\begin{vmatrix} 1 & 1 \\ 1 & 0 \end{vmatrix}}{1\begin{vmatrix} 1 & 0 \\ 0 & 1 \end{vmatrix} - (-2)\begin{vmatrix} 1 & 1 \\ 0 & 1 \end{vmatrix} + (-3)\begin{vmatrix} 1 & 1 \\ 1 & 0 \end{vmatrix}}$$

$$= \frac{180(1-0) + 6(1-0) + 6(0-1)}{1(1-0) + 2(1-0) - 3(0-1)}$$

$$= \frac{180(1) + 6(1) + 6(-1)}{1(1) + 2(1) - 3(-1)} = \frac{180 + 6 - 6}{1 + 2 + 3} = \frac{180}{6} = 30°$$

$$B = \frac{\begin{vmatrix} 1 & 180 & 1 \\ -2 & -6 & 0 \\ -3 & 6 & 1 \end{vmatrix}}{6} = \frac{1\begin{vmatrix} -6 & 0 \\ 6 & 1 \end{vmatrix} - (-2)\begin{vmatrix} 180 & 1 \\ 6 & 1 \end{vmatrix} + (-3)\begin{vmatrix} 180 & 1 \\ -6 & 0 \end{vmatrix}}{6}$$

$$= \frac{1(-6-0) + 2(180-6) - 3(0+6)}{6}$$

$$= \frac{1(-6) + 2(174) - 3(6)}{6} = \frac{-6 + 348 - 18}{6} = \frac{324}{6} = 54°$$

$$C = \frac{\begin{vmatrix} 1 & 1 & 180 \\ -2 & 1 & -6 \\ -3 & 0 & 6 \end{vmatrix}}{6} = \frac{1\begin{vmatrix} 1 & -6 \\ 0 & 6 \end{vmatrix} - (-2)\begin{vmatrix} 1 & 180 \\ 0 & 6 \end{vmatrix} + (-3)\begin{vmatrix} 1 & 180 \\ 1 & -6 \end{vmatrix}}{6}$$

$$= \frac{1(6-0) + 2(6-0) - 3(-6-180)}{6}$$

$$= \frac{1(6) + 2(6) - 3(-186)}{6} = \frac{6 + 12 + 558}{6} = \frac{576}{6} = 96°$$

Just as with two equations in two unknowns, the three equations must be in standard form before setting up the determinants. When any variable is missing from an equation, its coefficient is zero, and zero is used in the appropriate position in the determinants when finding the solution. If the determinant of the denominator is zero and that of the numerator is not zero, the system is inconsistent and there is no solution. If the determinants of both numerator and denominator are zero, the system is dependent and there is an infinite solution.

This method of evaluating third order determinants is called *expansion by minors*, and may be used to evaluate any determinant of order greater than two.

Exercises Evaluate each of the following determinants.

1. $\begin{vmatrix} 5 & -1 & 4 \\ 7 & 1 & 1 \\ 8 & -2 & -6 \end{vmatrix}$

2. $\begin{vmatrix} -3 & -8 & -4 \\ 5 & 10 & 0 \\ 2 & -1 & 1 \end{vmatrix}$

3. $\begin{vmatrix} 0 & 1 & -5 \\ -2 & 6 & -2 \\ -4 & 0 & 9 \end{vmatrix}$

4. $\begin{vmatrix} 8 & 9 & -6 \\ -4 & -2 & 5 \\ -3 & 7 & 2 \end{vmatrix}$

5. $\begin{vmatrix} 6 & 5 & 2 \\ 10 & -7 & 2 \\ -2 & 6 & -3 \end{vmatrix}$

6. $\begin{vmatrix} 1 & 4 & 2 \\ -1 & 3 & -6 \\ 0 & -7 & 0 \end{vmatrix}$

Solve each of the following systems of equations by using determinants.

7. $A + B - C = 1$
    $3A + B = 0$
    $B + 4C = 0$

8. $2X - 3Y + 3Z = 7$
    $3X - Y - 2Z = -11$
    $5X - 2Y + 4Z = 11$

9. $R - S = -6$
    $R + T = -9$
    $3S + T = 1$

10. $L + M - N = 4$
    $2L - 3M + N = 1$
    $L - 4M - 2N = -7$

11. $P + Q + 3R = 3$
    $2P - Q = 0$
    $4Q - R = 8$

12. $U + 2V + W = 9$
    $U + V - W = 12$
    $3U - V + 2W = 12$

13. $2A + B - 3C = 1$
    $A - B + 2C = 1$
    $A + 3B - C = 6$

14. $X - 2Y + Z = 5$
    $2X + 3Z = 4$
    $Y + 2Z = 2$

15. $R - S - T = 1$
    $2R + 3S + 2T = 8$
    $4R + 3S - 2T = -2$

16. $2L + M - 4N = 5$
    $3L - 2M + 3N = 9$
    $L + M - N = 0$

17. $2P - 3Q = 8 - 4R$
    $3P = 5R - 4Q - 4$
    $4P + 6R = 12 + 5Q$

18. $3U + 2V = W + 2$
    $U - 1 = V + 9W$
    $4U + 3W = -3V$

For each of the following, set up an appropriate system of equations and solve by using determinants.

19. A man invested $15,000. Part of this sum is invested at 5.5%, part at 6%, and the remainder at 6.5%. The total annual interest earned is $890. The interest earned from the 6% investment is $50 less than the combined interest earned from the other two investments. Find the amount invested at each rate.

20. The perimeter of a triangular shaped area is 120 feet. The longest side is

Sec. 6–9. Chapter Review   177

20 feet longer than the sum of the other two sides, while the second longest side is 10 feet longer than the shortest side. Find the lengths of the three sides.

6–9. CHAPTER REVIEW

Solve each of the following systems of equations graphically.

1. $2X + Y = 4$
   $X - Y = 5$

2. $A + B = 6$
   $3A - 2B = 3$

3. $5R - S = 2$
   $4R + S = 7$

4. $5P - 2Q = 6$
   $3P - 4Q = 12$

5. $3V - 5W = 13$
   $6V + 7W = -8$

6. $4T + 7U = 2$
   $3T + 5U = 1$

Solve each of the following systems of equations algebraically.

7. $2K - 3L = 1$
   $2K - 4L = 2$

8. $X + 3Y = -7$
   $2X - 3Y = 13$

9. $5A - 2B = 4$
   $A + B = 5$

10. $7M - 2N = 4$
    $9M - 3N = 3$

11. $5R + 3S = 3$
    $3R - 2S = 17$

12. $3P - 2Q = 2$
    $6P - 5Q = -1$

13. $2V + 3W = 7$
    $3V + W = 7$

14. $-2X + Y = 9$
    $X + Y = 3$

15. $B = 2A + 3$
    $A = 2B - 3$

16. $3K = 3 - 5L$
    $9K - 2 = -L$

Evaluate each of the following second-order determinants.

17. $\begin{vmatrix} 1 & 3 \\ 4 & -2 \end{vmatrix}$

18. $\begin{vmatrix} 5 & 9 \\ -7 & 6 \end{vmatrix}$

19. $\begin{vmatrix} -3 & 2 \\ -5 & -4 \end{vmatrix}$

20. $\begin{vmatrix} -2 & 2 \\ -3 & 8 \end{vmatrix}$

21. $\begin{vmatrix} -6 & -3 \\ -8 & -4 \end{vmatrix}$

22. $\begin{vmatrix} 9 & 3 \\ -6 & 2 \end{vmatrix}$

Solve each of the following systems of equations by using determinants.

23. $3X - 5Y = 3$
    $4X - 3Y = 15$

24. $5P + 3Q = 7$
    $9P - Q = 3$

25. $15R + 7S = 37$
    $9R - 10S = 8$

26. $20A + 23B = 11$
    $14A - 13B = 95$

27. $11T + 5U = 53$
    $31T - 14U = 37$

28. $6K + 3L = 9$
    $8K + 4L = 7$

Solve each of the following systems of equations by first making the substitutions: $U = \dfrac{1}{A}$ and $V = \dfrac{1}{B}$.

29. $\dfrac{2}{A} + \dfrac{3}{B} = 2$

$\dfrac{4}{A} - \dfrac{3}{B} = 1$

30. $\dfrac{7}{A} - \dfrac{4}{B} = 5$

$\dfrac{5}{A} - \dfrac{2}{B} = 4$

31. $\dfrac{2}{A} - \dfrac{1}{B} = 8$

$\dfrac{3}{A} - \dfrac{2}{B} = 19$

32. $\dfrac{3}{A} - \dfrac{2}{B} = 4$

$\dfrac{2}{A} + \dfrac{4}{B} = 1$

Solve each of the following systems of equations algebraically.

33. $A + B + 2C = 9$
$A - B + C = 5$
$3A + 2B - C = 12$

34. $2K - L - M = 8$
$4K + 3L = 0$
$5K - 7L + 2M = -16$

35. $R - 2S + 3T = -1$
$2R - 3S + T = 11$
$3R - S + 2T = 8$

36. $2X + 3Y + Z = 4$
$X - 2Y - 3Z = -1$
$3X + Y + Z = 3$

37. $2U + V = -1$
$ 2V + 3W = -1$
$4U + 2W = 4$

38. $3I + 4J - K = -3$
$4I + 5J + 7K = 14$
$I - 2J - 2K = 9$

Evaluate each of the following determinants by expanding along any row or column.

39. $\begin{vmatrix} 1 & 2 & -1 \\ -3 & -5 & 2 \\ 4 & 1 & -3 \end{vmatrix}$

40. $\begin{vmatrix} 3 & 2 & -1 \\ 7 & 4 & -1 \\ 2 & -3 & 1 \end{vmatrix}$

41. $\begin{vmatrix} -1 & -7 & 3 \\ 4 & -2 & 4 \\ 0 & -3 & 5 \end{vmatrix}$

42. $\begin{vmatrix} 6 & -5 & -7 \\ -1 & 2 & 4 \\ 2 & -3 & 1 \end{vmatrix}$

43. $\begin{vmatrix} 5 & -1 & 0 \\ 2 & 6 & -2 \\ 3 & -4 & 1 \end{vmatrix}$

44. $\begin{vmatrix} -1 & 1 & -2 \\ 3 & -2 & 0 \\ -2 & 0 & -1 \end{vmatrix}$

Solve each of the following systems of equations by using determinants.

45. $2X - 3Y + 4Z = 8$
$3X + 4Y - 5Z = -4$
$4X - 5Y + 6Z = 12$

46. $A - B + C = 1$
$2A + 3B + 2C = 8$
$4A + 3B - 2C = -2$

Sec. 6–9. Chapter Review

47. $2R - S + 3T = 4$
    $R + 3S + 3T = -2$
    $3R + 2S - 6T = 6$

48. $2K + 2L + 3M = 0$
    $3K + L + 4M = 21$
    $-K - 3L + 7M = 15$

49. $2I + J + K = 1$
    $I + 2J + K = -5$
    $I + J + 2K = 4$

50. $6U + 4V + 3W = 36$
    $6U - 5V + 3W = 9$
    $18U + 16V - 15W = 24$

For each of the following, set up an appropriate system of equations and solve by any appropriate method.

51. A motorboat travels 6 miles down a stream in 20 minutes and requires 30 minutes for the return trip upstream. What is the rate of the boat in still water, and what is the rate of the current?

52. A man owns two stores. Two years ago one store earned 10% of the investment in it while the other lost 5%, giving the man a net gain of $6,250. Last year the first store earned 8% while the second earned 6%, giving the man a net gain of $9,500. What is the value of each store?

53. The length of each leg of an isosceles triangle is $2\frac{1}{4}$ times the length of the base. What is the length of each side of the triangle if the perimeter is 38 inches?

54. The total number of businesses that failed during the past two years is 29,460. If there were 460 more failures last year than the year before, how many businesses failed in each of the two years?

55. A man has two investments, one paying 5% interest and the other 6%. His annual return on these investments is $210. If the 5% investment paid 6% and vice versa, the man would have an annual income of $230. How much money is invested at each rate?

56. A shipment of 3 dozen resistors and 5 dozen capacitors costs $189. Another shipment of 4 dozen resistors and 7 dozen capacitors costs $261.60. What is the cost of a single resistor and capacitor?

57. An equation of the type $AX + BY + CZ = D$ represents a plane when graphed. Three planes can intersect in a single point, a line, or a plane. This intersection would be the solution of the system of three equations which represent the three planes. Find the point of intersection of:

$$5X + 4Y - 2Z = 7$$
$$2X - 3Y + Z = 0$$
$$X + 8Y - 5Z = 4$$

58. Kirchhoff's laws may be applied to different electric circuits. If these laws are applied to the circuit shown, the following system of equations is obtained (I_1, I_2, and I_3 represent currents in amperes):

$$I_1 - I_2 - I_3 = 0$$
$$4I_1 + 5I_2 = -2$$
$$ -5I_2 + 3I_3 = 2$$

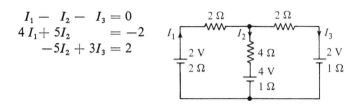

Find the three currents.

59. In a certain electric circuit there are three resistors. The sum of their resistances is 22 ohms. The sum of the first two resistances is 2 ohms more than the third resistance. Three times the first resistance is 1 ohm more than twice the second resistance. Find the resistances of the three resistors.

60. To minimize costs and complete a 35-mile section of roadway as soon as possible, the work was subcontracted to three different construction companies. The portion assigned company A was $\frac{1}{2}$ the length of the portion assigned company B. The portion assigned company C was 5 miles shorter than the portion assigned company B. What length of roadway was assigned to each company?

chapter 7

Introduction to Trigonometry

7–1. Angles and Their Measure

The word *trigonometry* literally means "triangle measure." Thus, the field of trigonometry is concerned with the properties, measure, and uses of different types of triangles. This field of study is important to us since many technological and scientific problems can be represented using some type of triangle, and hence, depend for their solution on our ability to work with these triangles. Before we can solve such problems, however, we must be thoroughly familiar with some more basic concepts. In this section we will look at one of these concepts—the concept of an angle.

An angle is generated by the rotation of a ray about its endpoint from an initial (*starting*) *position to a terminal* (*ending*) *position*. A ray is that part of a line to one side of a fixed point (including the point) on the line. The initial position of the ray is called the *initial side* of the angle and the terminal position of the ray is called the *terminal side* of the angle. The fixed point about which the rotation occurs is called the vertex (see Figure 7–1.1).

In generating angles, the direction of the rotation is extremely important. If the rotation is counterclockwise (↺), the angle is *positive*. If the rotation is clockwise (↻), the angle is *negative*.

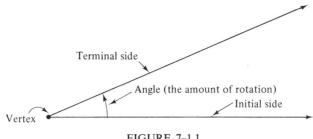

FIGURE 7-1.1

Example A:
1.

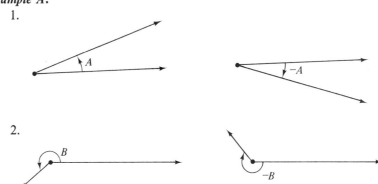

2.

Angles may be measured using either *degress* or *radians*. A degree (°) is defined as $\frac{1}{360}$ of a complete rotation. Therefore, one complete rotation from some initial position back to that same position is equal to 360°.

Example B:

Degrees may be further broken down into minutes and seconds. A *minute* (') is defined as $\frac{1}{60}$ of a degree. Thus, 1° = 60'. A *second* (") is defined as $\frac{1}{60}$ of a minute. Thus, 1' = 60". Combining these two results, we obtain 1° = 60' = 3600". To change from a larger unit to a smaller unit (i.e., degrees to minutes or minutes to seconds), we multiply by 60. To change from a smaller unit to a larger unit (i.e., seconds to minutes or minutes to degrees), we divide by 60.

Example C:
1. Change .7° to minutes.

$$.7 \times 60 = 42$$

Therefore, .7° = 42'.

Sec. 7–1. Angles and Their Measure   183

2. Change 3' to seconds.
$$3 \times 60 = 180$$
Therefore, $3' = 180''$.

3. In the angle 41.6°, change .6° to minutes.
$$.6 \times 60 = 36$$
Therefore, $41.6° = 41°36'$.

4. Change 48" to minutes.
$$48 \div 60 = .8$$
Therefore, $48'' = .8'$.

5. Change 24.3' to degrees.
$$24.3 \div 60 = .405$$
Therefore, $24.3' = .405°$.

6. In the angle 26°15', change 15' to degrees.
$$15 \div 60 = .25$$
Therefore, $26°15' = 26.25°$.

A *radian* is defined as the measure of an angle, with its vertex at the center of a circle, that intercepts an arc on the circle equal in length to the radius. Since the circumference of a circle equals $2\pi R$ (2π times the radius), and since each time the radius is measured on the circumference we obtain an angle of 1 radian, one complete rotation is equal to 2π radians (see Figure 7–1.2). π is

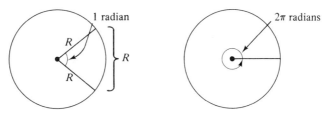

FIGURE 7–1.2

an irrational number approximately equal to 3.1416. Therefore, 2π is approximately equal to 6.2832. The property of radians that makes them extremely useful for measuring angles, especially in applied problems, is that they are just numbers with no associated dimension (such as degrees, minutes, or seconds). Therefore, operations involving radians do not affect the dimensional units in any problem.

Since one complete rotation is equal to 360° or $2\pi(6.283)$ radians, we have the basic equation:

$$360° = 2\pi \text{ radians}$$

If we divide both sides of this equation by 2π, we obtain: 1 radian $= \dfrac{360°}{2\pi} = \dfrac{180°}{\pi} \approx 57.3°$. Therefore, to change from degrees to radians, we divide by $57.3° \left(\dfrac{180°}{\pi}\right)$. To change from radians to degrees, we multiply by $57.3° \left(\dfrac{180°}{\pi}\right)$.

Example D:
1. Change 72.8° to radians.

$$72.8° \div 57.3° = 1.27$$

Therefore, 72.8° = 1.27 radians.
2. Change 265.7° to radians.

$$265.7° \div 57.3° = 4.64$$

Therefore, 265.7° = 4.64 radians.
3. Change 315° to radians in terms of π.

$$315° \div \frac{180°}{\pi} = 315° \times \frac{\pi}{180°} = \frac{315}{180}\pi = 1.75\pi$$

Therefore, 315° = 1.75π radians.
4. Change 163.5° to radians in terms of π.

$$163.5° \div \frac{180°}{\pi} = 163.5° \times \frac{\pi}{180°} = \frac{163.5}{180}\pi = .91\pi$$

Therefore, 163.5° = .91π radians.

Example E:
1. Change 3 radians to degrees.

$$3 \times 57.3° = 171.9°$$

Therefore, 3 radians = 171.9°.
2. Change 5.9 radians to degrees.

$$5.9 \times 57.3° = 338.07°$$

Therefore, 5.9 radians = 338.07°.
3. Change .7π radians to degrees.

$$.7\pi \times \frac{180°}{\pi} = .7 \times 180° = 126°$$

Therefore, .7π radians = 126°.

Sec. 7-1. Angles and Their Measure

4. Change 4.1π radians to degrees.

$$4.1\pi \times \frac{180°}{\pi} = 4.1 \times 180° = 738°.$$

Therefore, 4.1π radians $= 738°$.

Figure 7-1.3 shows the relationship between degrees and radians for one complete rotation.

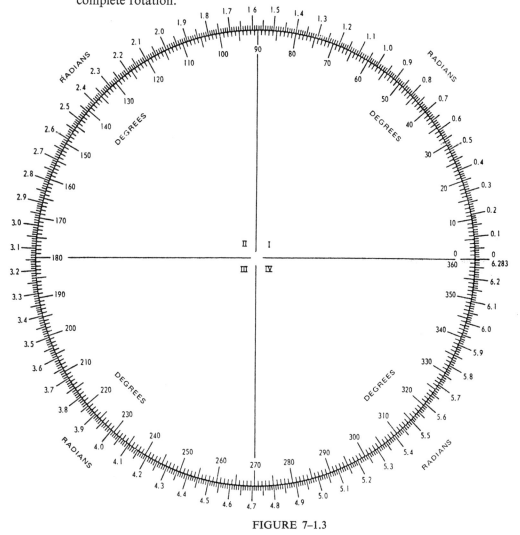

FIGURE 7-1.3

Besides understanding the basic concept of what an angle is and how it may be measured, we should be aware of the different types of angles that might be encountered. Following is a list of different types of angles with a brief description of each.

Coterminal Angles: Angles having the same initial and terminal sides.
A and *B* are coterminal.

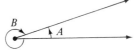

A and *B* are coterminal.

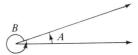

Acute Angle: An angle between 0° and 90° $\left[\text{between 0 and } 1.57\left(\frac{\pi}{2}\right) \text{ radians}\right]$.
Angle *A* is acute.

Obtuse Angle: An angle between 90° and 180° $\left[\text{between } 1.57\left(\frac{\pi}{2}\right) \text{ and } 3.14(\pi) \text{ radians}\right]$.
Angle *B* is obtuse.

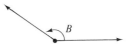

Right Angle: An angle equal to 90° $\left[\text{equal to } 1.57\left(\frac{\pi}{2}\right) \text{ radians}\right]$.
Angle *C* is a right angle.

Straight Angle: An angle equal to 180° [equal to 3.14(π) radians].
Angle *D* is a straight angle.

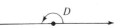

Adjacent Angles: Angles having a common side and vertex.
A and *B* are adjacent angles.

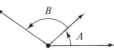

Vertical Angles: Angles formed by intersecting lines. They are equal.
X and *Y* are vertical angles.

Sec. 7–1. Angles and Their Measure

Alternate Interior Angles: Angles formed by a line intersecting a pair of parallel lines. They are equal.

P and Q are alternate interior angles.

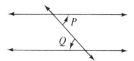

Complementary Angles: Two acute angles whose sum is 90° $\left(1.57 \text{ or } \frac{\pi}{2} \text{ radians}\right)$.

A and B are complementary angles.

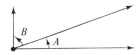

Supplementary Angles: Two positive angles whose sum is 180° (3.14 or π radians).

X and Y are supplementary angles.

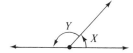

Example F:

1. Find the smallest positive and smallest negative angles that are coterminal with 72°. The smallest positive coterminal angle is obtained by adding 360° to the given angle. Therefore, 360° + 72° = 432°. The smallest negative coterminal angle is obtained by subtracting the given angle from 360° and expressing the answer as a negative angle. Therefore, 360° − 72° = 288°, which we express as −288°. The answers are 432° and −288°. That is, the angles 72°, 432°, and −288° all have the same initial and terminal sides.

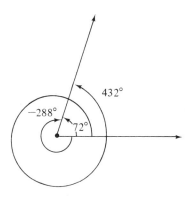

2. Find the smallest positive and smallest negative angles that are coterminal with 2.1 (radians). 6.28 + 2.1 = 8.38. 6.28 − 2.1 = 4.18 which we express as −4.18. Therefore, the answers are 8.38 and −4.18. That is, the angles 2.1, 8.38, and −4.18 (all in radians) have the same initial and terminal sides.

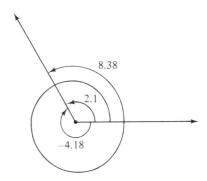

Example G:

Lines 1 and 2 are parallel.

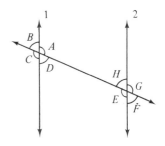

In this diagram:
1. Angles B, D, F, and H are acute.
2. Angles A, C, E, and G are obtuse.
3. Angles A and B, B and C, C and D, D and A, E and F, F and G, G and H, and H and E are pairs of adjacent angles.
4. Angles A and C, B and D, E and G, and F and H are pairs of vertical angles.
5. Angles A and E and D and H are pairs of alternate interior angles.
6. Angles A and B, B and C, C and D, D and A, E and F, F and G, G and H, and H and E are pairs of supplementary angles. (There are others).

Exercises

1. Change .32° to minutes.
2. Change .86° to minutes.
3. Change .6 minutes to seconds.
4. Change 13 minutes to seconds.

Sec. 7-1. Angles and Their Measure

5. In the angle 142.8°, change .8° to minutes.
6. In the angle 274.2°, change .2° to minutes.
7. Change 14′ to degrees.
8. Change 72.9′ to degrees.
9. Change 27″ to minutes.
10. Change 48″ to minutes.
11. In the angle 56°24′, change 24′ to degrees.
12. In the angle 192°40′, change 40′ to degrees.

Express each of the following angles in degrees.

13. .8   14. .94   15. 3.7   16. 4.3
17. 5.28   18. 2.71   19. .6π   20. 1.3π
21. $\frac{1}{5}\pi$ 22. $\frac{2}{3}\pi$ 23. 4.61π   24. 3.59π

Express each of the following angles in radians without using π in the expression.

25. 72.6°   26. 98.3°   27. 131.4°
28. 220.5°   29. 308.2°   30. 154.7°

Express each of the following angles in radians in terms of π.

31. 225°   32. 130°   33. 75°
34. 405°   35. 41.3°   36. 191.8°

Find the smallest positive and smallest negative coterminal angle for each of the following.

37. 74.5°   38. 123.2°   39. 98°20′   40. 210°35′
41. 4.83   42. 6.12   43. .5π   44. .1π
45.

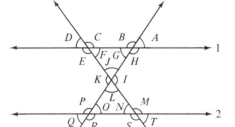

Lines 1 and 2 are parallel.

From the diagram on the preceding page, list:
- (a) 3 acute angles.
- (b) 3 obtuse angles.
- (c) 3 pair of adjacent angles.
- (d) 3 pair of vertical angles.
- (e) 3 pair of alternate interior angles.
- (f) 3 pair of supplementary angles.

46.

Line segments 1 and 2 are parallel.

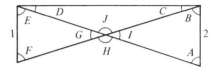

From the above diagram, list:
- (a) 3 acute angles.
- (b) 2 obtuse angles.
- (c) 3 pair of adjacent angles.
- (d) 2 pair of vertical angles.
- (e) 2 pair of alternate interior angles.
- (f) 3 pair of complementary angles.

7–2. Introduction to Triangles

At the beginning of the last section we stated that, basically, trigonometry involves the properties and uses of triangles. In this section we will look at the different types of triangles and some general properties that apply to all triangles.

A *triangle* is a geometric figure of three sides, all of which are connected, and three angles whose sum is 180°. Following is a list of the different types of triangles.

Equilateral Triangle: Triangle in which all three sides and all three angles are equal.

Isosceles Triangle: Triangle in which two angles are equal and the sides opposite those angles are equal.

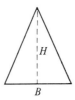

Right Triangle: Triangle in which one angle is 90°. If a triangle is not a right triangle it is called *oblique*.

Pythagorean Theorem: $\left.\begin{array}{l} C^2 = A^2 + B^2 \\ C = \sqrt{A^2 + B^2} \end{array}\right\}$ equivalent

Scalene Triangle: Triangle in which no two angles or sides are equal.

 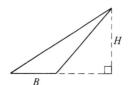

Similar Triangles: Triangles in which corresponding angles are equal and corresponding sides are proportional.

Congruent Triangles: Triangles in which corresponding angles and sides are equal.

For all triangles in a plane: The sum of the three interior angles is 180°. Area $= \frac{1}{2}BH$.

Example A: In the triangle given, find C.

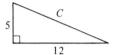

Using the Pythagorean Theorem, $C = \sqrt{5^2 + 12^2}$
$= \sqrt{25 + 144}$
$= \sqrt{169}$
$= 13$

Example B: In the similar triangles given, find the missing side in each triangle.

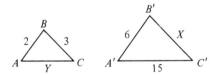

In similar triangles, corresponding sides are proportional. Therefore,

$$\frac{6}{2} = \frac{X}{3} = \frac{15}{Y}.$$

Since $\frac{6}{2} = 3$, then $X = 9$ since $\frac{9}{3} = 3$, and $Y = 5$ since $\frac{15}{5} = 3$.

Example C: In the triangle given, find the area.

Area $= \frac{1}{2} BH$
$= \frac{1}{2}(4'')(7'')$
$= 14$ sq. in.

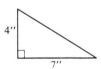

Example D: In the triangle given, find the area.

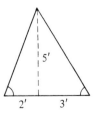

Base $= 2' + 3' = 5'$
Area $= \frac{1}{2} BH = \frac{1}{2}(5')(5') = 12.5$ sq ft.

Sec. 7–2. Introduction to Triangles

Exercises

1. In the isosceles triangle given, find angle X and side B.

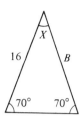

2. In the right triangle given, find angles A and B and side X.

3. In the equilateral triangle given, find all the remaining parts.

4. In the similar triangles given, find the two remaining sides in the larger triangle.

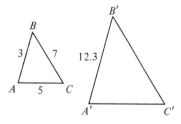

5. In the right triangle given, find side C and angle X.

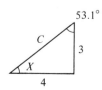

6. In the isosceles triangle given, find angle X and side A.

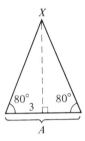

For each of the following triangles, find the area.

7.

8.

9.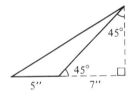

10.

11. This triangle is equilateral.

12. This triangle is isosceles.

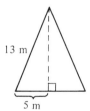

7–3. Angles in the Rectangular Coordinate System

Many problems that we encounter and which involve angles and triangles are often set up and analyzed with reference to the rectangular coordinate system. When angles are measured in the rectangular coordinate system, we usually start from what is called *standard position*. An angle is in standard position if its vertex is at the origin and its initial side lies along the positive horizontal axis. The position of the terminal side then determines the angle. If the terminal side lies within one of the quadrants, the angle is referred to as a first-, second-, third-, or fourth-quadrant angle, depending on where the terminal side lies. If the

Sec. 7-3. Angles in the Rectangular Coordinate System

terminal side lies along one of the axes, the angle is called a *quadrantal angle*. From this definition we should realize that quadrantal angles are 0°, 90°, 180°, 270°, and 360°, and any multiples of these.

Example A:

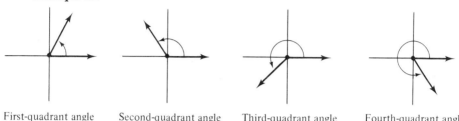

First-quadrant angle   Second-quadrant angle   Third-quadrant angle   Fourth-quadrant angle

All of the above angles are in standard position.

Example B:

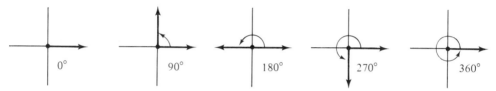

0°   90°   180°   270°   360°

Each of the above angles is a quadrantal angle.

From Example B we can determine the range of values for angles in each quadrant. When measured from standard position, the values of first-quadrant angles are between 0° and 90°, the values of second-quadrant angles are between 90° and 180°, the values of third-quadrant angles are between 180° and 270°, and the values of fourth-quadrant angles are between 270° and 360°. Figure 7-3.1 shows these "boundary values" for each quadrant in both degrees and radians.

Knowing these ranges of values for angles in each quadrant, we can quickly represent (at least approximately) any angle in standard position.

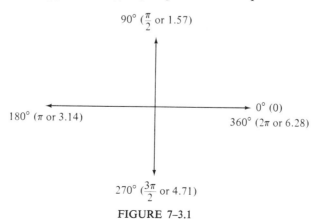

90° ($\frac{\pi}{2}$ or 1.57)

180° (π or 3.14)

0° (0)

360° (2π or 6.28)

270° ($\frac{3\pi}{2}$ or 4.71)

FIGURE 7-3.1

Example C: Represent each of the following angles in the rectangular coordinate system: 45°, 160°, 450°, 210°, −60°, −90°, 2.5, 5.1.

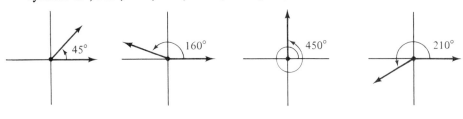

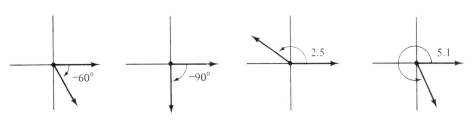

Exercises Represent each of the following angles in the rectangular coordinate system.

1. 30°   2. 110°   3. 320°   4. 245°
5. 420° 6. 190°   7. −75°   8. −120°
9. −200° 10. −350° 11. .6   12. 4.2
13. 2.9 14. 7.85 15. −2.5 16. −3.6
17. $\frac{\pi}{4}$ 18. $\frac{5\pi}{3}$ 19. $\frac{7\pi}{6}$ 20. 3π

Determine the approximate value of each of the following angles. (Express answers in both degrees and radians).

21. 22.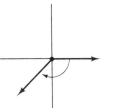 23.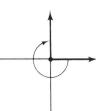

24.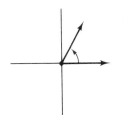 25.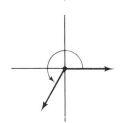 26.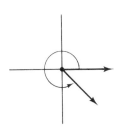

7–4. The Trigonometric Functions

If an angle is placed in standard position, and lines are drawn perpendicular to the horizontal axis from two different points on the terminal side of the angle, the diagram in Figure 7–4.1 is obtained. In this diagram, we have two similar

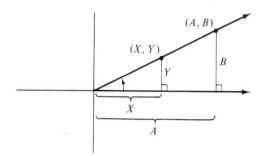

FIGURE 7–4.1

triangles, both with one vertex at the origin. We have previously pointed out that corresponding parts of similar triangles are proportional. Thus, ratios involving corresponding sides of these triangles will be equal. For example, $\frac{Y}{X} = \frac{B}{A}$. This will be true for the ratio of any two sides of these triangles. What this means is that no matter which point we pick on the terminal side of the angle from which to construct a line perpendicular to the horizontal axis, the ratios of corresponding sides of the different triangles formed will be equal.

Example A:

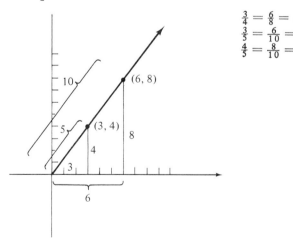

$\frac{3}{4} = \frac{6}{8} = .75$
$\frac{3}{5} = \frac{6}{10} = .6$
$\frac{4}{5} = \frac{8}{10} = .8$

From Example A, we should see that since the triangles formed are right triangles, the Pythagorean Theorem may be used to find the third side in each triangle (5 and 10). This third side, the line from the origin to a point on the terminal side of an angle, is called the *radius vector* and is denoted by R.

For any angle in standard position, then, we may select any point on the terminal side of that angle, construct a line from that point perpendicular to the horizontal axis, and obtain the diagram in Figure 7–4.2.

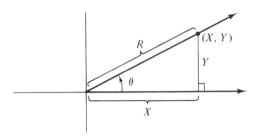

FIGURE 7–4.2

NOTE: While the terminal side is drawn in the first quadrant, it may, in fact, lie anywhere in the coordinate system.

From the right triangle in Figure 7–4.2, we may set up six ratios, each involving two sides of the triangle. These ratios are $\frac{Y}{R}, \frac{X}{R}, \frac{Y}{X}, \frac{R}{Y}, \frac{R}{X}$, and $\frac{X}{Y}$. The values of these ratios depend on the terminal side of the angle and hence, on the angle itself. Also, for any given angle, these values will be unique (i.e., there is no more than one value for any given ratio once an angle is specified). Therefore, according to our definition in Chapter 4, these ratios are functions of the angle. These functions are called the *trigonometric functions* and are defined as follows:

$$\frac{Y}{R} = \frac{\text{ordinate of point } P}{\text{radius vector of point } P} = \text{Sine } \theta \ (\text{Sin } \theta)$$

$$\frac{X}{R} = \frac{\text{abscissa of point } P}{\text{radius vector of point } P} = \text{Cosine } \theta \ (\text{Cos } \theta)$$

$$\frac{Y}{X} = \frac{\text{ordinate of point } P}{\text{abscissa of point } P} = \text{Tangent } \theta \ (\text{Tan } \theta)$$

$$\frac{R}{Y} = \frac{\text{radius vector of point } P}{\text{ordinate of point } P} = \text{Cosecant } \theta \ (\text{Csc } \theta)$$

$$\frac{R}{X} = \frac{\text{radius vector of point } P}{\text{abscissa of point } P} = \text{Secant } \theta \ (\text{Sec } \theta)$$

$$\frac{X}{Y} = \frac{\text{abscissa of point } P}{\text{ordinate of point } P} = \text{Cotangent } \theta \ (\text{Cot } \theta)$$

In these ratios, R is always considered positive and the ratios are undefined if the denominators are equal to zero.

Sec. 7-4. The Trigonometric Functions

So if we have an angle in standard position and know a point on the terminal side of the angle, we may find any or all of the trigonometric functions for that angle.

Example B: Find the trigonometric functions of the angle with a terminal side passing through the point (4, 3).

$X = 4$ and $Y = 3$
$R = \sqrt{3^2 + 4^2} = \sqrt{9 + 16}$
$ = \sqrt{25} = 5$
Sin $\theta = \frac{3}{5}$ Csc $\theta = \frac{5}{3}$
Cos $\theta = \frac{4}{5}$ Sec $\theta = \frac{5}{4}$
Tan $\theta = \frac{3}{4}$ Cot $\theta = \frac{4}{3}$

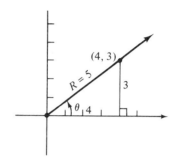

Example C: Find the trigonometric functions of the angle with a terminal side passing through the point (5, −12).

$X = 5$ and $Y = -12$
$R = \sqrt{5^2 + (-12)^2} = \sqrt{25 + 144}$
$ = \sqrt{169} = 13$
Sin $\theta = -\frac{12}{13}$ Csc $\theta = -\frac{13}{12}$
Cos $\theta = \frac{5}{13}$ Sec $\theta = \frac{13}{5}$
Tan $\theta = -\frac{12}{5}$ Cot $\theta = -\frac{5}{12}$

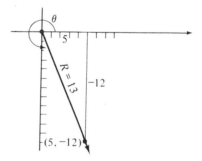

From our definition of the trigonometric functions and from Examples B and C, we should note that the Sin and Csc functions, the Cos and Sec functions, and the Tan and Cot functions are *reciprocals* of one another. That is:

$$\text{Sin } \theta = \frac{1}{\text{Csc } \theta} \quad \text{or} \quad \text{Csc } \theta = \frac{1}{\text{Sin } \theta}$$

$$\text{Cos } \theta = \frac{1}{\text{Sec } \theta} \quad \text{or} \quad \text{Sec } \theta = \frac{1}{\text{Cos } \theta}$$

$$\text{Tan } \theta = \frac{1}{\text{Cot } \theta} \quad \text{or} \quad \text{Cot } \theta = \frac{1}{\text{Tan } \theta}$$

Example D: If Sin $\theta = \frac{5}{6}$, find Tan θ.

Since $\sin \theta = \dfrac{Y}{R} = \dfrac{5}{6}$, $Y = 5$, and $R = 6$, $X = \sqrt{6^2 - 5^2} = \sqrt{36 - 25} = \sqrt{11}$. Therefore, $\tan \theta = \dfrac{Y}{X} = \dfrac{5}{\sqrt{11}}$ or 1.51.

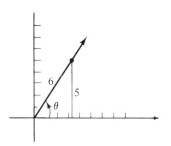

Example E: If $\tan \theta = \dfrac{7}{10}$, find $\csc \theta$.

Since $\tan \theta = \dfrac{Y}{X} = \dfrac{7}{10}$, $Y = 7$, and $X = 10$, $R = \sqrt{7^2 + 10^2} = \sqrt{49 + 100} = \sqrt{149}$. Therefore, $\csc \theta = \dfrac{R}{Y} = \dfrac{\sqrt{149}}{7}$ or 1.744.

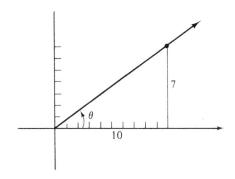

Exercises

For each of the following, the points listed are on the terminal side of an angle. Show that the indicated functions are the same for each point.

1. (4, 3) and (8, 6), $\cos \theta$ and $\tan \theta$.
2. (5, 7) and (15, 21), $\sin \theta$ and $\sec \theta$.
3. (6, 9) and (9, 13.5), $\tan \theta$ and $\csc \theta$.
4. (8, 5) and (20, 12.5), $\cos \theta$ and $\cot \theta$.

For each of the following, find the trigonometric functions of the angles whose terminal sides pass through the given points.

5. (3, 4)
6. (5, 12)
7. (7, 2)
8. (4, 9)
9. (−4, 3)
10. (2, −5)
11. (−3, −6)
12. (−12, −5)
13. (−1, 1)
14. (3, −2)

For each of the following, use the given trigonometric functions to find the required trigonometric functions.

15. If $\sin \theta = \dfrac{4}{9}$, find $\sec \theta$.
16. If $\cos \theta = \dfrac{6}{11}$, find $\tan \theta$.

Sec. 7-5. *Functions of Acute Angles*

17. If $\tan \theta = \frac{9}{2}$, find $\csc \theta$.
18. If $\csc \theta = \frac{5}{3}$, find $\cot \theta$.
19. If $\sec \theta = \frac{12}{7}$, find $\sin \theta$.
20. If $\cot \theta = \frac{1}{10}$, find $\cos \theta$.

7-5. Functions of Acute Angles

At this point we are able to calculate the trigonometric functions of any angle that is in standard position and for which we know one point on the terminal side. In the majority of the problems we encounter, however, we are usually given an angle in degrees or radians and required to find one or more functions of this angle. Therefore, we must be able to find the trigonometric functions of angles given in degrees or radians. In this section, we will discuss this procedure for *acute angles*. These are angles between 0° (0 radians) and 90° $\left(\frac{\pi}{2}\right.$ or 1.57 radians).

One method of finding trigonometric functions of acute angles is to start from standard position and use a protractor to make a scale drawing. We would then select a point on the terminal side of the angle, measure the lengths of X, Y, and R, and then use the proper ratios to determine the functions.

Example A: Find the trigonometric functions of 40°. Using a protractor to measure the angle, we select a point on the terminal side and then by measurement obtain $X = 1.2''$, $Y = 1''$, and $R = 1.56''$.

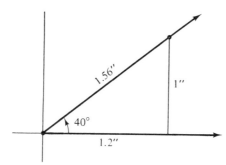

$$\sin 40° = \frac{1}{1.56} = .641 \qquad \csc 40° = \frac{1.56}{1} = 1.56$$

$$\cos 40° = \frac{1.2}{1.56} = .769 \qquad \sec 40° = \frac{1.56}{1.2} = 1.3$$

$$\tan 40° = \frac{1}{1.2} = .833 \qquad \cot 40° = \frac{1.2}{1} = 1.2$$

This procedure is impractical because it is time-consuming to make a scale drawing each time we are trying to find a trigonometric function, and even if a complete table of functions of acute angles were determined by this method, the results would be only approximate. Therefore, other methods for finding trigonometric functions are necessary.

We should realize that whatever method is used, two requirements must be met. There should be a complete list of the functions of acute angles so that they do not have to be re-calculated with each problem, and the values should be as precise as is necessary in a given situation. Such a list appears at the end of this text in Table 2. This list includes all angles at 10' intervals from 0° to 90° inclusive. These values were obtained by using *power series* and are much more precise than values obtained by using a scale drawing. While we have listed the functions with four-decimal-place precision, they may be expressed with as many decimal places as are necessary.

When this table is used, angles from 0° to 45° inclusive are listed on the left and read down, while the functions for these angles are listed along the top. Angles from 45° to 90° inclusive are listed on the right and read up, while the functions for these angles are listed along the bottom. We should also note that each angle is given in both degrees and radians.

Example B:
1. Find Sin 23°. Locating 23° on the left, under the Sin column at the top, we read Sin 23° = .3907.
2. Find Tan 57°. Locating 57° on the right, above the Tan column at the bottom, we read Tan 57° = 1.540.
3. Find Cos 41°20'. Locating 41°20' on the left, under the Cos column at the top, we read Cos 41°20' = .7509.
4. Find Sin 1.213. Locating 1.213 on the right, above the Sin column at the bottom, we read Sin 1.213 = .9367.
5. Find Csc .192. Locating .192 on the left, under the Csc column at the top, we read Csc .192 = 5.241.

If a function is given and the corresponding angle required, the procedure would be reversed.

Example C:
1. If Cos θ = .9703, find θ. Locating .9703 in the Cos column at the top, we read the corresponding angle on the left. Thus, Cos 14° (.2443) = .9703.
2. If Tan θ = .6330, find θ. Locating .6330 in the Tan column at the top, we read the corresponding angle on the left. Thus, Tan 32°20' (.5643) = .6330.
3. If Sec θ = 1.509, find θ. Locating 1.509 in the Sec column at the bottom, we read the corresponding angle on the right. Thus, Sec 48°30' (.8465) = 1.509.

Sec. 7–5. Functions of Acute Angles

If functions of acute angles that are not listed in the table are desired, a method called *interpolation* must be used. This method assumes that if a particular angle lies between two of those listed in the table, then the functions of that angle are at the same proportional distance between the functions listed and vice versa.

Example D: Find Sin 36°25′. We find this value by setting up and solving a proportion obtained from the following diagram.

$$10'\left[5'\left[\begin{array}{l}\text{Sin } 36°20' = .5925 \\ \text{Sin } 36°25' = \ ? \end{array}\right]x\right].0023$$
$$\text{Sin } 36°30' = .5948$$

$$\frac{5}{10} = \frac{x}{.0023} \longrightarrow 10X = .0115 \longrightarrow X = .00115.$$

Therefore, Sin 36°25′ = .5925 + .00115 = .59365. In other words, since 36°25′ is one half ($\frac{5}{10}$) of the way between 36°20′ and 36°30′, we assume that Sin 36°25′ is one half of the way between Sin 36°20′ and Sin 36°30′.

Example E: Find Cos 41°43′. Again we use a proportion.

$$10'\left[3'\left[\begin{array}{l}\text{Cos } 41°40' = .7470 \\ \text{Cos } 41°43' = \ ? \end{array}\right]x\right].0029$$
$$\text{Cos } 41°50' = .7451$$

$$\frac{3}{10} = \frac{x}{.0029} \longrightarrow 10X = .0087 \longrightarrow X = .00087$$

Therefore, Cos 41°43′ = .7470 − .00087 = .74613. In other words, since 41°43′ is three tenths of the way between 41°40′ and 41°50′, we assume that Cos 41°43′ is three tenths of the way between Cos 41°40′ and Cos 41°50′.

Note that when a function is *increasing* (Example D), the calculated value is *added* to the smaller value from the table, whereas when a function is *decreasing* (Example E), the calculated value is *subtracted* from the larger value from the table.

Example F: If Tan θ = 2.418, find θ.

$$10'\left[x\left[\begin{array}{l}\text{Tan } 67°30' = 2.414 \\ \qquad\quad ? \ \ = 2.418 \end{array}\right].004\right].020$$
$$\text{Tan } 67°40' = 2.434$$

$$\frac{x}{10} = \frac{.004}{.02} \longrightarrow .02X = .04 \longrightarrow X = \frac{.04}{.02} = 2$$

Therefore, 67°30′ + 2′ = 67°32′. Tan 67°32′ = 2.418.

Example G: If $\cos \theta = .8352$, find θ.

$$10' \left[x \left[\begin{array}{l} \cos 33°20' = .8355 \\ ? = .8352 \\ \cos 33°30' = .8339 \end{array} \right] .0003 \right] .0016$$

$$\frac{x}{10} = \frac{.0003}{.0016} \longrightarrow .0016X = .003 \longrightarrow X = \frac{.003}{.0016} = 1.875 = 2$$

Therefore, $33°20' + 2' = 33°22'$. $\cos 33°22' = .8352$.

Trigonometric functions of acute angles may also be found by using a slide rule or a hand calculator (see Appendix B). These instruments also make use of the tables to which we have just referred. In the case of the slide rule, the tables are "listed" on the different scales. With calculators, the tables are "programmed" into the calculator.

Exercises

Use a scale drawing to find the trigonometric functions of the following angles.

1. 30°   2. 45°   3. 65°   4. 20°

For each of the following, find the value of the trigonometric function.

5. Sin 16°
6. Tan 53°
7. Csc 27°40'
8. Cos 70°10'
9. Cot 7°30'
10. Sec 58°20'
11. Tan 83°50'
12. Csc 22°40'
13. Sin 41.5°
14. Cos 62.5°
15. Tan .448
16. Sec .861
17. Sin 39°47'
18. Cos 56°18'
19. Sec 11°13'
20. Csc 30°16'
21. Sin 66.7°
22. Tan 15.4°
23. Cot .6
24. Cos 1.3

For each of the following, find θ.

25. $\sin \theta = .3256$
26. $\sec \theta = 1.108$
27. $\tan \theta = .3217$
28. $\cos \theta = .8073$
29. $\csc \theta = 1.072$
30. $\tan \theta = .6412$
31. $\sin \theta = .4669$
32. $\cos \theta = .5175$
33. $\cot \theta = 1.213$
34. $\sin \theta = .7585$
35. $\cos \theta = .9942$
36. $\csc \theta = 3.822$
37. $\sin \theta = .7612$
38. $\tan \theta = .6070$
39. $\csc \theta = 1.460$
40. $\cos \theta = .4882$

41. $\sin \theta = .1518$ **42.** $\tan \theta = 2.714$
43. $\sec \theta = 3.542$ **44.** $\cot \theta = 13.41$
45. The formula for finding the direction (θ) of a resultant vector is $\tan \theta = \frac{Y}{X}$, where Y is the vertical component and X is the horizontal component of the vector. Find θ when $Y = 12$ pounds and $X = 5$ pounds.
46. The acceleration of a block sliding down a frictionless inclined plane is given by the equation $a = g \sin \theta$, where g is the acceleration due to gravity and θ is the angle the plane makes with the horizontal. Find a if $g = 32$ ft/sec² and $\theta = 37°45'$.

7–6. Problems Involving Right Triangles

A large number of problems in the technical field either directly or indirectly depend on the solution to a right triangle. This section will deal with the terminology and techniques necessary to solve problems involving a right triangle.

Every triangle has six parts, three angles and three sides (see Figure 7–6.1). When one of the angles is a 90° angle (a right angle), the triangle is called a *right triangle*. The angles (or vertices) of the triangle are usually labeled with large letters, A, B, and C (or any three letters), while the sides opposite these angles are labeled with small letters a, b, and c, respectively. The right angle is usually labeled with a large C, while the side opposite the right angle is called the *hypotenuse* and labeled with a small c.

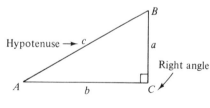

FIGURE 7–6.1

The sum of the angles of a triangle in a plane is 180°. Since one of these angles is 90°, the remaining two angles are acute angles whose sum is 90°. When the sum of two acute angles is 90° the angles are called *complementary angles*. Thus, every right triangle has two complementary angles.

The six trigonometric functions have already been defined in terms of an angle in standard position in the rectangular coordinate system. It is now necessary to define the six trigonometric functions in relation to the sides of a right triangle and still be consistent with the definitions as they have been stated already.

Basic Definition	New Definition
$\sin A = \dfrac{Y}{R}$	$\sin A = \dfrac{a}{c} = \dfrac{\text{side opposite angle } A}{\text{hypotenuse}}$
$\cos A = \dfrac{X}{R}$	$\cos A = \dfrac{b}{c} = \dfrac{\text{side adjacent to angle } A}{\text{hypotenuse}}$
$\tan A = \dfrac{Y}{X}$	$\tan A = \dfrac{a}{b} = \dfrac{\text{side opposite angle } A}{\text{side adjacent to angle } A}$
$\csc A = \dfrac{R}{Y}$	$\csc A = \dfrac{c}{a} = \dfrac{\text{hypotenuse}}{\text{side opposite angle } A}$
$\sec A = \dfrac{R}{X}$	$\sec A = \dfrac{c}{b} = \dfrac{\text{hypotenuse}}{\text{side adjacent to angle } A}$
$\cot A = \dfrac{X}{Y}$	$\cot A = \dfrac{b}{a} = \dfrac{\text{side adjacent to angle } A}{\text{side opposite angle } A}$

The opposite side and the adjacent side are relative to the acute angle with which we are working, while the hypotenuse is always the side opposite the right angle (see Figure 7–6.2).

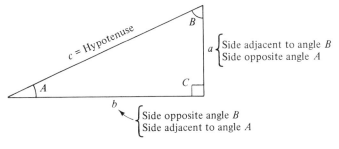

FIGURE 7–6.2

Using our new definition of the trigonometric functions (in terms of opposite, adjacent, and hypotenuse) makes it unnecessary to place an angle in standard position in order to find the trigonometric functions of that angle. Thus, from the above diagram:

Sec. 7-6. Problems Involving Right Triangles

$$\text{Sin } A = \frac{a}{c} \qquad \text{Sin } B = \frac{b}{c}$$

$$\text{Cos } A = \frac{b}{c} \qquad \text{Cos } B = \frac{a}{c}$$

$$\text{Tan } A = \frac{a}{b} \qquad \text{Tan } B = \frac{b}{a}$$

$$\text{Csc } A = \frac{c}{a} \qquad \text{Csc } B = \frac{c}{b}$$

$$\text{Sec } A = \frac{c}{b} \qquad \text{Sec } B = \frac{c}{a}$$

$$\text{Cot } A = \frac{b}{a} \qquad \text{Cot } B = \frac{a}{b}$$

Examining these trigonometric functions for the angles A and B, we should observe another fact. Note that Sin A = Cos B and Cos A = Sin B, Tan A = Cot B and Cot A = Tan B, and Csc A = Sec B and Sec A = Csc B. A and B are complementary angles. Also, the *S*ine and *Co*sine, *T*angent and *Co*tangent, *S*ecant and *Co*secant are all pairs of *cofunctions*. Therefore, we can state that *cofunctions of acute complementary angles are equal*. In other words, Sin A = Cos $(90° - A)$ and Cos A = Sin $(90° - A)$, Tan A = Cot $(90° - A)$ and Cot A = Tan $(90° - A)$, and Sec A = Csc $(90° - A)$ and Csc A = Sec $(90° - A)$.

Example A:
1. Sin $40°$ = Cos $(90° - 40°)$ = Cos $50°$ = .6428
2. Cot $62°$ = Tan $(90° - 62°)$ = Tan $28°$ = .5317
3. Sec $17°$ = Csc $(90° - 17°)$ = Csc $73°$ = 1.046

When solving a right triangle (or any triangle) it is necessary to know two parts of the triangle other than the right angle, and at least one of these two other parts must be a side. When this information is given it is possible to find the remaining three parts of the triangle.

Example B: Solve the right triangle with $A = 34°$ and $b = 16$.
  1. Draw a right triangle with the given values.

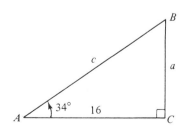

2. Side *a* can be found by using the tangent function since we know the measure of angle *A* and the side adjacent to angle *A*, and *a* is the opposite side.

$$\text{Tan } 34° = \frac{a}{16}$$

$$.6745 = \frac{a}{16}$$

$$.6745(16) = a$$

$$10.792 = a$$

3. Side *c* can be found by using the cosine function since we know the measure of angle *A* and the side adjacent to angle *A*, and *c* is the hypotenuse of the triangle.

$$\text{Cos } 34° = \frac{16}{c}$$

$$.8290 = \frac{16}{c}$$

$$c = \frac{16}{.8290}$$

$$= 19.300$$

NOTE: Side *c* could also be found by using the Pythagorean Theorem.

4. Angle *B* is complementary to *A*.

$$B = 90° - 34°$$

$$= 56°$$

Example C: Solve the right triangle with $c = 10$ and $b = 8$.

1. Draw a right triangle with the given values.

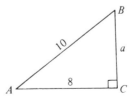

2. Angle *A* can be found by using the cosine function since the two sides that are given are the side adjacent to angle *A* and the hypotenuse.

$$\text{Cos } A = \tfrac{8}{10} = .8000$$

$$A = 36.87°$$

Sec. 7-6. Problems Involving Right Triangles

3. Angle B is complementary to A.

$$B = 90° - 36.87°$$
$$= 53.13°$$

4. Side a can be determined by the tangent function since angle A is known and the adjacent side is also known.

$$\text{Tan } 36.87° = \frac{a}{8}$$
$$.7499 = \frac{a}{8}$$
$$.7499(8) = a$$
$$5.999 = a$$

NOTE: Other trigonometric functions could have been used to find the length of side a, as well as the Pythagorean Theorem.

Example D: A guy wire forms an angle of .75 radians with a vertical pole and exerts a force of 100 kilograms. What is the horizontal force on the pole?

1. Construct a triangle representative of the conditions given in the statement.

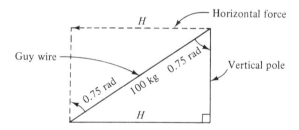

2. The Sine function can be used since the 100-kilogram force represents the hypotenuse of the right triangle and the horizontal force (H) represents the side opposite the acute angle of .75 radians. Thus,

$$\text{Sin } .75 = \frac{H}{100 \text{ kg}}$$
$$.6816(100 \text{ kg}) = H$$
$$68.16 \text{ kg} = H$$

NOTE: It is only necessary to find one part of the right triangle in this problem.

Two angles of importance in applied problems are the angle of elevation and the angle of depression. An *angle of elevation* is an angle formed by a horizontal line and a line of sight above the horizontal. An *angle of depression* is an angle formed by a horizontal line and a line of sight below the horizontal (see Figure 7–6.3).

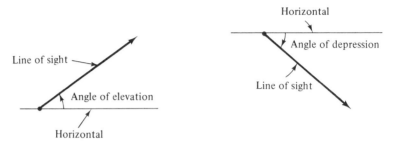

FIGURE 7–6.3

Example E: An angle of depression of a ship sighted from a lighthouse is 15° and the height of the lighthouse is 154 meters. What is the distance from the ship to the lighthouse?

1. Make a representative drawing. Let D represent the distance from the ship to the lighthouse.

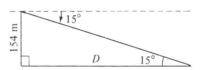

2. Since the side opposite the 15° angle is known to be 154 meters and the side adjacent to the 15° angle is the side the problem calls for, the tangent function can be used to solve this problem.

$$\text{Tan } 15° = \frac{154 \text{ m}}{D}$$

$$D = \frac{154 \text{ m}}{.2679}$$

$$D = 574.84 \text{ m}$$

Exercises

Solve the right triangles with the following given parts.

1. $b = 2.5$, $a = 6$
2. $B = 41°$, $b = 11$
3. $A = .73$, $c = 151$
4. $b = 13$, $a = 12$

Sec. 7-6. Problems Involving Right Triangles

5. $A = 33°12'$, $b = 3.89$
6. $B = 1.71$, $a = 2.4$
7. $c = .0138$, $b = .0871$
8. $a = 1238$, $B = 14°31'$
9. Determine a_3, α, and β.

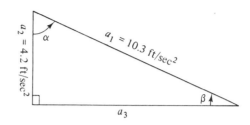

10. Find the distance across the river.

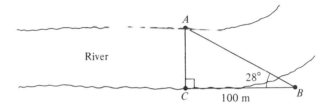

11. Find the distance between the monuments.

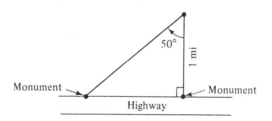

12. An airplane with an air speed of 120 miles per hour is headed directly north. The wind is from the east at 30 miles per hour. What are the ground speed and direction of the airplane?

13. A pilot sights an object on the ground and notes that the angle of depression is 30° and his distance is 1500 meters from the object. Fifteen minutes later the pilot sights the same object and notes that the angle of depression is 45° and his distance is 700 meters from the object. The object was in front of the pilot for both sightings. How far has the pilot traveled between sightings and how fast is the pilot traveling?

14. Determine V_b.

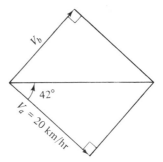

15. The base of an isosceles triangle is 20.3 centimeters and the base angles are 52.3°. Find the equal sides and the altitude.

16. To measure the height of a certain building, a distance of 50 meters is measured from the base of the building, and from this point the angle of elevation of the top of the building is measured to be 54.7°. What is the height of the building?

7–7. Functions of Angles That Are Not Acute

This section will deal with angles that are not acute; that is, angles whose terminal side lies in quadrants II, III, or IV. These angles may be positive or negative.

The six trigonometric functions are basically defined in terms of X, Y, and R. The algebraic signs of X and Y depend on the quadrant in which a point lies, while R is always a positive number ($R > 0$).

The algebraic sign for any one of the six trigonometric functions can be determined by referring to its basic definition as outlined in the following table.

BASIC DEFINITION OF FUNCTION	SIGN OF COMPONENTS IN THE FOUR QUADRANTS				SIGN OF FUNCTION IN THE FOUR QUADRANTS			
	I	II	III	IV	I	II	III	IV
$\sin \theta = \dfrac{Y}{R}$	$Y>0$ $R>0$	$Y>0$ $R>0$	$Y<0$ $R>0$	$Y<0$ $R>0$	+	+	−	−
$\cos \theta = \dfrac{X}{R}$	$X>0$ $R>0$	$X<0$ $R>0$	$X<0$ $R>0$	$X>0$ $R>0$	+	−	−	+
$\tan \theta = \dfrac{Y}{X}$	$Y>0$ $X>0$	$Y>0$ $X<0$	$Y<0$ $X<0$	$Y<0$ $X>0$	+	−	+	−
$\csc \theta = \dfrac{R}{Y}$	$R>0$ $Y>0$	$R>0$ $Y>0$	$R>0$ $Y<0$	$R>0$ $Y<0$	+	+	−	−
$\sec \theta = \dfrac{R}{X}$	$R>0$ $X>0$	$R>0$ $X<0$	$R>0$ $X<0$	$R>0$ $X>0$	+	−	−	+
$\cot \theta = \dfrac{X}{Y}$	$X>0$ $Y>0$	$X<0$ $Y>0$	$X<0$ $Y<0$	$X>0$ $Y<0$	+	−	+	−

From this table we can see that each of the six trigonometric function values is positive in the first quadrant and one other quadrant and negative in the remaining two quadrants. The Sin and Csc are positive in the first and second quadrants and negative in the third and fourth. The Cos and Sec are positive in the first and fourth quadrants and negative in the second and third. The Tan and Cot are positive in the first and third quadrants and negative in the second and fourth.

Consider an angle of 135° which is a quadrant II angle (see Figure 7-7.1). The supplement of 135° is 45°. Thus, a point on the terminal side of the angle

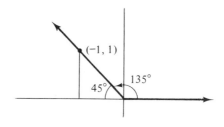

FIGURE 7-7.1

would be $(-1, 1)$. In Section 7-4 it was pointed out that the numerical value of a trigonometric function can be determined if a point on the terminal side of an angle is known. Thus,

$$\text{Tan } 135° = \frac{1}{-1} = -1$$

It should be noted that the absolute value of Tan 135° is equal to Tan 45°, and it should also be noted that the supplementary angle of 135° is 45°. In general, a trigonometric function can be found for a quadrant II angle by finding the trigonometric function of the supplementary angle. That is, if θ_2 represents a second quadrant angle and α represents its supplement, determine the trigonometric function of α and affix the appropriate algebraic sign. If F represents any one of the six basic trigonometric functions, the reference angle for quadrant II is $\alpha = 180° - \theta_2$ and $\mathbf{F}(\boldsymbol{\theta_2}) = \pm \mathbf{F}(180° - \boldsymbol{\theta_2}) = \pm \mathbf{F}(\boldsymbol{\alpha})$ (see Figure 7-7.2).

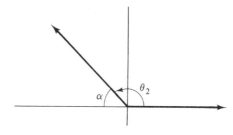

FIGURE 7-7.2

Consider a quadrant III angle of 225° (see Figure 7-7.3). This angle would be a straight angle (180°) plus 45°. Thus, a point on the terminal side of the

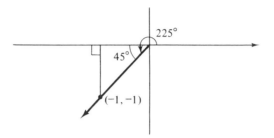

FIGURE 7–7.3

angle would be $(-1, -1)$. Thus,

$$\text{Tan } 225° = \frac{-1}{-1} = 1$$

It should be noted that the absolute value of Tan 225° is equal to Tan 45°. In general, a trigonometric function can be found for a quadrant III angle by finding the trigonometric function of the reference angle (α) and affixing the appropriate algebraic sign. The reference angle for quadrant III is $\alpha = \theta_3 - 180°$ and $F(\theta_3) = \pm F(\theta_3 - 180°) = \pm F(\alpha)$ (see Figure 7-7.4).

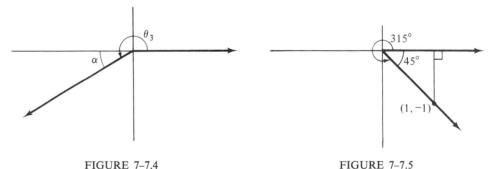

FIGURE 7–7.4   FIGURE 7–7.5

Consider a quadrant IV angle of 315° (see Figure 7-7.5). This angle would be short of a complete revolution by 45°. Thus, a point on the terminal side would be $(1, -1)$. Thus,

$$\text{Tan } 315° = \frac{-1}{1} = -1$$

It should be noted that the absolute value of Tan 315° is equal to Tan 45°. In general, a trigonometric function can be found for a quadrant IV angle by finding the trigonometric function of the reference angle (α) and affixing the appropriate algebraic sign. The reference angle for quadrant IV is $\alpha = 360° - \theta_4$ and $F(\theta_4) = \pm F(360° - \theta_4) = \pm F(\alpha)$ (see Figure 7-7.6).

Sec. 7–7. Functions of Angles That Are Not Acute 215

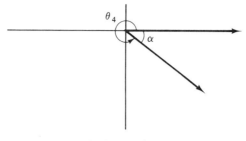

FIGURE 7–7.6

Example A: Determine the trigonometric functions of an angle of 120°.

$$\alpha = 180° - 120° = 60°$$

Sin 120° = Sin 60° = .8660
Cos 120° = −1·Cos 60° = −1·.5 = −.5
Tan 120° = −1·Tan 60° = −1·1.732 = −1.732
Csc 120° = Csc 60° = 1.155
Sec 120° = −1·Sec 60° = −1·2 = −2
Cot 120° = −1·Cot 60° = −1·.5774 = −.5774

NOTE: −1 is the algebraic sign of the Cos, Tan, Sec, and Cot functions for all second quadrant angles.

Example B: Determine the trigonometric functions of an angle of 210°.

$$\alpha = 210° - 180° = 30°$$

Sin 210° = −1·Sin 30° = −1·.5 = −.5
Cos 210° = −1·Cos 30° = −1·.8660 = −.8660
Tan 210° = Tan 30° = .5774
Csc 210° = −1·Csc 30° = −1·2 = −2
Sec 210° = −1·Sec 30° = −1·1.155 = −1.155
Cot 210° = Cot 30° = 1.732

Example C: Determine the trigonometric functions of an angle of 315°.

$$\alpha = 360° - 315° = 45°$$

Sin 315° = −1·Sin 45° = −1·.707 = −.707
Cos 315° = Cos 45° = .707
Tan 315° = −1·Tan 45° = −1·1 = −1
Csc 315° = −1·Csc 45° = −1·1.414 = −1.414

$$\text{Sec } 315° = \text{Sec } 45° = 1.414$$
$$\text{Cot } 315° = -1 \cdot \text{Cot } 45° = -1 \cdot 1 = -1$$

The trigonometric functions of a negative angle are determined by considering a point on the terminal side of that angle.

Example D: Determine the trigonometric functions of $-45°$.

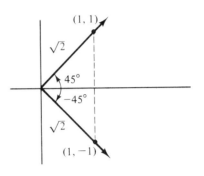

$$\text{Sin}(-45°) = \frac{-1}{\sqrt{2}} = -\frac{\sqrt{2}}{2} = -.707$$
$$\text{Cos}(-45°) = \frac{1}{\sqrt{2}} = \frac{\sqrt{2}}{2} = .707$$
$$\text{Tan}(-45°) = \frac{-1}{1} = -1$$
$$\text{Csc}(-45°) = \frac{\sqrt{2}}{-1} = -\sqrt{2} = -1.414$$
$$\text{Sec}(-45°) = \frac{\sqrt{2}}{1} = 1.414$$
$$\text{Cot}(-45°) = \frac{1}{-1} = -1$$

NOTE: $\text{Sin}(-45°) = -1 \cdot \text{Sin } 45°$
$\text{Cos}(-45°) = 1 \cdot \text{Cos } 45°$
$\text{Tan}(-45°) = -1 \cdot \text{Tan } 45°$
$\text{Csc}(-45°) = -1 \cdot \text{Csc } 45°$
$\text{Sec}(-45°) = 1 \cdot \text{Sec } 45°$
$\text{Cot}(-45°) = -1 \cdot \text{Cot } 45°$

In general, trigonometric functions of negative angles are found by using the following formulas. These formulas may be verified by referring to Figure 7–7.7.

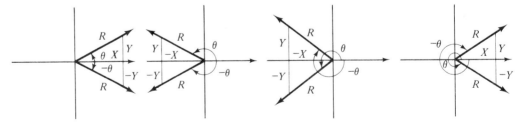

FIGURE 7–7.7

$$\text{Sin}(-\theta) = -\text{Sin } \theta \qquad \text{Csc}(-\theta) = -\text{Csc } \theta$$
$$\text{Cos}(-\theta) = \text{Cos } \theta \qquad \text{Sec}(-\theta) = \text{Sec } \theta$$
$$\text{Tan}(-\theta) = -\text{Tan } \theta \qquad \text{Cot}(-\theta) = -\text{Cot } \theta$$

Sec. 7–7. Functions of Angles That Are Not Acute

Example E:
1. $\text{Sin}(-20°) = -\text{Sin } 20° = -.342$
2. $\text{Cos}(-150°) = \text{Cos } 150° = -\text{Cos } 30° = -.8660$
3. $\text{Tan}(-50°) = -\text{Tan } 50° = -1.192$
4. $\text{Csc}(-250°) = -\text{Csc } 250° = -(-\text{Csc } 70°) = 1.064$
5. $\text{Sec}(-300°) = \text{Sec } 300° = \text{Sec } 60° = 2$
6. $\text{Cot}(-80°) = -\text{Cot } 80° = -.1763$

The trigonometric functions of the quadrantal angles can be determined by using the definitions.

Functions   **Reciprocal Functions**

$\text{Sin } 0° = \frac{0}{1} = 0$ $\qquad$ $\text{Csc } 0° = \frac{1}{0}$ (undefined)

$\text{Cos } 0° = \frac{1}{1} = 1$ $\qquad$ $\text{Sec } 0° = \frac{1}{1} = 1$ $\qquad$ (1, 0), $Y = 0$, $X = 1$, $R = 1$

$\text{Tan } 0° = \frac{0}{1} = 0$ $\qquad$ $\text{Cot } 0° = \frac{1}{0}$ (undefined)

$\text{Sin } 90° = \frac{1}{1} = 1$ $\qquad$ $\text{Csc } 90° = \frac{1}{1} = 1$

$\text{Cos } 90° = \frac{0}{1} = 0$ $\qquad$ $\text{Sec } 90° = \frac{1}{0}$ (undefined) $\qquad$ (0, 1), $Y = 1$, $X = 0$, $R = 1$

$\text{Tan } 90° = \frac{1}{0}$ (undefined) $\qquad$ $\text{Cot } 90° = \frac{0}{1} = 0$

$\text{Sin } 180° = \frac{0}{1} = 0$ $\qquad$ $\text{Csc } 180° = \frac{1}{0}$ (undefined)

$\text{Cos } 180° = \frac{-1}{1} = -1$ $\qquad$ $\text{Sec } 180° = \frac{1}{-1} = -1$ $\qquad$ (−1, 0), $Y = 0$, $X = -1$, $R = 1$

$\text{Tan } 180° = \frac{0}{-1} = 0$ $\qquad$ $\text{Cot } 180° = \frac{-1}{0}$ (undefined)

$\text{Sin } 270° = \frac{-1}{1} = -1$ $\qquad$ $\text{Csc } 270° = \frac{1}{-1} = -1$

$\text{Cos } 270° = \frac{0}{1} = 0$ $\qquad$ $\text{Sec } 270° = \frac{1}{0}$ (undefined) $\qquad$ $Y = -1$, $X = 0$, $R = 1$

$\text{Tan } 270° = \frac{-1}{0}$ (undefined) $\qquad$ $\text{Cot } 270° = \frac{0}{-1} = 0$

(0, −1)

Values of the functions of 360° are the same as the values of the functions of 0°.

Example F: We may now find any of the six trigonometric functions for any angle whatsoever, whether that angle is given in radians or degrees.
1. Sin 240° = −Sin 60° = −.8660
2. Tan 228° = Tan 48° = 1.111
3. Csc (−123°) = −Csc 123° = −Csc 57° = −1.192
4. Cot 270° = 0
5. Cos 430° = Cos (430° − 360°) = Cos 70° = .3420
6. Sin 2 = Sin (2·57.3°) = Sin 114.6° = Sin 65.4° = .9093
7. Tan .5 = Tan (.5·57.3°) = Tan 28.65° = .5467
8. Cos (−180°) = Cos 180° = −1

If a function is given and the corresponding angle desired, the previous procedure would be reversed.

Example G: Find θ such that $0° \leq \theta < 360°$.
1. Tan θ = −1. Locating 1 in the Tan column we read the corresponding angle (45°) on the left. This angle is the reference angle for θ. The Tan function is negative in quadrants II and IV. Therefore, θ will be either of two angles. In quadrant II, $\theta = 180° − 45° = 135°$. In quadrant IV, $\theta = 360° − 45° = 315°$.
2. Cos θ = −.8660. Locating .8660 in the Cos column we read the corresponding angle (30°) on the left. This angle is the reference angle for θ. Since the Cos function is negative in quadrants II and III, θ can be determined for quadrant II by $180° − 30° = 150°$ and for quadrant III by $180° + 30° = 210°$.
3. Sin θ = .8746. Locating .8746 in the Sin column we read the corresponding angle (61°) on the right. This angle is the reference angle for θ. The Sin function is positive in quadrants I and II. In quadrant I, θ is the reference angle. Therefore, $\theta = 61°$. In quadrant II, $\theta = 180° − 61° = 119°$.

If θ is to be expressed in radian measure, then each angle should be converted to radians using the method described in Section 7–1.

Exercises

Determine the trigonometric functions for each of the following.

1. 135°
2. 140°
3. 2
4. 175°
5. 200°
6. 305°
7. 2.1
8. 253°
9. 183°
10. 1.75
11. 5.1
12. 3.85
13. 370°
14. 7.0
15. 142°
16. 189°
17. 600°
18. 415°

Determine the trigonometric functions for each negative angle.

19. $-15°$ **20.** $-135°$ **21.** $-235°$
22. -1.5 **23.** -4.85 **24.** $-700°$

Find θ such that $0° \leq \theta < 360°$ (give answers in degrees).

25. $\tan\theta = -1$ **26.** $\cos\theta = .707$
27. $\sec\theta = -.5367$ **28.** $\sin\theta = -.5$
29. $\csc\theta = 2.41$ **30.** $\cot\theta = -324$

Find θ such that $0 \leq \theta < 2\pi$ (give answers in radians).

31. $\sin\theta = .8680$ **32.** $\cos\theta = -.8002$
33. $\tan\theta = 0.0$ **34.** $\sin\theta = -.85$
35. $\cos\theta = .98$ **36.** $\cot\theta = 3.21$

7-8. Applications of Radian Measure

We have seen that angles may be measured in both degrees and radians. While we are probably more familiar with degrees, there are many problems in which radian measure is more appropriate. In this section, we will look at some of these types of problems.

In geometry we learned that the length of arc on a circle is proportional to the central angle. If we let S represent arc length, then $S = 2\pi R$ for a complete circle. That is, the arc length of a complete circle (the circumference) is equal to the central angle for a complete circle (2π radians) times the radius. Thus *for any part of a circle, the length of arc equals the central angle (in radians) times the radius* (see Figure 7-8.1).

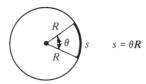

FIGURE 7-8.1

Example A: In a given circle, the radius is 5 inches and the central angle is 60°. Find the length of arc intercepted.

$$60° \div 57.3° = 1.05 \text{ rad}$$
$$s = \theta R = 1.05(5) = 5.25 \text{ in.}$$

Example B: In a given circle, an angle of 120° intercepts an arc of 9.8 feet. Find the radius.

$$120° \div 57.3° = 2.09 \text{ rad}$$

$$s = \theta R \quad \text{or} \quad R = \frac{s}{\theta} = \frac{9.8}{2.09} = 4.69 \text{ ft}$$

We should recall that a *sector* of a circle is formed by two radii and part of the circumference. In geometry we also learned that areas of sectors of circles are proportional to their central angles. For a complete circle, $A = \pi R^2 = \frac{1}{2}(2\pi)R^2$. That is, the area of a complete circle (the largest possible sector) is equal to $\frac{1}{2}$ times the central angle for a complete circle (2π radians) times the square of the radius. Thus *for any sector, the area equals $\frac{1}{2}$ times the central angle (in radians) times the square of the radius.*

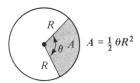

FIGURE 7-8.2

Example C: In a circle with a radius of 3 feet, a sector is formed by a central angle of 50°. Find the area of this sector.

$$50° \div 57.3° = .87 \text{ rad}$$

$$A = \tfrac{1}{2}\theta R^2 = \tfrac{1}{2}(.87)(9 \text{ sq ft}) = 3.92 \text{ sq ft}$$

Example D: In a circle with a radius of 8 decimeters, the area of a certain sector is 22.3 sq dec. Find the central angle for this sector.

$$A = \frac{1}{2}\theta R^2 \quad \text{or} \quad \theta = \frac{2A}{R^2} = \frac{2(22.3 \text{ sq dec})}{64 \text{ sq dec}} = .7 \text{ rad } (40.1°)$$

The next type of problem involves velocity. Average velocity is equal to distance divided by time $\left(v = \frac{s}{t}\right)$. For example, if we travel 200 miles in 4 hours, our average velocity is $\frac{200 \text{ miles}}{4 \text{ hours}} = 50$ miles per hour. For an object moving in a *circular* path with constant velocity, this equation becomes $v = \frac{\theta R}{t}$ or $v = \frac{\theta}{t}R$. $\frac{\theta}{t}$ is the ratio of an angle to time (radians per second or radians per minute). Such a ratio is called *angular velocity* and is usually represented by the Greek letter ω (omega). Our formula thus becomes **$v = \omega R$**, and expresses

Sec. 7–8. Applications of Radian Measure

a relationship between *linear velocity* (v) and *angular velocity* $\left(\omega = \dfrac{\theta}{t}, \text{ where } \theta \right.$ must be expressed in radians).

Example E: The propeller of a small airplane is 6 feet in diameter and rotates at 2400 revolutions per minute. What is the linear velocity of the tip of the propeller?

$$(2400 \text{ rev/min}) \times (6.28 \text{ rad/rev}) = 15072 \text{ rad/min}$$
$$v = \omega R = (15072 \text{ rad/min}) \times (3 \text{ ft}) = 45216 \text{ ft/min}$$

NOTE: Revolutions (or cycles) must be changed to radians by using 1 rev = 6.28 rad.

Example F: For an automobile traveling at 55 miles per hour, what is the angular velocity of the tires which are 26 inches in diameter?

$$\dfrac{55 \text{ mi}}{1 \text{ hr}} \times \dfrac{1 \text{ hr}}{3600 \text{ sec}} \times \dfrac{5280 \text{ ft}}{1 \text{ mi}} = 80.67 \text{ ft/sec}$$

similar units

radius of tires = 13 in. ÷ 12 = 1.08 ft

$$v = \omega R \quad \text{or} \quad \omega = \dfrac{v}{R} = \dfrac{80.67 \text{ ft/sec}}{1.08 \text{ ft}} = 74.7 \text{ rad/sec}$$

In working with AC circuits, the equations $i = I \sin \omega t$ and $v = V \cos \omega t$ are used to represent the flow of current (i) and voltage (v), respectively, under certain conditions at time t. In both of these equations, when ω is expressed in radians, we may calculate either current (i) or voltage (v).

Example G: If $i = I \sin \omega t$, find i if I = 6 amps, ω = 60 cycles/sec and t = .01 sec.

$$(60 \text{ cycles/sec}) \times (6.28 \text{ rad/cycle}) = 376.8 \text{ rad/sec}$$
$$i = (6 \text{ amps}) \sin (376.8 \times .01)$$
$$= (6 \text{ amps}) \sin (3.768)$$
$$= (6 \text{ amps})(-.586)$$
$$= -3.52 \text{ amps}$$

Exercises

For each of the following, use the formula $S = \theta R$ to find the missing quantity.
1. $\theta = 75°$ and R = 9 in.
2. $\theta = 220°$ and R = 4.2 ft
3. S = 12 ft and R = 5.2 ft
4. S = 51 in. and R = 74 in.
5. S = 18 ft and θ = 2.6
6. S = 63 in. and θ = 18°

For each of the following, use the formula $A = \frac{1}{2}\theta R^2$ to find the missing quantity.

7. $\theta = 40°$ and $R = 62$ cm
8. $\theta = 110°$ and $R = 3.4$ m
9. $A = 48$ sq ft and $R = 5.2$ ft
10. $A = 192$ sq cm and $R = 16$ cm
11. $A = 72$ sq in. and $\theta = 4$
12. $A = 14$ sq m and $\theta = 30°$

For each of the following, use the formula $v = \omega R$ to find the missing quantity.

13. $\omega = 4$ rad/sec and $R = 3.5$ m
14. $\omega = 60$ cycles/sec and $R = 4$ in.
15. $v = 40$ mph and $R = 1$ ft
16. $v = 66$ ft/sec and $R = 1.2$ ft
17. $v = 88$ ft/sec and $\omega = 12$ rev/sec
18. $v = 30$ m/sec and $\omega = 11.2$ rad/sec
19. A 4-foot pendulum oscillates through an angle of 6°. What is the distance through which the end of the pendulum swings as it goes from one extreme position to the other?
20. To measure the length of a certain proposed curved section of road, a surveyor selects a reference point that enables him to consider the section of road as part of a circular arc. From this reference point, the distance to both ends of the road section is 200 meters and the angle between these two distances is 130°. What is the length of road to be constructed?
21. A man wishes to build a semicircular walk. From a point that is exactly midway between the ends of the walk, the distance to the near side of the walk is 8 feet and to the far side 10 feet. What is the area (in square feet) of the walk?
22. From a circular piece of flexible material which is 16 feet in diameter, a circular sector with a central angle of 200° is cut. A cone is then formed by bringing the two radii of the sector together. What is the outside surface area of the cone?
23. The distance from the earth to the moon is approximately 240,000 miles. If it takes the moon about 28 days to revolve around the earth once, what is the linear velocity of the moon?
24. If an automobile is traveling 60 miles per hour and the tires are 28 inches in diameter, what is the angular velocity of the tires in revolutions per second?
25. If $i = I \operatorname{Sin} \omega t$, find i if $I = 4$ amps, $\omega = 30$ cycles/sec and $t = .02$ sec.
26. If $v = V \operatorname{Cos} \omega t$, find v if $V = 110$ volts, $\omega = 60$ cycles/sec and $t = .01$ sec.

7-9. CHAPTER REVIEW

Express each of the following angles in degrees.

1. $.8\pi$
2. $\frac{\pi}{12}$
3. $.62$
4. 2.3

Express each of the following angles in radians in terms of π.

5. $32°$
6. $160°$

Express each of the following angles in radians without using π.

7. $126.9°$
8. $312.8°$

Find the area of each of the following tirangles.

9.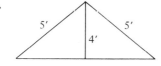
10.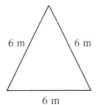

11.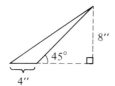
12.

Find the six trigonometric functions of the angle whose terminal side passes through each given point.

13. $(3, -4)$
14. $(-12, -5)$
15. $(1, 6)$
16. $(-5, 2)$

For each of the following, find the value of the trigonometric function.

17. Sin $31°$
18. Tan $68°$
19. Cos $47.6°$
20. Cot $81.3°$
21. Sin 1.2
22. Sec $25°42'$
23. Csc $52°58'$
24. Sin $\frac{\pi}{8}$
25. Cos 3.6
26. Tan 3.7π
27. Tan $193°$
28. Sin $248°$
29. Csc $305.7°$
30. Tan $.52\pi$
31. Sin $(-61°)$

32. Cos .2
33. Tan 6.1
34. Sin 122.8°
35. Cos 340°
36. Sin (−4.2)

For each of the following, find θ as a positive acute angle in degrees.

37. Sin θ = .7623
38. Cos θ = .2182
39. Tan θ = 2.539
40. Csc θ = 5.487
41. Sec θ = 1.567
42. Cot θ = 3.962

For each of the following, find θ as a positive acute angle in radians without using π.

43. Sin θ = .1045
44. Cos θ = .9822
45. Tan θ = .7860
46. Csc θ = 1.045
47. Sec θ = 8.016
48. Cot θ = .3249

For each of the following, find all values for A between 0° and 360° that satisfy the given functional values. Give answers in degrees.

49. Sin A = .5491
50. Tan A = 17.82
51. Cos A = −.7242
52. Sec A = −6.483

For each of the following, find all values for A between 0 and 6.28 that satisfy the given functional values. Give answers in radians.

53. Tan A = −4.196
54. Cos A = .9237
55. Sin A = −.3647
56. Cot A = .0198

In each of the following triangles, find all unknown parts.

57.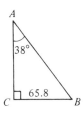
58.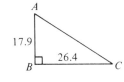
59.
60.

61. From an observation tower 35 meters high, a forest ranger spots a fire at an angle of depression of 26°. Assuming the terrain to be level, how far from the tower is the fire?

62. If a jet cruises at 700 miles per hour and climbs at an angle of 18°, what is its gain in altitude in 5 minutes?

63. To determine the height of a tree, a distance of 50 feet is measured from the base of the tree. From this point, the angle of elevation of the top of the tree is 62°. What is the height of the tree?

64. A regular pentagon is inscribed in a circle of diameter 72 centimeters. Find the length of the sides of the pentagon.

65. Find the distance between cities A and B if angle ACB is a central angle.

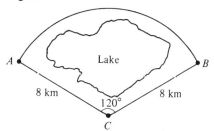

66. On a certain meter, a 15 centimeter needle deflects through an angle of 2.3 radians on a circular scale. What is the length of the scale?

67. On a rescue mission, the pilot of a certain plane is to search an area that is a circular sector. If this area has a radius of 50 miles and a central angle of 18°, how many square miles must this pilot cover in his search?

68. A certain plot of land is a circular sector. The area of the plot is 30 acres and the sides (radii) intersect at an angle of 50°. What is the length in feet of the sides?

69. An automobile is traveling at 80 kilometers per hour. If the tires are 71 centimeters in diameter, what is the angular velocity of the tires in revolutions per second?

70. In a certain electric circuit, the current (i) is given as a function of the time (t) by the equation $i = I \sin \omega t$. Find i if $I = 3.4$ amps, $\omega = 60$ cycles/sec, and $t = .001$ sec.

71. In a certain electric circuit, the voltage (v) is given as a function of the time (t) by the equation $v = V \cos \omega t$. Find v if $V = 220$ volts, $\omega = 30$ cycles/sec, and $t = .03$ sec.

chapter 8

Vectors and Oblique Triangles

8–1. Introduction to Vectors

Problem-solving in the field of technology utilizes the techniques used in setting up and solving vector problems.

Physical quantities such as force, displacement, velocity, and acceleration can be represented by a vector. A *vector* is a quantity having both magnitude and direction. Graphically, a vector is represented by an arrow that defines the direction (see Figure 8–1.1). The magnitude is indicated by the length of the arrow. Point O is the initial point of the vector, while point P is the head, or terminal, point of the vector.

A vector may also be identified by its head point. A vector is an ordered pair of real numbers, (A, B), where the head point is the point (A, B) and the initial point is at the origin of the rectangular coordinate system (see Figure 8–1.2).

It is very convenient to represent vectors this way in the rectangular coordinate system, since by letting standard position (0°) correspond to an arrow pointing east, we can represent both the magnitude and direction in a familiar setting.

Vectors will be represented in this text by an arrow over a letter: $\vec{A}$. The magnitude of the vector is indicated by $|A|$.

Sec. 8–1. Introduction to Vectors   227

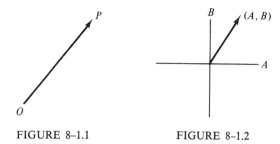

FIGURE 8–1.1   FIGURE 8–1.2

Example A: An airplane is heading northeast at 130 miles per hour. What is its magnitude?

$$|130| = 130$$

Besides vector quantities which require magnitude and direction there are quantities that possess only magnitude, such as temperature, time, length, and mass. These quantities can be represented by real numbers and are called *scalars*. Operations with scalars follow the same rules as in algebra. These operations were dealt with in detail in the discussion of dimensional analysis in Chapter Three. The algebra that this section will deal with is vector addition and subtraction.

1. Two vectors $\vec{A}$ and $\vec{B}$ are equal if they have the same magnitude and direction. Since $\vec{A}$ and $\vec{B}$ are parallel to each other and have the same magnitude, then $\vec{A} = \vec{B}$. (See Figure 8-1.3.)

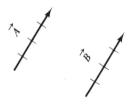

FIGURE 8–1.3

2. If a second vector has the same magnitude as $\vec{A}$, but its direction is opposite to that of $\vec{A}$, it is denoted by $-\vec{A}$. (See Figure 8-1.4.)

FIGURE 8–1.4

3. The sum of $\vec{A}$ and $\vec{B}$ is found by positioning the initial point of $\vec{B}$ at the head point of $\vec{A}$. The sum is the vector determined by the initial point of $\vec{A}$ and the head point of $\vec{B}$. This is represented by $\overrightarrow{A + B}$ and the procedure is called *adding vectors by the triangle method.* (See Figure 8-1.5.)

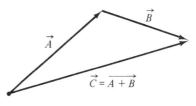

FIGURE 8-1.5

Example B: Add the following displacements:

$$\vec{A} = 10 \text{ ft northwest}$$
$$\vec{B} = 20 \text{ ft } 30° \text{ north of east}$$
$$\vec{C} = 35 \text{ ft due south}$$

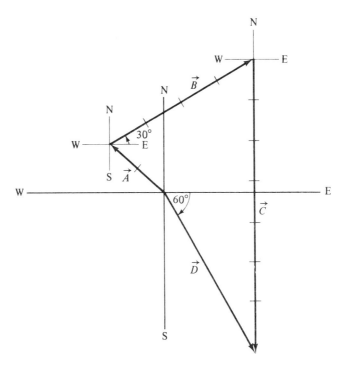

$\vec{D}$ is formed by making its origin the initial point of $\vec{A}$ and its terminal point the head point of $\vec{C}$.

Sec. 8-1. Introduction to Vectors

The *parallelogram method* of adding two vectors is equivalent to adding vectors by the triangle method.

Example C: Add $\vec{A}$ and $\vec{B}$ by both the parallelogram method and the triangle method. $\vec{A} = 30$ at $75°$ and $\vec{B} = 45$ at $15°$.

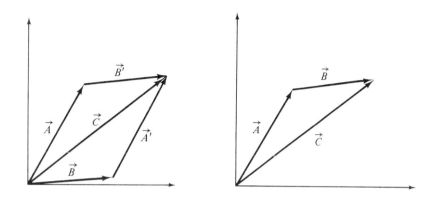

NOTE: $\vec{A} = \vec{A}'$ and $\vec{A}' \parallel \vec{A}$, $\vec{B} = \vec{B}'$ and $\vec{B}' \parallel \vec{B}$. Therefore, the quadralateral is a parallelogram. $\vec{C}$ is the diagonal of the parallelogram. $\vec{C}$ in the left diagram is equal to $\vec{C}$ in the right diagram.

Regardless of how we add or subtract vectors, the sum or result is called the *resultant vector*, while the individual vectors being combined are called *components*.

Any vector in a plane has a horizontal and vertical component. These components can be found by using right triangle trigonometry.

Example D: Find the horizontal and vertical components of $\vec{A}$.

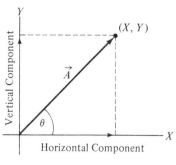

$$\text{Cos } \theta = \frac{X}{|A|} \quad \text{or} \quad X = |A| \text{ Cos } \theta$$

Therefore, the horizontal component can be determined by finding the product of the magnitude and Cos θ.

$$\text{Sin } \theta = \frac{Y}{|A|} \quad \text{or} \quad Y = |A| \text{ Sin } \theta$$

Therefore, the vertical component can be determined by finding the product of the magnitude and Sin θ.

Example E: Find the components of $\vec{B}$ with a magnitude of 100 and an angle in standard position of 225°.

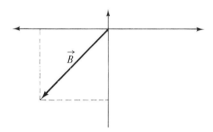

$$\begin{aligned}
\text{Horizontal component} &= (100)(\cos 225°) \\
&= (100)(-.707) \\
&= -70.7 \\
\text{Vertical component} &= (100)(\sin 225°) \\
&= 100(-.707) \\
&= -70.7
\end{aligned}$$

If two component vectors are given, the resultant vector can be determined by using the Pythagorean Theorem and the tangent function.

Example F: Find the sum or the resultant vector of a horizontal vector ($\vec{A}$) and a vertical vector ($\vec{B}$). Since $\vec{R}$ is the hypotenuse of a right triangle formed

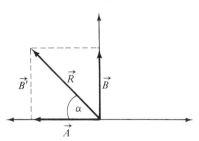

Sec. 8–1. Introduction to Vectors   231

by $\vec{A}$ and $\vec{B}'$, the Pythagorean Theorem is applied to find the magnitude of $\vec{R}$. Thus

$$|R| = \sqrt{|B|^2 + |A|^2}$$

Angle α of the right triangle is found by using the tangent function. Thus,

$$\text{Tan } \alpha = \frac{\text{vertical vector } (\vec{B})}{\text{horizontal vector } (\vec{A})}$$

Example G: Find the resultant vector if $\vec{A}$ has a magnitude of 10 and a direction of 180°, and $\vec{B}$ has a magnitude of 12 and a direction of 270°.

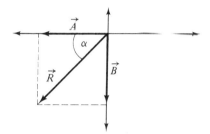

$$|R| = \sqrt{12^2 + 10^2} = \sqrt{144 + 100} = \sqrt{244} = 15.6$$
$$\text{Tan } \alpha = \frac{-12}{-10} = \frac{6}{5} = 1.2; \alpha = 50.19°$$

θ, an angle in standard position, is equal to

$$\alpha + 180° = 50.19° + 180° = 230.19°$$

Summarizing the formulas for vector analysis:

1. The components of any vector can be determined by:
    (a) Horizontal component of vector $\vec{A} = |A| \cos \theta$.
    (b) Vertical component of vector $\vec{A} = |A| \sin \theta$, where θ is an angle in standard position.
2. The magnitude of the resultant vector or sum of a vertical ($\vec{B}$) and horizontal ($\vec{A}$) vector can be found by:

$$|R| = \sqrt{|A|^2 + |B|^2}$$

3. The angle in standard position can be determined by

$$\text{Tan } \theta = \frac{\text{vertical vector } (\vec{B})}{\text{horizontal vector } (\vec{A})}$$

When two or more vectors are given, the sum or resultant vector can be found by finding the algebraic sum of the horizontal and vertical components of all the vectors and then using the appropriate formulas for finding the magnitude and direction of the resultant vector.

Find the resultant vector of two vectors ($\vec{P}$ and $\vec{Q}$) acting on an object from a single point. (See Figure 8-1.6.)

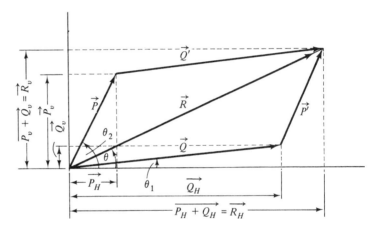

FIGURE 8–1.6

Vectors	COMPONENTS	
	Horizontal	Vertical
$\vec{P}$	$\vec{P_H} = \|P\| \text{Cos } \theta_2$	$\vec{P_V} = \|P\| \text{Sin } \theta_2$
$\vec{Q}$	$\vec{Q_H} = \|Q\| \text{Cos } \theta_1$	$\vec{Q_V} = \|Q\| \text{Sin } \theta_1$
$\vec{R}$	$\vec{P_H} + \vec{Q_H} = \vec{R_H}$	$\vec{P_V} + \vec{Q_V} = \vec{R_V}$

$$|R| = \sqrt{|R_H|^2 + |R_V|^2}$$

$$\text{Tan } \theta = \frac{\vec{R_V}}{\vec{R_H}}$$

NOTE: This procedure can be extended to more than two vectors.

Sec. 8–1. Introduction to Vectors

Example H: Find the resultant force of the following two forces.

$$\vec{P}: |P| = 13.6 \text{ kg at } 75°$$
$$\vec{Q}: |Q| = 21.3 \text{ kg at } 15°$$

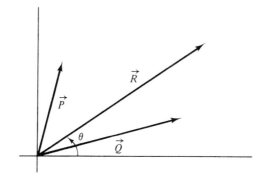

COMPONENTS

Vectors	Horizontal	Vertical
$\vec{P}$	(13.6)(Cos 75°) = 3.52	(13.6)(Sin 75°) = 13.14
$\vec{Q}$	(21.3)(Cos 15°) = 20.57	(21.3)(Sin 15°) = 5.51
$\vec{R}$	24.09	18.65

$$|R| = \sqrt{24.09^2 + 18.65^2}$$
$$= \sqrt{580.3 + 347.8}$$
$$= \sqrt{928.1}$$
$$= 30.46 \text{ kg}$$

$$\text{Tan } \theta = \frac{18.65}{24.09} = .7742$$

$$\theta = 37.75°$$

Example I: Find the sum of the following three vectors.

$$\vec{A}: \ |A| = 8.6, \ A_\theta = 60°$$
$$\vec{B}: \ |B| = 10.1, \ B_\theta = 105°$$
$$\vec{C}: \ |C| = 6.4, \ C_\theta = 225°$$

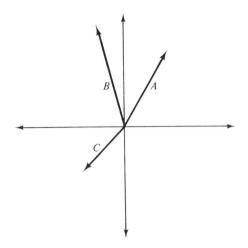

	COMPONENTS	
Vectors	Horizontal	Vertical
$\vec{A}$	(8.6) Cos 60° = 4.30	(8.6) Sin 60° = 7.44
$\vec{B}$	(10.1) Cos 105° = −2.61	(10.1) Sin 105° = 9.76
$\vec{C}$	(6.4) Cos 225° = −4.53	(6.4) Sin 225° = −4.53
$\vec{R}$	−2.84	12.67

$$|R| = \sqrt{(2.84)^2 + 12.67^2}$$
$$= \sqrt{8.06 + 160.5}$$
$$= \sqrt{168.56}$$
$$= 12.98$$
$$\text{Tan } \theta = \frac{12.67}{-2.84} = -4.46$$
$$\theta = 180° - 77.4° = 102.6°$$

Exercises   State which of the following are scalars and which are vectors.

1. 150 kg
2. 600 cal
3. density
4. 13 cm³
5. 17 mph northeast
6. 10 mi
7. energy
8. velocity
9. momentum
10. 17 ft/sec² due to gravity

Sec. 8–1. Introduction to Vectors

Represent graphically:
11. A force of 75 lb, 30° north of east.
12. A force of 10 kg, .75 south of west.
13. A force of 150 g at an angle of 135°.
14. A force of 35 t at an angle of −30°.
15–18. Represent the opposite of the vectors in problems 11–14.

Find the sum of the following displacements graphically.
19. $\vec{P}$, 56 ft southwest
    $\vec{Q}$, 143 ft east
    $\vec{R}$, 200 ft northwest
20. $\vec{A}$, 24 m west
    $\vec{B}$, 38 m southeast
    $\vec{C}$, 56 m northeast

Given vectors $\vec{X}$, $\vec{Y}$, and $\vec{Z}$, construct

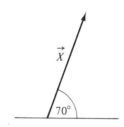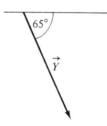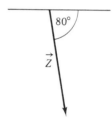

21. $\vec{X} + \vec{Y} + \vec{Z}$
22. $\vec{X} - (\vec{Y} + \vec{X})$
23. $2\vec{Y} + \vec{Z} - \tfrac{1}{2}\vec{X}$
24. $\vec{Z} - 2\vec{Y} + 3\vec{X}$

Determine the horizontal and vertical components of each of the following.
25. $\vec{F}, |F| = 17, \theta = 47°$
26. $\vec{C}, |C| = 156, \theta = 32°$
27. $\vec{A}, |A| = .0861, \theta = 163°$
28. $\vec{D}, |D| = 2186, \theta = 225°$
29. $\vec{P}, |P| = 12, \theta = 90°$
30. $\vec{Q}, |Q| = 1780, \theta = 270°$
31. $\vec{Z}, |Z| = 34, \theta = 360°$
32. $\vec{K}, |K| = .742, \theta = 180°$

Determine the resultant vector for each of the following:
33. $\vec{A}$: $|A| = 13, \theta_A = 90°$
    $\vec{B}$: $|B| = 19, \theta_B = 180°$
34. $\vec{X}$: $|X| = 136, \theta_X = 35°$
    $\vec{Y}$: $|Y| = 215, \theta_Y = 135°$
35. $\vec{F}$: $|F| = .891, \theta_F = -60°$
    $\vec{G}$: $|G| = 1.21, \theta_G = 165°$
36. $\vec{F}_1$: $|F_1| = 15.71, \theta_{F_1} = 200°$
    $\vec{F}_2$: $|F_2| = 21.82, \theta_{F_2} = 265°$

37. $\vec{A}_1$: $|A_1| = 18.9, \theta_{A_1} = 35°$
   $\vec{A}_2$: $|A_2| = 21.0, \theta_{A_2} = 125°$
   $\vec{A}_3$: $|A_3| = 29.5, \theta_{A_3} = 235°$

38. $\vec{T}_1$: $|T_1| = 2490, \theta_{T_1} = 125°$
   $\vec{T}_2$: $|T_2| = 2200, \theta_{T_2} = -55°$
   $\vec{T}_3$: $|T_3| = 3620, \theta_{T_3} = 200°$

8-2. Problems Involving Vectors

When solving an applied problem that involves vector quantities, the verbal statement and/or the drawing should be translated into vector quantities in the rectangular coordinate system. This procedure will make the problem easier to analyze and thus, easier to solve. This section will deal only with static (stationary) forces in a single plane.

Example A: A projectile fired from the ground is traveling at 2000 feet per second at an angle of elevation of 23°. What are the horizontal and vertical components of its velocity? This problem calls for finding the components of a single vector with a magnitude of 2000 feet per second at an angle of 23°.

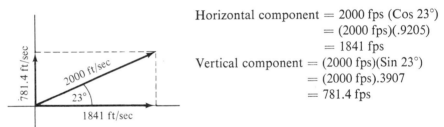

Horizontal component = 2000 fps (Cos 23°)
  = (2000 fps)(.9205)
  = 1841 fps
Vertical component = (2000 fps)(Sin 23°)
  = (2000 fps).3907
  = 781.4 fps

Example B: A boat heads north across a lake at 8 miles per hour. The current flows east at 3 miles per hour. What is the velocity of the boat in relation to the water? Neglect the wind. Since these vectors are at right angles to each other the resultant vector will be the velocity of the boat in relation to the water.

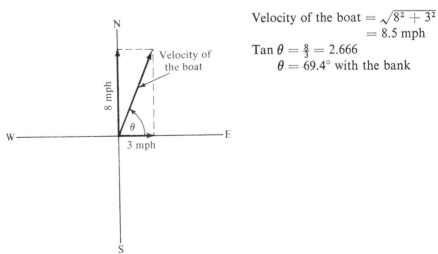

Velocity of the boat = $\sqrt{8^2 + 3^2}$
  = 8.5 mph
Tan $\theta = \frac{8}{3} = 2.666$
$\theta = 69.4°$ with the bank

Example C: A 200-pound object hangs away from a wall, supported by a cable and a metal brace as shown. What are the vectors representing the cable and brace?

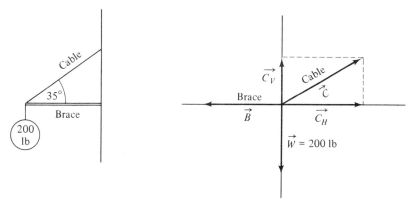

Picture of problem

Vector diagram: all forces and their components.

The solution to this problem depends on Newton's Law. This principle states that the resultant force on a body in equilibrium is zero. This means that the vector sum of all horizontal forces must be zero and the vector sum of all vertical forces must be zero.

The vector representing the cable in the diagram is situated as it is because the cable would have a tendency to pull up on the weight; the vector representing the metal brace is situated as it is because the brace would be holding the weight away from the wall.

Find the sum of the vertical components:

$\vec{C}_v - \vec{W} = 0$ (The minus sign indicates that $\vec{W}$ is headed in the negative direction.)

$|C| \cdot \text{Sin } 35° - 200 \text{ lb} = 0$

$$|C| = \frac{200 \text{ lb}}{\text{Sin } 35°} = \frac{200 \text{ lb}}{.5736}$$

$$|C| = 348.7 \text{ lb}$$

Find the sum of the horizontal components:

$\vec{C}_H - \vec{B} = 0$ (The minus sign indicates that $\vec{B}$ is headed in the negative direction.)

$|C| \cdot \text{Cos } 35° - |B| = 0$

$$|B| = |C| \cdot \text{Cos } 35° = 348.7(.8192)$$

$$|B| = 285.7 \text{ lb}$$

Example D: An 800-pound block is resting on a plane that forms an angle of 25° with the horizontal. What force would have to be exerted on the block to prevent the block from sliding down the plane? What is the friction between the block and the plane?

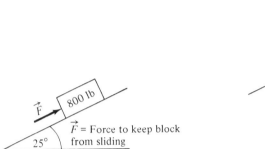
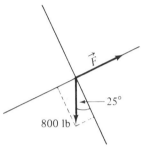

The weight of the block is exerted straight down in a direction perpendicular to the horizontal, but only that part of the weight parallel to the incline and opposite to $\vec{F}$ will cause the block to slide down the incline.

$$|F| - (800)(\sin 25°) = 0$$
$$|F| = (800)(\sin 25°) = 800(.4226)$$
$$= 338.1 \text{ lb}$$

Therefore, 338.1 lb is the force necessary to prevent the block from sliding down the incline. The force representing the friction between the block and the plane is perpendicular to the plane. Therefore, the friction force equals $800 \cdot \cos 25°$ $= 800(.9063) = 725$ lb.

Exercises

1. An automobile travels 54 miles per hour in a southwest direction. What are the westerly and southerly components of its velocity?

2. Two forces are acting on a body from the same point. One force is 78 pounds, and the other is 84 pounds. These forces are at right angles to each other. What is the resultant of these two forces?

3. If an airplane heads due east at 1500 miles per hour and a wind is blowing at 45 miles per hour from the north, what is the velocity of the airplane in relation to the ground?

4. What force is needed to hold a 156-kilogram weight in place on an incline of 45°?

5. What force is needed to hold a 156-kilogram weight in place on an incline of 60°?

Sec. 8–2. Problems Involving Vectors   239

6. What force is needed to hold a 156-kilogram weight in place on an incline of 30°?

7. A 200-pound object is supported away from a wall by a cable and metal brace from the top of the object. If the cable forms an angle of elevation of 25°, what is the magnitude of the vectors representing the cable and the brace?

8. What forces represent the cable and brace in problem 7 if the angle of elevation is increased to 45°?

9. If a boy weighing 125 pounds sits on a swing, what tension is in each rope of the swing?

10. If the boy in problem 9 is pulled horizontally by a force of 60 pounds, what is the tension in each rope of the swing?

11. A boat is heading due north across a lake at a speed of 12 miles per hour, the current is heading 35° east of north at 4 miles per hour, and the wind is from the northwest at 8 miles per hour. What is the velocity of the boat?

12. The following diagram shows a wind force on the sail of a sailboat. The components of this vector force are the drive force and tip force on the sail boat. Determine each of these component forces if the velocity of the wind is 15 miles per hour.

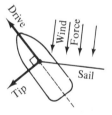
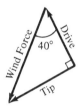

13. A truck weighing 5 tons is parked on a hill which rises 2 feet for every 50 feet of road. What force tends to make the truck roll down the hill?

14. A 2-ton weight is to be raised by a crane. The following diagram shows the forces between the weight, the cable, and the crane boom. Find the thrust force of the boom.

15. A boy wants to pull a lawn roller up a step. The lawn roller weighs 30 pounds. If the boy pulls with a force of 80 pounds at a 38° angle with the horizontal, will there be enough lifting force to pull the roller up the step?

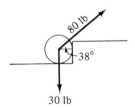

8–3. The Law of Sines

Since not all triangles are right triangles, it is necessary to develop two basic formulas that can be used to solve triangles that are oblique. *Oblique* triangles have no right angles.

The first of these formulas is the law of sines. The derivation of this formula follows (see Figure 8-3.1).

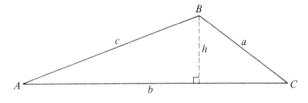

FIGURE 8–3.1

Applying right triangle relationships:

$$\text{Sin } A = \frac{h}{c} \quad \text{and} \quad \text{Sin } C = \frac{h}{a}$$

Solving both equations for h:

$$h = c \cdot \text{Sin } A \quad \text{and} \quad h = a \cdot \text{Sin } C$$

By the transitive property of an equality:

$$c \text{ Sin } A = a \text{ Sin } C$$

or

$$\frac{\text{Sin } A}{a} = \frac{\text{Sin } C}{c}$$

Sec. 8–3. The Law of Sines

If one of the remaining two altitudes of the triangle is drawn, it can be shown that $\dfrac{\sin A}{a} = \dfrac{\sin B}{b}$.

These two formulas can be put together into one statement to form the *law of sines*, a statement of proportionality between the lengths of the sides of a triangle and the sine function of the angle opposite each side. That is:

$$\frac{\sin A}{a} = \frac{\sin B}{b} = \frac{\sin C}{c}$$

When the given information consists of two sides and the angle opposite one of these sides or two angles and one side, then the law of sines can be used to solve the triangle.

NOTE: The case of an oblique triangle containing an obtuse angle will be considered in problem (1) of the exercises.

Example A: Given a triangle where angle $A = 37°$, angle $B = 100°$, and $a = 10$, find b and c.

$$\frac{\sin 100°}{b} = \frac{\sin 37°}{10}$$

$$b = \frac{10 \sin 100°}{\sin 37°} = 16.4$$

Angle $C = 180° - (37° + 100°) = 43°$

$$\frac{\sin 37°}{10} = \frac{\sin 43°}{c}$$

$$c = \frac{10(\sin 43°)}{\sin 37°} = 11.3$$

Example B: A ship travels 40 miles due east from its home port and then heads 43° north of east. After 2 hours of traveling in this direction its displacement from home port is 105 miles. What was the velocity of the ship after it turned on the course of 43° north of east?

$$\frac{\sin 137°}{105} = \frac{\sin B}{40}$$

$$\sin B = \frac{(40)(\sin 137°)}{105}$$

$$\sin B = .2598$$

$$B = 15.06°$$

$$A = 180° - (137° + 15.06°)$$
$$= 27.94°$$
$$\frac{\sin 27.94°}{a} = \frac{\sin 137°}{105}$$
$$a = \frac{(105)(\sin 27.94°)}{\sin 137°} = \frac{(105)(.4685)}{.6820} = 72.13 \text{ mi.}$$
$$\text{velocity} = \frac{\text{distance}}{\text{time}} = \frac{72.13 \text{ mi.}}{2 \text{ hr.}} = 36.065 \text{ mph}$$

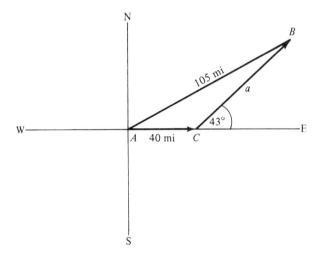

When the information given for a triangle consists of two sides and the angle opposite one of these sides there are a number of possibilities for triangles. There may be no triangles, one, or two triangles. This situation will be demonstrated in the following diagrams.

1.

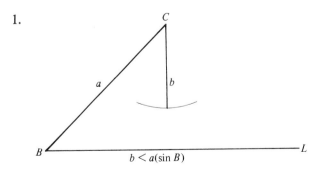

An arc with radius b and center point C might not intersect L to form a triangle. Thus, in this situation no triangle exists.

2.

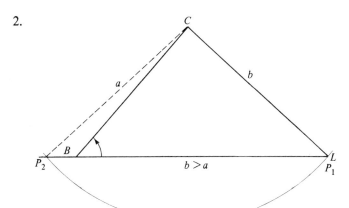

An arc with radius b and center point C could intersect L at two points. Point P_2 would not be a vertex of a triangle with angle B as one of its angles. Point P_1 would be a vertex of a triangle with the given parts. Thus, there is one triangle.

3.

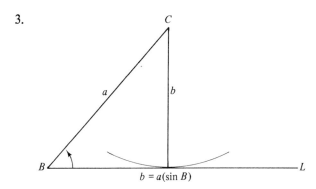

There may be only one point of intersection of the arc and the line L. In this case there is only one triangle and it is a right triangle.

4.

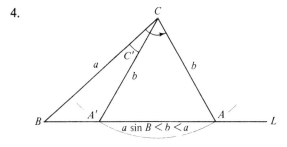

An arc with radius b and center point C could intersect L at two points. Two triangles can be formed with the given information. These two triangles are BAC and $BA'C'$.

Example C: Solve the triangle with $a = 5.2$, $b = 4.3$, and $B = 48°$.

Sin $B \cdot a < b < a$
$3.86 < 4.3 < 5.2$

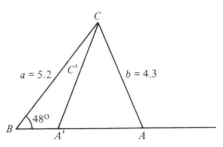

Thus, there will be two triangles.

$$\frac{\text{Sin } 48°}{4.3} = \frac{\text{Sin } A}{5.2} \longrightarrow \text{Sin } A = \frac{(5.2)(\text{Sin } 48°)}{4.3} = .8986$$

$$A = 64°$$
$$C = 180° - (48° + 64°) = 68°$$

$$\frac{\text{Sin } 48°}{4.3} = \frac{\text{Sin } 68°}{c} \longrightarrow c = \frac{(4.3)(\text{Sin } 68°)}{\text{Sin } 48°} = \frac{3.86}{.743}$$

$$c = 5.365$$

In triangle BCA: $B = 48°$ $b = 4.3$
                        $A = 64°$ $a = 5.2$
                        $C = 68°$ $c = 5.4$

In triangle $BA'C'$: $A' = 180° - 64 = 116°$
                         $C' = 180° - (48° + 116°) = 16°$

$$\frac{\text{Sin } 48°}{4.3} = \frac{\text{Sin } 16°}{c'}$$

$$c' = \frac{(4.3)(\text{Sin } 16°)}{\text{Sin } 48°} = 1.59 = 1.6$$

Exercises

1. Show that the law of sines also holds for an oblique triangle containing an obtuse angle.

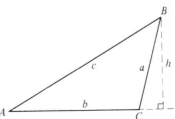

Sec. 8–3. The Law of Sines

Solve each of the following triangles for the remaining parts.

2. $c = 700, B = 120°, C = 43°$
3. $a = 87.3, c = 38, C = 21°$
4. $a = 44.9, b = 126, A = 64°$
5. $b = .74, B = 50°, C = 4°$
6. $b = 45, a = 50, B = 30°$
7. Determine the length of the upper support (L).

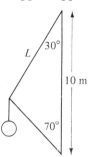

8. Resolve $\vec{V}$ with a magnitude of 135 into two components $\vec{U}$ and $\vec{W}$.

9. Three forces are represented by the sides of a triangle. One of the forces is 1000 kilograms. The other two forces make angles of 33° and 42° with the given force. Determine each of the unknown forces.

10. The velocity of a boat is 20 kilometers per hour in a direction 50° north of east. The resultant force is 24 kilometers per hour in a direction 40° north of east. If the wind is blowing from the west, what is its velocity?

11. A forester measures the angle of elevation to the top of a tree to be 29°. He then moves 25 feet farther away and measures the angle of elevation to be 21°10′. How tall is the tree?

12. A 3-ton weight is to be raised by a crane. The following diagram shows the forces between the weight cable and crane boom. Find the thrust of the boom and the force on the tie cable. [Refer to problem (14) of Section 8–2.]

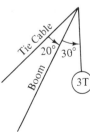

8–4. The Law of Cosines

If the given information for an oblique triangle consists of two sides and the included angle or three sides it is necessary to use the law of cosines to solve the triangle.

The derivation of the law of cosines follows. (Refer to Figure 8-4.1.)

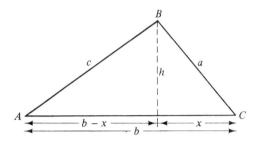

FIGURE 8-4.1

h is an altitude of triangle ABC. Applying the Pythagorean Theorem:

$$c^2 = h^2 + (b - x)^2 \qquad \text{and} \qquad a^2 = h^2 + x^2$$
$$c^2 = h^2 + b^2 - 2bx + x^2 \qquad\qquad h^2 = a^2 - x^2$$
$$c^2 - b^2 + 2bx - x^2 = h^2$$

By the transitive property of equality:

$$c^2 - b^2 + 2bx - x^2 = a^2 - x^2$$

Solving for c^2:

$$c^2 = a^2 + b^2 - 2bx$$

Since $\cos C = \dfrac{x}{a}$,

$$x = a \cdot \cos C$$

Substituting $a \cdot \cos C$ in place of x,

$$c^2 = a^2 + b^2 - 2ab \cos C$$

This last statement is the *law of cosines*. There are two other forms of this law that relates any one side of a triangle to the remaining two sides and the cosine of the angle between the two sides. The other forms are:

$$b^2 = a^2 + c^2 - 2ac \cos B$$
$$a^2 = b^2 + c^2 - 2bc \cos A$$

Both of these formulas could be arrived at by constructing the other two altitudes of the triangle ABC.

Sec. 8-4. The Law of Cosines

NOTE: The case of an oblique triangle containing an obtuse angle will be considered in problem (1) of the Exercises.

Example A: Solve the triangle with $a = 30$, $b = 41.6$, and $C = 134°$. Side c can be found by using the law of cosines.

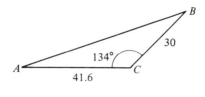

$$c^2 = a^2 + b^2 - 2ab \cos C$$
$$= 30^2 + 41.6^2 - 2(30)(41.6)(\cos 134°)$$
$$= 900 + 1730.56 + 1733.97$$
$$c = \sqrt{4364.53}$$
$$c = 66.06$$

The law of sines can be used to find the remaining parts.

$$\frac{\sin 134°}{66.06} = \frac{\sin A}{30}$$

$$\sin A = \frac{30(\sin 134°)}{66.06} = .3266$$

$$A = 19.067°$$

$$B = 180° - (134° + 19.067°) = 26.93°$$

Example B: Solve the triangle with sides $a = 3.14$, $b = 4.69$, and $c = 6.01$. This triangle can be solved using the law of cosines when it is solved for $\cos A$, $\cos B$, or $\cos C$.

$$a^2 = b^2 + c^2 - 2bc \cos A$$
$$2bc \cos A = b^2 + c^2 - a^2$$
$$\cos A = \frac{b^2 + c^2 - a^2}{2bc}$$
$$= \frac{4.69^2 + 6.01^2 - (3.14)^2}{2(4.69)(6.01)}$$
$$= \frac{21.99 + 36.12 - 9.86}{56.37}$$
$$= \frac{48.25}{56.37}$$
$$= .8560$$

$$A = 31.12°$$

The law of sines can be used to determine one other angle.

$$\frac{\sin 31.12°}{3.14} = \frac{\sin B}{4.69}$$

$$\sin B = \frac{(4.69)(\sin 31.12°)}{3.14}$$

$$= .7719$$

$$B = 50.53°$$

$$C = 180° - (31.12° + 50.53°)$$

$$= 98.35°$$

Example C: A boy bicycles 3 kilometers due west; then he hikes 2 kilometers in a direction 40° east of north. How far is he from his point of departure? Since two sides and the included angle are given, the third side, which is his displacement from the point of departure, can be determined, using the law of cosines.

$$c^2 = a^2 + b^2 - 2ab \cos C$$
$$= 3^2 + 2^2 - 2(2)(3) \cos 50°$$
$$= 9 + 4 - 12(.6428)$$
$$= 9 + 4 - 7.71$$
$$= 5.29$$
$$c = \sqrt{5.29} \text{ km}$$
$$= 2.3 \text{ km}$$

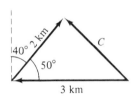

Exercises

1. Show that the law of cosines also holds for an oblique triangle containing an obtuse angle.

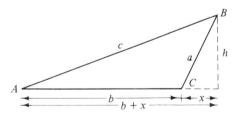

Solve each of the following triangles for the missing parts.

2. $a = 10, b = 12, C = 52°$
3. $A = 131°, c = 5, b = 7$
4. $a = 5, b = 4, c = 6$
5. $a = 30, b = 20, c = 30$
6. $B = 10°, a = 1, c = .68$
7. Find the resultant force of $\vec{A}$ and $\vec{B}$.

Sec. 8–4. The Law of Cosines 249

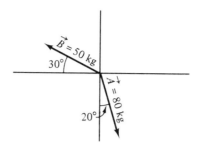

8. Find θ_A and θ_B if the tension in $\vec{A}$ is 54 pounds and the tension in $\vec{B}$ is 80 pounds.

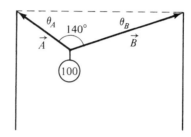

9. Find the measure of α in the accompanying diagram.

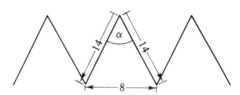

10. If $\vec{V}_1 = 7.5$ ft/sec, $\vec{V}_2 = 10$ ft/sec, $\vec{V}_3 = 5.2$ ft/sec, find B.

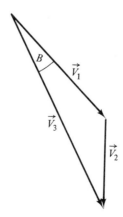

8-5. CHAPTER REVIEW

In problems 1–4, determine the horizontal and vertical components of the given vectors.

1. $\vec{A}$: $|A| = 2.5$, $\theta_A = -120°$
2. $\vec{B}$: $|B| = .75$, $\theta_B = 135°$
3.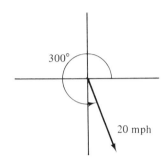

4. The velocity of an object is 1500 kilometers per hour in a direction southwest.

5. Determine the magnitude and direction of the resultant vector.

   $|A| = 18.1$, $\theta_A = 30°$
   $|B| = 9.3$, $\theta_B = -100°$

   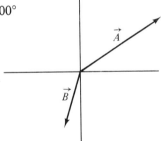

6. A horizontal force on an object is 15 kilograms and a vertical force on the same object is 45 kilograms. What is the resultant force?

Solve each of the following problems.

7. A jet travels on a course due east at 850 kilometers per hour. A wind blows from the southwest at 45 kilometers per hour. What is the velocity of the jet with respect to the ground?

8. A force of 15 kilograms acts on an object. At a point 60° from this force, another force of 20 kilograms acts on the same object. What is the resultant force?

9. A car leaves a city and travels 4 kilometers northwest. A second car leaves the same city and travels 10 kilometers due east. What is the diplacement between the two cars?

Sec. 8–5. Chapter Review

10. A single force of 120 kilograms is replacing two forces of 100 kilograms and 30 kilograms. What is the angle between the two original forces?

11. Determine the displacement from A to C in the following diagram.

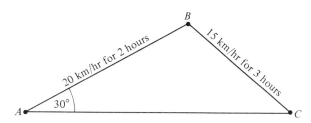

12. A pilot flying on a straight course sights an object on the ground at an angle of depression of 23°. After flying 1000 meters the object is sighted again at an angle of depression of 35°. If both observations were taken while the pilot was flying on the same side of the object, what was the pilot's distance from the object after the second sighting?

13. The navigator of a ship sights a lighthouse at 300° when the ship is traveling due north. One hour later he sights the same lighthouse at 210° while on the same course. After the second sighting, the distance from the lighthouse to the ship is 99 kilometers. How far has the ship sailed since the first sighting?

14. The following triangle was laid out in order to determine the distance from a point A on the bank of a river to an object B on the opposite side of the river. Determine the distance from A to B.

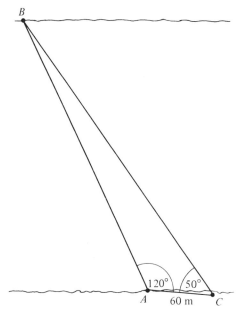

15. A ship is sailing due east when an obstruction is observed at 63° east of north. After the ship has traveled 2200 meters the obstruction is observed at 48° east of north. If the ship stays on the same course, how close will the ship approach the obstruction?

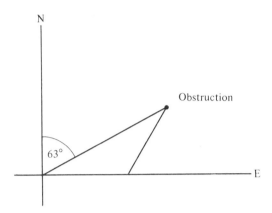

16. Solve the triangle *ABC*, given $a = 31$, $b = 52$, and $A = 33°$.

17. Two forces of 16.8 kilograms and 22.6 kilograms act on a body. If these forces make an angle of 50.2° with each other, find the magnitude of their resultant and the angle that it makes with each force.

18. A tree leaning 5.6° from a vertical position casts a shadow 31 meters long when the angle of elevation of the sun is 43.2°. Find the height of the tree.

19. A force of 80 kilograms is applied to the handle of a lawn mower. The angle between the handle and the ground is 38°. Resolve the force into its component parts.

20. A sign weighing 100 kilograms is supported away from a wall by a rigid bracket and a chain from the end of the bracket to the wall. The angle between the bracket and the chain is 35°. Calculate the force along the chain and the bracket.

21. A 500-kilogram load is being rolled up an inclined plane. The plane is 10 meters long and 4 meters high. What force tends to make the load roll back down the plane?

22. A 100-kilogram weight is suspended by two wires, each making an angle of 45° with the horizontal. Find the tension in each wire.

23. A weight of 20 kilograms rests on an inclined plane that is 1.83 meters high and 3.05 meters long. Find: (a) the force pushing the block against the plane, and (b) the force tending to make it slide down the plane.

24.

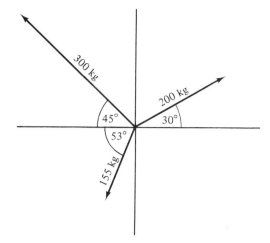

(a) Find the resultant of the three forces.
(b) Find the magnitude and direction of a fourth force that must be added to make the resultant force zero.

chapter 9

Graphs of the Trigonometric Functions

9-1. Graphs of the Trigonometric Functions

As we pointed out in Chapter 4, one of the best ways of analyzing functions is by examining their graphs. Constructing the graph of a function gives us a picture of the relationship between the variables involved. In this chapter we will construct the basic graphs of the trigonometric functions and see what effect certain important concepts have on these graphs. We will also analyze the properties of the individual trigonometric functions and relationships between different functions from their graphs. Finally, we will see some applications of the trigonometric graphs.

In constructing the graphs of the trigonometric functions, several ideas should be kept in mind. The angle will be considered as the independent variable and the calculated values for the functions will be the values for the dependent variable. We have already seen that the values of the trigonometric functions repeat every 360° or 2π radians. Therefore, we need only take values of the angle from 0° to 360° or from 0 to 2π radians, inclusive. The angle for each function will be expressed in radians. This is because the angle and functional value will then both be numbers and these numbers may be assigned different units of measurement in applied problems.

Sec. 9–1. Graphs of the Trigonometric Functions

Example A: Construct the graph of $Y = \text{Sin } X$.

	0°	30°	60°	90°	120°	150°	180°	210°	240°	270°	300°	330°	360°
X	0	$\frac{\pi}{6}$	$\frac{\pi}{3}$	$\frac{\pi}{2}$	$\frac{2\pi}{3}$	$\frac{5\pi}{6}$	π	$\frac{7\pi}{6}$	$\frac{4\pi}{3}$	$\frac{3\pi}{2}$	$\frac{5\pi}{3}$	$\frac{11\pi}{6}$	2π
Y	0	.5	.87	1	.87	.5	0	−.5	−.87	−1	−.87	−.5	0

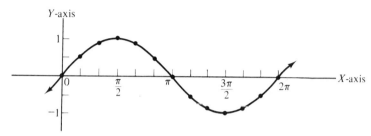

NOTES: The maximum value for the basic sine function is 1 $\left(\text{reached at } 90° \text{ or } \frac{\pi}{2}\right)$, while the minimum value is −1 $\left(\text{reached at } 270° \text{ or } \frac{3\pi}{2}\right)$. The functional values repeat every 360° and we have indicated this by extending the graph beyond the values listed in the table. This graph represents one cycle of the basic sine function. This basic shape should be readily recognized since it will always be the same.

Example B: Construct the graph of $Y = \text{Cos } X$.

X	0	$\frac{\pi}{6}$	$\frac{\pi}{3}$	$\frac{\pi}{2}$	$\frac{2\pi}{3}$	$\frac{5\pi}{6}$	π	$\frac{7\pi}{6}$	$\frac{4\pi}{3}$	$\frac{3\pi}{2}$	$\frac{5\pi}{3}$	$\frac{11\pi}{6}$	2π
Y	1	.87	.5	0	−.5	−.87	−1	−.87	−.5	0	.5	.87	1

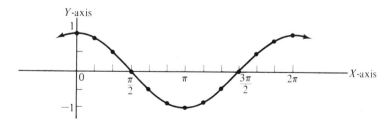

NOTES: The maximum value for the basic cosine function is 1 (reached at 0° and 360° or 0 and 2π), while the minimum value is −1 (reached at 180° or π).

The functional values repeat every 360° and this graph represents one cycle of the basic cosine function. This basic shape should be readily recognized since it will always be the same.

Example C: Construct the graph of $Y = \text{Tan } X$.

X	0	$\frac{\pi}{6}$	$\frac{\pi}{3}$	$\frac{\pi}{2}$	$\frac{2\pi}{3}$	$\frac{5\pi}{6}$	π	$\frac{7\pi}{6}$	$\frac{4\pi}{3}$	$\frac{3\pi}{2}$	$\frac{5\pi}{3}$	$\frac{11\pi}{6}$	2π
Y	0	.58	1.7	undefined	−1.7	−.58	0	.58	1.7	undefined	−1.7	−.58	0

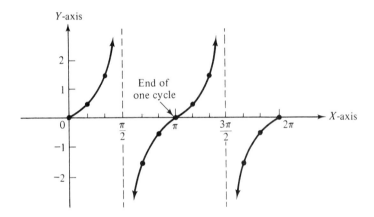

NOTES: The tangent function has no maximum or minimum values. The functional values repeat every 180° and thus, this graph represents two cycles of the basic tangent function. The tangent function is undefined for angles of 90° or $\frac{\pi}{2}$, and 270° or $\frac{3\pi}{2}$, and we have indicated this fact with asymptotes. Examining the tables for values of the trigonometric functions leads us to the fact that the tangent function becomes very large as the angle approaches 90° $\left(\frac{\pi}{2}\right)$. This basic shape should be readily recognized since it will always be the same.

Example D: Construct the graph of $Y = \text{Csc } X$.

X	0	$\frac{\pi}{6}$	$\frac{\pi}{3}$	$\frac{\pi}{2}$	$\frac{2\pi}{3}$	$\frac{5\pi}{6}$	π	$\frac{7\pi}{6}$	$\frac{4\pi}{3}$	$\frac{3\pi}{2}$	$\frac{5\pi}{3}$	$\frac{11\pi}{6}$	2π
Y	undefined	2	1.15	1	1.15	2	undefined	−2	−1.15	−1	−1.15	−2	undefined

Sec. 9–1. Graphs of the Trigonometric Functions

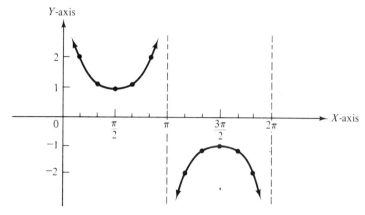

NOTES: The cosecant function has no maximum or minimum values. It does not assume any values between 1 and −1. The functional values repeat every 360° and this graph represents one cycle of the basic cosecant function. The cosecant function is undefined for angles of 0° or 0, 180° or π, and 360° or 2π, and we have indicated this fact with asymptotes. Since $\text{Csc } \theta = \dfrac{1}{\text{Sin } \theta}$, the functional values in the table above are obtained by finding the reciprocals of the sine values for the same angles. This basic shape should be readily recognized since it will always be the same.

Example E: Construct the graph of $Y = \text{Sec } X$.

X	0	$\frac{\pi}{6}$	$\frac{\pi}{3}$	$\frac{\pi}{2}$	$\frac{2\pi}{3}$	$\frac{5\pi}{6}$	π	$\frac{7\pi}{6}$	$\frac{4\pi}{3}$	$\frac{3\pi}{2}$	$\frac{5\pi}{3}$	$\frac{11\pi}{6}$	2π
Y	1	1.15	2	undefined	−2	−1.15	−1	−1.15	−2	undefined	2	1.15	1

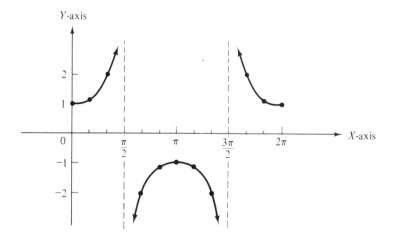

NOTES: The secant function has no maximum or minimum values. It does not assume any values between 1 and −1. The functional values repeat every 360° and this graph represents one cycle of the basic secant function. The secant function is undefined for angles of 90° or $\frac{\pi}{2}$, and 270° or $\frac{3\pi}{2}$, and we have indicated this fact with asymptotes. Since $\text{Sec } \theta = \frac{1}{\text{Cos } \theta}$, the functional values in the table above are obtained by finding the reciprocals of the cosine values for the same angles. This basic shape should be readily recognized since it will always be the same.

Example F: Construct the graph of $Y = \text{Cot } X$.

X	0	$\frac{\pi}{6}$	$\frac{\pi}{3}$	$\frac{\pi}{2}$	$\frac{2\pi}{3}$	$\frac{5\pi}{6}$	π	$\frac{7\pi}{6}$	$\frac{4\pi}{3}$	$\frac{3\pi}{2}$	$\frac{5\pi}{3}$	$\frac{11\pi}{6}$	2π
Y	undefined	1.73	.58	0	−.58	−1.73	undefined	1.73	.58	0	−.58	−1.73	undefined

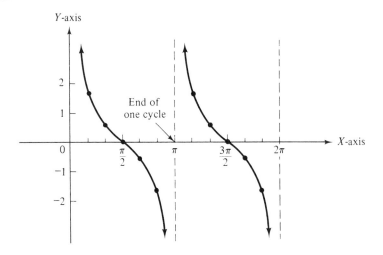

NOTES: The cotangent function has no maximum or minimum values. The functional values repeat every 180° and thus, this graph represents two cycles of the basic cotangent function. The cotangent function is undefined for angles of 0° or 0, 180° or π, and 360° or 2π, and we have indicated this fact with asymptotes. Since $\text{Cot } \theta = \frac{1}{\text{Tan } \theta}$, the functional values in the table above are obtained by finding the reciprocals of the tangent values for the same angles. This basic shape should be readily recognized since it will always be the same.

Sec. 9–1. Graphs of the Trigonometric Functions

From these graphs, the following conclusions may be reached:

1. The *functional values* of the sine, cosine, cosecant, and secant functions repeat every 360°. Those of the tangent and cotangent functions repeat every 180°.
2. The sine and cosine functions are defined for all angles, while the tangent, cosecant, secant, and cotangent functions are undefined for certain angles.
3. The sine and cosine functions have maximum values of 1 and minimum values of -1. The tangent, cosecant, secant, and cotangent functions have no maximum or minimum values. The cosecant and secant functions have no values between 1 and -1.
4. The *sine function* equals zero at the beginning, end, and midpoint of one normal cycle. Its maximum and minimum values are reached halfway between these points.
5. The *cosine function* equals zero one quarter and three quarters of the way through one normal cycle. Its maximum and minimum values are reached at the beginning, end, and midpoint of the cycle.
6. The *tangent function* equals zero at the beginning and end of one normal cycle. It is undefined at the midpoint of the cycle.
7. The *cosecant function* is undefined at the beginning, end, and midpoint of one normal cycle. It reaches its "turning points" one quarter and three quarters of the way through the cycle.
8. The *secant function* is undefined one quarter and three quarters of the way through one normal cycle. It reaches its "turning points" at the beginning, end, and midpoint of the cycle.
9. The *cotangent function* equals zero at the midpoint of one normal cycle. It is undefined at the beginning and end of the cycle.

Exercises

1. Complete the following table for the functions $Y = \text{Sin } X$, $Y = \text{Cos } X$, and $Y = \text{Tan } X$. Then draw the graphs of these functions.

X	$-\pi$	$\frac{-3\pi}{4}$	$\frac{-\pi}{2}$	$\frac{-\pi}{4}$	0	$\frac{\pi}{4}$	$\frac{\pi}{2}$	$\frac{3\pi}{4}$	π
Sin X									
Cos X									
Tan X									

2. Complete the following table for the functions $Y = \text{Sin } X$, $Y = \text{Cos } X$, and $Y = \text{Tan } X$. Then draw the graphs of these functions.

X	0	1	2	3	4	5	6	7
Sin X								
Cos X								
Tan X								

For each of the following functions, construct a table of values by using values from $-180°$ $(-\pi)$ to $360°$ (2π) at intervals of $30°$ $\left(\dfrac{\pi}{6}\right)$. Then draw the graph of each of these functions.

3. $Y = \text{Sin } X$ 4. $Y = \text{Cos } X$ 5. $Y = \text{Tan } X$
6. $Y = \text{Csc } X$ 7. $Y = \text{Sec } X$ 8. $Y = \text{Cot } X$

9-2. Amplitude, Period, and Displacement

In this section we will see how the trigonometric graphs, while remaining unchanged in their basic shapes and characteristics, may be modified or affected by certain concepts.

The first of these concepts is called the *amplitude*. If $Y = AF(X)$, where F represents any of the six trigonometric functions, each of the normal functional values of $F(X)$ is to be multiplied by A. The absolute value of this number A is the amplitude of the function $F(X)$.

Example A:
1. If $Y = 3 \text{ Sin } X$, the amplitude of this function is $|3| = 3$.
2. If $Y = -2 \text{ Cos } X$, the amplitude of this function is $|-2| = 2$.
3. If $Y = 5 \text{ Sec } X$, the amplitude of this function is $|5| = 5$.

Example B: Construct the graph of $Y = 3 \text{ Sin } X$.

X	0	$\dfrac{\pi}{6}$	$\dfrac{\pi}{3}$	$\dfrac{\pi}{2}$	$\dfrac{2\pi}{3}$	$\dfrac{5\pi}{6}$	π	$\dfrac{7\pi}{6}$	$\dfrac{4\pi}{3}$	$\dfrac{3\pi}{2}$	$\dfrac{5\pi}{3}$	$\dfrac{11\pi}{6}$	2π
Y	0	1.5	2.61	3	2.61	1.5	0	-1.5	-2.61	-3	-2.61	-1.5	0

Sec. 9–2. Amplitude, Period, and Displacement

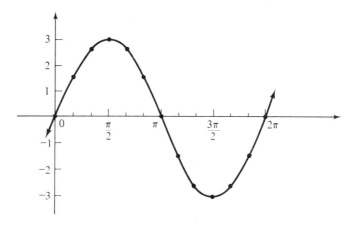

Each functional value of the normal sine graph has been multiplied by 3.

Example C: Construct the graph of $Y = -2 \cos X$.

X	0	$\frac{\pi}{6}$	$\frac{\pi}{3}$	$\frac{\pi}{2}$	$\frac{2\pi}{3}$	$\frac{5\pi}{6}$	π	$\frac{7\pi}{6}$	$\frac{4\pi}{3}$	$\frac{3\pi}{2}$	$\frac{5\pi}{3}$	$\frac{11\pi}{6}$	2π
Y	−2	−1.74	−1	0	1	1.74	2	1.74	1	0	−1	−1.74	−2

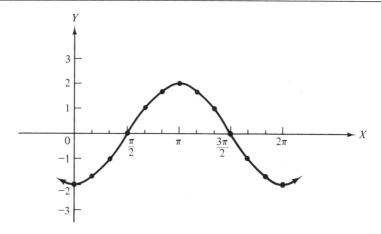

Each functional value of the normal Cos graph has been multiplied by −2. A negative value for A will *invert* the normal graph of any trigonometric function.

Example D: Construct the graph of $Y = 5 \sec X$.

X	0	$\frac{\pi}{6}$	$\frac{\pi}{3}$	$\frac{\pi}{2}$	$\frac{2\pi}{3}$	$\frac{5\pi}{6}$	π	$\frac{7\pi}{6}$	$\frac{4\pi}{3}$	$\frac{3\pi}{2}$	$\frac{5\pi}{3}$	$\frac{11\pi}{6}$	2π
Y	5	5.75	10	undefined	−10	−5.75	−5	−5.75	−10	undefined	10	5.75	5

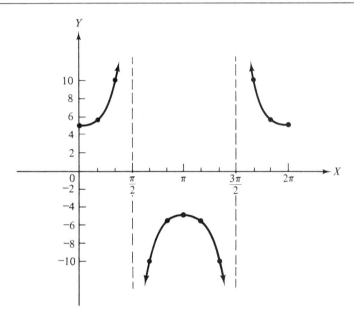

Each functional value of the normal secant graph has been multiplied by 5.

Another change that may be made in the normal trigonometric graphs is in the frequency with which the functional values repeat. The trigonometric functions are examples of what are called *periodic functions*. That is, their functional values repeat at regular intervals. The *period* of such functions is the horizontal distance between any point of the graph of the function and the next corresponding point for which the functional values start repeating. From the previous section, we should realize that the sine, cosine, cosecant, and secant functions have normal periods of 360° (2π), while the tangent and cotangent functions have normal periods of 180° (π). In general, if $Y = F(X)$ represents any of the six trigonometric functions with a normal period of P, then $Y = F(BX)$ represents the same trigonometric function with a period of $\frac{P}{B}$.

Example E:

1. The period of $Y = \text{Sin}(2X)$ is $\frac{2\pi}{2} = \pi$.

Sec. 9–2. Amplitude, Period, and Displacement

2. The period of $Y = \text{Tan}(3X)$ is $\frac{\pi}{3}$.
3. The period of $Y = \text{Csc}(\pi X)$ is $\frac{2\pi}{\pi} = 2$.

Example F: Construct the graph of $Y = \text{Sin}(2X)$. In this problem, we will use multiples of $\frac{\pi}{8}$ (22.5°) from 0 (0°) to π (180°) since we will be multiplying the angle by 2.

X	0	$\frac{\pi}{8}$	$\frac{\pi}{4}$	$\frac{3\pi}{8}$	$\frac{\pi}{2}$	$\frac{5\pi}{8}$	$\frac{3\pi}{4}$	$\frac{7\pi}{8}$	π
2X	0	$\frac{\pi}{4}$	$\frac{\pi}{2}$	$\frac{3\pi}{4}$	π	$\frac{5\pi}{4}$	$\frac{3\pi}{2}$	$\frac{7\pi}{4}$	2π
Y	0	.7	1	.7	0	$-.7$	-1	$-.7$	0

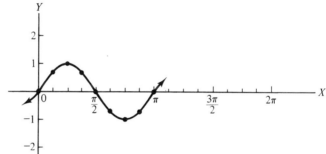

The period has been altered from the normal period for the sine function (2π) to a new period $\left(\pi = \frac{2\pi}{2}\right)$.

Example G: Construct the graph of $Y = \text{Tan}(3X)$. In this problem we will use multiples of $\frac{\pi}{12}$ (15°) from 0 (0°) to $\frac{2\pi}{3}$ (120°) since we will be multiplying the angle by 3.

X	0	$\frac{\pi}{12}$	$\frac{\pi}{6}$	$\frac{\pi}{4}$	$\frac{\pi}{3}$	$\frac{5\pi}{12}$	$\frac{\pi}{2}$	$\frac{7\pi}{12}$	$\frac{2\pi}{3}$
3X	0	$\frac{\pi}{4}$	$\frac{\pi}{2}$	$\frac{3\pi}{4}$	π	$\frac{5\pi}{4}$	$\frac{3\pi}{2}$	$\frac{7\pi}{4}$	2π
Y	0	1	undefined	-1	0	1	undefined	-1	0

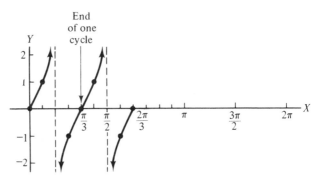

The period has been altered from the normal period for the tangent function (π) to a new period $\left(\dfrac{\pi}{3}\right)$.

Example H: Construct the graph of $Y = \text{Csc}\,(\pi X)$. In this problem, we will use multiples of $\frac{1}{4}$ (14.3°) from 0 (0°) to 2 (114.4°) since we will be multiplying the angle by π.

X	0	$\dfrac{1}{4}$	$\dfrac{1}{2}$	$\dfrac{3}{4}$	1	$\dfrac{5}{4}$	$\dfrac{3}{2}$	$\dfrac{7}{4}$	2
πX	0	$\dfrac{\pi}{4}$	$\dfrac{\pi}{2}$	$\dfrac{3\pi}{4}$	π	$\dfrac{5\pi}{4}$	$\dfrac{3\pi}{2}$	$\dfrac{7\pi}{4}$	2π
Y	undefined	1.4	1	1.4	undefined	-1.4	-1	-1.4	undefined

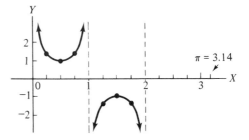

The period has been altered from the normal period for the cosecant function (2π) to a new period $\left(2 = \dfrac{2\pi}{\pi}\right)$.

One other change that may be made in the basic trigonometric graphs is to move or shift the curve either left or right. If $Y = F(X + C)$, where F represents any of the six trigonometric functions, C is called a *phase angle* and indicates a shift or *displacement* of the curve $Y = F(X)$. This displacement is to the left if C is positive and to the right if C is negative. The numerical value of the displace-

Sec. 9–2. Amplitude, Period, and Displacement

ment is equal to $\frac{C}{B}$. If $B = 1$, the numerical value of the displacement is equal to C.

Example I:

1. If $Y = \text{Sin}\left(X + \frac{\pi}{4}\right)$, the phase angle is $\frac{\pi}{4}$ and the displacement is $\frac{\pi/4}{1} = \frac{\pi}{4}$ to the left.

2. If $Y = \text{Cot}\left(X - \frac{\pi}{2}\right)$, the phase angle is $-\frac{\pi}{2}$ and the displacement is $\frac{\pi/2}{1} = \frac{\pi}{2}$ to the right.

3. If $Y = \text{Cos}\left(2X - \frac{\pi}{2}\right)$, the phase angle is $-\frac{\pi}{2}$ and the displacement is $\frac{\pi/2}{2} = \frac{\pi}{4}$ to the right.

Example J: Construct the graph of $Y = \text{Sin}\left(X + \frac{\pi}{4}\right)$. In this problem, we will use multiples of $\frac{\pi}{4}$ from $-\frac{\pi}{4}$ to $\frac{7\pi}{4}$ since we will be adding $\frac{\pi}{4}$ to each angle.

X	$\frac{-\pi}{4}$	0	$\frac{\pi}{4}$	$\frac{\pi}{2}$	$\frac{3\pi}{4}$	π	$\frac{5\pi}{4}$	$\frac{3\pi}{2}$	$\frac{7\pi}{4}$
$X + \frac{\pi}{4}$	0	$\frac{\pi}{4}$	$\frac{\pi}{2}$	$\frac{3\pi}{4}$	π	$\frac{5\pi}{4}$	$\frac{3\pi}{2}$	$\frac{7\pi}{4}$	2π
Y	0	.7	1	.7	0	−.7	−1	−.7	0

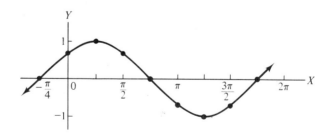

The normal sine curve has been displaced $\frac{\pi}{4}$ units to the left.

Example K: Construct the graph of $Y = \text{Cot}\left(X - \frac{\pi}{2}\right)$. In this problem, we will use multiples of $\frac{\pi}{4}$ from $\frac{\pi}{2}$ to $\frac{5\pi}{2}$ since we will be subtracting $\frac{\pi}{2}$ from each angle.

X	$\frac{\pi}{2}$	$\frac{3\pi}{4}$	π	$\frac{5\pi}{4}$	$\frac{3\pi}{2}$	$\frac{7\pi}{4}$	2π	$\frac{9\pi}{4}$	$\frac{5\pi}{2}$
$X - \frac{\pi}{2}$	0	$\frac{\pi}{4}$	$\frac{\pi}{2}$	$\frac{3\pi}{4}$	π	$\frac{5\pi}{4}$	$\frac{3\pi}{2}$	$\frac{7\pi}{4}$	2π
Y	undefined	1	0	−1	undefined	1	0	−1	undefined

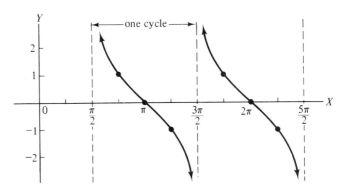

The normal cotangent curve has been displaced $\frac{\pi}{2}$ units to the right.

Example L: Construct the graph of $Y = \text{Cos}\left(2X - \frac{\pi}{2}\right)$. In this problem, we will use multiples of $\frac{\pi}{8}$ from $\frac{\pi}{4}$ to $\frac{5\pi}{4}$. Also, we will have a new period of $\frac{2\pi}{2} = \pi$, as well as a displacement.

X	$\frac{\pi}{4}$	$\frac{3\pi}{8}$	$\frac{\pi}{2}$	$\frac{5\pi}{8}$	$\frac{3\pi}{4}$	$\frac{7\pi}{8}$	π	$\frac{9\pi}{8}$	$\frac{5\pi}{4}$
$2X - \frac{\pi}{2}$	0	$\frac{\pi}{4}$	$\frac{\pi}{2}$	$\frac{3\pi}{4}$	π	$\frac{5\pi}{4}$	$\frac{3\pi}{2}$	$\frac{7\pi}{4}$	2π
Y	1	.7	0	−.7	−1	−.7	0	.7	1

Sec. 9-2. Amplitude, Period, and Displacement

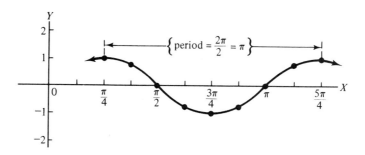

The normal cosine curve has been displaced $\dfrac{\pi}{4}\left(\dfrac{\frac{\pi}{2}}{2}\right)$ units to the right.

We can now summarize the 3 changes that may be made in any of the normal trigonometric graphs. If $Y = AF(BX + C)$, where F represents any of the six trigonometric functions, then:

1. The amplitude is $|A|$.
2. The period is the normal period (2π or π) divided by B.
3. The displacement is $\dfrac{C}{B}$ and is to the left if C is positive and to the right if C is negative. C itself is a phase angle.

Example M:

1. If $Y = 2 \operatorname{Sin} (3X - \pi)$, the amplitude is 2, the period is $\dfrac{2\pi}{3}$, and the displacement is $\dfrac{\pi}{3}$ units to the right.
2. If $Y = -3 \operatorname{Cos} (\pi X - \pi)$, the amplitude is $|-3| = 3$, the period is $2\left(\dfrac{2\pi}{\pi}\right)$, and the displacement is $1\left(\dfrac{\pi}{\pi}\right)$ unit to the right.
3. If $Y = 4 \operatorname{Tan}\left(2X + \dfrac{\pi}{2}\right)$, the amplitude is 4, the period is $\dfrac{\pi}{2}$, and the displacement is $\dfrac{\pi}{4}\left(\dfrac{\frac{\pi}{2}}{2}\right)$ unit to the left.

Combining the significance of the amplitude, period, and displacement with the basic shape and characteristics of the trigonometric graphs pointed out in Section 9-1, we are now able to sketch the graph of any trigonometric function of the type $Y = AF(BX + C)$ without making a table of values.

Example N: Sketch the graph of $Y = 2 \sin(3X - \pi)$, with amplitude = 2, period = $\frac{2\pi}{3}$, and displacement = $\frac{\pi}{3}$ to the right.

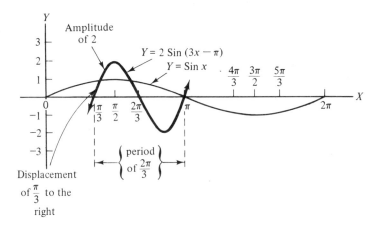

Example O: Sketch the graph of $Y = -3 \cos(\pi X - \pi)$, with amplitude = $|-3| = 3$, period = $\frac{2\pi}{\pi} = 2$, and displacement = $\frac{\pi}{\pi} = 1$ to the right.

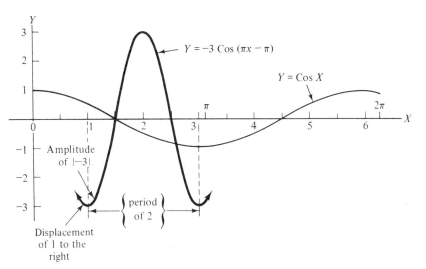

Example P: Sketch the graph of $Y = 4 \tan\left(2X + \frac{\pi}{2}\right)$, with amplitude = 4, period = $\frac{\pi}{2}$, and displacement = $\frac{\frac{\pi}{2}}{2} = \frac{\pi}{4}$ to the left.

Sec. 9–2. Amplitude, Period, and Displacement

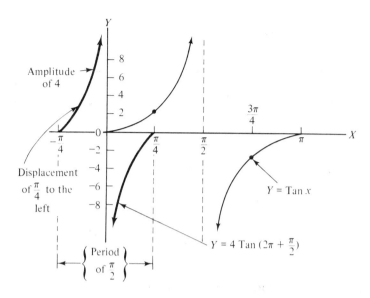

Exercises

For each of the following, list the amplitude, period, and displacement.

1. $Y = -2 \text{ Tan }(4X)$
2. $Y = \text{Cos }(3X + \pi)$
3. $Y = 6 \text{ Csc}\left(X - \dfrac{\pi}{2}\right)$
4. $Y = 5 \text{ Cot}\left(X + \dfrac{1}{2}\right)$
5. $Y = \text{Sec }(2\pi X - \pi)$
6. $Y = 4 \text{ Sin}\left(5X + \dfrac{\pi}{2}\right)$
7. $Y = -5 \text{ Cos }(2X - 6)$
8. $Y = 2 \text{ Tan}\left(\pi X + \dfrac{\pi}{4}\right)$
9. $Y = -\text{Cot}\left(4X + \dfrac{\pi}{2}\right)$
10. $Y = -3 \text{ Sin}\left(\dfrac{1}{3}X - \pi\right)$

For each of the following, determine the amplitude, period, and displacement, sketch one cycle of the graph, and label the starting and ending points of the cycle. (Do not make out a table of values).

11. $Y = 5 \text{ Cos } X$
12. $Y = -3 \text{ Tan } X$
13. $Y = 2 \text{ Csc } X$
14. $Y = -2 \text{ Cot } X$
15. $Y = \text{Sin }(4X)$
16. $Y = \text{Cos }(3X)$
17. $Y = \text{Tan}\left(\dfrac{1}{2}X\right)$
18. $Y = \text{Csc}\left(\dfrac{X}{3}\right)$
19. $Y = \text{Cot}\left(X - \dfrac{\pi}{4}\right)$
20. $Y = \text{Sin}\left(X + \dfrac{\pi}{3}\right)$

21. $Y = \text{Tan}(X - 1)$

22. $Y = \text{Sec}\left(X + \dfrac{1}{2}\right)$

23. $Y = 3 \text{Sin}\left(2X + \dfrac{\pi}{2}\right)$

24. $Y = -4 \text{Cos}\left(\dfrac{1}{2}X - \pi\right)$

25. $Y = 2 \text{Tan}\left(\dfrac{1}{3}X\right)$

26. $Y = -\text{Csc}(2X + \pi)$

27. $Y = \text{Cot}\left(2X - \dfrac{\pi}{4}\right)$

28. $Y = -2 \text{Sec}\left(X - \dfrac{\pi}{2}\right)$

29. $Y = -\text{Sin}\left(3X - \dfrac{\pi}{2}\right)$

30. $Y = 5 \text{Cos}\left(\dfrac{X}{3} + \pi\right)$

31. $Y = 4 \text{Tan}(2\pi X - 2\pi)$

32. $Y = -3 \text{Sin}(\pi X + 2\pi)$

33. $Y = 3 \text{Tan}\left(\dfrac{\pi}{2}X + \pi\right)$

34. $Y = 2 \text{Cos}\left(2X - \dfrac{\pi}{3}\right)$

9–3. Composite Trigonometric Functions

When the trigonometric functions are used in technical and scientific areas, very often an expression involving more than one function is encountered. For example, (Sin X + 2 Cos X), (Cos X − Sin $2X$), (3 Tan X + Sin X), or other expressions similar to these may appear in a problem. Our objective in this section is to construct the graphs of such expressions. This is done by sketching the graph of each function separately and then graphically adding the functional values (values of the dependent variable). This procedure is usually referred to as the *addition of ordinates*.

Example A: Sketch the graph of $Y = \text{Sin } X + 2 \text{ Cos } X$ by using the addition of ordinates method.

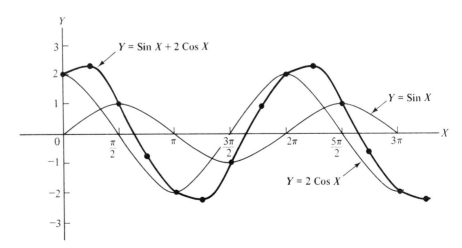

NOTE: The graphs of $Y = \operatorname{Sin} X$ and $Y = 2 \operatorname{Cos} X$ are sketched on the same set of axes (light curves on the graph). The functional values of these two curves are then added to obtain the points on the resulting curve (heavy curve on the graph). Where one curve crosses the horizontal axis, the point on the resulting curve will correspond to the value of the other curve. Care and neatness will insure an accurate result.

Example B: Sketch the graph of $Y = \operatorname{Cos} X - \operatorname{Sin}(2X)$ by using the addition of ordinates method.

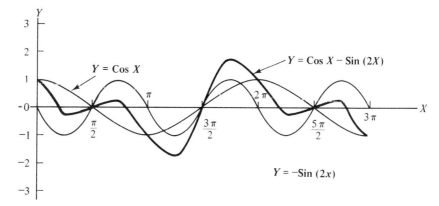

Example C: Sketch the graph of $Y = 2 \operatorname{Cos}(2X) - 3 \operatorname{Sin} X$ by using the addition of ordinates method.

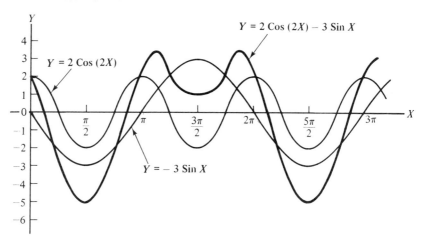

Other functions may be combined with the trigonometric functions in the same way.

Example D: Sketch the graph of $Y = 2X + \operatorname{Cos} X$ by using the addition of ordinates method.

Ch. 9. Graphs of the Trigonometric Functions

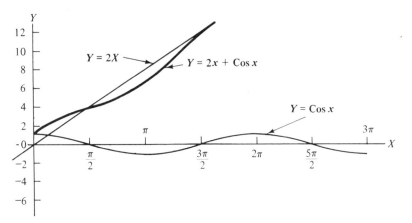

Example E: Sketch the graph of $Y = X - 2 \cos(\pi X)$ by using the addition of ordinates method.

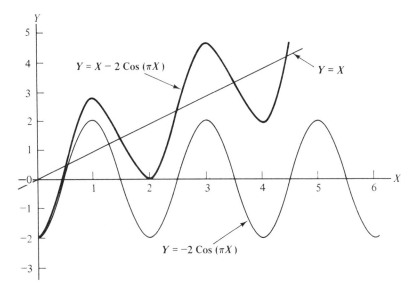

Exercises

Sketch each of the following curves by using the addition of ordinates method.

1. $Y = 2 \sin X + \cos X$
2. $Y = \cos(2X) - 3 \sin\left(\dfrac{X}{2}\right)$
3. $Y = -2 \cos X - 2 \sin(2X)$
4. $Y = 2 \sin(\pi X) + \cos(2\pi X)$
5. $Y = \sin(3X) - \cos\left(X + \dfrac{\pi}{4}\right)$
6. $Y = 3 \cos X + \sin\left(X - \dfrac{\pi}{3}\right)$
7. $Y = 2 \cos(\pi X) - \sin\left(\dfrac{\pi}{2} X\right)$
8. $Y = 2 \sin\left(X + \dfrac{\pi}{3}\right) - 2 \cos X$

Sec. 9–4. Applications of the Graphs of the Trigonometric Functions

9. $Y = \operatorname{Sin}\left(X + \frac{\pi}{4}\right) + \operatorname{Cos}\left(X - \frac{\pi}{3}\right)$

10. $Y = 2 \operatorname{Cos}(X - \pi) - \operatorname{Cos} X$

11. $Y = 3 \operatorname{Sin}\left(X - \frac{\pi}{2}\right) + 2 \operatorname{Sin}\left(X - \frac{\pi}{4}\right)$

12. $Y = \operatorname{Cos}\left(\frac{1}{2}X - \frac{\pi}{4}\right) + \operatorname{Cos}\left(2X - \frac{\pi}{2}\right)$

13. $Y = X - \operatorname{Cos} X$ 14. $Y = X + 2 \operatorname{Sin} X$

15. $Y = X - 2 \operatorname{Sin}(\pi X)$ 16. $Y = 3 \operatorname{Cos}(2X) - X$

17. $Y = \frac{X}{2} + 3 \operatorname{Sin}(2X)$ 18. $Y = \frac{X}{2} - \operatorname{Sin}(X + \pi)$

19. $Y = 2X + \operatorname{Cos}(3X + \pi)$ 20. $Y = \operatorname{Cos}\left(\pi X - \frac{\pi}{2}\right) - 2X$

21. The electromotive force (E) in a certain electric circuit is given by the equation $E = IR \operatorname{Sin}(\omega t) - \frac{I}{\omega C} \operatorname{Cos}(\omega t)$. Sketch the curve of this equation for $I = 2$, $R = 55$, $\omega = 60$ cycles/sec., and $C = .000014$.

22. In the field of optics, two waves interfere destructively if, when they pass through the same medium, the amplitude of their sum is zero. Sketch the curve representing the sum of $Y = \operatorname{Sin} X$ and $Y = \operatorname{Cos}\left(X + \frac{\pi}{2}\right)$, and determine whether destructive interference occurs.

9–4. Applications of the Graphs of the Trigonometric Functions

In this section we will look at a few very important technical applications of the trigonometric graphs. Previously we have talked (in Section 7–8) about the velocity of an object traveling in a circular path. If this angular velocity for a given object is constant, the projection of the object on a diameter of the circle about which the motion occurs is called *simple harmonic motion*. Examples of such motion are the horizontal or vertical position of the end of a spoke on a wheel which is turning, the movement of an object on the end of a spring, or any other uniform "back-and-forth" motion. As the object P moves about the circle with constant velocity, its position may be projected on the diameter D. The projections of some of the positions of P are shown in Figure 9-4.1. From this figure we should see that as P moves about the circle, the projection moves back and forth on the diameter D. This back-and-forth motion of the projection is what we mean by simple harmonic motion.

From Figure 9-4.2, we note that the length of the projection (d) of any position will be given by $d = R \operatorname{Sin} \theta$, where R is the radius of the circle and θ is

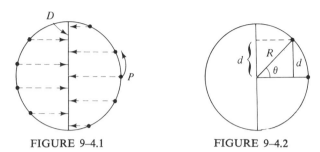

FIGURE 9-4.1   FIGURE 9-4.2

the angle giving the position of the object measured from a horizontal position (standard position).

In Section 7-8 we saw that $\omega = \dfrac{\theta}{t}$. Therefore, $\theta = \omega t$. Thus, the length of the projection in terms of time (t) is given by $d = R \operatorname{Sin}(\omega t)$. So the simple harmonic motion of the projection may be represented graphically by a Sin curve with an amplitude of R and a period of $\dfrac{2\pi}{\omega}$ seconds. This graphical representation is shown in Figure 9-4.3.

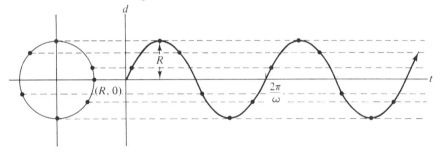

FIGURE 9-4.3

Example A: Sketch one cycle of the curve of the simple harmonic motion represented by $d = R \operatorname{Sin}(\omega t)$ when $R = 3$ and $\omega = 4$ rad/sec.

Therefore, $d = 3 \operatorname{Sin}(4t)$, with amplitude = 3, period = $\dfrac{2\pi}{4} = \dfrac{\pi}{2}$, and displacement = 0.

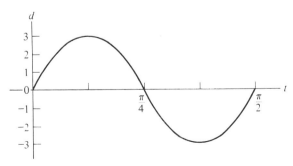

Sec. 9–4. Applications of the Graphs of the Trigonometric Functions   **275**

Example B: Sketch one cycle of the curve of the simple harmonic motion represented by $d = R \sin(\omega t)$, when $R = 2$ and $\omega = 30$ rpm. Since 1 rpm = 2π rad/min, 30 rpm = $30(2\pi) = 60\pi$ rad/min.

Therefore, $d = 2 \sin(60\pi t)$, with amplitude = 2, period = $\frac{2\pi}{60\pi} = \frac{1}{30}$ min, and displacement = 0.

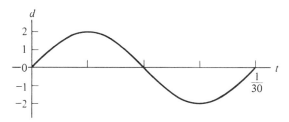

If the object starts from a position other than $(R, 0)$ (i.e., other than standard position), then we will have a phase angle, since the position $(R, 0)$ corresponds to an angle of $0°$. For example, if we start at $(.7R, .7R)$, the phase angle would be $\frac{\pi}{4}$ (45°). If we start at $(0, R)$, the phase angle would be $\frac{\pi}{2}$ (90°). In these situations the simple harmonic motion would be represented by $d = R \sin(\omega t + C)$ where C is the phase angle.

Example C: Sketch one cycle of the curve of the simple harmonic motion represented by $d = R \sin(\omega t + C)$ when $R = 2$, $\omega = 1$ rad/sec, and $C = \frac{\pi}{2}$.

Therefore, $d = 2 \sin\left(t + \frac{\pi}{2}\right)$, with amplitude = 2, period = $\frac{2\pi}{1} = 2\pi$, and displacement = $\frac{\pi/2}{1} = \frac{\pi}{2}$ left.

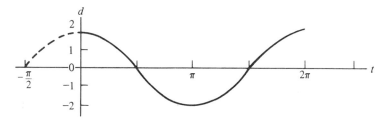

NOTE: In this case we actually end up with a cosine curve. In general, when $d = R \sin(\omega t + C)$, if $C = \frac{\pi}{2}$ when $t = 0$, then $d = R \cos(\omega t)$. Also, t cannot be negative, as is indicated by the dotted part of the graph.

Example D: Sketch one cycle of the curve of the simple harmonic motion represented by $d = R \sin(\omega t + C)$ when $R = 5$, $\omega = 60$ cycles/sec, and $C = \frac{\pi}{4}$.

Since 1 cycle/sec = 2π rad/sec, 60 cycles/sec = 60 (2π) = 120π rad/sec. Therefore, $d = 5 \text{ Sin} \left(120\pi t + \frac{\pi}{4}\right)$, with amplitude = 5, period = $\frac{2\pi}{120\pi} = \frac{1}{60}$ sec, and displacement = $\frac{\pi/4}{120\pi} = \frac{1}{480}$ sec left.

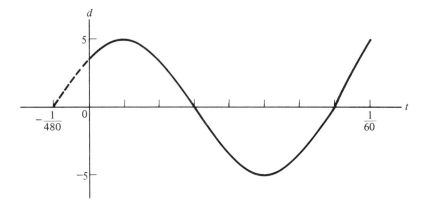

Graphs of the trigonometric functions are also used in the field of electronics where alternating current is involved. Alternating current is induced (caused) in a coil of wire which rotates in a magnetic field. If this rotation is with a constant angular velocity (ω), then the current (i) in the wire at any time (t) is given by the equation $i = I \text{ Sin } (\omega t + C)$, where I is the maximum possible current and C is the phase angle.

Example E: Sketch two cycles of the curve represented by $i = I \text{ Sin } (\omega t + C)$ when $I = 5$ amps, $\omega = 60$ cycles/sec, and $C = 0$.

Therefore, $i = 5 \text{ Sin } (120\pi t)$, with amplitude = 5, period = $\frac{2\pi}{120\pi} = \frac{1}{60}$ sec, and displacement = 0.

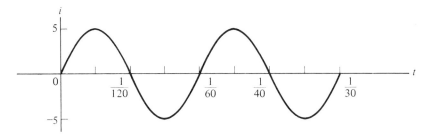

Example F: Sketch two cycles of the curve represented by $i = I \text{ Sin } (\omega t + C)$ when $I = 3$ amps, $\omega = 60$ cycles/sec, and $C = \frac{\pi}{3}$.

Therefore, $i = 3 \sin\left(120\pi t + \frac{\pi}{3}\right)$, with amplitude = 3, period = $\frac{2\pi}{120\pi}$ = $\frac{1}{60}$ sec, and displacement = $\frac{\pi/3}{120\pi} = \frac{1}{360}$ sec left.

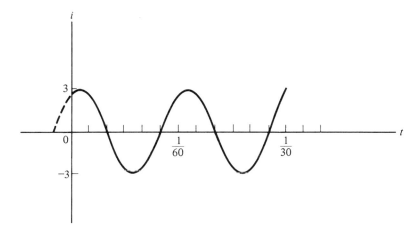

In the same type of electrical circuit, the voltage (v) at any time (t) is given by the equation $v = V \cos(\omega t + C)$, where V is the maximum possible voltage and C is the phase angle.

Example G: Sketch two cycles of the curve represented by $v = V \cos(\omega t + C)$ when $V = 110$ volts, $\omega = 60$ cycles/sec, and $C = 0$.

Therefore, $v = 110 \cos(120\pi t)$, with amplitude = 110, period = $\frac{2\pi}{120\pi}$ = $\frac{1}{60}$ sec, and displacement = 0.

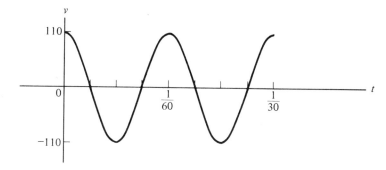

Example H: Sketch two cycles of the curve represented by $v = V \cos(\omega t + C)$ when $V = 220$ volts, $\omega = 60$ cycles/sec, and $C = \frac{\pi}{4}$.

Therefore, $v = 220 \cos\left(120\pi t + \dfrac{\pi}{4}\right)$, with amplitude = 220, period = $\dfrac{2\pi}{120\pi} = \dfrac{1}{60}$ sec, and displacement = $\dfrac{\pi/4}{120\pi} = \dfrac{1}{480}$ sec left.

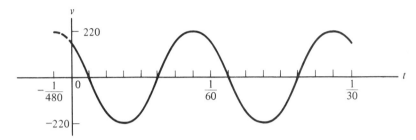

From the previous examples, we see that the current and voltage both take on positive and negative values. This simply means that they move in opposite directions and are, in fact, "alternating."

If we are given a value for current (voltage) or time, we can find the corresponding value for the other variable from the graph.

Example I: If $i = 3 \sin(\pi t)$:
1. What is the current after 1.2 seconds?
2. After how many seconds is the current .5 amperes?

Amplitude = 3, period = $\dfrac{2\pi}{\pi} = 2$ sec, and displacement = 0.

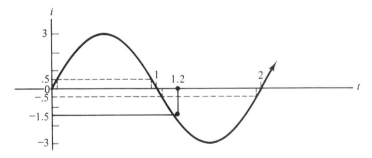

1. After 1.2 seconds, the current is 1.5 amperes.
2. The current is .5 ampere after .1 second, .9 second, 1.1 seconds, and 1.9 seconds, and any of these values $+2K$, where K is a positive integer.

Example J: If $v = V \cos(\omega t + C)$, sketch one cycle of the curve of this relationship when $V = 220$ volts, $\omega = 60$ cycles/sec, and $C = \dfrac{\pi}{3}$.

Sec. 9–4. Applications of the Graphs of the Trigonometric Functions 279

1. What is the voltage after $\frac{1}{120}$ second?
2. After how many seconds is the voltage 60 volts?

$v = 220 \cos\left(120\pi t + \frac{\pi}{3}\right)$, with amplitude $= 220$, period $= \frac{2\pi}{120\pi} = \frac{1}{60}$ sec, and displacement $= \frac{\pi/3}{120\pi} = \frac{1}{360}$ sec left.

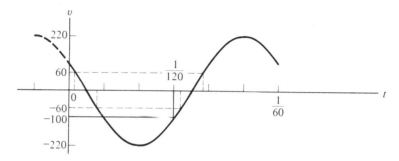

1. After $\frac{1}{120}$ second, the voltage is 100 volts.
2. The voltage is 60 volts after .001 second, .002 second, .009 second, and .01 second, and any of these values $+\frac{K}{60}$, where K is a positive integer.

Exercises

Sketch one cycle of the curve of the relationship $d = R \sin(\omega t)$ for each of the following sets of values.

1. $R = 2$ and $\omega = 6$ rad/sec
2. $R = 5$ and $\omega = 4\pi$ rad/sec
3. $R = -2$ and $\omega = 60$ cycles/sec
4. $R = 4$ and $\omega = 60$ rpm

Sketch one cycle of the curve of the relationship $d = R \sin(\omega t + C)$ for each of the following sets of values.

5. $R = 3$, $\omega = 5$ rad/sec, and $C = \frac{\pi}{2}$.

6. $R = -1$, $\omega = 6$ rad/sec, and $C = -\frac{\pi}{4}$.

7. $R = -3$, $\omega = 60$ cycles/sec, and $C = \frac{\pi}{3}$.

8. $R = 4$, $\omega = 2\pi$ rad/sec, and $C = \frac{\pi}{6}$.

Sketch two cycles of the curve of the relationship $i = I \sin(\omega t + C)$ for each of the following sets of values.

9. $I = 3$ amps, $\omega = 8$ rad/sec, and $C = 0$.
10. $I = .5$ amps, $\omega = 30$ cycles/sec, and $C = 0$.
11. $I = 2$ amps, $\omega = 60$ cycles/sec, and $C = \frac{\pi}{3}$.
12. $I = 6$ amps, $\omega = 3\pi$ rad/sec, and $C = -\frac{\pi}{2}$.

Sketch two cycles of the curve of the relationship $v = V \cos(\omega t + C)$ for each of the following sets of values.

13. $V = 110$ volts, $\omega = 30$ cycles/sec, and $C = 0$.
14. $V = 220$ volts, $\omega = 2\pi$ rad/sec, and $C = 0$.
15. $V = 220$ volts, $\omega = 60$ rpm, and $C = -\frac{\pi}{4}$.
16. $V = 110$ volts, $\omega = 4$ rad/sec, and $C = \frac{\pi}{3}$.
17. If $i = I \sin(\omega t + C)$, sketch one cycle of the curve of this relationship when $I = 5$ amps, $\omega = 60$ cycles/sec, and $C = -\frac{\pi}{4}$.

    (a) What is the current flowing after $\frac{1}{240}$ second?
    (b) What is the current flowing after $\frac{1}{480}$ second?
    (c) After how many seconds will 3 amperes of current be flowing?
    (d) After how many seconds will 1.2 amperes of current be flowing?

18. If $v = V \cos(\omega t + C)$, sketch one cycle of the curve of this relationship when $V = 110$ volts, $\omega = 60$ cycles/sec, and $C = \frac{\pi}{3}$.

    (a) What is the voltage after $\frac{1}{180}$ second?
    (b) What is the voltage after $\frac{1}{90}$ second?
    (c) After how many seconds will the voltage be 100 volts?
    (d) After how many seconds will the voltage be 40 volts?

19. When a wave travels in a string, each point on the string will oscillate in simple harmonic motion. At any time (t), the displacement (d) of any point on the string from its equilibrium position is given by the equation $d = A \cos\left(\frac{2\pi}{\lambda} t\right)$, where A is the amplitude and λ is the wavelength. Sketch two cycles of the curve of this relationship when $A = .1$ in. and $\lambda = .1$ sec.

20. In certain types of electric circuits containing a capacitor, the charge (q) on the capacitor in terms of the time (t) is given by $q = -\frac{I_0}{\omega} \cos(\omega t)$. Sketch two cycles of the curve of this relationship for $I_0 = 6$ amps and $\omega = 12$ rad/sec.

9–5. CHAPTER REVIEW

For each of the following, determine the amplitude, period, and displacement, sketch one cycle of the graph, and determine the starting and ending points of the cycle. Do not make out a table of values.

1. $Y = 2 \sin(3X)$
2. $Y = -2 \cos(2X)$
3. $Y = 3 \tan(2X)$
4. $Y = -4 \cos\left(\frac{X}{2}\right)$
5. $Y = 3 \sin(X - \pi)$
6. $Y = -\cos\left(X + \frac{\pi}{4}\right)$
7. $Y = 2 \sin\left(X + \frac{1}{2}\right)$
8. $Y = -2 \csc\left(X + \frac{\pi}{2}\right)$
9. $Y = \cos(3X - \pi)$
10. $Y = \sin\left(2X - \frac{\pi}{2}\right)$
11. $Y = \sec(\pi X - \pi)$
12. $Y = \cot\left(\frac{X}{2} - \frac{\pi}{4}\right)$
13. $Y = 2 \sin\left(X - \frac{\pi}{2}\right)$
14. $Y = -3 \cos\left(2X + \frac{\pi}{4}\right)$
15. $Y = -\sin(\pi X + \pi)$
16. $Y = 2 \cos\left(\frac{X}{2} - \frac{\pi}{3}\right)$
17. $Y = 4 \tan\left(3X + \frac{\pi}{4}\right)$
18. $Y = -2 \csc\left(2X - \frac{\pi}{6}\right)$
19. $Y = 3 \sin\left(\frac{\pi X}{2} - \frac{\pi}{2}\right)$
20. $Y = 5 \cos\left(3X - \frac{\pi}{2}\right)$

For each of the following, sketch the curve by using the addition of ordinates method.

21. $Y = \sin\left(X + \frac{\pi}{2}\right) + \cos(2X - \pi)$
22. $Y = 3 \cos(\pi x) - 2 \sin\left(\pi X + \frac{\pi}{2}\right)$
23. $Y = 2 \sin\left(X - \frac{\pi}{3}\right) + 3 \sin\left(X + \frac{\pi}{4}\right)$

24. $Y = -\text{Cos}\left(X - \frac{\pi}{4}\right) - 2\,\text{Sin}\left(2X + \frac{\pi}{3}\right)$

25. $Y = X + 4\,\text{Cos}\left(\pi X - \frac{\pi}{2}\right)$

26. $Y = X - 2 + \text{Sin}\left(\frac{X}{2} - \frac{\pi}{3}\right)$

27. $Y = 2X - \text{Sin}\left(3X - \frac{\pi}{2}\right)$

28. $Y = 3\,\text{Cos}\left(2X + \frac{\pi}{3}\right) - X + 1$

29. The electric current in a certain circuit is given by $i = I\,\text{Sin}\,(\omega t + C)$. Sketch two cycles of the curve of this relationship for $I = 5$ amps, $\omega = 60$ cycles/sec, and $C = \frac{\pi}{4}$.

30. The voltage in a certain electric circuit is given by $v = V\,\text{Cos}\,(\omega t + C)$. Sketch two cycles of the curve of this relationship for $V = 110$ volts, $\omega = 2\pi$ rad/sec, and $C = -\frac{\pi}{2}$.

31. For a certain object oscillating at the end of a spring, the displacement (d) from equilibrium as a function of time (t) is given by the equation $d = .6\,\text{Cos}\left(4t + \frac{\pi}{3}\right)$. Sketch three cycles of the graph of this equation. (d is measured in feet and t is measured in seconds.)

32. The angular displacement (A) of a certain pendulum bob is given in terms of its initial displacement (A_0) by the equation $A = A_0\,\text{Sin}\,(\omega t + B)$. Sketch two cycles of this equation for $A_0 = \frac{\pi}{25}$ rad, $\omega = 2$ rad/sec, and $B = -\frac{\pi}{2}$.

33. The voltage in a certain electric circuit is given by the equation

$$v = 40\,\text{Cos}\,30\pi t + 70\,\text{Cos}\,50\pi t$$

Sketch the curve of this equation using the addition of ordinates method.

34. Under certain conditions, the displacement (d) from equilibrium of an object oscillating on a spring as a function of time (t) is given by the equation $d = .3\,\text{Sin}\,4t + .3\,\text{Cos}\,6t$. Sketch the curve of this equation using the addition of ordinates method. (d is measured in feet and t is measured in seconds.)

chapter 10

Trigonometric Identities, Equations, and Inverse Relations

10–1. Additional Angle Formulas

Examining the definition of the trigonometric functions that was given in Section 7–4, we should note two facts. First, there are definite relationships that exist among the trigonometric functions. Some of these relationships are obvious and others will be derived in this chapter. Second, the definitions do not depend on the angle θ, and thus are valid for *any angle*.

From the basic definitions in terms of X, Y, and R, the following relationships are obtained.

(1) $\quad \mathrm{Sin}\,\theta = \dfrac{1}{\mathrm{Csc}\,\theta} \quad$ or $\quad \mathrm{Csc}\,\theta = \dfrac{1}{\mathrm{Sin}\,\theta} \quad$ or $\quad \mathrm{Sin}\,\theta\,\mathrm{Csc}\,\theta = 1$

(2) $\quad \mathrm{Cos}\,\theta = \dfrac{1}{\mathrm{Sec}\,\theta} \quad$ or $\quad \mathrm{Sec}\,\theta = \dfrac{1}{\mathrm{Cos}\,\theta} \quad$ or $\quad \mathrm{Cos}\,\theta\,\mathrm{Sec}\,\theta = 1$

(3) $\quad \mathrm{Tan}\,\theta = \dfrac{1}{\mathrm{Cot}\,\theta} \quad$ or $\quad \mathrm{Cot}\,\theta = \dfrac{1}{\mathrm{Tan}\,\theta} \quad$ or $\quad \mathrm{Tan}\,\theta\,\mathrm{Cot}\,\theta = 1$

(4) $\quad \dfrac{\mathrm{Sin}\,\theta}{\mathrm{Cos}\,\theta} = \dfrac{Y/R}{X/R} = \dfrac{Y}{X} = \mathrm{Tan}\,\theta$

(5) $\quad \dfrac{\mathrm{Cos}\,\theta}{\mathrm{Sin}\,\theta} = \dfrac{X/R}{Y/R} = \dfrac{X}{Y} = \mathrm{Cot}\,\theta$

From the basic definitions and the Pythagorean Theorem ($Y^2 + X^2 = R^2$), we obtain the following relationships.

(6) $\quad \dfrac{Y^2}{R^2} + \dfrac{X^2}{R^2} = \dfrac{R^2}{R^2} \quad$ or $\quad \text{Sin}^2\,\theta + \text{Cos}^2\,\theta = 1$

Also,
$$\text{Sin}^2\,\theta = 1 - \text{Cos}^2\,\theta$$
$$\text{Cos}^2\,\theta = 1 - \text{Sin}^2\,\theta$$

(7) $\quad \dfrac{Y^2}{X^2} + \dfrac{X^2}{X^2} = \dfrac{R^2}{X^2} \quad$ or $\quad \text{Tan}^2\,\theta + 1 = \text{Sec}^2\,\theta$

Also,
$$\text{Tan}^2\,\theta = \text{Sec}^2\,\theta - 1$$
$$\text{Sec}^2\,\theta - \text{Tan}^2\,\theta = 1$$

(8) $\quad \dfrac{Y^2}{Y^2} + \dfrac{X^2}{Y^2} = \dfrac{R^2}{Y^2} \quad$ or $\quad 1 + \text{Cot}^2\,\theta = \text{Csc}^2\,\theta$

Also,
$$\text{Cot}^2\,\theta = \text{Csc}^2\,\theta - 1$$
$$\text{Csc}^2\,\theta - \text{Cot}^2\,\theta = 1$$

These eight relationships are usually referred to as the *fundamental*, or *basic*, *trigonometric identities*. They should be thoroughly known and understood before proceeding to the remainder of this section and chapter.

We will now develop formulas for the sine and cosine of the sum and difference of two angles, using the following diagram (Figure 10-1.1). Angle A and angle RSO are equal since they are alternate interior angles, formed by two parallel lines cut by a transversal. Angle RSO and angle RTS are equal since they are both complementary to angle TSR.

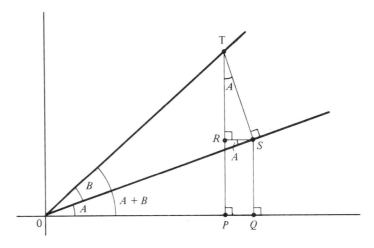

FIGURE 10-1.1

Sec. 10-1. Additional Angle Formulas

(9) $\text{Sin}(A+B) = \dfrac{PT}{OT} = \dfrac{PR+TR}{OT} = \dfrac{SQ+TR}{OT} = \dfrac{SQ}{OT} + \dfrac{TR}{OT}$

$= \dfrac{SQ}{OS} \cdot \dfrac{OS}{OT} + \dfrac{TR}{TS} \cdot \dfrac{TS}{OT}$

$= \text{Sin } A \text{ Cos } B + \text{Cos } A \text{ Sin } B$

(10) $\text{Cos}(A+B) = \dfrac{OP}{OT} = \dfrac{OQ-PQ}{OT} = \dfrac{OQ-RS}{OT} = \dfrac{OQ}{OT} - \dfrac{RS}{OT}$

$= \dfrac{OQ}{OS} \cdot \dfrac{OS}{OT} - \dfrac{RS}{TS} \cdot \dfrac{TS}{OT}$

$= \text{Cos } A \text{ Cos } B - \text{Sin } A \text{ Sin } B$

We have previously seen that $\text{Sin}(-B) = -\text{Sin } B$ and $\text{Cos}(-B) = \text{Cos } B$. Using these two facts, we obtain the following relationships.

(11) $\text{Sin}(A-B) = \text{Sin}[A+(-B)] = \text{Sin } A \text{ Cos}(-B) + \text{Cos } A \text{ Sin}(-B)$

$= \text{Sin } A \text{ Cos } B - \text{Cos } A \text{ Sin } B$

(12) $\text{Cos}(A-B) = \text{Cos}[A+(-B)] = \text{Cos } A \text{ Cos}(-B) - \text{Sin } A \text{ Sin}(-B)$

$= \text{Cos } A \text{ Cos } B + \text{Sin } A \text{ Sin } B$

Some of the most important relationships among the trigonometric functions are those involving twice an angle or one-half an angle. We will develop these relationships next.

(13) $\text{Sin}(2A) = \text{Sin}(A+A) = \text{Sin } A \text{ Cos } A + \text{Cos } A \text{ Sin } A$

$= 2 \text{ Sin } A \text{ Cos} A$

(14) $\text{Cos}(2A) = \text{Cos}(A+A) = \text{Cos } A \text{ Cos } A - \text{Sin } A \text{ Sin } A$

$= \text{Cos}^2 A - \text{Sin}^2 A$

$= (1 - \text{Sin}^2 A) - \text{Sin}^2 A = 1 - 2 \text{ Sin}^2 A$

$= \text{Cos}^2 A - (1 - \text{Cos}^2 A) = 2 \text{ Cos}^2 A - 1$

Therefore,

$$\text{Cos}(2A) = \text{Cos}^2 A - \text{Sin}^2 A = 1 - 2 \text{ Sin}^2 A = 2 \text{ Cos}^2 A - 1$$

Since $\text{Cos}(2A) = 1 - 2 \text{ Sin}^2 A$ or $2 \text{ Cos}^2 A - 1$, we may state this relationship as $\text{Cos } A = 1 - 2 \text{ Sin}^2\left(\dfrac{A}{2}\right)$ or $2 \text{ Cos}^2\left(\dfrac{A}{2}\right) - 1$ and still maintain the same basic relationship between the involved angles.

(15) $$\text{Cos } A = 1 - 2 \text{ Sin}^2\left(\dfrac{A}{2}\right)$$

$$2 \operatorname{Sin}^2 \left(\frac{A}{2}\right) = 1 - \operatorname{Cos} A$$

$$\operatorname{Sin}^2 \left(\frac{A}{2}\right) = \frac{1 - \operatorname{Cos} A}{2}$$

$$\operatorname{Sin} \left(\frac{A}{2}\right) = \pm \sqrt{\frac{1 - \operatorname{Cos} A}{2}}$$

(16)
$$2 \operatorname{Cos}^2 \left(\frac{A}{2}\right) - 1 = \operatorname{Cos} A$$

$$2 \operatorname{Cos}^2 \left(\frac{A}{2}\right) = 1 + \operatorname{Cos} A$$

$$\operatorname{Cos}^2 \left(\frac{A}{2}\right) = \frac{1 + \operatorname{Cos} A}{2}$$

$$\operatorname{Cos} \left(\frac{A}{2}\right) = \pm \sqrt{\frac{1 + \operatorname{Cos} A}{2}}$$

In these two formulas, the appropriate sign is determined by the quadrant in which the angle $\frac{A}{2}$ lies.

These 16 formulas involving the trigonometric functions are the most useful and important for our purposes. Other formulas may be derived, but we will not consider them at this point. A complete list of trigonometric formulas may be found in Table 6.

The following examples should help explain and further clarify these formulas.

Example A: Verify formula (1) for $\theta = 70°$.

$$\operatorname{Sin} 70° = .9397 \quad \frac{1}{\operatorname{Csc} 70°} = \frac{1}{1.064} = .9397$$

$$(.9397)(1.064) = 1$$

Example B: Verify formula (4) for $\theta = 50°$.

$$\frac{\operatorname{Sin} 50°}{\operatorname{Cos} 50°} = \frac{.7660}{.6428} = 1.1917 = \operatorname{Tan} 50°$$

Example C: Verify formula (10) for $A = 40°$ and $B = 25°$.

$$\operatorname{Cos} (40° + 25°) = \operatorname{Cos} 65° = .4226$$
$$\operatorname{Cos} 40° \operatorname{Cos} 25° - \operatorname{Sin} 40° \operatorname{Sin} 25° = .7660(.9063) - .6428(.4226)$$
$$= .6942 - .2716 = .4226$$

Sec. 10–1. Additional Angle Formulas

Example D: Verify formula (15) for $A = 80°$.

$$\text{Sin}\left(\frac{80°}{2}\right) = \text{Sin } 40° = .6428$$

$$\sqrt{\frac{1 - \text{Cos } 80°}{2}} = \sqrt{\frac{1 - .1736}{2}} = \sqrt{\frac{.8264}{2}} = \sqrt{.4132} = .6428$$

Note that the positive square root is taken since $\frac{80°}{2} = 40°$ is in the first quadrant and the sine function is positive there.

Example E: Find Sin 75° by using $75° = 45° + 30°$.

$$\text{Sin } 75° = \text{Sin } (45° + 30°) = \text{Sin } 45° \text{ Cos } 30° + \text{Cos } 45° \text{ Sin } 30°$$
$$= .7071(.8660) + .7071(.5000) = .6123 + .3536$$
$$= .9659$$

Example F: Find Cos 30° by using $30° = 90° - 60°$.

$$\text{Cos } 30° = \text{Cos } (90° - 60°) = \text{Cos } 90° \text{ Cos } 60° + \text{Sin } 90° \text{ Sin } 60°$$
$$= 0(.5000) + 1(.8660) = 0 + .8660$$
$$= .8660$$

Example G: Find Sin 60° by using $60° = 2(30°)$.

$$\text{Sin } 60° = \text{Sin } [2(30°)] = 2 \text{ Sin } 30° \text{ Cos } 30°$$
$$= 2(.5)(.8660)$$
$$= .8660$$

Example H: Find Cos 120° by using $120° = \frac{1}{2}(240°)$.

$$\text{Cos } 120° = \text{Cos }\left(\frac{240°}{2}\right) = -\sqrt{\frac{1 + \text{Cos } 240°}{2}} = -\sqrt{\frac{1 - .5}{2}} = -\sqrt{\frac{.5}{2}}$$
$$= -\sqrt{.25} = -.5$$

Note that the negative square root is taken since $\frac{240°}{2} = 120°$ is in the second quadrant and the cosine function is negative there.

Example I: If $\text{Sin } A = \frac{5}{13}$ in the second quadrant and $\text{Cos } B = \frac{4}{5}$ in the first quadrant, find $\text{Cos } (A + B)$. Using the Pythagorean Theorem, we find that $\text{Cos } A = -\frac{12}{13}$ and $\text{Sin } B = \frac{3}{5}$.

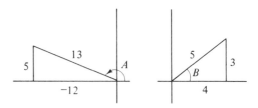

$$\text{Cos }(A+B) = \text{Cos }A\text{ Cos }B - \text{Sin }A\text{ Sin }B = -\tfrac{12}{13}\cdot\tfrac{4}{5} - \tfrac{5}{13}\cdot\tfrac{3}{5}$$
$$= -\tfrac{48}{65} - \tfrac{15}{65} = -\tfrac{63}{65} = -.9692$$

Example J: Find Sin $(2T)$ if Tan $T = .5$ in the third quadrant. Using the Pythagorean Theorem, we find that Sin $T = .4472$ and Cos $T = .8944$.

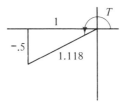

$$\text{Sin }(2T) = 2\text{ Sin }T\text{ Cos }T = 2(-.4472)(-.8944) = .7999$$

Exercises

1. Verify formula (2) for $\theta = 52°$.
2. Verify formula (5) for $\theta = 1$ radian.
3. Verify formula (6) for $\theta = 2$ radians.
4. Verify formula (8) for $\theta = 220°$.
5. Verify formula (9) for $A = 27°$ and $B = 79°$.
6. Verify formula (12) for $A = 312°$ and $B = 210°$.
7. Verify formula (13) for $A = 35°$.
8. Verify formula (16) for $A = 4$ radians.
9. Find Sin 105°, using $105° = 60° + 45°$.
10. Find Sin 90° using $90° = 45° + 45°$.
11. Find Cos 75°, using $75° = 45° + 30°$.
12. Find Cos 90°, using $90° = 60° + 30°$.
13. Find Sin 15°, using $15° = 60° - 45°$.
14. Find Sin 30°, using $30° = 60° - 30°$.
15. Find Cos 0°, using $0° = 30° - 30°$.

16. Find Cos 15°, using 15° = 45° − 30°.
17. Find Sin 120°, using 120° = 2(60°).
18. Find Sin 180°, using 180° = 2(90°).
19. Find Cos 270°, using 270° = 2(135°).
20. Find Cos 60°, using 60° = 2(30°).
21. Find Sin 45°, using 45° = ½(90°).
22. Find Sin 30°, using 30° = ½(60°).
23. Find Cos 15°, using 15° = ½(30°).
24. Find Cos 22.5°, using 22.5° = ½(45°).
25. If Sin $A = \frac{4}{5}$ in the second quadrant and Cos $B = -\frac{12}{13}$ in the second quadrant, find Sin $(A + B)$.
26. If Sin $A = \frac{3}{5}$ in the first quadrant and Cos $B = \frac{4}{5}$ in the fourth quadrant, find Cos $(A - B)$.
27. If Cos $X = \frac{3}{5}$ in the first quadrant, find Sin $(2X)$.
28. If Sin $Y = -\frac{4}{5}$ in the third quadrant, find Cos $(2Y)$.

10–2. Trigonometric Identities

In Section 10–1 we listed the basic trigonometric identities. More complicated identities may be handled by making use of these basic ones plus the additional formulas which were developed from them.

We should recall from Section 2–6 that an identity is an equation that is true for all values of the variable(s) involved. A *trigonometric identity*, then, is an equation involving trigonometric functions that is true for all values of the involved variable (the angle). To *prove* such identities, we can start with one side only and by appropriate substitutions and algebraic manipulations show that it equals the other side. Or we can work with both sides separately and, using the same type of procedures just mentioned, show that they equal the same quantity. The important thing to remember is that we are not solving equations and thus we cannot employ equation-solving techniques (i.e., performing the same operation on both members simultaneously) in working with identities. We are actually trying to prove that we have an equation, and until this is actually done, we cannot treat the relationship as an equation.

The following examples will illustrate the correct techniques to be used. We should keep in mind, however, that each identity is different and a procedure that works for one may not work for another. The best tool we have is to be thoroughly familiar with the basic identities and with the other formulas that have been derived, and be able to quickly recognize the different forms.

Example A: Prove the identity:

$$\text{Sec } X - \text{Sin } X \text{ Tan } X = \text{Cos } X$$

Here we will work with the left-hand side.

$$\begin{aligned}
\text{Sec } X - \text{Sin } X \text{ Tan } X &= \frac{1}{\text{Cos } X} - \text{Sin } X \frac{\text{Sin } X}{\text{Cos } X} && \text{formulas (2) and (4)} \\
&= \frac{1}{\text{Cos } X} - \frac{\text{Sin}^2 X}{\text{Cos } X} && \text{algebra} \\
&= \frac{1 - \text{Sin}^2 X}{\text{Cos } X} && \text{algebra} \\
&= \frac{\text{Cos}^2 X}{\text{Cos } X} && \text{formula (6)} \\
&= \frac{\text{Cos } X \text{ Cos } X}{\text{Cos } X} && \text{algebra} \\
&= \text{Cos } X && \text{algebra}
\end{aligned}$$

Example B: Prove the identity:

$$\text{Csc } T \text{ Sec } T = \text{Cot } T + \text{Tan } T$$

Here we will work with the right-hand side.

$$\begin{aligned}
\text{Cot } T + \text{Tan } T &= \frac{\text{Cos } T}{\text{Sin } T} + \frac{\text{Sin } T}{\text{Cos } T} && \text{formulas (4) and (5)} \\
&= \frac{\text{Cos}^2 T + \text{Sin}^2 T}{\text{Sin } T \text{ Cos } T} && \text{algebra} \\
&= \frac{1}{\text{Sin } T \text{ Cos } T} && \text{formula (6)} \\
&= \text{Csc } T \text{ Sec } T && \text{formulas (1) and (2)}
\end{aligned}$$

Example C: Prove the identity:

$$\text{Sin} \left(\frac{\pi}{2} - B \right) = \text{Cos } B$$

Here we will work with the left-hand side.

$$\begin{aligned}
\text{Sin} \left(\frac{\pi}{2} - B \right) &= \text{Sin } \frac{\pi}{2} \text{ Cos } B - \text{Cos } \frac{\pi}{2} \text{ Sin } B && \text{formula (11)} \\
&= (1) \text{ Cos } B - (0) \text{ Sin } B && \text{Sin } \frac{\pi}{2} = 1 \text{ and Cos } \frac{\pi}{2} = 0 \\
&= \text{Cos } B - 0 && \text{algebra} \\
&= \text{Cos } B && \text{algebra}
\end{aligned}$$

Sec. 10–2. Trigonometric Identities

Example D: Prove the identity:

$$\text{Sec}^2\ Y = 2\ \text{Csc}\ (2Y)\ \text{Tan}\ Y$$

Here we will work with the right-hand side.

$$2\ \text{Csc}\ (2Y)\ \text{Tan}\ Y = \frac{2\ \text{Sin}\ Y}{\text{Sin}\ (2Y)\ \text{Cos}\ Y} \qquad \text{formulas (1) and (4)}$$

$$= \frac{2\ \text{Sin}\ Y}{2\ \text{Sin}\ Y\ \text{Cos}\ Y\ \text{Cos}\ Y} \qquad \text{formula (13)}$$

$$= \frac{1}{\text{Cos}\ Y\ \text{Cos}\ Y} \qquad \text{algebra}$$

$$= \frac{1}{\text{Cos}^2\ Y} \qquad \text{algebra}$$

$$= \text{Sec}^2\ Y \qquad \text{formula (2)}$$

Example E: Prove the identity:

$$\text{Cos}\ \theta = 1 - 2\ \text{Sin}^2\left(\frac{\theta}{2}\right)$$

Here we will work with the right-hand side.

$$1 - 2\ \text{Sin}^2\left(\frac{\theta}{2}\right) = 1 - 2\left(\pm\sqrt{\frac{1 - \text{Cos}\ \theta}{2}}\right)^2 \qquad \text{formula (15)}$$

$$= 1 - 2\left(\frac{1 - \text{Cos}\ \theta}{2}\right) \qquad \text{algebra}$$

$$= 1 - (1 - \text{Cos}\ \theta) \qquad \text{algebra}$$

$$= 1 - 1 + \text{Cos}\ \theta \qquad \text{algebra}$$

$$= \text{Cos}\ \theta \qquad \text{algebra}$$

Example F: Prove the identity:

$$2\ \text{Sin}^2\ X = 1 - \text{Cos}\ (2X)$$

Here we will work with the right-hand side.

$$1 - \text{Cos}\ (2X) = 1 - (1 - 2\ \text{Sin}^2\ X) \qquad \text{formula (14)}$$

$$= 1 - 1 + 2\ \text{Sin}^2\ X \qquad \text{algebra}$$

$$= 2\ \text{Sin}^2\ X \qquad \text{algebra}$$

Example G: Prove the identity:

$$\text{Sin}\ (3A) = 3\ \text{Sin}\ A - 4\ \text{Sin}^3\ A$$

Here we will work with the left-hand side.

$$\begin{aligned}
\operatorname{Sin}(3A) &= \operatorname{Sin}(2A + A) & &\text{algebra} \\
&= \operatorname{Sin}(2A)\operatorname{Cos} A + \operatorname{Cos}(2A)\operatorname{Sin} A & &\text{formula (9)} \\
&= 2\operatorname{Sin} A \operatorname{Cos} A \operatorname{Cos} A + (\operatorname{Cos}^2 A - \operatorname{Sin}^2 A)\operatorname{Sin} A & &\text{formulas (13) and (14)} \\
&= 2\operatorname{Sin} A \operatorname{Cos}^2 A + \operatorname{Sin} A \operatorname{Cos}^2 A - \operatorname{Sin}^3 A & &\text{algebra} \\
&= 3\operatorname{Sin} A \operatorname{Cos}^2 A - \operatorname{Sin}^3 A & &\text{algebra} \\
&= 3\operatorname{Sin} A(1 - \operatorname{Sin}^2 A) - \operatorname{Sin}^3 A & &\text{formula (6)} \\
&= 3\operatorname{Sin} A - 3\operatorname{Sin}^3 A - \operatorname{Sin}^3 A & &\text{algebra} \\
&= 3\operatorname{Sin} A - 4\operatorname{Sin}^3 A & &\text{algebra}
\end{aligned}$$

Example H: Prove the identity:

$$\operatorname{Sin}[90° - (X + Y)] = \operatorname{Cos} X \operatorname{Cos} Y - \operatorname{Sin} X \operatorname{Sin} Y$$

1. Working with the left-hand side:

$$\begin{aligned}
\operatorname{Sin}[90° - (X + Y)] &= \operatorname{Sin} 90° \operatorname{Cos}(X + Y) & &\text{formula (11)} \\
&\quad - \operatorname{Cos} 90° \operatorname{Sin}(X + Y) & & \\
&= (1)\operatorname{Cos}(X + Y) & &\operatorname{Sin} 90° = 1 \text{ and} \\
&\quad - (0)\operatorname{Sin}(X + Y) & &\operatorname{Cos} 90° = 0 \\
&= \operatorname{Cos}(X + Y) - 0 & &\text{algebra} \\
&= \operatorname{Cos}(X + Y) & &\text{algebra}
\end{aligned}$$

2. Working with the right-hand side:

$$\operatorname{Cos} X \operatorname{Cos} Y - \operatorname{Sin} X \operatorname{Sin} Y = \operatorname{Cos}(X + Y) \qquad \text{formula (10)}$$

Since both sides are equal to the same quantity, they are equal to each other.

Exercises Prove each of the following identities.
1. $\operatorname{Cos} X \operatorname{Cot} X - \operatorname{Csc} X = -\operatorname{Sin} X$
2. $\operatorname{Tan} X \operatorname{Cos} X - \operatorname{Sin} X = 0$
3. $\operatorname{Csc}^2 A (1 - \operatorname{Cos}^2 A) = 1$
4. $(1 - \operatorname{Sin}^2 B)(1 + \operatorname{Tan}^2 B) = 1$
5. $\operatorname{Cos}(90° + T) = -\operatorname{Sin} T$
6. $\operatorname{Sin}\left(\frac{3\pi}{2} - Y\right) = -\operatorname{Cos} Y$
7. $\operatorname{Sin}^2 A - \operatorname{Sin}^2 B = \operatorname{Sin}(A + B) \times \operatorname{Sin}(A - B)$

Sec. 10–2. Trigonometric Identities

8. $\text{Cos } X \text{ Cos } Y = \frac{1}{2}[\text{Cos }(X+Y) + \text{Cos }(X-Y)]$
9. $1 + \text{Cos }(2\theta) = 2 \text{ Cos}^2 \theta$
10. $4 \text{ Cos}^3 \theta - 3 \text{ Cos } \theta = \text{Cos }(3\theta)$
11. $\text{Cos}^4 R - \text{Sin}^4 R = \text{Cos } 2R$
12. $1 + \text{Sin }(2\theta) = (\text{Sin } \theta + \text{Cos } \theta)^2$
13. $2 \text{ Sin}\left(\frac{U}{2}\right) \text{Cos}\left(\frac{U}{2}\right) = \text{Sin } U$
14. $\text{Cos } V = 1 - 2 \text{ Sin}^2\left(\frac{V}{2}\right)$
15. $2 \text{ Cos}^2\left(\frac{V}{2}\right) - \text{Cos } V = 1$
16. $\text{Cos }(R+S) \text{ Cos } S + \text{Sin }(R+S) \text{ Sin } S = \text{Cos } R$
17. $\text{Sin } A \text{ Sin } B = \frac{1}{2}[\text{Cos }(A-B) - \text{Cos }(A+B)]$
18. $\text{Cos }(X + 2\pi) + \text{Cos } X = 2 \text{ Cos } X$
19. $\dfrac{\text{Sin }(2\theta)}{2 \text{ Sin } \theta} + \dfrac{\text{Cos }(2\theta)}{2 \text{ Cos } \theta} + \dfrac{1}{2 \text{ Cos } \theta} = 2 \text{ Cos } \theta$
20. $\dfrac{\text{Cos } X}{\text{Sec } X + \text{Tan } X} = 1 - \text{Sin } X$
21. Using the formulas for $\text{Sin }(A+B)$ and $\text{Cos }(A+B)$, and then dividing both numerator and denominator by $\text{Cos } A \text{ Cos } B$, show that $\text{Tan }(A+B) = \dfrac{\text{Tan } A + \text{Tan } B}{1 - \text{Tan } A \text{ Tan } B}$.
22. Employing the same procedure, show that $\text{Tan }(A-B) = \dfrac{\text{Tan } A - \text{Tan } B}{1 + \text{Tan } A \text{ Tan } B}$.
23. Using the formulas for $\text{Sin }(2A)$ and $\text{Cos }(2A)$, and then dividing both numerator and denominator by $\text{Cos}^2 A$, show that $\text{Tan }(2A) = \dfrac{2 \text{ Tan } A}{1 - \text{Tan}^2 A}$.
24. Using the formulas for $\text{Sin}\left(\frac{A}{2}\right)$ and $\text{Cos}\left(\frac{A}{2}\right)$, show that $\text{Tan}\left(\frac{A}{2}\right) = \pm\sqrt{\dfrac{1 - \text{Cos } A}{1 + \text{Cos } A}}$.
25. If $Y = \text{Csc}^2 \theta$ and $X = \text{Cot } \theta$, express Y in terms of X.
26. If $Y = \text{Tan}^2 \theta$ and $X = \text{Sec } \theta$, express Y in terms of X.
27. In studying the motion of a system of particles, the trigonometric identity
$$\text{Tan}\left(\frac{\theta}{2}\right) = \dfrac{2 \text{ Sin}\left(\frac{\theta}{2}\right) \text{Cos}\left(\frac{\theta}{2}\right)}{2 \text{ Cos}^2\left(\frac{\theta}{2}\right)}$$
appears. Verify this identity.

28. In the field of electromagnetics, the following identity appears: $\text{Sin } \theta + \text{Sin } (\theta - 120°) + \text{Sin } (\theta + 120°) = 0$. Verify this identity.

10–3. Basic Trigonometric Equations

A trigonometric equation is a statement of equality containing at least one trigonometric function. There are two types of trigonometric equations, the trigonometric identity and the conditional trigonometric equation.

A *trigonometric identity* is a statement of equality that is true for any value of the angle involved. This type of trigonometric equation was studied in the previous section.

A *conditional trigonometric equation* is a statement of equality that is true for certain replacements of the angle or angles involved.

Example A: $\text{Tan } B - 1 = 0$ is an example of a conditional trigonometric equation. If $B = 45° + K(360°)$ or $225° + K(360°)$ where K is an integer, the equation is true. If any other replacement is substituted for B, the equation would be false. B could also be expressed as $\frac{\pi}{4} + K(2\pi)$ and $\frac{5\pi}{4} + K(2\pi)$.

This section will deal with the solution of conditional trigonometric equations. Furthermore, the solutions will be limited to angles between 0° and 360° inclusive or in terms of radians between 0 and 2π inclusive. It is important to realize that a *trigonometric function represents a real number*, and thus, a term in an equation consisting of a trigonometric function can be treated the same as a real number.

The following examples will illustrate the solution of trigonometric equations as compared to the solution of similar algebraic equations. Each solution to the trigonometric equations will be stated in radian measure.

Trigonometric Equation	Algebraic Equation

Example B:

$$2 \text{ Cos } X + 1 = 0 \qquad\qquad 2X + 1 = 0$$
$$2 \text{ Cos } X = -1 \qquad\qquad 2X = -1$$
$$\text{Cos } X = -\tfrac{1}{2} \qquad\qquad X = -\tfrac{1}{2}$$
$$X = \tfrac{2\pi}{3}, \tfrac{4\pi}{3}$$

Check: $2 \text{ Cos } \left(\tfrac{2\pi}{3}\right) + 1 = 0 \qquad\qquad$ *Check:* $(-\tfrac{1}{2}) + 1 = 0$
$$2(-\tfrac{1}{2}) + 1 = 0 \qquad\qquad\qquad -1 + 1 = 0$$
$$-1 + 1 = 0 \qquad\qquad\qquad\qquad 0 = 0$$
$$0 = 0$$

Sec. 10-3. Basic Trigonometric Equations

Trigonometric Equation

$2 \cos\left(\frac{4\pi}{3}\right) + 1 = 0$
$2(-\frac{1}{2}) + 1 = 0$
$-1 + 1 = 0$
$0 = 0$

Algebraic Equation

Example C:

$\tan A - \sqrt{3} = 0$
$\tan A = \sqrt{3}$
$A = \frac{\pi}{3}, \frac{4\pi}{3}$

Check: $\tan\left(\frac{\pi}{3}\right) - \sqrt{3} = 0$
$\sqrt{3} - \sqrt{3} = 0$
$0 = 0$
$\tan\left(\frac{4\pi}{3}\right) - \sqrt{3} = 0$
$\sqrt{3} - \sqrt{3} = 0$
$0 = 0$

$A - \sqrt{3} = 0$
$A = \sqrt{3}$

Check: $\sqrt{3} - \sqrt{3} = 0$
$0 = 0$

Another procedure that is used in solving equations is based on the theorem that states *if a product of real numbers is equal to zero, then at least one of the numbers must be equal to zero.*

The following examples will show how this theorem is used in solving equations.

Trigonometric Equation

Algebraic Equation

Example D:

$\sin B \cos B = 0$
$\sin B = 0$ and/or $\cos B = 0$
$B = 0, \pi$ and/or $B = \frac{\pi}{2}, \frac{3\pi}{2}$

Check: $\sin 0 \cos 0 = 0$
$0(1) = 0$
$0 = 0$
$\sin \pi \cos \pi = 0$
$0(-1) = 0$
$0 = 0$
$\sin\left(\frac{\pi}{2}\right) \cos\left(\frac{\pi}{2}\right) = 0$
$(1)0 = 0$
$0 = 0$

$(B - 2)(B + 4) = 0$
$B - 2 = 0$ and/or $B + 4 = 0$
$B = 2$ and/or $B = -4$

Check: $(2 - 2)(2 + 4) = 0$
$0(6) = 0$
$0 = 0$

$(-4 - 2)(-4 + 4) = 0$
$(-6)(0) = 0$
$0 = 0$

Trigonometric Equation	Algebraic Equation

$$\text{Sin}\left(\frac{3\pi}{2}\right) \text{Cos}\left(\frac{3\pi}{2}\right) = 0$$
$$(-1)0 = 0$$
$$0 = 0$$

Example E:

$(\text{Sin } \alpha + 2)(\text{Sin } \alpha - 1) = 0$	$(X + 3)(X + 2) = 0$
$\text{Sin } \alpha + 2 = 0$ and/or	$X + 3 = 0$ and/or $X + 2 = 0$
$\text{Sin } \alpha - 1 = 0$	
$\text{Sin } \alpha = -2$ and/or $\text{Sin } \alpha = 1$	$X = -3$ and/or $X = -2$
$\alpha = \dfrac{\pi}{2}$	

NOTE: Since the Sine function cannot be greater than 1, there are no angles satisfying $\text{Sin } \alpha = -2$.

Check: $\left[\text{Sin}\left(\dfrac{\pi}{2}\right) + 2\right]\left[\text{Sin}\left(\dfrac{\pi}{2}\right) - 1\right]$	Check: $(-3 + 3)(-3 + 2) = 0$
$= 0$	$(0)(-1) = 0$
$(1 + 2)(1 - 1) = 0$	$0 = 0$
$(3)(0) = 0$	$(-2 + 3)(-2 + 2) = 0$
$0 = 0$	$(1)(0) = 0$
	$0 = 0$

For the following trigonometric equations there are no counterparts in the set of algebraic equations. Each solution depends upon the substitution of a trigonometric identity.

Example F:

$\text{Sin }(2B) + \text{Cos } B = 0$	Since $\text{Sin }(2B) = 2 \text{ Sin } B \text{ Cos } B$, the right member of this identity can be substituted into the original equation.
$2 \text{ Sin } B \text{ Cos } B + \text{Cos } B = 0$	
$\text{Cos } B(2 \text{ Sin } B + 1) = 0$	
$\text{Cos } B = 0$ and/or $2 \text{ Sin } B + 1 = 0$	
$\text{Cos } B = 0$ and/or $\text{Sin } B = -\frac{1}{2}$	
$B = \dfrac{\pi}{2}, \dfrac{3\pi}{2}, \dfrac{7\pi}{6}, \dfrac{11\pi}{6}$	

NOTE: All four of these roots check in the original equation.

Sec. 10–3. Basic Trigonometric Equations

Example G:

$$\text{Cos}\left(\frac{X}{2}\right) = -\frac{\sqrt{2}}{2}$$

$$\pm\sqrt{\frac{1+\text{Cos } X}{2}} = -\frac{\sqrt{2}}{2}$$

$$\left(\pm\sqrt{\frac{1+\text{Cos } X}{2}}\right)^2 = \left(-\frac{\sqrt{2}}{2}\right)^2$$

$$\frac{1+\text{Cos } X}{2} = \frac{2}{4} = \frac{1}{2}$$

$$1+\text{Cos } X = 1$$

$$\text{Cos } X = 0$$

$$X = \frac{\pi}{2}, \frac{3\pi}{2}$$

Since $\text{Cos}\left(\frac{X}{2}\right) = \pm\sqrt{\frac{1+\text{Cos } X}{2}}$, the right member of this identity can be substituted into the original equation.

Check:
$$\text{Cos}\left(\frac{\pi/2}{2}\right) = -\frac{\sqrt{2}}{2}$$

$$\text{Cos}\left(\frac{\pi}{4}\right) = -\frac{\sqrt{2}}{2}$$

$$\frac{\sqrt{2}}{2} \neq -\frac{\sqrt{2}}{2}$$

$$X \neq \frac{\pi}{2}$$

$$\text{Cos}\left(\frac{3\pi/2}{2}\right) = -\frac{\sqrt{2}}{2}$$

$$\text{Cos}\left(\frac{3\pi}{4}\right) = -\frac{\sqrt{2}}{2}$$

$$-\frac{\sqrt{2}}{2} = -\frac{\sqrt{2}}{2}$$

$$X = \frac{3\pi}{2}$$

Since both members of the equation were squared, it is necessary to check each root.

Example H:

$$\text{Cos }(2A) + \text{Cos } A + 1 = 0$$
$$2\text{ Cos}^2 A - 1 + \text{Cos } A + 1 = 0$$
$$2\text{ Cos}^2 A + \text{Cos } A = 0$$
$$\text{Cos } A\,(2\text{ Cos } A + 1) = 0$$
$$\text{Cos } A = 0 \text{ and/or } 2\text{ Cos } A + 1 = 0$$
$$\text{Cos } A = 0 \text{ and/or } \text{Cos } A = -\tfrac{1}{2}$$
$$A = \frac{\pi}{2}, \frac{3\pi}{2}, \frac{2\pi}{3}, \frac{4\pi}{3}$$

$$\text{Cos }(2A) = \text{Cos}^2 A - \text{Sin}^2 A$$
$$= 2\text{ Cos}^2 A - 1$$
$$= 1 - 2\text{ Sin}^2 A$$

Of these three, $2\text{ Cos}^2 A - 1$ would be useful in solving the equation.

The check is left to the student. All roots do check.

The following trigonometric equations include multiple angles.

Example I:

$$\text{Sin}(3B) = \frac{\sqrt{2}}{2}$$

Since the value of B must be $0 \leq B < 2\pi$, the value of $3B$ must be $0 \leq 3B < 6\pi$.

$$3B = \frac{\pi}{4}, \frac{3\pi}{4}, \frac{9\pi}{4}, \frac{11\pi}{4}, \frac{17\pi}{4}, \frac{19\pi}{4}$$

and

$$B = \frac{\pi}{12}, \frac{\pi}{4}, \frac{3\pi}{4}, \frac{11\pi}{12}, \frac{17\pi}{12}, \frac{19\pi}{12}$$

The check is left to the student. All roots do check.

Example J:

$$\text{Cos}\left(\frac{\alpha}{2}\right) = \frac{1}{2} \qquad \text{Since the value of } \alpha \text{ must be } 0 \leq \alpha < 2\pi, \frac{\alpha}{2} \text{ must be}$$

$$\frac{\alpha}{2} = \frac{2\pi}{3} \qquad 0 \leq \frac{\alpha}{2} < \pi.$$

$$\alpha = \frac{4\pi}{3}$$

The check is left to the student. The root does check.

Exercises Solve each of the following trigonometric equations.

1. $2 \text{ Cos } A + 1 = 0$
2. $-2 \text{ Sin } B = 1$
3. $3 \text{ Tan } \theta + \sqrt{3} = 0$
4. $\text{Sin } \alpha \text{ Cos } \alpha = 0$
5. $2 \text{ Sin } B \text{ Cos } B - \text{Cos } B = 0$
6. $\text{Tan } \theta \text{ Sin } \theta - \text{Tan } \theta = 0$
7. $\text{Sin }(2A) = -\frac{\sqrt{3}}{2}$
8. $\text{Sin}\left(\frac{B}{2}\right) = \frac{1}{2}$
9. $\text{Cot}\left(\frac{A}{3}\right) + \frac{1}{\sqrt{3}} = 0$
10. $\text{Tan }(2X) = -1$
11. $\text{Sin}^2 \theta = \frac{3}{2}$
12. $\text{Cos}^2 A = -\frac{1}{2}$
13. $(\text{Cos } A - 1)(2 \text{ Sin } A + 1) = 0$
14. $(\text{Tan } B - 1)(4 \text{ Sin}^2 B + 3) = 0$
15. $\text{Sec}^2 X - 2 = 0$
16. $2 \text{ Csc}^2 \alpha + 8 = 0$
17. $\text{Sin }(2X) + 2 \text{ Cos } X = 0$
18. $2 \text{ Cos } A \text{ Csc } A = \sqrt{3} \text{ Csc } A$
19. $2 \text{ Cos}^4 B - 3 \text{ Cos}^2 B + 1 = 0$

Sec. 10-4. The Inverse Trigonometric Relations

20. $\cot^2 X - \csc X = 1$
21. The vertical displacement (s) of a weight on the end of a spring is given by $s = \sin 4t - \sin 2t$. Find the values of t (in seconds) for which $s = 0$.
22. Current as a function of time in a particular circuit is $i = \cos 2t - \cos t$. Find the time t (in seconds) when $i = 0$.

10-4. The Inverse Trigonometric Relations

A procedure for finding an *inverse* of a given relation consists of two steps:

1. Interchange the dependent and independent variables in the original equation.
2. Solve the resulting equation for the dependent variable.

Example A: Find the inverse of $Y = 4X - 3$.
1. Interchange the variables.
$$X = 4Y - 3$$
2. Solve the resulting equation for Y.
$$Y = \frac{X+3}{4} = \frac{1}{4}X + \frac{3}{4}$$

The graphs of $Y = 4X - 3$ and $Y = \dfrac{X+3}{4}$ are shown below.

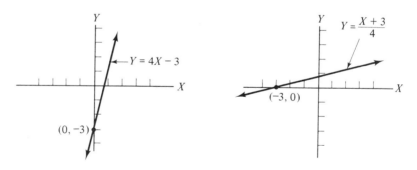

NOTE: The point $(0, -3)$ on $Y = 4X - 3$ has a corresponding point $(-3, 0)$ on $Y = \dfrac{X+3}{4}$.

In general, since the dependent and independent variables were interchanged, it follows that any point on the graph of the original relation will have a corresponding point on the graph of the inverse relation. This point can be found by interchanging the ordered pair of numbers corresponding to the point on the graph of the original relation.

Example B: Find the inverse of $A = 2B^2 - 1$.

1. Interchange the variables.

$$B = 2A^2 - 1$$

2. Solve the resulting equation for A.

$$A = \pm\sqrt{\frac{B+1}{2}}$$

The graphs of $A = 2B^2 - 1$ and $A = \pm\sqrt{\frac{B+1}{2}}$ are shown below.

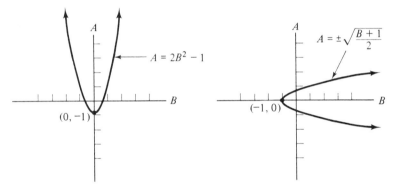

Recall from Section 4–2 our definition of a function as a relationship such that for each and every value of the independent variable there is no more than one value for the dependent variable. Also, no vertical line will intersect the graph of a function in more than one point. If we refer to Examples A and B, we should note that in both cases the original relations are functions. Therefore, the notation $f(X)$ and $f(B)$ may be used to describe these relations. In Example A, the inverse relation is also a function. When this occurs (i.e., *when the inverse relation is a function*), we use the notation f^{-1} to describe the relationship. Thus, $f(X) = 4X - 3$ and $f^{-1}(X) = \frac{1}{4}X + \frac{3}{4}$. In Example B, the inverse relation is not a function and hence, we may not use this notation. Thus, $f(B) = 2B^2 - 1$ and $A = \pm\sqrt{\frac{B+1}{2}}$.

Our primary concern in this section is to find the *inverse trigonometric relations*. To do this, we will use the procedure just described.

To find the inverse of $Y = \text{Sin } X$:

1. Interchange the dependent and independent variables: $X = \text{Sin } Y$.
2. Solve the resulting equation for Y.

Sec. 10–4. The Inverse Trigonometric Relations

There are no properties of an equation that will allow us to solve for Y. Therefore, it is necessary to introduce a new notation: $Y = \arcsin X$. The meaning of this notation is "Y is the angle whose sine is X."

Example C: What values will satisfy the equation $Y = \arcsin \frac{1}{2}$? This means, "Which angle or angles have a Sine function of $\frac{1}{2}$?" Since $\sin 30° = \frac{1}{2}$ and $\sin 150° = \frac{1}{2}$, the value of Y would be $30° + K(360°)$ or $150° + K(360°)$, where K is an integer.

NOTE: For a given value of X there are infinitely many values of Y. Thus, arcsin X is not a function. This also follows since the domain and range of $Y = \sin X$ are interchanged for $Y = \arcsin X$. Recall the graph of $Y = \sin X$.

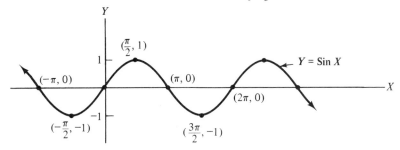

When each ordered pair of numbers from $Y = \sin X$ are interchanged we obtain the graph of $Y = \arcsin X$. It can be seen from this graph that $Y = 30° + K(360°)$ or $150° + K(360°)$ when $X = \frac{1}{2}$.

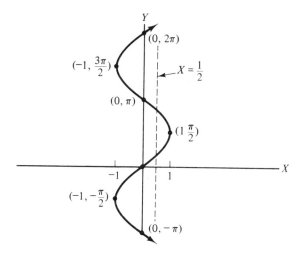

The remaining inverse trigonometric relations are derived in the same manner. Thus, the relations $Y = \arccos X$, $Y = \arctan X$, $Y = \text{arccsc } X$, $Y = \text{arcsec } X$, $Y = \text{arccot } X$ are handled in the same manner as $Y = \arcsin X$.

Example D: Find the smallest positive angle for each of the following.
1. $Y = \arccos(-\frac{1}{2})$. Since $\cos[120° + K(360°)] = -\frac{1}{2}$ and $\cos[240° + K(360°)] = -\frac{1}{2}$, $Y = 120°$ or $\frac{2\pi}{3}$.
2. $T = \arctan 1$. Since $\tan[45° + K(360°)] = 1$ and $\tan[225° + K(360°)] = 1$, $T = 45°$ or $\frac{\pi}{4}$.
3. $M = \text{arcsec } 2$. Since $\sec[60° + K(360°)] = 2$ and $\sec[300° + K(360°)] = 2$, $M = 60°$ or $\frac{\pi}{3}$.

Example E: For each of the following find θ, such that $0° \leq \theta < 360°$.
1. $\theta = \arcsin \frac{1}{2}$. Since $\sin[30° + K(360°)] = \frac{1}{2}$ and $\sin[150° + K(360°)] = \frac{1}{2}$, $\theta = 30°, 150°$.
2. $\theta = \text{arccsc } 1$. Since $\csc[90° + K(360°)] = 1$, $\theta = 90°$.
3. $\theta = \arctan 14.3$. Since $\tan[86° + K(360°)] = 14.3$ and $\tan[266° + K(360°)] = 14.3$, $\theta = 86°, 266°$.

Example F: Solve for X: $Y = \cos(3X)$. This means that $3X$ is the angle whose cosine is Y. Therefore, $3X = \arccos Y$ or $X = \frac{\arccos Y}{3}$.

Example G: Solve for T: $3V = \arctan(2T)$. This means that $3V$ is the angle whose tangent is $2T$. Therefore, $2T = \tan(3V)$ or $T = \frac{\tan(3V)}{2}$.

Exercises

Solve each of the following for X.
1. $Y = \tan(2X)$
2. $Y = 3 \csc X$
3. $Y = 5 \sin(X+1)$
4. $Y = 4 \cos(2X - 3)$
5. $Y = \arcsin(5X)$
6. $Y = 2 \arctan(3X)$
7. $Y + 1 = 3 \arccos(2X)$
8. $Y = 5 \text{ arcsec}(3X + 2)$

Find the smallest positive angle for each of the following.
9. $Y = \arcsin 0$
10. $T = \arccos(-1)$
11. $K = \arctan \sqrt{3}$
12. $X = \text{arccot } \sqrt{3}$
13. $\theta = \text{arcsec } 2$
14. $\alpha = \arccos 0$
15. $R = \text{arccot } 0$
16. $P = \arcsin .3333$
17. $M = \text{arcsec } 1.0324$
18. $S = \text{arccot}(-.7340)$

For each of the following, find θ such that $0° \leq \theta < 360°$.
19. $\theta = \arcsin(-\frac{1}{2})$
20. $\theta = \text{arcsec}(-2)$

Sec. 10–5. *The Inverse Trigonometric Functions and Their Graphs*   303

21. $\theta = \text{arccot } 1$
22. $\theta = \text{arccos }.3420$
23. $\theta = \text{arccsc}(-.8660)$
24. $\theta = \text{arctan }.8098$
25. $\theta = \text{arcsin }.1736$
26. $\theta = \text{arctan}(-3.487)$
27. $\theta = 2 \text{ arccos}(-\frac{1}{2})$
28. $\theta = 3 \text{ arctan } 1$
29. $\theta = \frac{1}{2} \text{ arcsin }.6691$
30. $\theta = 4 \text{ arccsc}(-1.305)$

10–5. The Inverse Trigonometric Functions and Their Graphs

In order for $Y = \arcsin X$ to be a function it is necessary to restrict the range so that for any value in the domain there is not more than one corresponding value in the range. If the range is restricted to $-\frac{\pi}{2} \leq Y \leq \frac{\pi}{2}$ then $Y = \arcsin X$ will be a function and therefore, for any value in the domain $(-1 \leq X \leq 1)$, there is not more than one corresponding value of Y. When this restriction is placed on Y, thus making $Y = \arcsin X$ a function, we use the notation

$$Y = \text{Arcsin } X$$

That is, the capital "A" denotes the *inverse function* while the small "a" denotes the *inverse relation*. When the capital A is used and we obtain a function, we refer to $Y = \text{Arcsin } X$ as the *principal value* of $Y = \arcsin X$.

The graphs of $Y = \arcsin X$ and $Y = \text{Arcsin } X$ are pictured in Figure 10-5.1.

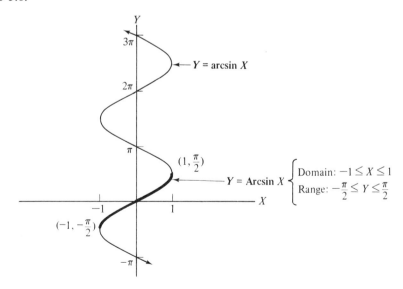

FIGURE 10–5.1

In order for the remaining inverse trigonometric relations to become functions, it is necessary to restrict their ranges in a similar manner. These restrictions together with the graphs of the remaining inverse relations and functions are given in Figures 10-5.2 through 10-5.6.

By observing the graphs of the inverse functions and their ranges, we should see that the *principal values* for the inverse trigonometric relations which, in effect, make them functions are as follows:

$$\frac{-\pi}{2} \leq \text{Arcsin } X \leq \frac{\pi}{2} \qquad \frac{-\pi}{2} \leq \text{Arccsc } X \leq \frac{\pi}{2} \text{ (except 0)}$$

$$0 \leq \text{Arccos } X \leq \pi \qquad 0 \leq \text{Arcsec } X \leq \pi \left(\text{except } \frac{\pi}{2}\right)$$

$$\frac{-\pi}{2} < \text{Arctan } X < \frac{\pi}{2} \qquad 0 < \text{Arccot } X < \pi$$

We should note that these particular ranges are chosen for the principal values in order to give us two consecutive quadrants for each function, one in which the basic trigonometric function is positive and one in which it is negative. Thus for Arcsine, Arccosecant, and Arctangent we have chosen the first and fourth quadrants. For Arccosine, Arcsecant, and Arccotangent we have chosen the first and second quadrants.

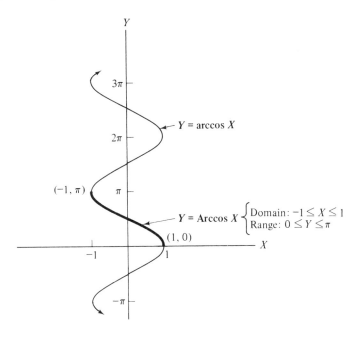

FIGURE 10-5.2

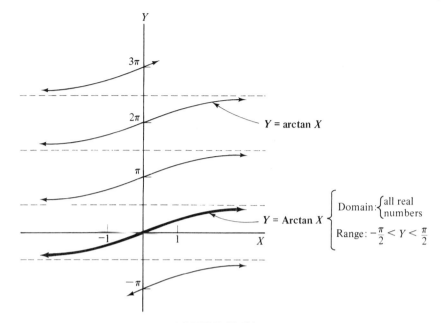

FIGURE 10–5.3

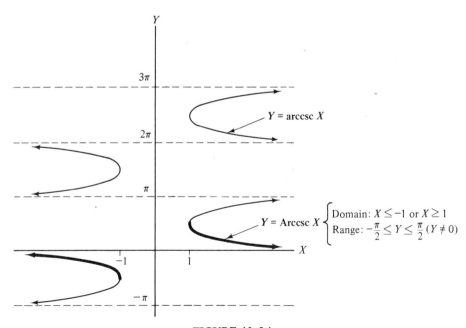

FIGURE 10–5.4

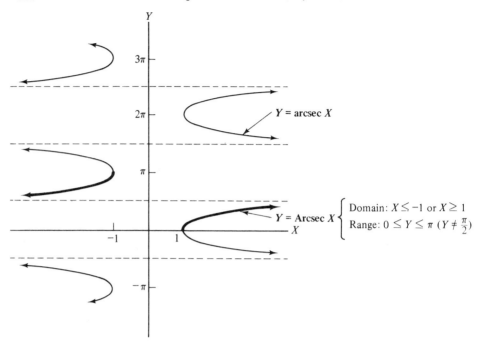

FIGURE 10-5.5

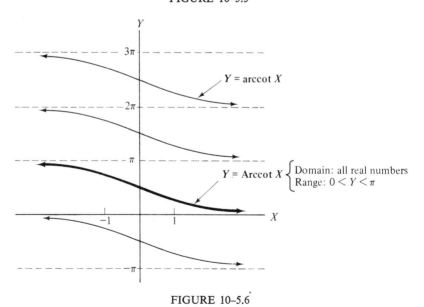

FIGURE 10-5.6

Example A: Find Y, when $Y = \arcsin \frac{1}{2}$ and $Y = \text{Arcsin } \frac{1}{2}$.
If $Y = \arcsin \frac{1}{2}$, then $Y = 30° + K(360°)$ or $150° + K(360°)$.
If $Y = \text{Arcsin } \frac{1}{2}$, then $Y = 30°$.

Sec. 10–5. The Inverse Trigonometric Functions and Their Graphs

Example B: Find θ, when $\theta = \arctan(-1)$ and $\theta = \text{Arctan}(-1)$.
If $\theta = \arctan(-1)$, then $\theta = 135° + K(360°)$ or $315° + K(360°)$.
If $\theta = \text{Arctan}(-1)$, then $\theta = 315°$.

From Examples A and B, we should note that when working with inverse relations (a) we obtain an infinite number of values for the range. When working with inverse functions (A) we obtain a single value for the range.

Example C: Find Sin (Arccos .3420).

$$\text{Arccos } .3420 = 70°$$
$$\text{Sin } 70° = .9397$$

Therefore, Sin (Arccos .3420) = .9397.

Example D: Find Csc [Arctan (−1.732)].

$$\text{Arctan}(-1.732) = 300°$$
$$\text{Csc } 300° = -\text{Csc } 60° = -1.155$$

Therefore, Csc [Arctan (−1.732)] = −1.155.

Example E: Find an algebraic expression for Sin (Arctan X). Arctan X means "the angle whose tangent is X." Since $\text{Tan } \theta = \dfrac{Y}{X} = \dfrac{X}{1}$, $R = \sqrt{X^2 + 1}$.

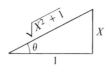

Therefore, Sin (Arctan X) = Sin $\theta = \dfrac{X}{\sqrt{X^2 + 1}}$.

Example F: Find an algebraic expression for Tan (Arccos X). Arccos X means "the angle whose cosine is X." Since $\text{Cos } \theta = \dfrac{X}{R} = \dfrac{X}{1}$, $Y = \sqrt{1 - X^2}$.

Therefore, Tan (Arccos X) = Tan $\theta = \dfrac{X}{\sqrt{1 - X^2}}$.

Exercises For each of the following find θ.

1. $\theta = \text{Arctan } 1$
2. $\theta = \text{Arccos } (-\tfrac{1}{2})$
3. $\theta = \text{Arccsc } (-2)$
4. $\theta = \text{Arcsin } .7071$
5. $\theta = \text{Arccsc } 1.494$
6. $\theta = \text{Arccot } (-1.540)$
7. $\theta = \text{Arcsec } (-1.305)$
8. $\theta = \text{Arcsin } .4067$
9. $\theta = \text{Arctan } (-.0524)$
10. $\theta = \text{Arccos } .9848$

Evaluate each of the following.

11. Sin (Arctan 2.145)
12. Tan [Arcsec (-2)]
13. Cot (Arccos .3256)
14. Cos [Arcsin $(-.9511)$]
15. Csc [Arccot $(-.8391)$]
16. Sec (Arccos .9925)

Find an algebraic expression for each of the following.

17. Csc (Arccos X)
18. Sec (Arccot T)
19. Tan (Arccsc K)
20. Cot (Arcsin M)

10–6. CHAPTER REVIEW

1. Find Sin 90°, using 90° = 45° + 45°.
2. Find Cos 120°, using 120° = 90° + 30°.
3. Find Sin 60°, using 60° = 90° − 30°.
4. Find Cos 15°, using 15° = 60° − 45°.
5. Find Sin 60°, using 60° = 2(30°).
6. Find Cos 120°, using 120° = 2(60°).
7. Find Sin 15°, using 15° = $\tfrac{1}{2}$(30°).
8. Find Cos 22.5°, using 22.5° = $\tfrac{1}{2}$(45°).

Prove each of the following identities.

9. $\text{Tan}^2 B - \text{Sec}^2 B = -1$
10. $\text{Csc } A - \text{Sin } A = \text{Cot } A \text{ Cos } A$
11. $\text{Sin } \theta \text{ Tan } \theta + \text{Cos } \theta = \text{Sec } \theta$
12. $\text{Sec}^2 X - \text{Sec}^2 X \text{ Sin}^2 X = 1$
13. $\text{Cos } (2T) + 1 - \text{Cos}^2 T = \text{Cos}^2 T$
14. $\text{Cos} \left(\dfrac{\pi}{2} + Y \right) = -\text{Sin } Y$

Sec. 10–6. Chapter Review

15. $\dfrac{\operatorname{Cos} X \operatorname{Tan} X + \operatorname{Sin} X}{\operatorname{Tan} X} = 2 \operatorname{Cos} X$

16. $\operatorname{Cos}^2 R - \operatorname{Sin}^2 T = \operatorname{Cos}(R+T) \operatorname{Cos}(R-T)$

17. $\operatorname{Sin}\left(\dfrac{A}{2}\right) \operatorname{Cos}\left(\dfrac{A}{2}\right) = \dfrac{\operatorname{Sin} A}{2}$

18. $\operatorname{Cos} B (\operatorname{Sec} B - \operatorname{Cos} B) = \operatorname{Sin}^2 B$

19. $\operatorname{Tan}\left(\dfrac{U}{2}\right) = \dfrac{\operatorname{Sin} U}{1 + \operatorname{Cos} U}$

20. $\operatorname{Tan}\theta + \operatorname{Cot}\theta = \dfrac{2}{\operatorname{Sin}(2\theta)}$

Solve each of the following equations.

21. $2 \operatorname{Cos}\theta + 1 = 0$
22. $5 \operatorname{Tan}\theta - 3 = 0$.
23. $\operatorname{Csc} X \operatorname{Sec} X = 0$
24. $(\operatorname{Tan} A - 2)(2 \operatorname{Sin} A + 1) = 0$
25. $\operatorname{Cos}^2 T + \operatorname{Cos} T = 0$
26. $5 \operatorname{Sin}^2 Y = 4 \operatorname{Sin} Y$
27. $\operatorname{Tan}^2 U - 2 \operatorname{Tan} U - 3 = 0$
28. $\operatorname{Sec}^2 V + 3 \operatorname{Sec} V = -2$
29. $\operatorname{Cos}(2X) - \operatorname{Sin} X - 1 = 0$
30. $2 \operatorname{Sin}^2 B = \operatorname{Cos}^2 B$

Evaluate each of the following.

31. Arccos (-1)
32. Arctan 2
33. Arcsin $(.5736)$
34. Arccsc (1.058)
35. Sin (Arctan 3)
36. Cos (Arcsin .1736)

Solve each of the following for T.

37. $A = \operatorname{Sin}(2T)$
38. $Y = 3 \operatorname{Tan}(4T)$
39. $U = .6 \operatorname{Cos}(5T)$
40. $Y + 1 = \operatorname{Sin}\left(T + \dfrac{\pi}{2}\right)$
41. $Y = 2 \operatorname{arcsec}(3T)$
42. $R = 5 \operatorname{arccot}(8T)$
43. $M - 3 = \operatorname{arcsin}(T + 2)$
44. $K + 1 = 4 \operatorname{arctan}(T - 3)$

Find an algebraic expression for each of the following.

45. Sin (2 Arccos X)
46. Tan [Arcsec $(2X)$]

47. Engineers who are studying stress use the expression $A \operatorname{Cos}^2 \theta$. Show that this can be written as $\dfrac{A}{2} + \dfrac{A}{2} \operatorname{Cos}(2\theta)$.

48. In mechanics, a certain force is represented as $2F \operatorname{Sin}^2\left(\dfrac{\theta}{2}\right)$. Show that this can be written as $F(1 - \operatorname{Cos}\theta)$.

49. For a bullet fired with a muzzle velocity of v at an angle of elevation θ, the maximum range is given by $R = \dfrac{2v^2}{g} \operatorname{Sin} \theta \operatorname{Cos} \theta$ (we neglect friction). Show that this expression can be written as $R = \dfrac{v^2}{g} \operatorname{Sin}(2\theta)$.

50. The equation
$$C = g\left(\dfrac{\operatorname{Tan} A + \operatorname{Tan} B}{1 - \operatorname{Tan} A \operatorname{Tan} B}\right)$$
states the relationship between the bank angle (A), coefficient of friction ($\operatorname{Tan} B$), and centrifugal acceleration (C) of a race car on a banked track. Show that this equation can be written as $C = g \operatorname{Tan}(A + B)$.

51. To integrate $\operatorname{Sin} A \operatorname{Sin} B$, we write it as $\tfrac{1}{2}[\operatorname{Cos}(A - B) - \operatorname{Cos}(A + B)]$. Verify this change.

52. To integrate
$$\dfrac{1 - X^2}{1 + X^2}$$
we let $X = \operatorname{Tan} \theta$ and change
$$\dfrac{1 - X^2}{1 + X^2}$$
to $\operatorname{Cos}(2\theta)$. Perform this change.

53. In the study of simple harmonic motion, the relationship $d = R \operatorname{Sin}(\omega t + C)$ is found. Solve for t.

54. The voltage in a certain electric circuit is given by the equation $v = V[\operatorname{Cos}(\omega t) \operatorname{Cos} C - \operatorname{Sin}(\omega t) \operatorname{Sin} C]$. Solve for t.

55. The Law of Cosines may be written as $a^2 = b^2 + c^2 - 2bc \operatorname{Cos} A$. Solve for A.

56. The equation $M = I \operatorname{Cos} \theta \operatorname{Sin} \theta$ gives the magnitude of a ray of a certain type of light. Solve for θ.

chapter 11

Imaginary and Complex Numbers

11–1. Imaginary Numbers

In Chapter 2, we looked very carefully at roots of quantities. At that time we defined the square root of a number A ($\sqrt{A}$) to be a quantity that, when multiplied times itself, would yield the number A. For example, $\sqrt{4} = 2$ since $2 \times 2 = 4$. From our knowledge of the rules of signs (listed in Chapter 1), we should realize that any number (positive or negative) times itself will yield a positive result. For example, $2 \cdot 2 = 4$ and $(-2) \cdot (-2) = 4$. Thus our definition of square root does not apply to square roots of negative numbers (e.g., $\sqrt{-1}$, $\sqrt{-2}$, $\sqrt{-9}$). However, we must be able to work with such numbers in order to solve certain types of equations and solve problems in many different areas. Therefore, in this chapter we will take a look at these numbers that are called *imaginary numbers*.

The basic quantity for this class of numbers is $\sqrt{-1}$ which is called the *imaginary unit* and which will be represented by the symbol j. Every imaginary number can then be expressed in terms of j.

Example A:
1. $\sqrt{-4} = \sqrt{4 \cdot (-1)} = \sqrt{4} \cdot \sqrt{-1} = 2j$
2. $\sqrt{-25} = \sqrt{25 \cdot (-1)} = \sqrt{25} \cdot \sqrt{-1} = 5j$

3. $-\sqrt{-9} = -\sqrt{9\cdot(-1)} = -\sqrt{9}\cdot\sqrt{-1} = -3j$
4. $\sqrt{-8} = \sqrt{8\cdot(-1)} = \sqrt{8}\cdot\sqrt{-1} = 2\sqrt{2}\,j$

Using the definitions of exponents and of j, we may raise the quantity j to different powers.

$$j^1 = j$$
$$j^2 = (\sqrt{-1})^2 = -1$$
$$j^3 = j^2 \cdot j^1 = (-1)j = -j$$
$$j^4 = j^3 \cdot j^1 = (-j)j = -j^2 = -(-1) = 1$$
$$j^5 = j^4 \cdot j^1 = 1(j) = j$$
$$j^6 = j^4 \cdot j^2 = 1(-1) = -1$$
$$j^7 = j^4 \cdot j^3 = 1(-j) = -j$$
$$j^8 = j^4 \cdot j^4 = 1(1) = 1$$

The integral powers of j will continue to repeat the cycle of values: j, -1, $-j$, 1. This is because any integral power of j may be written as some power of j^4 (which will be equal to 1) times j, j^2, or j^3. Knowing this fact, we may very quickly raise j to any integral power.

Example B:
1. $j^{13} = j^{12} \cdot j^1 = (j^4)^3 \cdot j^1 = 1 \cdot j = j$
2. $j^{127} = j^{124} \cdot j^3 = (j^4)^{31} \cdot j^3 = 1 \cdot j^3 = -j$
3. $j^{-18} = \dfrac{1}{j^{18}} = \dfrac{1}{j^{16}\cdot j^2} = \dfrac{1}{(j^4)^4 \cdot j^2} = \dfrac{1}{1 \cdot j^2} = \dfrac{1}{-1} = -1$
4. $j^{-36} = \dfrac{1}{j^{36}} = \dfrac{1}{(j^4)^9} = \dfrac{1}{1} = 1$

Exercises

Express each of the following in terms of j.

1. $\sqrt{-9}$
2. $\sqrt{-16}$
3. $-\sqrt{-25}$
4. $-\sqrt{-36}$
5. $\sqrt{-7}$
6. $\sqrt{-11}$
7. $-\sqrt{-5}$
8. $-\sqrt{-13}$
9. $\sqrt{-12}$
10. $\sqrt{-18}$
11. $\sqrt{-.01}$
12. $\sqrt{-.25}$
13. $-\sqrt{-24}$
14. $-\sqrt{-20}$
15. $-\sqrt{-.04}$
16. $-\sqrt{-.16}$

Simplify each of the following.

17. j^{17}
18. j^{89}
19. j^{102}

Sec. 11–2. Complex Numbers   313

20. j^{64} 21. j^{22} 22. j^{962}
23. j^0 24. j^{228} 25. j^{-12}
26. j^{-38} 27. j^{-57} 28. j^{-75}

11–2. Complex Numbers

If we wish to examine further the properties and uses of imaginary numbers, we must introduce a new type of number called complex. A *complex number* is one that may be written in the form $A + Bj$, where A and B are real numbers and $j = \sqrt{-1}$. This form $(A + Bj)$ is called the *rectangular form* of a complex number. If $B = 0$, we have a pure real number (A) or if $A = 0$, we have a pure imaginary number (Bj). Thus, the set of complex numbers includes all real numbers, all imaginary numbers, and all combinations of real and imaginary numbers. Reference to the diagram at the end of Section 1–1 will point out exactly how these and other classes of numbers are related in our number system.

Example A: $2, -5, 4 + 3j, -7 + 2j, -3 - 9j, \sqrt{2} - \sqrt{5}j$ and $\pi - \frac{7}{2}j$ are all examples of complex numbers.

Complex numbers are *neither positive nor negative* in the ordinary sense since we do not know how to interpret j in terms of a sign. However, the individual parts may be positive or negative.

Example B:
1. In the complex number $-2 + 3j$, the real part (-2) is negative and the imaginary part $(+3j)$ is positive.
2. In the complex number $5 - 4j$, the real part (5) is positive and the imaginary part $(-4j)$ is negative.

Every complex number has a *conjugate*, which is obtained by changing the sign of the imaginary part of the number.

Example C:
1. The conjugate of $2 - 3j$ is $2 + 3j$.
2. The conjugate of $-5 + 4j$ is $-5 - 4j$.
3. The conjugate of $6j$ is $-6j$.
4. The conjugate of -7 is -7.

Two complex numbers are equal if and only if their corresponding parts are equal. That is:

$$A + Bj = C + Dj \text{ if and only if } A = C \text{ and } B = D.$$

Example D:
1. $X + Yj = 2 + 9j$ if and only if $X = 2$ and $Y = 9$.
2. $R + Sj = -4 + 6j$ if and only if $R = -4$ and $S = 6$.

Example E: Solve the equation: $7A - 2Bj = 4j + 21$.

$$7A - 2Bj = 21 + 4j$$

Therefore,

$$7A = 21 \text{ and } -2B = 4$$
$$A = 3 \text{ and } \quad B = -2$$

Example F: Solve the equation: $6Rj + 2R + 3 = 7j + Sj - S$.

$$(2R + 3) + (6R)j = -S + (7 + S)j$$

Therefore,

$$2R + 3 = -S \text{ and } 6R = 7 + S$$
$$2R + S = -3$$
$$\underline{6R - S = 7}$$
$$8R = 4$$
$$R = \tfrac{1}{2} \qquad\qquad 2R + 3 = -S$$
$$2(\tfrac{1}{2}) + 3 = -S$$
$$4 = -S$$
$$-4 = S$$
$$R = \tfrac{1}{2} \text{ and } S = -4$$

Exercises

Express each of the following in the form, $A + Bj$.

1. $7 + \sqrt{-9}$
2. $-6 - \sqrt{-16}$
3. $4j^3 - 3$
4. $5j - 2j^2$
5. $32 - \sqrt{-8}$
6. $-\sqrt{12} - \sqrt{-27}$

Find the conjugate of each of the following.

7. $7 + j$
8. $8 - 3j$
9. $-1 + 8j$
10. $-11 - 9j$
11. -17
12. $-12j$

Sec. 11-3. Polar Form of a Complex Number

Solve each of the following equations for the variables involved.

13. $2A - 5Bj = -8 - 15j$
14. $-3X + 2j = -12 - 6Yj$
15. $2P - 4 - 2j = -2Pj - Qj$
16. $X - 1 + j = Y - Yj - Xj$
17. $R + 3S + 3Rj = -j + 5 - Sj$
18. $3K + L + 2j = 12 + Lj - 2Kj$

11-3. Polar Form of a Complex Number

In this section we will see how to represent complex numbers graphically and use this representation to obtain an alternative way of writing complex numbers.

Since a complex number consists of two parts (the real part and the imaginary part), we will use the rectangular coordinate system to represent these numbers graphically. As we pointed out in Chapter 4, this sytem allows us to represent graphically two sets of numbers and the relationship between them. We must make a few changes in terminology, however. The independent (horizontal) axis will be called the *real axis*, the dependent (vertical) axis will be called the *imaginary axis*, the abscissa of a point will correspond to the real part of a complex number, and the ordinate of a point will correspond to the imaginary part of a complex number. This arrangement is referred to as the *complex plane* (see Figure 11-3.1).

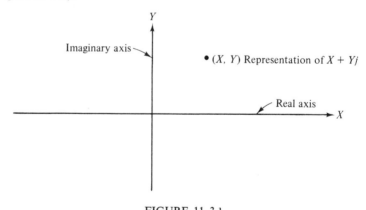

FIGURE 11-3.1

The coordinates of any point in the complex plane, then, correspond to the real and imaginary parts of a complex number. We should note that when complex numbers are represented this way, the vertical coordinate (the ordinate) is automatically multiplied by *j*.

Example A: Represent the following numbers in the complex plane:

$$3 + 4j, -5 + j, 2 - 3j, -3, 3j$$

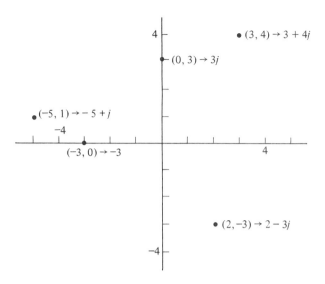

If a complex number is represented graphically, a vector is drawn from the origin to the point corresponding to the complex number, and a line segment is drawn from that point perpendicular to the horizontal axis, the following diagram is obtained (see Figure 11-3.2).

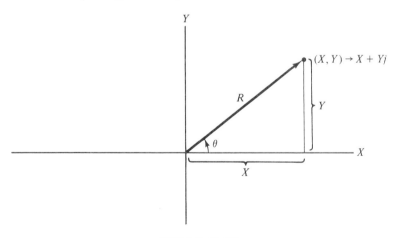

FIGURE 11-3.2

From this diagram, we should observe that $\frac{Y}{R} = \text{Sin } \theta$ or $Y = R \text{ Sin } \theta$ and that $\frac{X}{R} = \text{Cos } \theta$ or $X = R \text{ Cos } \theta$. Therefore,

$$X + Yj = (R \text{ Cos } \theta) + (R \text{ Sin } \theta)j = (R \text{ Cos } \theta) + j(R \text{ Sin } \theta)$$
$$= R(\text{Cos } \theta + j \text{ Sin } \theta)$$

Sec. 11–3. Polar Form of a Complex Number

The last expression is called the *polar form* of a complex number. Thus any complex number may be expressed in terms of rectangular coordinates (X and Y) or in terms of polar coordinates (R and θ). A shorthand notation, $R/\underline{\theta}$, is often used in place of $R(\cos\theta + j\sin\theta)$. In this polar form, R is called the *modulus* or *absolute value* and θ is called the *argument* of the complex number.

Example B:
1. $2(\cos 30° + j \sin 30°) = 2/\underline{30°}$
2. $7(\cos 240° + j \sin 240°) = 7/\underline{240°}$
3. $5/\underline{160°} = 5(\cos 160° + j \sin 160°)$
4. $\sqrt{3}/\underline{2.1} = \sqrt{3}(\cos 2.1 + j \sin 2.1)$

Knowing the range of angles for each quadrant in the coordinate system allows us to represent graphically complex numbers given in polar form.

Example C:
1. Represent $5/\underline{70°}$ graphically.

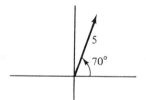

2. Represent $2/\underline{210°}$ graphically.

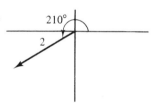

We may change from rectangular form to polar form and vice versa by making use of the following formulas which we first encountered in Section 8–1 and which may be verified by a reference to Figure 11–3.2.

Rectangular form to polar form: $R = \sqrt{X^2 + Y^2}$
$$\tan\theta = \frac{Y}{X}$$

Polar form to rectangular form: $X = R\cos\theta$
$$Y = R\sin\theta$$

Example D: Change $3 + 4j$ to polar form.

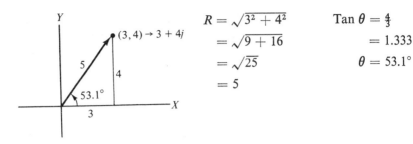

Therefore, $3 + 4j = 5\underline{/53.1°}$.

Example E: Change $3 - 2j$ to polar form.

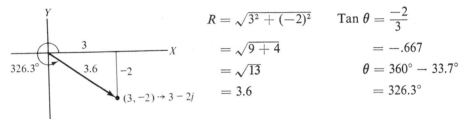

Therefore, $3 - 2j = 3.6\underline{/326.3°}$.

Example F: Change $4\underline{/225°}$ to rectangular form.

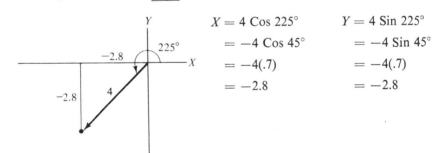

Therefore, $4\underline{/225°} = -2.8 - 2.8j$.

Example G: Change $7\underline{/156°}$ to rectangular form.

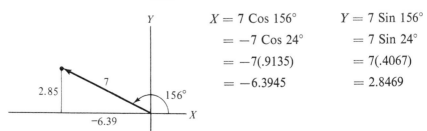

Sec. 11-3. Polar Form of a Complex Number

Therefore, $7\underline{/156°} = -6.39 + 2.85j$.

Pure real and pure imaginary numbers may also be expressed in polar form. Knowing that such numbers lie on one of the axes, we obtain the following:

For any real number: $A = A\underline{/0°}$
$-A = A\underline{/180°}$

For any imaginary number: $Bj = B\underline{/90°}$
$-Bj = B\underline{/270°}$

Example H: Represent each of the following in polar form: $3, -3, 3j, -3j$.

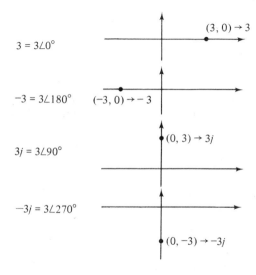

$3 = 3\angle 0°$

$-3 = 3\angle 180°$

$3j = 3\angle 90°$

$-3j = 3\angle 270°$

NOTE: There is a third way of writing a complex number. It is called the *exponential form* and is expressed as $Re^{j\theta}$. However, we will not deal with this form here.

Exercises

Represent each of the following complex numbers graphically.

1. $2 + 5j$
2. $5 - 7j$
3. $-4 + 6j$
4. $-1 - j$
5. $\sqrt{3} + j$
6. $-\pi + 2j$
7. $6j$
8. -5
9. $3\underline{/60°}$
10. $7\underline{/312°}$
11. $10\underline{/270°}$
12. $\sqrt{2}\underline{/110°}$

Express each of the following in polar form.

13. $-3 + 4j$
14. $5 + 12j$
15. $-3 - 8j$
16. $7 - j$
17. $\sqrt{3} + 2j$
18. $-2 - 2j$

320 Ch. 11. Imaginary and Complex Numbers

19. $-1 + 5j$ 20. $\sqrt{2} + \sqrt{5}j$ 21. -9
22. 16 23. $4j$ 24. $-7j$

Express each of the following in rectangular form.

25. $3/\underline{40°}$ 26. $5/\underline{110°}$ 27. $6/\underline{305°}$
28. $2/\underline{262°}$ 29. $4/\underline{3.1}$ 30. $5/\underline{\dfrac{\pi}{6}}$
31. $\sqrt{2}/\underline{0°}$ 32. $15/\underline{\pi}$ 33. $12/\underline{\dfrac{3\pi}{2}}$
34. $\sqrt{7}/\underline{90°}$ 35. $3/\underline{192°}$ 36. $2/\underline{4.9}$

11–4. Operations with Complex Numbers

Operations with complex numbers may be performed in either rectangular or polar form. However, as we will see in this section, certain operations are handled more easily in one form or the other.

To *add or subtract complex numbers*, we observe the basic rule governing addition or subtraction of any quantities. That is, *only like quantities may be combined*. Thus, in rectangular form, we add or subtract the real and imaginary parts separately, while in polar form we may add or subtract only if the angles are identical.

In rectangular form: $(A + Bj) \pm (C + Dj) = (A \pm C) + (B \pm D)j$.

In polar form: $R_1/\underline{\theta} \pm R_2/\underline{\theta} = (R_1 \pm R_2)/\underline{\theta}$.

Example A:
1. $(2 + 3j) + (5 - 4j) = (2 + 5) + (3 - 4)j = 7 - j$
2. $(-3 + 2j) - (2 - j) = (-3 - 2) + [2 - (-1)]j = -5 + 3j$
3. $(4j - 2) - (2j - j^2) = (-2 + 4j) - [2j - (-1)]$
   $\qquad\qquad = (-2 + 4j) - (1 + 2j) = (-2 - 1) + (4 - 2)j$
   $\qquad\qquad = -3 + 2j$
4. $(2 + 4j) + (6 - 3j) - (7 + 6j) = (2 + 6 - 7) + (4 - 3 - 6)j$
   $\qquad\qquad = 1 - 5j$

Example B:
1. $2/\underline{30°} + 7/\underline{30°} = 9/\underline{30°}$
2. $12/\underline{210°} - 9/\underline{210°} = 3/\underline{210°}$
3. $4/\underline{45°} - 10/\underline{45°} = -6/\underline{45°} = 6/\underline{225°}$

NOTE: In this case, recalling that a negative number is the opposite of a positive number, -6 in a direction of 45° is the same as 6 in the opposite direction, which is 225° (45° + 180° = 225°).

4. $3/\underline{120°} - 8/\underline{120°} = -5/\underline{120°} = 5/\underline{300°}$ (120° + 180° = 300°)

Sec. 11–4. Operations with Complex Numbers

We may also add two complex numbers graphically. If we draw a vector from the origin to the point corresponding to each complex number and draw two additional sides to complete a parallelogram, the diagonal of the parallelogram which starts at the origin will be a vector which ends at the point corresponding to the sum of the original complex numbers. In other words, complex numbers may be added diagrammatically in the same way as vectors.

Example C:
1. $(4 + 2j) + (1 + 5j) = 5 + 7j$

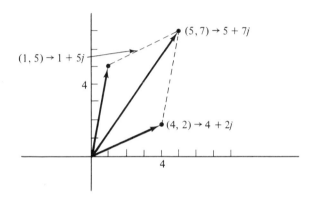

2. $(6 + 3j) - (2 - 4j) = (6 + 3j) + (-2 + 4j) = 4 + 7j$

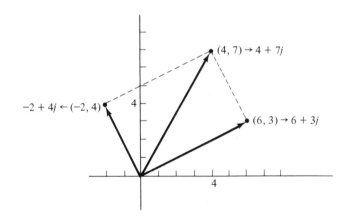

To *multiply complex numbers*, we apply the basic rule for multiplying algebraic expressions. That is, *each term in one expression must be multiplied by each term in the other expression* as we multiply them two at a time. We must also keep in mind that $j^2 = (-1)$.

In rectangular form: $(A + Bj)(C + Dj) = AC + ADj + BCj + BD(j^2)$
$$= (AC - BD) + (AD + BC)j$$

In polar form: $(R_1 \underline{/\theta_1})(R_2 \underline{/\theta_2})$
$= [R_1(\cos\theta_1 + j\sin\theta_1)][R_2(\cos\theta_2 + j\sin\theta_2)]$
$= R_1 R_2[\cos\theta_1 \cos\theta_2 + \cos\theta_1 \sin\theta_2 j + \sin\theta_1 \cos\theta_2 j + \sin\theta_1 \sin\theta_2 (j^2)]$
$= R_1 R_2[(\cos\theta_1 \cos\theta_2 - \sin\theta_1 \sin\theta_2) + j(\sin\theta_1 \cos\theta_2 + \sin\theta_2 \cos\theta_1)]$
$= R_1 R_2[\cos(\theta_1 + \theta_2) + j\sin(\theta_1 + \theta_2)]*$
$= R_1 R_2 \underline{/\theta_1 + \theta_2}$

Example D:
1. $(2 + 4j)(5 - 6j) = 10 - 12j + 20j - 24(j^2) = 10 + 24 - 12j + 20j$
$= 34 + 8j$
2. $(7 - 3j)(-2 + j) = -14 + 7j + 6j - 3(j^2) = -14 + 3 + 7j + 6j$
$= -11 + 13j$

Example E:
1. $(3\underline{/120°})(5\underline{/72°}) = 15\underline{/192°}$
2. $(4\underline{/210°})(2\underline{/170°}) = 8\underline{/380°} = 8\underline{/20°}$

NOTE: $380° = 380° - 360° = 20°$.

A special case of multiplication of complex numbers is *multiplication by* j. The following examples illustrate that multiplying a complex number by j has the same effect as taking the complex number and rotating it through an angle of 90°. For this reason, the quantity j is often referred to as the j-*operator*.

Example F:
1. $j(2) = 2j$
2. $j(2j) = 2j^2 = -2$
3. $j(-2) = -2j$
4. $j(-2j) = -2j^2 = 2$

Example G: $j(2 + 3j) = 2j + 3j^2 = -3 + 2j$

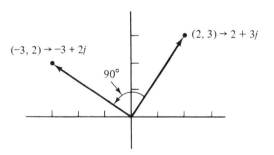

*Refer to trigonometric identities in Chapter 10.

Sec. 11-4. Operations with Complex Numbers

Example H: $j(-1-j) = -j - j^2 = 1 - j$

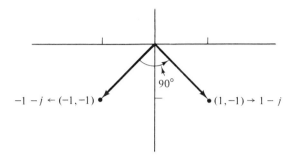

To *divide complex numbers*, we multiply both numerator and denominator by the conjugate of the denominator. This procedure is equivalent to multiplying by 1 (hence, the value of the expression remains unchanged), and also enables us to obtain a single complex number for our answer.

In rectangular form:

$$\frac{(A + Bj)}{(C + Dj)} = \frac{(A + Bj) \times (C - Dj)}{(C + Dj) \times (C - Dj)} = \frac{AC - ADj + BCj - BD(j^2)}{C^2 - CDj + CDj - D^2(j^2)}$$

$$= \frac{(AC + BD) + (BC - AD)j}{C^2 + D^2}$$

In polar form:

$$\frac{R_1 \underline{/\theta_1}}{R_2 \underline{/\theta_2}} = \frac{R_1(\cos \theta_1 + j \sin \theta_1)}{R_2(\cos \theta_2 + j \sin \theta_2)}$$

$$= \frac{R_1(\cos \theta_1 + j \sin \theta_1) \times R_2(\cos \theta_2 - j \sin \theta_2)}{R_2(\cos \theta_2 + j \sin \theta_2) \times R_2(\cos \theta_2 - j \sin \theta_2)}$$

$$= \frac{R_1 R_2 [\cos \theta_1 \cos \theta_2 - \sin \theta_2 \cos \theta_1 j + \sin \theta_1 \cos \theta_2 j - \sin \theta_1 \sin \theta_2 (j^2)]}{(R_2)^2 [\cos \theta_2 \cos \theta_2 - \sin \theta_2 \cos \theta_2 j + \sin \theta_2 \cos \theta_2 j - \sin \theta_2 \sin \theta_2 (j^2)]}$$

$$= \frac{R_1 [(\cos \theta_1 \cos \theta_2 + \sin \theta_1 \sin \theta_2) + j(\sin \theta_1 \cos \theta_2 - \sin \theta_2 \cos \theta_1)]}{R_2 [\cos^2 \theta_2 + \sin^2 \theta_2]}$$

$$= \frac{R_1}{R_2} [\cos (\theta_1 - \theta_2) + j \sin (\theta_1 - \theta_2)]*$$

$$= \frac{R_1}{R_2} \underline{/\theta_1 - \theta_2}$$

Example I:

1. $\dfrac{2 + 4j}{5 - 2j} = \dfrac{(2 + 4j) \times (5 + 2j)}{(5 - 2j) \times (5 + 2j)} = \dfrac{10 + 4j + 20j + 8j^2}{25 + 10j - 10j - 4(j^2)} = \dfrac{2 + 24j}{29}$

*Refer to trigonometric identities in Chapter 10.

2. $\dfrac{5-4j}{2j} = \dfrac{(5-4j)\times(-2j)}{(2j)\times(-2j)} = \dfrac{-10j+8j^2}{-4j^2} = \dfrac{-8-10j}{4} = \dfrac{2(-4-5j)}{4}$
$= \dfrac{-4-5j}{2}$

3. $\dfrac{2+j}{j^2-j} = \dfrac{2+j}{-1-j} = \dfrac{(2+j)\times(-1+j)}{(-1-j)\times(-1+j)} = \dfrac{-2+2j-j+j^2}{1-j+j-j^2}$
$= \dfrac{-3+j}{2}$

Example J:
1. $\dfrac{12\underline{/72°}}{6\underline{/30°}} = 2\underline{/42°}$

2. $\dfrac{18\underline{/120°}}{3\underline{/200°}} = 6\underline{/-80°} = 6\underline{/280°}$

3. $\dfrac{2\underline{/260°}}{5\underline{/100°}} = .4\underline{/160°}$

Finding *powers or roots of complex numbers* is a very important procedure. In rectangular form, *integral powers* of complex numbers may be found by successive multiplication. With the exception of pure real numbers, roots of complex numbers usually cannot be found when the numbers are in rectangular form.

Example K:
1. $(5+4j)^2 = (5+4j)(5+4j) = 25+20j+20j+16j^2 = 9+40j$
2. $(2-3j)^3 = (2-3j)(2-3j)(2-3j) = (4-6j-6j+9j^2)(2-3j)$
$= (-5-12j)(2-3j) = -10+15j-24j+36j^2$
$= -46-9j$
3. $(-3+j)^{-2} = \dfrac{1}{(-3+j)^2} = \dfrac{1}{(-3+j)(-3+j)} = \dfrac{1}{9-3j-3j+j^2}$
$= \dfrac{1}{8-6j} = \dfrac{1}{8-6j} \times \dfrac{8+6j}{8+6j} = \dfrac{8+6j}{64+48j-48j-36j^2}$
$= \dfrac{2(4+3j)}{100} = \dfrac{4+3j}{50}$

In polar form, we make use of the definition of multiplication to develop a general rule for finding both powers and roots.

$(R\underline{/\theta})^2 = (R\underline{/\theta})(R\underline{/\theta}) = R^2\underline{/\theta+\theta} = R^2\underline{/2\theta}$
$(R\underline{/\theta})^3 = (R\underline{/\theta})(R\underline{/\theta})(R\underline{/\theta}) = R^3\underline{/\theta+\theta+\theta} = R^3\underline{/3\theta}$
$(R\underline{/\theta})^4 = (R\underline{/\theta})(R\underline{/\theta})(R\underline{/\theta})(R\underline{/\theta}) = R^4\underline{/\theta+\theta+\theta+\theta} = R^4\underline{/4\theta}$

In general, for any value of N,

$$(R\underline{/\theta})^N = R^N\underline{/N\theta}.$$

Sec. 11–4. Operations with Complex Numbers 325

This rule is called *DeMoivre's Theorem*, and may be used to find both powers (if N is an integer) and roots (if N is a fraction) of complex numbers if they are in polar form.

Example L:
1. $(2\underline{/30°})^5 = 2^5\underline{/5(30°)} = 32\underline{/150°}$
2. $(5\underline{/200°})^3 = 5^3\underline{/3(200°)} = 125\underline{/600°} = 125\underline{/240°}$
3. $(4\underline{/72°})^{-4} = \dfrac{1}{(4\underline{/72°})^4} = \dfrac{1\underline{/0°}}{4^4\underline{/4(72°)}} = \dfrac{1\underline{/0°}}{256\underline{/288°}} = .004\underline{/-288°}$
$= .004\underline{/72°}$

If a complex number in rectangular form is to be raised to a large integral power, it is usually easier to change the number to polar form and use DeMoivre's Theorem.

Example M:
1. Evaluate $(3 + 2j)^7$.

   $R = \sqrt{9+4}$ $\text{Tan } \theta = \tfrac{2}{3}$
   $= \sqrt{13}$ $= .667$
   $= 3.6$ $\theta = 33.7°$

   $(3 + 2j)^7 = (3.6\underline{/33.7°})^7 = 3.6^7\underline{/7(33.7°)} = 7836.4\underline{/235.9°}$

2. Evaluate $(4 - j)^{10}$.

   $R = \sqrt{16+1}$ $\text{Tan } \theta = \dfrac{-1}{4}$
   $= \sqrt{17}$ $= -.25$
   $= 4.12$ $\theta = 360° - 14.1° = 345.9°$

   $(4 - j)^{10} = (4.12\underline{/345.9°})^{10} = 4.12^{10}\underline{/10(345.9°)} = 1409198\underline{/3459°}$
   $= 1409198\underline{/219°}$

3. Evaluate $\dfrac{(-2+j)^6}{(3+3j)^4}$

   (a) $-2 + j$: $R = \sqrt{4+1}$ $\text{Tan } \theta = \dfrac{1}{-2}$
   $= \sqrt{5}$ $= -.5$
   $= 2.24$ $\theta = 180° - 26.6° = 153.4°$

   (b) $3 + 3j$: $R = \sqrt{9+9}$ $\text{Tan } \theta = \tfrac{3}{3}$
   $= \sqrt{18}$ $= 1$
   $= 4.24$ $\theta = 45°$

   (c) $\dfrac{(-2+j)^6}{(3+3j)^4} = \dfrac{(2.24\underline{/153.4°})^6}{(4.24\underline{/45°})^4} = \dfrac{2.24^6\underline{/6(153.4°)}}{4.24^4\underline{/4(45°)}} = \dfrac{126.3\underline{/920.4°}}{323.2\underline{/180°}}$
   $= .39\underline{/740.4°} = .39\underline{/20.4°}$

Example N: Find the cube root of -8. We know that $(-2)(-2)(-2) = -8$ and therefore, $\sqrt[3]{-8} = -2$. Let us verify this result by using DeMoivre's Theorem.

$$\sqrt[3]{-8} = (-8)^{1/3} = (8\underline{/180°})^{1/3} = 8^{1/3}\underline{/\tfrac{1}{3}(180°)} = 2\underline{/60°}$$

Examining the result ($2\underline{/60°}$), we find

$$X = 2\cos 60° = 2(.5) = 1 \quad \text{and} \quad Y = 2\sin 60° = 2(.866) = 1.732$$

Therefore, $\sqrt[3]{8} = 2\underline{/60°} = 1 + 1.732j$, which is not the result we expected. However, since adding $360°$ to any angle gives us an angle with the same terminal side and hence the same functional values, let us try this:

$$180° + 360° = 540°$$

Therefore,

$$(8\underline{/180°})^{1/3} = (8\underline{/540°})^{1/3} = 8^{1/3}\underline{/\tfrac{1}{3}(540°)} = 2\underline{/180°}$$

When $2\underline{/180°}$ is put in rectangular form, we obtain the number 2, which is what we originally expected. If we repeat this procedure, $540° + 360° = 900°$. Therefore,

$$(8\underline{/180°})^{1/3} = (8\underline{/540°})^{1/3} = (8\underline{/900°})^{1/3}$$
$$(8\underline{/900°})^{1/3} = 8^{1/3}\underline{/\tfrac{1}{3}(900°)} = 2\underline{/300°} = 1 - 1.732j$$

in rectangular form. If we try this once more, $900° + 360° = 1260°$.

$$(8\underline{/1260°})^{1/3} = 8^{1/3}\underline{/\tfrac{1}{3}(1260°)} = 2\underline{/420°} = 2\underline{/60°}$$

which is the same as our first answer. Therefore, adding more multiples of $360°$ will yield the same answers we have already obtained.

Thus, we have found three cube roots of -8: $1 + 1.732j$, -2, and $1 - 1.732j$. If any of these numbers are cubed, we will get a result of -8. Thus, they are all true cube roots of -8.

What we have done in the previous example may be generalized to the fact that *every complex number has* **N** *Nth roots*. This fact may be proven, but we will accept it here without a formal proof. That is, every complex number has 2 square roots, 3 cube roots, 4 fourth roots, etc. To find these roots, the complex number must be in polar form ($R\underline{/\theta}$). We then apply DeMoivre's Theorem N times using θ to obtain the first root and adding $(N - 1)$ multiples of $360°$ to θ to obtain the remaining roots. Also, when the roots are in polar form, the angles differ by $\dfrac{360°}{N}$. Thus, in the previous example, the 3 cube roots ($2\underline{/60°}$, $2\underline{/180°}$, and $2\underline{/300°}$) differ by $\dfrac{360°}{3} = 120°$.

Sec. 11–4. Operations with Complex Numbers

Example O: Find the fourth roots of $3j$.

$$3j = 3\underline{/90°}$$
$$\sqrt[4]{3j} = (3j)^{1/4} = (3\underline{/90°})^{1/4} = 1.32\underline{/22.5°}$$
$$= (3\underline{/450°})^{1/4} = 1.32\underline{/112.5°}$$
$$= (3\underline{/810°})^{1/4} = 1.32\underline{/202.5°}$$
$$= (3\underline{/1170°})^{1/4} = 1.32\underline{/292.5°}$$

The 4 fourth roots of $3j$.

NOTE: The 4 roots are $\dfrac{360°}{4} = 90°$ apart.

Example P: Find the sixth roots of $(2 - 2j)$.

$$R = \sqrt{4+4} \qquad \text{Tan } \theta = \dfrac{-2}{2}$$
$$= \sqrt{8} \qquad\qquad\quad = -1$$
$$= 2.83 \qquad\qquad \theta = 360° - 45° = 315°$$
$$\sqrt[6]{2-2j} = (2-2j)^{1/6} = (2.83\underline{/315°})^{1/6} = 1.19\underline{/52.5°}$$
$$= (2.83\underline{/675°})^{1/6} = 1.19\underline{/112.5°}$$
$$= (2.83\underline{/1035°})^{1/6} = 1.19\underline{/172.5°}$$
$$= (2.83\underline{/1395°})^{1/6} = 1.19\underline{/232.5°}$$
$$= (2.83\underline{/1755°})^{1/6} = 1.19\underline{/292.5°}$$
$$= (2.83\underline{/2115°})^{1/6} = 1.19\underline{/352.5°}$$

The 6 sixth roots of $(2 - 2j)$.

NOTE: The 6 roots are $\dfrac{360°}{6} = 60°$ apart.

Once the N Nth roots of a complex number are obtained, they may be expressed in either polar or exponential form. However, unless there is a specific reason for doing otherwise, it is normally easier to leave them in polar form.

Exercises

For each of the following, perform the indicated operation(s).

1. $(5 + 2j) + (-4 + 8j)$
2. $(7 - 3j) + (2j - 6)$
3. $(4 + j) - (5j + 1) - (2j - j^2)$
4. $(j^3 + 1) - 4j + (3j^2)$
5. $3\underline{/60°} + 5\underline{/60°}$
6. $4\underline{/110°} + 7\underline{/110°} - 3\underline{/110°}$
7. $6\underline{/220°} - 9\underline{/220°}$
8. $5\underline{/90°} - 7\underline{/90°}$
9. $(2 + 5j)(3 - j)$
10. $(-4 + 3j)(-2 - 7j)$
11. $(-1 - j)(2 - 6j)3j$
12. $(7 + 3j)(j^2 - 1)(-2 - 3j)$
13. $(5\underline{/20°})(12\underline{/142°})$
14. $(3\underline{/143°})(6\underline{/289°})$
15. $(2\underline{/18°})(5\underline{/71°})(9\underline{/145°})$
16. $(7\underline{/220°})(3\underline{/95°})(4\underline{/110°})$
17. $\dfrac{2 - 3j}{-1 - j}$
18. $\dfrac{2j}{5 - 4j}$

19. $\dfrac{3-6j}{j^2+j}$ 20. $\dfrac{-1+3j}{4j}$

21. $\dfrac{8\underline{/145°}}{4\underline{/77°}}$ 22. $\dfrac{25\underline{/274°}}{10\underline{/98°}}$

23. $\dfrac{12\underline{/210°}}{4\underline{/330°}}$ 24. $\dfrac{8\underline{/45°}}{20\underline{/276°}}$

25. $(2-j)^3$ 26. $(5-5j)^2$

27. $(j^2-2j^3)^2$ 28. $(5j-3)^3$

29. $(5\underline{/20°})^4$ 30. $(2\underline{/220°})^3$

31. $(4\underline{/143°})^6$ 32. $(3\underline{/310°})^5$

For each of the following, perform the indicated operations graphically.

33. $(4+5j)+(2-j)$ 34. $(-3+2j)+(-1-j)$
35. $(-4+j)-(5-2j)$ 36. $(3-6j)-(1-j)$

In each of the following, change the numbers to polar form before performing the indicated operations. Leave answers in polar form.

37. $(5+j)^7(3-4j)^6$ 38. $(-2+3j)^{10}(4-6j)^4$
39. $\dfrac{(-1-j)^6}{(-5+2j)^5}$ 40. $\dfrac{(6-2j)^7}{(-4-7j)^8}$

Use DeMoivre's Theorem to find each of the following indicated roots. Leave answers in polar form.

41. $\sqrt[4]{16}$ 42. $\sqrt[3]{-8j}$ 43. $\sqrt[5]{4+4j}$
44. $\sqrt{5-12j}$ 45. $\sqrt[3]{-7+3j}$ 46. $\sqrt[6]{1-2j}$

11–5. An Application to Vectors

In Section 11–3 we saw that complex numbers could be written in either rectangular or polar form, and in Section 11–4 we saw that when complex numbers are written in rectangular form they may be added graphically in the same manner as vectors. In Section 8–1 we defined a vector as a quantity that has both a magnitude and a direction. By comparison, we should see that the *magnitude and direction* (R *and* θ) *of a vector* correspond to the *modulus and argument* (R *and* θ) *of a complex number*. Thus, when we are given a vector, this corresponds to being given a complex number in polar form. Therefore, a method of adding vectors, an alternative to the method described in Chapter 8, is to consider them as complex numbers in polar form, change them to rectangular form, and add. The answer may be expressed in either rectangular or polar form.

Sec. 11–5. An Application to Vectors

Example A: Add the following vectors.

$$\vec{R}: \ |R| = 30, \text{ direction} = 40°.$$
$$\vec{S}: \ |S| = 50, \text{ direction} = 160°.$$
$$R_x = 30 \text{ Cos } 40° = 30(.7660) = 22.98$$
$$R_y = 30 \text{ Sin } 40° = 30(.6428) = 19.28$$

Therefore, $\vec{R} = 22.98 + 19.28j$.

$$S_x = 50 \text{ Cos } 160° = 50(-\text{Cos } 20°) = 50(-.9397) = -46.98$$
$$S_y = 50 \text{ Sin } 160 = 50(\text{Sin } 20°) = 50(.3420) = 17.10$$

Therefore, $\vec{S} = -46.98 + 17.10j$.

$$\vec{R} + \vec{S} = (22.98 - 46.98) + (19.28 + 17.10)j = -24 + 36.38j$$

Example B: Add the following vectors.

$$\vec{A}: \ |A| = 26, \text{ direction} = 58°.$$
$$\vec{B}: \ |B| = 45, \text{ direction} = 215°.$$
$$\vec{C}: \ |C| = 38, \text{ direction} = 312°.$$
$$A_x = 26 \text{ Cos } 58° = 26(.5299) = 13.78$$
$$A_y = 26 \text{ Sin } 58° = 26(.8480) = 22.05$$

Therefore, $\vec{A} = 13.78 + 22.05j$.

$$B_x = 45 \text{ Cos } 215° = 45(-\text{Cos } 35°) = 45(-.8192) = -36.86$$
$$B_y = 45 \text{ Sin } 215° = 45(-\text{Sin } 35°) = 45(-.5736) = -25.81$$

Therefore, $\vec{B} = -36.86 - 25.81j$.

$$C_x = 38 \text{ Cos } 312° = 38(\text{Cos } 48°) = 38(.6691) = 25.43$$
$$C_y = 38 \text{ Sin } 312° = 38(-\text{Sin } 48°) = 38(-.7431) = -28.24$$

Therefore, $\vec{C} = 25.43 - 28.24j$.

$$\vec{A} + \vec{B} + \vec{C} = (13.78 - 36.86 + 25.43) + (22.05 - 25.81 - 28.24)j$$
$$= 2.35 - 32j$$

We should see from these examples that, not only do the magnitude and direction of a vector correspond to the modulus and argument of a complex number in polar form, but the horizontal and vertical components of a vector correspond to the real and imaginary parts of a complex number in rectangular form. Therefore, as we have mentioned previously, vectors may be used to represent complex numbers, and complex numbers may be used to represent vectors.

Exercises In each of the following problems, find the resultant for the given vectors by considering them as complex numbers. Leave answers in rectangular form for problems (1)–(10).

1. $\vec{A}$: $|A| = 45$, direction $= 50°$.
   $\vec{B}$: $|B| = 51$, direction $= 100°$.
2. $\vec{R}$: $|R| = 22$, direction $= 126°$.
   $\vec{S}$: $|S| = 39$, direction $= 260°$.
3. $\vec{P}$: $|P| = 98$, direction $= 92°$.
   $\vec{Q}$: $|Q| = 110$, direction $= 8°$.
4. $\vec{X}$: $|X| = 2650$, direction $= 170°$.
   $\vec{Y}$: $|Y| = 3200$, direction $= 305°$.
5. $\vec{J}$: $|J| = 16.8$, direction $= 86°$.
   $\vec{K}$: $|K| = 27.3$, direction $= 12°$.
6. $\vec{M}$: $|M| = 7.5$, direction $= 185°$.
   $\vec{N}$: $|N| = 12.1$, direction $= 220°$.
7. $\vec{T}$: $|T| = 61$, direction $= 345°$.
   $\vec{V}$: $|V| = 52$, direction $= 78°$.
8. $\vec{I}$: $|I| = 19.6$, direction $= 116°$.
   $\vec{J}$: $|J| = 24.3$, direction $= 195°$.
9. $\vec{A}$: $|A| = 150$, direction $= 65°$.
   $\vec{B}$: $|B| = 186$, direction $= 105°$.
   $\vec{C}$: $|C| = 162$, direction $= 253°$.
10. $\vec{R}$: $|R| = 17.6$, direction $= 100°$.
    $\vec{S}$: $|S| = 25.8$, direction $= 250°$.
    $\vec{T}$: $|T| = 19.1$, direction $= 350°$.
11. A plane is headed in a direction 35° north of west at a speed of 650 miles per hour. If the wind is from the southwest at 20 miles per hour, what is the resultant speed and direction of the plane?
12. After leaving port, a ship travels due west for 78 kilometers and then turns and heads 12° south of west for 24 kilometers. How far is the ship from port and in what direction?

11–6. An Application to AC Circuits

In this section we will see how complex numbers may be used in analyzing *AC circuits*. *AC* stands for *alternating current*, and in this type of electric circuit, the

Sec. 11-6. An Application to AC Circuits

current and voltage have different *phases*. This means that the current and voltage do not reach their peak (maximum) values at the same time. When complex numbers are used to represent current and voltage, the modulus (R) will correspond to the actual value of the current or voltage and the argument (θ) will be a measure of the phase. We will be considering circuits containing a resistance, a capacitance, and an inductance.

A *resistance* is an element in the circuit that tends to obstruct the flow of current. It is denoted by R, is measured in ohms, and is represented in circuit diagrams by -ww-. A *capacitance* is essentially two unconnected plates. No current flows across the gap between the plates, but in an AC circuit a charge is continually going to and from each plate so that current is not really blocked. A capacitance is denoted by C, is measured in farads, and is represented in circuit diagrams by -)(-. An *inductance* is a coil of wire in which a current is induced due to the changing current in the circuit. It is denoted by L, is measured in henrys, and is represented in circuit diagrams by -∞-. Each of these elements offers some type of resistance to the flow of current in an AC circuit. The effective resistance of any part of such a circuit is called the *reactance* and is denoted by X. The voltage across any part of a circuit that has a reactance of X is given by $V = IX$, where V is the voltage (in volts) and I is the current (in amperes). Therefore, across a resistor, $V_R = IX_R$, across a capacitor, $V_C = IX_C$, and across an inductor, $V_L = IX_L$.

To determine the voltage across a combination of these elements, we must take into account both the reactance and the phase of the voltage in that part of the circuit. In Chapter 9, we saw that both the current and voltage may be represented by a sine or cosine curve. Thus, each of these quantities periodically reaches a maximum value. If these quantities reach their maximum values at the same time, then they are *in phase*. Otherwise, they are out of phase and the voltage will either *lead* or *lag* the current, reaching its maximum value before or after the current reaches its maximum value. In the study of electricity and electric circuits, it can be shown that across a resistor the voltage and current are in phase, across a capacitor the voltage lags the current by $90° \left(\frac{\pi}{2}\right)$, and across an inductor the voltage leads the current by $90° \left(\frac{\pi}{2}\right)$.

Taking these phase differences into account, we shall represent X_R as a vector along the postive horizontal axis (voltage and current in phase with phase difference of 0°), X_C as a vector along the negative imaginary axis (voltage lags current by 90° with phase difference of $-90°$), and X_L as a vector along the positive imaginary axis (voltage leads current by 90° with phase difference of 90°). These vectors are called *phasors* (see Figure 11-6.1).*

---

*Reactances are not really vectors but are represented this way in order to be able to represent phase differences between voltage and current. Also, any position could be chosen as a starting point for X_R, but the positive horizontal axis is the most convenient.

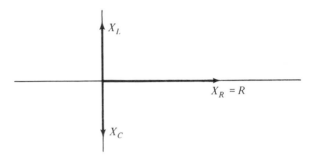

FIGURE 11-6.1

The voltage across a combination of a resistor, a capacitor, and an inductor will be given by:

$$V_{RLC} = V_R + V_L + V_C$$
$$= IX_R + IX_L j + IX_C(-j)$$
$$= IX_R + Ij(X_L - X_C)$$
$$= I[X_R + (X_L - X_C)j]$$
$$= IZ$$

$Z = X_R + (X_L - X_C)j$ is a complex number and is called the *impedance* of the circuit. It is the total effective resistance of the combination of the elements in the circuit. It has a magnitude, $|Z| = \sqrt{X_R^2 + (X_L - X_C)^2}$, and makes an angle with the positive horizontal axis which is given by $\text{Tan } \theta = \dfrac{X_L - X_C}{X_R}$. Since Z was obtained by taking into account the phase difference between voltage and current in each element, this angle θ represents the phase difference between voltage and current across the combination of elements (see Figure 11-6.2).

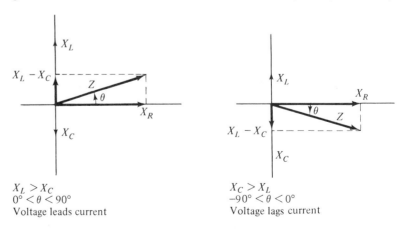

FIGURE 11-6.2

Sec. 11–6. An Application to AC Circuits

Example A: In a given circuit, $X_R(R) = 3$ ohms, $X_L = 8$ ohms, and $X_C = 4$ ohms. Find the impedance and the phase angle across the combination.

$X_R = 3$ ohms $X_L = 8$ ohms $X_C = 4$ ohms

$$Z = 3 + (8 - 4)j = 3 + 4j$$
$$|Z| = \sqrt{X_R^2 + (X_L - X_C)^2} = \sqrt{3^2 + (8 - 4)^2} = \sqrt{9 + 16}$$
$$= \sqrt{25} = 5 \text{ ohms}$$
$$\text{Tan } \theta = \frac{X_L - X_C}{X_R} = \frac{8 - 4}{3} = \frac{4}{3} = 1.33\overline{3}.$$

Therefore, $\theta = 53.15°$ and voltage leads current by $53.15°$.

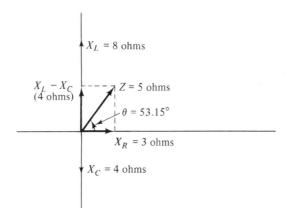

Example B: In a given circuit, $X_R(R) = 8$ ohms, $X_L = 4$ ohms, and $X_C = 10$ ohms. A current of 3 amperes is in the circuit. Find the voltage across each element, the impedance, the voltage across the combination, and the phase angle between the voltage and current.

A   B   C   D

$X_R = 8$ ohms $X_L = 4$ ohms $X_C = 10$ ohms

1. Voltage across the resistor (between A and B) is given by

$$V_R = IX_R = 3(8) = 24 \text{ V}$$

2. Voltage across the inductor (between B and C) is given by

$$V_L = IX_L = 3(4) = 12 \text{ V}$$

3. Voltage across the capacitor (between C and D) is given by

$$V_C = IX_C = 3(10) = 30 \ V$$

4. Impedance is given by

$$Z = 8 + (4 - 10)j = 8 - 6j$$
$$|Z| = \sqrt{X_R^2 + (X_L - X_C)^2} = \sqrt{8^2 + (4-10)^2}$$
$$= \sqrt{64 + 36} = \sqrt{100} = 10 \text{ ohms}$$

5. Voltage across the combination (between A and D) is given by

$$V = IZ = 3(10) = 30 \ V$$

6. $\text{Tan } \theta = \dfrac{X_L - X_C}{X_R} = \dfrac{4 - 10}{8} = \dfrac{-6}{8} = -.75$

Therefore, $\theta = -36.9°$ and voltage lags current by $36.9°$.

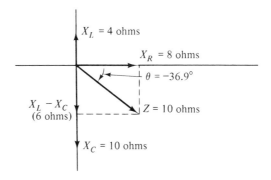

NOTE: The voltage across the combination is not the arithmetic sum of the individual voltages since the phase differences must be taken into account.

If I and Z are given as complex numbers, then we make use of our definition of multiplication of complex numbers to find V.

Example C: In a certain AC circuit, $I = 3 - 4j$ amperes and $Z = 5 + 2j$ ohms. Find V.

$$V = IZ = (3 - 4j)(5 + 2j) = 15 + 6j - 20j - 8j^2 = 15 - 14j + 8$$
$$= (23 - 14j) \ V$$
$$|V| = \sqrt{23^2 + (-14)^2} = \sqrt{529 + 196} = \sqrt{725} = 26.9 \ V$$

If a coil of wire rotates through a magnetic field, an alternating current

Sec. 11-6. An Application to AC Circuits

is produced. If the coil has a constant angular velocity ω (measured in radians per unit of time), the capactive and inductive reactances are given by

$$X_C = \frac{1}{\omega C} \quad \text{and} \quad X_L = \omega L$$

Example D: If $X_R = 15$ ohms, $L = .5$ henry, $C = .00021$ farad, and $\omega = 60$ radians per second, find the impedance and the phase difference between the voltage and current.

$$X_C = \frac{1}{\omega C} = \frac{1}{60(.00021)} = \frac{1}{.0126} = 79.4 \text{ ohms}$$

$$X_L = \omega L = 60(.5) = 30 \text{ ohms}$$

$$Z = 15 + (30 - 79.4)j = 15 - 49.4j$$

$$|Z| = \sqrt{X_R^2 + (X_L - X_C)^2} = \sqrt{15^2 + (30 - 79.4)^2}$$

$$= \sqrt{15^2 + (-49.4)^2} = \sqrt{225 + 2440.4} = \sqrt{2665.4}$$

$$= 51.6 \text{ ohms}$$

$$\text{Tan } \theta = \frac{X_L - X_C}{X_R} = \frac{30 - 79.4}{15} = \frac{-49.4}{15} = -3.29$$

Therefore, $\theta = -73.1°$ and voltage lags current by 73.1°.

Exercises

1. If $I = (5 - j)$ amperes and $Z = (2 + j)$ ohms, find V.
2. If $I = (6 + 2j)$ amperes and $Z = (3 - j)$ ohms, find V.
3. If $V = (8 + 3j)$ volts and $Z = (4 - 2j)$ ohms, find I.
4. If $V = (4 + 3j)$ volts and $Z = (1 + 3j)$ ohms, find I.
5. In a certain circuit, $X_R = 6$ ohms, $X_L = 5$ ohms, and $X_C = 9$ ohms. Find the impedance (and its magnitude) and phase angle across the combination.
6. In a certain circuit, $X_R = 12$ ohms, $X_L = 10$ ohms, and $X_C = 7$ ohms. Find the impedance (and its magnitude) and phase angle across the combination.
7. A current of 2 amperes flows in a certain circuit. If in the same circuit, $X_R = 8$ ohms, $X_L = 14$ ohms, and $X_C = 9$ ohms, find the voltage across each element, the impedance (and its magnitude), the voltage across the combination, and the phase angle between the voltage and current.
8. Four amperes of current flow in a given circuit. If the same circuit has reactances of $X_R = 16$ ohms, $X_L = 11$ ohms, and $X_C = 24$ ohms, find the voltage across each element, the impedance (and its magnitude), the voltage across the combination, and the phase angle between the voltage and current.

In a certain circuit, $X_R = 18$ ohms, $X_L = 15$ ohms, and $X_C = 8$ ohms. Find the impedance (and its magnitude) and the phase angle between the voltage and current for each of the following situations.

9. The inductor is removed.
10. The capacitor is removed.
11. The resistor is removed.
12. No element is removed.
13. A coil of wire rotates through a magnetic field at 60 cycles per second. If $X_R = 7$ ohms, $L = .03$ henry, and $C = .0005$ farad in the circuit, find the impedance (and its magnitude) and the phase angle between the voltage and current.
14. A coil of wire rotates through a magnetic field at 30 revolutions per second. In the circuit, $X_R = 2000$ ohms, $L = 6.2$ henrys, and $C = .00001$ farad. Find the impedance (and its magnitude) and the phase angle between the voltage and current.

11-7. CHAPTER REVIEW

Represent each of the following in $A + Bj$ form.

1. $5 - \sqrt{-36}$
2. $-3 + \sqrt{-121}$
3. $\sqrt{2} - \sqrt{-12}$
4. $\pi + \sqrt{-20}$

Solve each of the following for X and Y.

5. $X - 2 + 5Yj = Y - 3Xj - 2j$
6. $Y + 3 - 2Xj = X + 1 - 7Yj$

Express each of the following in polar form.

7. $5 - 4j$
8. $-1 - 2j$
9. $6 + 9j$
10. $-3 + 7j$

Express each of the following in rectangular form.

11. $5/220°$
12. $3/115°$
13. $7/295°$
14. $2/58°$

For each of the following, perform the indicated operations graphically.

15. $(5 + 2j) + (3 - j)$
16. $(4 - 3j) + (-1 + 2j)$
17. $(-3 + 5j) - (4 - 2j)$
18. $(-2 - j) - (-3 - 6j)$

For each of the following, perform the indicated operations.

19. $(3 - 5j) + (3j + 7)$
20. $(4 - 2j) + (6 - 3j) - (5 - 4j)$
21. $(2j - 3j^2) - (3j^4 + j^3)$
22. $(7 - 5j^3) - (j^2 + 3j)$

Sec. 11–7. Chapter Review

23. $3\underline{/70°} + 6\underline{/70°}$
24. $5\underline{/200°} - 3\underline{/200°}$
25. $8\underline{/120°} - 11\underline{/120°}$
26. $3\underline{/312°} - 9\underline{/312°}$
27. $(6 + 3j) \times (4 - j)$
28. $(2 - 5j) \times (3 + 2j)$
29. $(5j^2 - 2j) \times (6j - 4)$
30. $(7 + 3j^3) \times (5j - j^2)$
31. $(4 + 5j) \div (-1 - 3j)$
32. $(-1 + 4j) \div (2j^2 + 3j)$
33. $3\underline{/27°} \times 5\underline{/116°}$
34. $\sqrt{2}\underline{/145°} \times \sqrt{6}\underline{/210°}$
35. $12\underline{/168°} \div 6\underline{/91°}$
36. $4\underline{/245°} \div 18\underline{/358°}$
37. $(3 - 2j)^3$
38. $(4 + j)^4$
39. $(3\underline{/72°})^4$
40. $(2\underline{/305°})^5$

For each of the following, add the vectors by considering them as complex numbers.

41. $\vec{A}$: $|A| = 62$, direction $= 125°$
    $\vec{B}$: $|B| = 75$, direction $= 218°$

42. $\vec{R}$: $|R| = 136$, direction $= 73°$
    $\vec{S}$: $|S| = 210$, direction $= 305°$

43. $\vec{X}$: $|X| = 8.6$, direction $= 24°$
    $\vec{Y}$: $|Y| = 12.1$, direction $= 158°$
    $\vec{Z}$: $|Z| = 9.4$, direction $= 117°$

44. $\vec{A}$: $|A| = 54$, direction $= 344°$
    $\vec{B}$: $|B| = 35$, direction $= 41°$
    $\vec{C}$: $|C| = 46$, direction $= 198°$

45. In a certain AC circuit, $X_R = 20$ ohms, $X_L = 18$ ohms, $X_C = 30$ ohms, and $I = 5$ amperes. Find the voltage across each element, the magnitude of the impedance, the voltage across the combination, and the phase difference between the voltage and current.

46. A current of 3 amperes flows in a certain AC circuit. In this circuit $X_R = 16$ ohms, $X_L = 20$ ohms, and $X_C = 10$ ohms. Find the voltage across each element, the magnitude of the impedance, the voltage across the combination, and the phase difference between the voltage and current.

47. A coil of wire rotates through a magnetic field at 90 radians per second. In the circuit, $X_R = 50$ ohms, $L = .4$ henry, $C = .0002$ farad, and $I = 2$ amperes. Find the voltage across each element, the magnitude of the impedance, the voltage across the combination, and the phase difference between the voltage and current.

48. In a magnetic field, a coil of wire is rotating with an angular velocity of 60 revolutions per second. If $I = 6$ amperes, $X_R = 3000$ ohms, $L = 8$ henrys, and $C = .000002$ farad, find the voltage across each element, the magnitude of the impedance, the voltage across the combination, and the phase difference between the voltage and current.

chapter 12

Logarithms

12–1. Definition of a Logarithm

There are two very important functions that are useful in describing certain physical phenomena. These functions are the exponential and logarithmic functions.

An equation of the form $Y = \mathbf{b}^X$ is an *exponential function*, where X is the independent variable and Y is the dependent variable. The domain for this function is the set of all real numbers (X a real number) and the range is the set of all real numbers greater than 0 ($Y > 0$). The base must be greater than 0 and cannot equal 1 ($b > 0$ and $b \neq 1$).

The inverse of the exponential function can be determined in the following manner. (Refer to Section 10–5 for an explanation of inverse function.)

1. Interchange the dependent and independent variables in the exponential function $Y = b^X$. Thus, $X = b^Y$.
2. Solve $X = b^Y$ for Y. Since there are no algebraic techniques that will yield a form where Y is isolated on one side of the equation, it is necessary to define a new equation: $Y = \log_b X$, read "Y is the logarithm of X to the base b."

Therefore, in the exponential equation $Y = b^X$, X is called the *logarithm* of a number Y to the base b. We see then that a logarithm is an exponent.

Sec. 12–1. Definition of a Logarithm

The exponential and logarithmic equations listed above are equivalent to each other. Therefore, using this fact, any exponential equation can be stated as an equivalent logarithmic equation and vice versa.

$$\text{If } Y = b^X, \text{ then } X = \log_b Y.$$
$$\text{If } X = \log_b Y, \text{ then } Y = b^X.$$

This relationship is useful in evaluating both exponential and logarithmic expressions.

Example A:
1. Solve: $\log_3 81 = X$. Expressing this equation in its equivalent form, we have $3^X = 81$. Since $3^4 = 81$, $X = 4$.
2. Solve: $\log_b 4 = \frac{1}{2}$. Expressing this equation in its equivalent form we have $b^{1/2} = 4$. Therefore, $b = 4^2 = 16$.

NOTE: $b^{1/2} = 4$ was solved for b by squaring both members of the equation.

3. Solve: $\log_4 Y = \frac{1}{2}$. Expressing this equation in its equivalent form, we have $Y = 4^{1/2}$. Therefore, $Y = \sqrt{4} = 2$.

Exercises

Express each of the following as an equivalent logarithmic equation.

1. $5^3 = 125$
2. $16^{1/2} = 4$
3. $125^{2/3} = 25$
4. $10^3 = 1000$
5. $4^{-3} = \frac{1}{64}$
6. $(\frac{1}{2})^{-1} = 2$
7. $2^{-3} = \frac{1}{8}$
8. $2^3 = 8$
9. $.1^2 = .01$
10. $8^{1/3} = 2$

Express each of the following as an equivalent exponential equation.

11. $\log_3 9 = 2$
12. $\log_7 49 = 2$
13. $\log_4 16 = 2$
14. $\log_8 2 = \frac{1}{3}$
15. $\log_{1/3} 9 = -2$
16. $\log_8 4 = \frac{2}{3}$
17. $\log_{125} 25 = \frac{2}{3}$
18. $\log_{32} 2 = \frac{1}{5}$
19. $\log_2 32 = 5$
20. $\log_{.01} .1 = \frac{1}{2}$

Solve each of the following equations.

21. $\log_3 9 = A$
22. $\log_{10} 1 = B$
23. $\log_b 16 = 2$
24. $\log_{27} R = \frac{1}{3}$
25. $\log_{10} X = 0$
26. $\log_2 A = 2$
27. $\log_2 2 = Y$
28. $\log_b \sqrt{2} = \frac{1}{2}$
29. $\log_{16} \frac{1}{4} = A$
30. $\log_{10} \frac{Y}{5} = -3$

12-2. Basic Exponential and Logarithmic Graphs

The graph of an exponential function can be determined by using the graphing techniques of Chapter 4.

Construct a table of values where the independent variable is the exponent. Since b^x is a real number greater than zero for all real numbers X, the independent variable can be any real number.

Example A: Graph the equation $Y = 2^x$

X	−3	−2	−1	0	1	2	3
Y	$\frac{1}{8}$	$\frac{1}{4}$	$\frac{1}{2}$	1	2	4	8

Locating these points and drawing a smooth curve through these points will give the following graph.

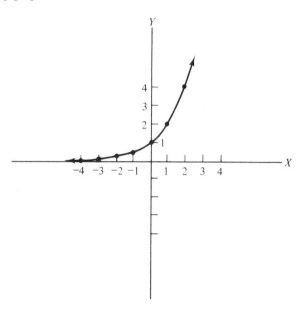

In general, any exponential function where $b > 1$ will have a graph that is increasing, will pass through the points $(0, 1)$ and $(1, b)$, and will have the independent axis as a horizontal asymptote.

Example B: Graph the equation $Y = (\frac{1}{3})^x$

X	−2	−1	0	1	2	3
Y	9	3	1	$\frac{1}{3}$	$\frac{1}{9}$	$\frac{1}{27}$

Sec. 12–2. Basic Exponential and Logarithmic Graphs

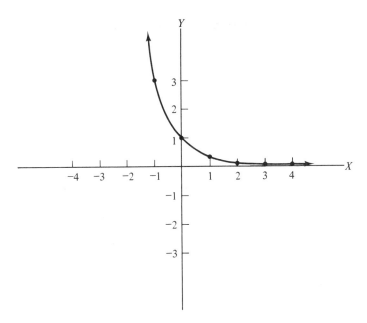

In general, if $0 < b < 1$, the graph of an exponential function will be decreasing, will pass through the points $(0, 1)$ and $(1, b)$, and will have the independent axis as a horizontal asymptote.

If $b = 1$, the graph would be a horizontal line passing through $(0, 1)$.

We should see that the exponential function is a continuous function. Therefore, any real number could be substituted for the independent variable in the table of values. For example, when $X = \pi$, Y would be equal to some real number, although it would be difficult to determine exactly the value of this number.

It should be noted that, since the dependent and independent variables were interchanged to determine the logarithmic function, the domain and range of the logarithmic function will be determined by interchanging the domain and range of the exponential function.

Thus, the domain for $Y = \log_b X$ is $X > 0$ and the range for $Y = \log_b X$ is the set of all real numbers. For any particular point of the graph of $Y = \log_b X$, simply interchange the ordered pair of numbers from a point on the graph of the corresponding exponential function. For example, the graphs of all exponential functions pass through $(0, 1)$. Therefore, the graphs of all logarithmic functions pass through $(1, 0)$.

Example C: Graph the equation $Y = \log_2 X$. Interchanging the values of X and Y from Example A, we obtain the following table.

X	$\frac{1}{8}$	$\frac{1}{4}$	$\frac{1}{2}$	1	2	4	8
Y	-3	-2	-1	0	1	2	3

Therefore, we get the following graph.

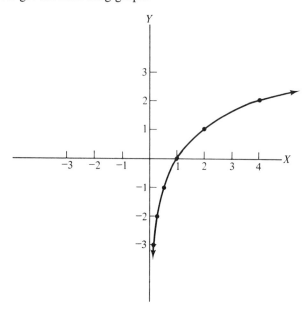

In general, any logarithmic function where $b > 1$ will have a graph that is increasing, will pass through the points $(1, 0)$ and $(b, 1)$, and will have the dependent axis as a vertical asymptote.

Example D: Graph the equation $Y = \log_{1/3} X$. Interchanging the values of X and Y from Example B, we obtain the following table.

X	9	3	1	$\frac{1}{3}$	$\frac{1}{9}$	$\frac{1}{27}$
Y	−2	−1	0	1	2	3

Therefore, we get the following graph.

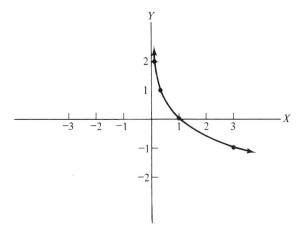

Sec. 12–2. Basic Exponential and Logarithmic Graphs

In general, if $0 < b < 1$, the graph of a logarithmic function will be decreasing, will pass through the points $(1, 0)$ and $(b, 1)$, and will have the dependent axis as a vertical asymptote.

In general, where $b > 1$ the graph of the logarithmic function is symmetrical to the increasing exponential function with respect to the line $Y = X$ (see Figure 12-2.1).

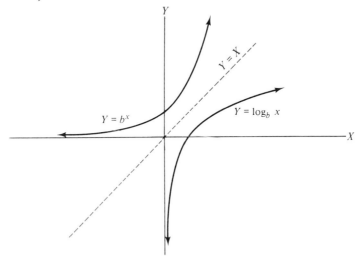

FIGURE 12–2.1

Example E: On the same set of axes, construct the graphs of $Y = 4^X$ and $Y = \log_4 X$.

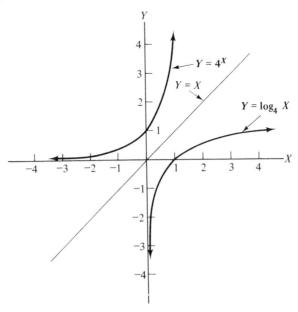

Another graphing technique that can be used in graphing the logarithmic function would be to change the logarithmic function to its equivalent exponential equation.

That is, $Y = \log_b X$ is equivalent to $X = b^Y$.

When graphing this equation, we substitute for Y and obtain corresponding values for X.

Example F: Graph $Y = \log_2 X$. First change $Y = \log_2 X$ to $2^Y = X$. Then set up a table of values and substitute for Y.

X	$\frac{1}{8}$	$\frac{1}{4}$	$\frac{1}{2}$	1	2	4	8
Y	−3	−2	−1	0	1	2	3

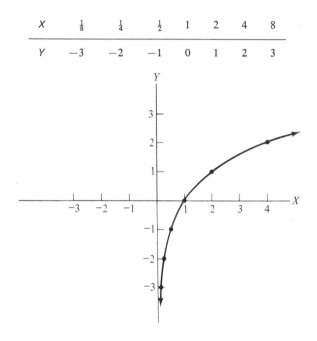

An *exponential* or *logarithmic equation* can be written from a set of data by noting the relationship among the data and putting it into the general form of the exponential or logarithmic equation.

Example G: Write an exponential equation from the following set of data.

A	−2	−1	0	1	2
B	$\frac{1}{2.25}$	$\frac{1}{1.5}$	1	1.5	2.25

Graph the set of data in the rectangular coordinate system.

Sec. 12-2. Basic Exponential and Logarithmic Graphs

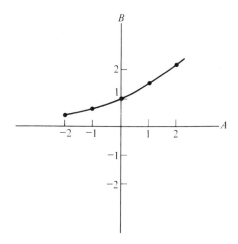

Note that the graph is increasing and that it passes through the points (1, 1.5) and (0, 1). With this information, it is determined that the graph is an increasing exponential function and its equation is of the form $Y = b^x$. Since it passes through the point (1, 1.5), $b = 1.5$. The independent variable is A and dependent variable is B.

Thus, the equation is $B = 1.5^A$.

Example H: Write a logarithmic equation from the following table of values.

S	$\frac{1}{64}$	$\frac{1}{4}$	1	4	16
T	-3	-1	0	1	2

Graph the set of data in the rectangular coordinate system.

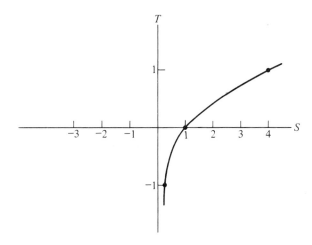

Note that the graph is increasing and that it passes through (1, 0) and (4, 1). With this information, it is determined that the graph is an increasing logarithmic function and its equation is of the form $Y = \log_b X$. Since it passes through the point (4, 1), the value of b is 4.

Thus, the equation is $T = \log_4 S$.

Exercises

Graph each of the following functions.

1. $Y = 2^x$
2. $Y = 3^x$
3. $Y = 4^x$
4. $Y = 2(3)^x$
5. $Y = 3(4)^x$
6. $Y = b^x$, where $b > 1$
7. $Y = (\frac{1}{2})^x$
8. $Y = (\frac{1}{3})^x$
9. $Y = 2(\frac{1}{2})^x$
10. $Y = b^x$, where $0 < b < 1$

Graph each of the following functions.

11. $Y = \log_2 X$
12. $Y = \log_3 X$
13. $Y = 3 \log_3 X$
14. $Y = \log_{10} X$
15. On the same set of axes, graph $Y = 2^x$ and $Y = \log_2 X$.
16. On the same set of axes, graph $Y = 3^x$ and $Y = \log_3 X$.
17. The following set of data was collected from the growth pattern of a plant over a period of five years. Write an empirical equation describing this growth pattern.

t (years)	0	1	2	3	4	5
Total Number of Plants	1	2	4	8	16	32

18. Write an equation for the relationship represented by the following graph.

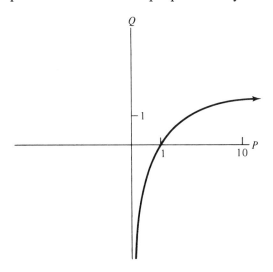

12-3. Properties of Logarithms

There are five properties of logarithms that are useful for solving logarithmic and exponential equations.

The derivation of these laws depends on the laws of exponents for multiplication and division.

1. *Logarithm of a product:*

Let
$$n = \log_b A \text{ and } m = \log_b B$$

Express each number in exponential form.
$$A = b^n \text{ and } B = b^m$$

Multiply.
$$A \cdot B = b^n \cdot b^m = b^{(n+m)}$$

Express $A \cdot B$ in logarithmic form.
$$\log_b (A \cdot B) = n + m$$

Substitute the logarithmic form of n and m.
$$\log_b A \cdot B = \log_b A + \log_b B$$

This last equation states that the logarithm of a product of two quantities is equal to the sum of the logarithms of the quantities. This property applies for any number of quantities.

2. *Logarithm of a quotient:*

Let
$$n = \log_b A \text{ and } m = \log_b B$$

Thus,
$$A = b^n \text{ and } B = b^m$$

Divide.
$$\frac{A}{B} = \frac{b^n}{b^m} = b^{(n-m)}$$

Express $\frac{A}{B}$ in logarithmic form.
$$\log_b \left(\frac{A}{B}\right) = n - m$$

Substitute the logarithmic form of n and m.
$$\log_b \left(\frac{A}{B}\right) = \log_b A - \log_b B$$

This last equation states that the logarithm of a quotient of two quantities is equal to the logarithm of the dividend minus the logarithm of the divisor.

3. *Logarithm of a number raised to a power:*

Let
$$n = \log_b A$$
Thus,
$$A = b^n$$
$$A^X = (b^n)^X = b^{nX}$$

Express A^X in logarithmic form.

$$\log_b A^X = n \cdot X$$

Substitute the logarithmic form of n.

$$\log_b A^X = X(\log_b A)$$

This last equation states that the logarithm of a quantity raised to a power is equal to the exponent times the logarithm of the quantity. The exponent can be any real number.

4. *Logarithm of the number 1:*

Let
$$Y = \log_b 1$$

Write this equation in exponential form.

$$b^Y = 1$$
Since
$$b^0 = 1, \; Y = 0$$
Therefore,
$$\log_b 1 = 0$$

This last equation states that the logarithm of 1 to any base is equal to 0.

5. *Logarithm of a number when that number is the base:*

Let
$$Y = \log_b b$$

Write this equation in exponential form.

$$b^Y = b$$
Since
$$b^1 = b, \; Y = 1$$
Therefore,
$$\log_b b = 1$$

This last equation states that the logarithm of a number, when that number is also the base, is equal to 1.

Sec. 12–3. Properties of Logarithms

The properties of logarithms may be used to simplify a logarithmic expression.

Example A: Simplify $\log_2 32$ using the properties of logarithms. Since 32 can be expressed in exponential form as 2^5, the $\log_2 32$ can be simplified in the following manner.

$$\log_2 32 = \log_2 2^5$$
$$= 5 \log_2 2$$
$$= 5(1) = 5$$

This example can be checked by writing $\log_2 32$ in its equivalent exponential form.

$$\log_2 32 = X \quad 2^X = 32$$
$$2^X = 2^5$$
$$X = 5$$

Example B: Simplify $\log_2 \frac{1}{2}$ using the properties of logarithms. Using property (2):

$$\log_2 \tfrac{1}{2} = \log_2 1 - \log_2 2.$$

According to property (4), $\log_2 1 = 0$, and according to property (5), $\log_2 2 = 1$. Therefore, $\log_2 \frac{1}{2} = 0 - 1 = -1$.

Example C: Condense $\log_3 8 - 3 \log_3 X$ into a single logarithmic statement. Since the multiple of a real number times a logarithm of another number implies that a number is being raised to a power, $3 \log_3 X$ can be expressed as $\log_3 X^3$. Since the difference between two logarithms implies the quotient of two numbers, $\log_3 8 - \log_3 X^3$ can be simplified to $\log_3 \left(\frac{8}{X^3}\right)$. Thus, $\log_3 8 - 3 \log_3 X = \log_3 \left(\frac{8}{X^3}\right)$.

The properties of logarithms may be used to solve some exponential and logarithmic equations.

Example D: Solve for Y.

$$\log_2 Y = \log_2 3 + \log_2 4$$

According to property (1), the right member of the equation can be restated.

$$\log_2 Y = \log_2 (3 \cdot 4) = \log_2 12$$

When the logarithms of two quantities to the same base are equal, the quantities themselves must be equal. Therefore, $Y = 12$.

Example E: Solve for B.

$$\log_3 B = 4 \log_3 2 + \log_3 4$$

According to properties (3) and (1), the equation can be restated as:

$$\log_3 B = \log_3 (2^4 \cdot 4) = \log_3 64$$

Since the logarithms of B and 64 to the same base are the same, $B = 64$.

Example F: Solve for Y in terms of X.

$$2 \log_a Y = \log_a 25 - \log_a X$$

According to properties (2) and (3), this equation can be restated as

$$\log_a Y^2 = \log_a \left(\frac{25}{X}\right)$$

Since a is the base for both members of the equation, the equation is

$$Y^2 = \frac{25}{X}$$

$$Y = \frac{5}{\sqrt{X}}$$

Exercises

Express each of the following as a sum, difference, or multiple of logarithms. Evaluate if possible.

1. $\log_2 8$
2. $\log_3 9$
3. $\log_{10} 100$
4. $\log_5 10$
5. $\log_2 25$

Condense each of the following expressions into a single logarithm.

6. $\log_b 4 + \log_b 5$
7. $\frac{1}{2} \log_2 5 - 3 \log_2 2$
8. $2 \log_a 3 + \log_a 4 - \log_a 18$
9. $5 - \log_2 4 + \log_3 \frac{1}{3}$
10. $2 \log_2 2 + \frac{1}{2} \log_3 81 - 4$

Solve each of the following equations for Y.

11. $\log_b Y = \log_b 10$
12. $\log_a Y = \log_a 4 + \log_a 6$
13. $\log_2 Y = \log_2 16 - \log_2 32$
14. $\frac{1}{2} \log_3 Y = \log_3 X + \log_3 1 - \log_3 2$
15. $\log_{10} Y = \frac{1}{3} \log_{10} 27 + 2 \log_{10} 5 - \log_{10} 6$

12–4. Common Logarithms

All positive real numbers can be expressed as some power of ten.

X (positive real number)	Y (a power of ten)
.001	10^{-3}
.1	10^{-1}
1.0	10^{0}
2.0	$10^{\cdot----}$
8.0	$10^{\cdot----}$
10.0	$10^{1.0}$
16.0	$10^{1.----}$
1000.0	10^{3}

Therefore, for all real numbers X, $X = 10^Y$ or $Y = \log_{10} X$. Some of the values of Y are easily determined, while others would have to be read from a table of logarithms when the base is ten. (Logarithms to the base ten are called *common logarithms*). The values of X and Y correspond to the values of X and Y on the graph of $Y = \log_{10} X$ (see Figure 12-4.1).

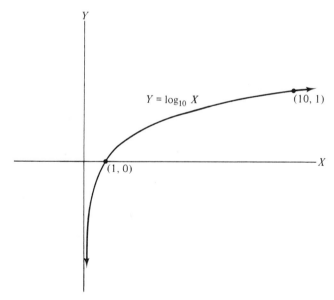

FIGURE 12–4.1

When referring to a common logarithm the subscript 10 is usually omitted. Therefore, $Y = \log_{10} X$ becomes **Y = log X**.

If a number is expressed in *Scientific Notation* ($N = P \times 10^K$, where $1 \leq P < 10$ and K is an integer) the value of Y (the logarithm of the number) can be easily determined.

Therefore:
$$N = P \times 10^K$$
$$\log N = \log(P \times 10^K)$$
$$\log N = \log P + K \log 10$$
$$\log N = \log P + K, \text{ since } \log 10 = 1$$

The value of K is called the *characteristic* of the number and is the power of ten when the number is written in scientific notation. The real number, $\log P$, is called the *mantissa* of the number and is theoretically a value of Y from the graph of $Y = \log X$ between 0 and 1, including 0. A table of values is used for the sake of convenience in finding mantissas.

Example A: Find the common logarithm of 48.3.

$$48.3 = 4.83 \times 10^1$$
$$\log 48.3 = \log 4.83 + 1$$
$$= .6839 + 1$$

NOTE: .6839 comes from the four-place table of common logarithms, Table 3. It may also be determined from a slide rule or from a hand calculator (see Appendix B).

$$\log 48.3 = 1.6839$$

Example B: Find the common logarithm of 2.58.

$$2.58 = 2.58 \times 10^0$$
$$\log 2.58 = \log 2.58 + 0$$
$$= .4116 + 0$$
$$= .4116$$

Example C: Find the common logarithm of .00692.

$$.00692 = 6.92 \times 10^{-3}$$
$$\log .00692 = \log 6.92 + (-3)$$
$$= .8401 + (-3)$$

NOTE: .8401 is a positive number; thus, the sum of $.8401 + (-3) = -2.1599$. It is sometimes convenient to express the log of a number by adding 10 and subtracting 10 in the following manner.

$$\begin{array}{rr} (-3) + .8401 & \\ 10 & -10 \\ \hline 7.8401 & -10 \end{array}$$

Therefore, the $\log .00692 = -2.1599$ or $7.8401 - 10$.

Sec. 12–4. Common Logarithms

The *antilogarithm* of a number indicates the reverse operation of finding the logarithm of a number. It means that a number must be found when the logarithm of the number is given.

Example D: If $\log N = 7.9258$, find N.

$$N = \text{antilog } 7.9258$$

1. The mantissa .9258 is located in the table of common logarithms (Table 3) and the number in *scientific notation* is read from the table.
2. The number in scientific notation must be multiplied by 10^7 since 7 is the characteristic.

Therefore, $N = 8.43 \times 10^7 = 84,300,000$.

Example E: If $\log N = 6.7404 - 10$, find N.

$$N = \text{antilog } (6.7404 - 10)$$

1. The mantissa .7404 yields 5.50 from the table of common logarithms.
2. The characteristic is $6 - 10 = -4$. Therefore, the power of ten is -4. Therefore, $N = 5.50 \times 10^{-4} = .00055$.

Example F: If $\log N = .6010$, find N.

$$N = \text{antilog } .6010$$

1. The mantissa .6010 yields 3.99 from the table of common logarithms.
2. The characteristic is 0. Therefore, the power of ten is 0. Therefore, $N = 3.99 \times 10^0 = 3.99$.

Example G: If $\log N = 2.3842$, find N.

$$N = \text{antilog } 2.3812$$

1. Since the mantissa is not in the table of common logarithms, but is in between a smaller and larger value, a proportion is set up and solved to determine a *number in scientific notation*. (This procedure, called *interpolation*, has been described previously in Section 7–5.)

Number in Scientific Notation		Mantissa	
2.42 ⎤ X ⎤		.3838 ⎤ .0004 ⎤	
2.42 ⎦ ⎦ .01		.3842 ⎦ ⎦ .0018	
2.43 ⎦		.3856 ⎦	

$$\frac{X}{.01} = \frac{.0004}{.0018}$$

$$X = \frac{(.0004)(.01)}{.0018} = .0022$$

$$2.42 + .0022 = 2.4222$$

2. Since the characteristic is 2, the power of ten is 2. Therefore, $N = 2.4222 \times 10^2 = 242.22$.

Exercises

Find the logarithm of each of the following numbers.

1. 49,000
2. 52.4
3. .00179
4. 982.1
5. 61.4×10^{-3}
6. .0702
7. .000009814
8. 18.6×10^4
9. .003967
10. 5.21

Find the number represented by each of the following antilogs.

11. $N = $ antilog 2.6010
12. $N = $ antilog 7.7619
13. $N = $ antilog $(7.8407 - 10)$
14. $N = $ antilog 3.0214
15. $N = $ antilog $(9.1067 - 10)$
16. $N = $ antilog $(3.8987 - 6)$
17. Find N, if $\log N = 1.5119$
18. Find N, if $\log N = 5.9069$
19. Find N, if $\log N = 2.9786 - 10$
20. Find N, if $\log N = 7.7404 - 10$

12–5. Computations with Logarithms

By combining the *properties* of logarithms (explained in Section 12–3) with the *methods* for finding common logarithms and antilogarithms (explained in Section 12–4), we may obtain the answers to *computational problems* involving the operations of multiplication, division, and finding powers or roots. This procedure is especially useful when the given computations or combination of computations are too difficult to be performed with an available slide rule or basic calculator.

The procedure for obtaining the correct answer to a computational problem by using logarithms is as follows:

1. Write the problem as an equation.
2. Take the common logarithm of both sides of the equation by making use of the properties of logarithms.
3. Evaluate the given logarithms, and combine to obtain a single result.
4. Take the common antilogarithm of this result which will be the answer to the problem.

Example A: Find the value of $(42.8)(156.3)$ by using logarithms.

$$N = (42.8)(156.3)$$

Sec. 12–5. Computations with Logarithms **355**

$$\log N = \log 42.8 + \log 156.3$$
$$= (1.6314) + (2.1939)$$
$$= 3.8253$$
$$N = 6.690 \times 10^3$$
$$= 6690$$

Example B: Find the value of $\frac{6.28}{72.4}$ by using logarithms.

$$N = \frac{6.28}{72.4}$$
$$\log N = \log 6.28 - \log 72.4$$
$$= (.7980) - (1.8597)$$
$$= (10.7980 - 10) - (1.8597)$$
$$= 8.9383 - 10$$
$$N = 8.676 \times 10^{-2}$$
$$= .08676$$

NOTE: $(10 - 10)$ was added to $.7980$ in order to obtain $(10.7980 - 10)$ and make the subtraction easier.

Example C: Find the value of $(6.2)^5$ by using logarithms.

$$N = (6.2)^5$$
$$\log N = 5 \log 6.2$$
$$= 5(.7924)$$
$$= 3.962$$
$$N = 9.162 \times 10^3$$
$$= 9162$$

Example D: Find the value of $\sqrt[5]{.872}$, using logarithms.

$$N = \sqrt[5]{.872} = (.872)^{1/5}$$
$$\log N = \tfrac{1}{5} \log .872$$
$$= \tfrac{1}{5}(49.9405 - 50)$$
$$= 9.9881 - 10$$
$$N = 9.73 \times 10^{-1}$$
$$= .973$$

NOTE: Since we had to divide by 5, the characteristic of log .872, which is -1, was written as $(49 - 50)$ to make the result easier to work with.

Example E: Find the value of $\left[\dfrac{(247)(\sqrt[3]{97})}{.738}\right]$, using logarithms.

$$N = \left[\dfrac{(247)(\sqrt[3]{97})}{.738}\right]$$

$\log N = \log 247 + \tfrac{1}{3} \log 97 - \log .738$

$ = (2.3927) + \tfrac{1}{3}(1.9868) - (9.8681 - 10)$

$ = (2.3927) + (.6623) - (9.8681 - 10)$

$ = (3.0550) - (9.8681 - 10)$

$ = (13.0550 - 10) - (9.8681 - 10)$

$ = 3.1869$

$N = 1.538 \times 10^3$

$ = 1538$

NOTE: $(10 - 10)$ was added to 3.0550 in order to obtain $(13.0550 - 10)$ and make the subtraction easier.

Example F: Find the value of $\left[\dfrac{\sqrt{86.2}}{(.541)(3.14)}\right]^3$, using logarithms.

$$N = \left[\dfrac{\sqrt{86.2}}{(.541)(3.14)}\right]^3$$

$\log N = 3[\tfrac{1}{2} \log 86.2 - (\log .541 + \log 3.14)]$

$ = 3\{\tfrac{1}{2}(1.9355) - [(9.7332 - 10) + .4969]\}$

$ = 3[.9678 - (10.2301 - 10)]$

$ = 3(.9678 - .2301)$

$ = 3(.7377)$

$ = 2.2131$

$N = 1.633 \times 10^2$

$ = 163.3$

NOTE: In this case, $(10 - 10)$ was dropped from $(10.2301 - 10)$ to obtain .2301 and make the subtraction easier.

Example G: The formula for the volume (V) of a sphere with radius (R) is $V = \tfrac{4}{3}\pi R^3$. Using logarithms, find the radius of a sphere that has a volume of 86.2 cm^3.

Sec. 12-6. Natural Logarithms

$$V = \frac{4}{3}\pi R^3 \text{ or } R^3 = \frac{3V}{4\pi}$$

$$R = \sqrt[3]{\frac{3V}{4\pi}} = \sqrt[3]{\frac{258.6}{12.57}}$$

$$\log R = \tfrac{1}{3}(\log 258.6 - \log 12.57)$$
$$= \tfrac{1}{3}(2.4126 - 1.1028)$$
$$= \tfrac{1}{3}(1.3098)$$
$$= .4366$$
$$R = 2.732 \times 10^0$$
$$= 2.732 \text{ cm}$$

Exercises Find the value of each of the following by using logarithms.

1. $(426)(91.3)$
2. $(.528)(.0327)$
3. $\dfrac{.861}{18.3}$
4. $\dfrac{7.26}{.0244}$
5. $(194)^3$
6. $(.831)^4$
7. $\sqrt[5]{452}$
8. $\sqrt[4]{.749}$
9. $\dfrac{(263)(76.5)}{483.6}$
10. $\dfrac{(.071)(4293)}{.418}$
11. $\dfrac{46}{(.069)(9180)}$
12. $\dfrac{.259}{(.095)(.694)}$
13. $\dfrac{\sqrt{641}}{\sqrt[3]{.916}}$
14. $\dfrac{(.328)^2}{\sqrt[4]{417}}$
15. $(72.48)^{4.2}$
16. $(3261)^{.6}$
17. $\dfrac{.518\sqrt{63.6}}{(18.3)^3}$
18. $\dfrac{\sqrt[3]{5823}}{528(.83)^2}$
19. $\left[\dfrac{2260\sqrt{.052}}{(4.7)^3 .0237}\right]^4$
20. $\sqrt[3]{\dfrac{(36)^2(.91)^3}{.841(58.2)^2}}$

21. The formula for the volume (V) of a right circular cylinder in terms of its radius (R) and height (H) is $V = \pi R^2 H$. Use logarithms to find the radius of a cylinder with a height of 12 inches and a volume of 408 cubic inches.

22. For an object falling under the influence of gravity, the distance(s) fallen is given by $s = \tfrac{1}{2} at^2$. If $a = 32$ ft/sec², use logarithms to determine how long it will take an object to fall a distance of 226 feet.

12-6. Natural Logarithms

There are situations in science and technology where numbers are expressed in a base other than base ten. The most important base other than base ten is base e, where $e \approx 2.718$. Whenever the logarithm of a number (N) is taken and the base is e, it is called the *natural*, or *Naperian, logarithm* of the number and is written, ln N.

It is difficult to have a table of natural logarithms of numbers because there are no two numbers that have the same mantissa as is the case with common logarithms. In general:

$$N = P \times 10^K$$
$$\ln N = \ln P + K \ln 10$$
$$\ln 10 = 2.3026 \ldots \quad \text{(This value comes from the table of natural logarithms, Table 4.)}$$

Therefore, $\ln N = \ln P + K(2.3026)$.

Example A: Find the ln 400.

$$400 = 4.0 \times 10^2$$
$$\ln 400 = \ln 4 + 2(2.3026)$$
$$= 1.3863 + 4.6052$$
$$= 5.9915$$

The table of natural logarithms (Table 4) was used to determine ln 4 = 1.3863.

Example B: Find the ln 40,000.

$$40,000 = 4.0 \times 10^4$$
$$\ln 40,000 = \ln 4 + 4 \ln 10$$
$$= 1.3863 + 4(2.3026)$$
$$= 1.3863 + 9.2104$$
$$= 10.5967$$

NOTE: The mantissa of 400 is not equal to the mantissa of 40,000. The common logarithms of these same two numbers would differ only by their characteristics.

Example C: Find the ln .78.

$$.78 = 7.8 \times 10^{-1}$$
$$\ln .78 = \ln 7.8 + -1(2.3026)$$
$$= 2.0541 - 2.3026$$
$$= -.2485$$

The table of natural logarithms (Table 4) is incomplete because no two numbers have the same mantissa, but the formula $\ln N = \ln P + K \ln 10$ will allow us to find the natural logarithms of any positive number.

Sec. 12–6. Natural Logarithms

The anti ln (natural antilogarithm) of a number can be found by breaking up the ln N into $2.3026K + \ln P$.

Example D: Find N if $\ln N = 6.4457$.

$$\ln N = \ln P + 2.3026K$$

1. Determine K such that $0 \leq \ln P < 2.3026$.

$$\ln N = 1.8405 + 2(2.3026)$$

2. Since $\ln P = 1.8405$, $P = 6.3$. This value of P is found from the table of mantissas of natural logarithms. Thus, $N = 6.3 \times 10^2 = 630$.

Example E: Find N if $\ln N = -6.5023$.

$$\ln N = .4055 + -3(2.3026)$$
$$N = 1.5 \times 10^{-3} = .0015$$

The logarithm of a number to a base other than base ten or base e can be found in the following manner:

Let $X = \log_b N$. Thus, $N = b^X$. Take the log of both members of the equation.

$$\log N = X \log b$$

Substitute for X.

$$\log N = (\log_b N)(\log b)$$

Solve for $\log_b N$.

$$\log_b N = \frac{\log N}{\log b}$$

This formula states that the logarithm of a number (N) to a base b may be found by dividing the common logarithm of N by the common logarithm of b. Furthermore, the natural logarithm of a number can be determined from a table of common logarithms by making use of this rule.

$$\ln N = \frac{\log N}{\log e} = \frac{\log N}{.4343}$$

Therefore, $\ln N = (2.3026) \log N$.

Example F: Find the ln 456.

$$\ln 456 = 2.3026(\log 456)$$
$$= 2.3026(2.6590)$$
$$= 6.1226$$

Example G: Find the ln .00819.

$$\ln .00819 = 2.3026(\log .00819)$$
$$= 2.3026(-2.0867)$$
$$= -4.8048$$

Example H: Determine N if $\ln N = 2.0694$. Use the formula $\log N = .4343 \ln N$, and substitute for $\ln N$.

$$\log N = .4343(2.0694)$$
$$= .8987$$
$$N = \text{antilog } .8987$$
$$N = 7.92$$

Example I: Determine N if $\ln N = -2.1302$.

$$\log N = .4343(-2.1302)$$
$$= -.9251$$
$$N = \text{antilog } (-.9251)$$
$$= \text{antilog } (9.0749 - 10)$$
$$= 1.188 \times 10^{-1} = .1188$$

Exercises

Find the natural logarithm of each number.

1. 31.6
2. .025
3. 1.4
4. 41×10^3
5. 1.7×10^{-1}
6. .00036
7. 1.17
8. 191
9. 48.2
10. .8156

Find the natural antilogarithm of each of the following numbers.

11. 2.2238
12. -1.3600
13. 4.0412
14. .1141
15. $-.8120$
16. 1.2104
17. 8.0040
18. -3.2781
19. $-.0010$
20. 2.5609

12-7. Basic Exponential and Logarithmic Equations

In Section 12-1 certain logarithmic and exponential equations were solved by applying the definition of a logarithm and the rules for working with exponents.

In this section we will extend these techniques so that a wider range of logarithmic and exponential equations can be solved.

Sec. 12–7. Basic Exponential and Logarithmic Equations

Exponential equations may be solved by taking the *common logarithm* of both members of the equation.

Example A: Solve the equation $2^X = 10$. Taking the common logarithm of both members, we obtain

$$X \log 2 = \log 10$$

To solve this equation for X, we recall that log 2 and log 10 are real numbers. Hence, the basic property for solving an equation with real numbers is used in solving this equation.

$$X = \frac{\log 10}{\log 2} = \frac{1}{.301} = 3.3222$$

Example B: Solve the equation $X^{2.12} = 14$. Taking the common logarithm of both members, we obtain

$$2.12 \log X = \log 14$$

$$\log X = \frac{\log 14}{2.12} = \frac{1.1461}{2.12} = .5406$$

$$X = \text{antilog } .5406$$

$$X = 3.472$$

Example C: Solve the equation $41^{.5} = X$. Taking the common logarithm of both members, we obtain

$$.5 \log 41 = \log X$$
$$.5(1.6128) = \log X$$
$$.8064 = \log X$$
$$\text{antilog } .8064 = X$$
$$6.403 = X$$

Logarithmic equations can be solved by simplifying the equation according to the basic definition and properties of logarithms.

Example D: Solve the equation $\log_2 (X + 2) = 3$. Write the equation in exponential form.

$$2^3 = X + 2$$
$$8 = X + 2$$
$$6 = X$$

Example E: Solve the equation $\log_4 (X + 2) + \log_4 (X - 1) = \log_4 4$. Using one of the basic properties of logarithms, we obtain

$$\log_4 (X+2)(X-1) = \log_4 4$$

From our basic definition of a logarithm, we should realize that when the logarithms of two quantities are equal, the quantities themselves are equal. Therefore,

$$X^2 + X - 2 = 4$$
$$X^2 + X - 6 = 0$$
$$(X+3)(X-2) = 0$$

$$X + 3 = 0 \qquad X - 2 = 0$$
$$X = -3 \qquad X = 2$$

When -3 is substituted into the original equation it will not check because the *logarithmic function is not defined for a negative number*.

$$\log_2 (-3+2) + \log_2 (-3-1) = \log_2 4$$
$$\log_2 (-1) + \log_2 (-4) = \log_2 4$$

This is not valid since $\log_2 (-1)$ and $\log_2 (-4)$ are undefined. When $X = 2$:

$$\log_2 (2+2) + \log_2 (2-1) = \log_2 4$$
$$\log_2 4 + \log_2 1 = \log_2 4$$
$$\log_2 4(1) = \log_2 4$$

Example F: Solve the equation $.2 = \log_4 (T - 1)$. Write this equation in its equivalent exponential form.

$$4^{.2} = T - 1$$

Solve $4^{.2}$ separately.

$$N = 4^{.2}$$
$$\log N = .2 \log 4 = .2(.6021) = .12042$$
$$N = \text{antilog} .12042 = 1.3195$$

Therefore,

$$1.3195 = T - 1$$
$$2.3195 = T$$

Example G: If a principal (P) is invested at a periodic interest rate (i) for a certain number of periods (N), it will amount to a certain sum (S) given by the formula $S = P(1 + i)^N$. Find S if $P = \$1000$, $i = 1.5\%$, and $n = 8$.

$$S = 1000(1 + .015)^8 = 1000(1.015)^8$$
Let $N = 1.015^8$

Sec. 12–8. Graphs on Logarithmic and Semilogarithmic Graph Paper

$$\log N = 8 \log 1.015 = 8(.0064) = .0516$$
$$N = \text{antilog } .0516 = 1.126$$

Therefore,
$$S = 1000(1.126)$$
$$= \$1126$$

Exercises Solve each of the following equations.

1. $3^{.2} = A$
2. $B^3 = 18$
3. $X^{.1} = 14$
4. $5^{(X+1)} = 10$
5. $.3^{2X} = 1$
6. $3^{(A+1)} = 4^A$
7. $8^{.3} = B$
8. $(3)4^X = 17$
9. $2.15^{.5} = B$
10. $.4(2.1)^3 = A$
11. $Y^{1.5} = 13.8$
12. $2A^{3/4} = 1.4$

Solve each of the following logarithmic equations.

13. $\log_2 (X + .1) = 4$
14. $\log (A + 2) + \log A = \log 8$
15. $\log (B - 1) - \log 5 = \log 19$
16. $\log_3 (Y + 4) = 1.5$
17. $3 \log T + \log 12 = \log 96$
18. $.2 \log A + \log 4 = \log 20$
19. $\log_4 X^2 - \log_4 2 = 1.2$
20. $.12 \log_5 X + \log_5 4 = 2.1$

21. The voltage decay across a capacitor that is discharged through a resistor is given by $E = E_0 e^{-t/(RC)}$. For a particular capacitor and resistor the formula is $E = 15e^{-t}$ ($e = 2.7182$). What is the value of E after 2 seconds? (t represents time in seconds.)

22. If $S = P(1 + i)^N$, find S if $P = \$600$, $i = 1\frac{1}{4}\%$, and $N = 12$.

12–8. Graphs on Logarithmic and Semilogarithmic Graph Paper

When the graph of an *exponential function* $Y = Ab^X$ is plotted on rectangular coordinate graph paper, the number of values of the domain and/or range which may be easily plotted is in most cases too small to get an accurate graph that would be useful in a detailed study of the function. Therefore, to plot such functions, we often use logarithmic or semilogarithmic graph paper.

Logarithmic graph paper (or log-log paper) has horizontal and vertical scales with intervals that are proportional to the logarithms of numbers from 1 to 10, rather than intervals uniformly spaced as they are on rectangular coordinate graph paper. Each series of such intervals is called a *cycle* and each cycle represents a different power of 10. Thus, in plotting numbers on this type of graph paper, we are actually plotting the common logarithm of these numbers (see Figure 12–8.1).

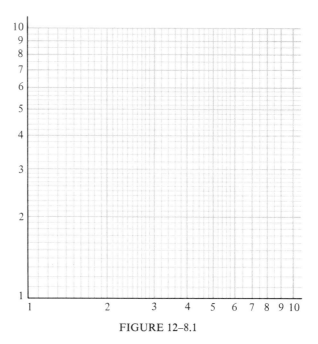

FIGURE 12–8.1

When the graph of an exponential function, $Y = Ab^x$, is plotted on log-log graph paper, the curve is simplified and we can include a much larger range of values for the variables.

The graph of $Y = 3^x$, plotted on rectangular coordinate graph paper, is an increasing function (see Figure 12–8.2). Because the graph increases so rapidly, it is difficult to determine a value of the dependent variable for increasing values of the independent variable. If this same function is graphed on log-log paper (see Figure 12–8.3), a corresponding value of the dependent variable for any value of the independent variable can be more easily determined.

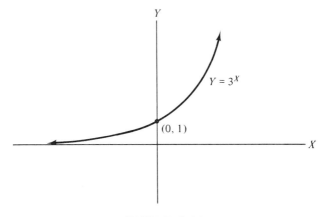

FIGURE 12–8.2

Sec. 12–8. Graphs on Logarithmic and Semilogarithmic Graph Paper

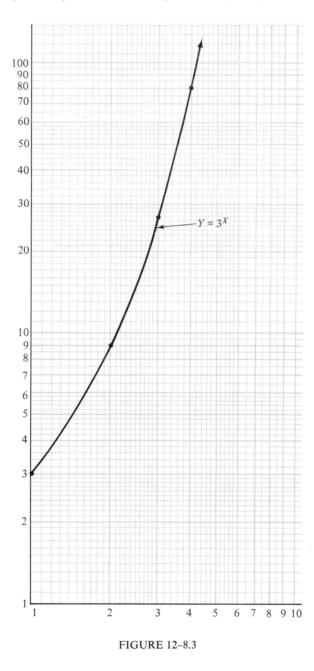

FIGURE 12–8.3

Example A: The following data was collected from an experimental situation relating power (P) transmitted by a steel shaft and the diameter (d) of the shaft. What horsepower might be expected with a shaft diameter of 20 centimeters?

d (cm)	3.8	5.1	6.3	7.6	8.9	10.0	11.4	12.7
P (hp)	11	27	50	91	140	213	308	415

1. Plot this data on logarithmic paper.

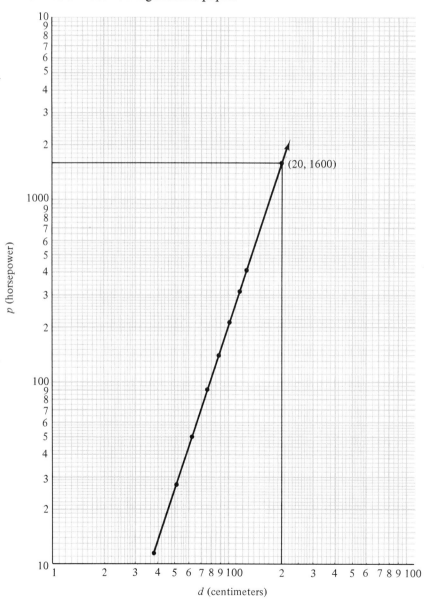

Sec. 12–8. Graphs on Logarithmic and Semilogarithmic Graph Paper

2. Read the horsepower from the graph for a diameter of 20 centimeters (1600 horsepower).

NOTE: If this data was plotted on rectangular coordinate graph paper it would be difficult or impossible to read the answer from the graph.

It should be noted that the answer to the question in Example A goes beyond the measured points that are given in the table. This process is known as *extrapolation*. The answer may not be in agreement with the actual results. Therefore, when a technician gets an answer to a problem by extrapolation, he or she should experiment to verify that answer.

If a *power function* (an equation of the form $Y = A \cdot X^M$) is graphed on logarithmic paper, the equation representing the data can be determined from the graph. ($Y = A \cdot X^M$ is a power function, whereas $Y = A \cdot b^X$ is an exponential function.)

A straight line on logarithmic graph paper will have an equation of the form $Y = A \cdot X^M$, and the values of A and M can be determined from the graph and the given information. The value of A can be read from the graph as the point of intersection of the vertical line $X = 1$ and the graph. This is true since $\log Y = \log A + M \log X$, and if $X = 1$, $\log 1 = 0$. Therefore, A would correspond to the vertical intercept of a straight line. The value of M, the slope of the line, is determined by using the slope formula as stated in Section 6–2. In using the slope formula, however, we must remember to take logarithms of the given numbers since these are the points we actually plot. If only one of the values of A or M can be found from the graph, the remaining one can be determined by substituting an ordered pair of numbers from the graph into the general power function equation and solving the resulting equation for the remaining variable.

Example B: The following data represents the pressure (P) of a solution related to the concentration (C) of the solution. Write the empirical equation representing this relationship.

C	1	2	3	4	5	6
P	27	58.3	91.4	125.8	161.1	197.3

The equation can be found by using the following procedure.
1. Plot this data on logarithmic graph paper.

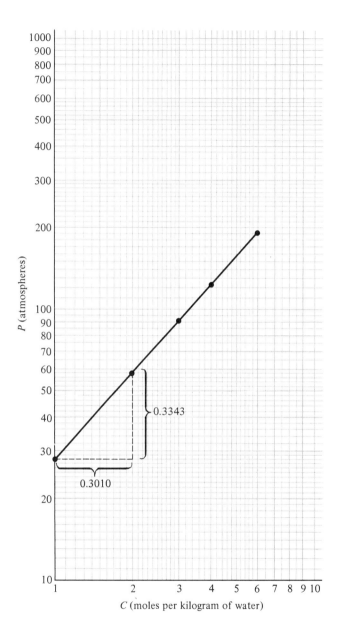

2. Determine the slope.

$$M = \frac{\log 58.3 - \log 27}{\log 2 - \log 1} = \frac{1.7657 - 1.4314}{.3010 - 0} = \frac{.3343}{.3010} = 1.11$$

3. On the vertical line $C = 1$ (since C is the independent variable), read the value of the vertical intercept, 27.

These values of the slope and intercept correspond to the values for the power function in logarithmic form, $\log P = \log A + M \log C$, and this is the logarithmic form of the equation of the power function $P = A \cdot C^M$. Therefore, the empirical equation is $P = 27C^{1.11}$.

Sec. 12-8. Graphs on Logarithmic and Semilogarithmic Graph Paper

Example C: The following data gives the relationship between voltage (E) and current (I) for a particular element. Determine the empirical equation.

E (volts)	20	25	30	35	40	45	50
I (amps)	.25	.40	.59	.82	1.08	1.39	1.73

1. Plot this data on logarithmic graph paper.

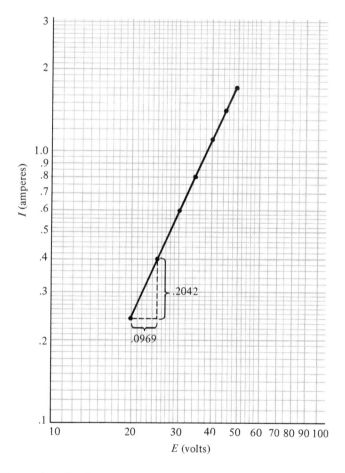

2. Determine the slope.

$$M = \frac{\log .40 - \log .25}{\log 25 - \log 20} = \frac{-.3979 - (-.6021)}{1.3979 - 1.3010} = \frac{.2042}{.0969} = 2.11$$

3. Since the independent variable (E) does not have a value of 1, the value of A in the equation $\log I = \log A + M \log E$ must be determined by substituting an ordered pair of numbers into the equation and solving the resulting equation for A.

$$\log I = \log A + M \log E$$
$$\log .25 = \log A + 2.11 \log 20$$
$$-.6021 = \log A + 2.11(1.3010)$$
$$-.6021 = \log A + 2.745$$
$$-3.347 = \log A$$

Therefore,
$$A = \text{antilog } -3.347$$
$$= .00045$$

Therefore, the empirical equation is $I = .00045 E^{2.11}$.

When the range of values of one variable is much greater than the range of values of the other variable for an exponential or a power function, the graph can be plotted on *semilogarithmic graph paper*. On semilogarithmic paper one scale is uniform as it is on rectangular coordinate paper, and the other scale is a logarithmic scale (see Figure 12–8.4).

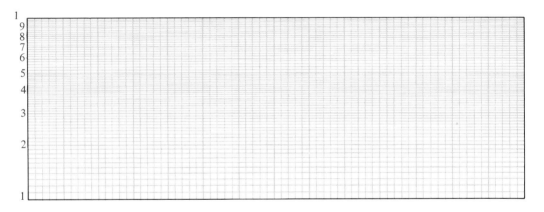

FIGURE 12–8.4

Example D: Plot the graph of $Y = 5(3)^X$ on semilog paper.

X	−1	0	1	2	3
Y	1.67	5	15	45	135

1. Since the range for the independent variable (X) is relatively small compared to the range of the dependent variable (Y), which is approximately 133, semilogarithmic graph paper should be used.
2. Three cycles of a logarithmic scale will be needed on the dependent axis in order to plot all given points: 1.67 and 5 would fit in a logarithmic cycle of 1–10, 15 and 45 would fit in a logarithmic cycle of 10–100, and 135 would fit in a logarithmic cycle of 100–1000.

Sec. 12–8. Graphs on Logarithmic and Semilogarithmic Graph Paper 371

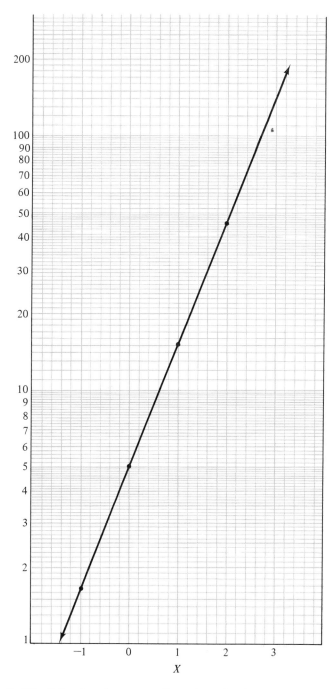

Example E: The following data was collected as a liquid was discharged through a capillary in a tank. The remaining volume in the tank was measured at 15-second intervals. From a graph, determine the volume at 20 seconds.

t (sec)	0	15	30	45	60	75	90
V (liters)	14	12.95	12.00	11.12	10.29	9.55	8.85

1. Since the time intervals are uniform, they can be plotted on a uniform scale along the horizontal axis. The volume can be plotted in two cycles of a logarithmic scale. These scales would be 1–10 and 10–100.

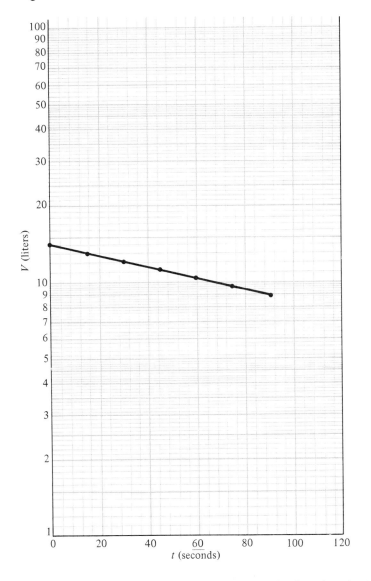

2. From this graph the answer to the question can be found to be 12.5 liters.

Example F: The formula describing the decay of Uranium U^{239} is

$$A = A_0 e^{-.0289t}$$

where A_0 is the original amount of Uranium U^{239} and A is the amount after t minutes. Graph A vs. t for 20 grams of Uranium U^{239} and determine the time for the half-life of Uranium U^{239}. (That is, determine the time when the amount is one-half of the original amount.) Make a table of values for $t > 0$ (see p. 374) such that A will decrease below 10 grams, and plot the graph of this information on semilog graph paper.

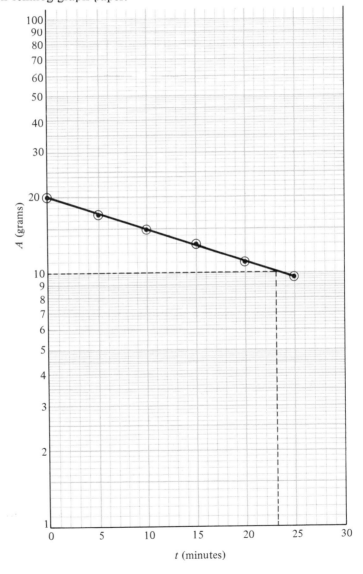

t (min)	0	5	10	15	20	25
A (gm)	20	17.0	14.8	13	11	9.6

Where the horizontal line at 10 grams on the vertical scale intersects the graph, we can determine the appropriate value of time. Therefore, the half-life of 20 gm of U^{239} is 23.2 minutes.

When the graph of an exponential function, $Y = Ab^X$, is plotted on semilog graph paper, the graph will be a straight line. Taking the logarithm of both sides of this equation, we obtain $\log Y = \log A + X \log b$. Log A and log b are constants, and since we are using semilog graph paper, log Y is plotted automatically. Therefore, this equation is of the form $U = MX + B$, where $U = \log Y$, $M = \log b$, and $B = \log A$. We should recognize this as the slope-intercept form of the equation of a straight line (see Section 6–2).

Example G: The following table of data is the watt density as a function of air temperature for a particular heating element. Plot this information on semilog graph paper (see p. 375) and determine the empirical equation from the graph.

Air Temperature (°C)	−12.2	80	204
watts/cm²	156	118	81

Refer to graph on p. 375. The vertical intercept is read from the graph:

$$\text{when } T = 0, \ W = 150.$$

$$M = \log b = \frac{\log 118 - \log 156}{80 - (-12.2)} = \frac{2.0719 - 2.1931}{92.2} = \frac{-.1212}{92.2} = -.00131$$

$$b = \text{antilog}(-.00131) = .997$$

Therefore, the empirical equation is

$$W = 150(.997)^T$$

In conclusion, both semilog and log-log graph papers allow for a wider range of values for one or both variables, result in simpler curves, and make the analysis of the relationships graphed much easier.

Sec. 12–8. Graphs on Logarithmic and Semilogarithmic Graph Paper

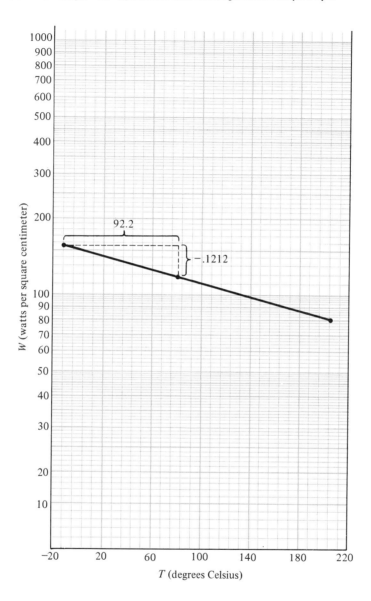

Exercises Plot each of the following on logarithmic graph paper.

1. $A = \dfrac{8}{B}$
2. $S^2 = \dfrac{25}{T^2}$
3. $M^2 N^3 = 4$

Plot each of the following on semilogarithmic graph paper.

4. $Y = 4^X$
5. $A = 4(3)^B$
6. $S = .5(2)^T$

7. Plot the following on logarithmic graph paper.

t (sec)	1.55	1.27	.31	.15
L (cm)	.92	.61	.88	.63

8. Plot the following data on log-log graph paper and determine the equation relating the volume (V) and pressure (P) of a gas at a constant temperature.

V (m³)	1	10	100	1000
P (atm)	4	.4	.04	.004

9. A charged capacitor was allowed to discharge through a resistor while the residual charge of the capacitor was measured by the deflection, D (centimeters) of a galvanometer. Plot a graph of this data on semilog graph paper, and determine the empirical equation.

t (sec)	0	25	50	75	100	125
D (cm)	45.1	41.8	38.8	36.0	33.4	30.9

10. Plot the following data relating flux density (D) as a function of magnetic intensity (I).

I	50	100	150	200
D	.04	.68	1.02	1.20

11. A formula that describes the relationship of the intensity of light passing through a transparent material is

$$I = I_0 e^{-KT}$$

where I_0 is the original intensity, K is a constant, and T is the thickness of the material. Plot the graph on semilog graph paper when $I_0 = 125$ candle power, $K = .45$, and $.5 \text{ cm} \leq T \leq 2.5 \text{ cm}$.

12. The tension T_1 in the driven side of a belt passing over a pulley, and the maximum tension T_2 that can occur in the driving side, are related by the equation

$$T_2 = T_1 e^{2.4\mu}$$

where μ is the coefficient of friction between belt and pulley. Let $T_1 = 60$. Plot the graph of T_2 vs. μ on semilog graph paper for a range of $.1 \leq \mu \leq .6$. From this graph determine T_2 when $\mu = .3$.

12–9. CHAPTER REVIEW

Solve each of the following logarithmic equations.

1. $\log A = 1$
2. $3 \log_2 2 = B$
3. $\log_b 64 = \frac{2}{3}$
4. $\log_b .02 = -2$
5. $\log_b 27 = -\frac{1}{3}$
6. $\log_{12} 1728 = X$
7. $\log 0 = A$
8. $3 \log_b 10 = 3$
9. $-\frac{1}{3} \log_{1/2} 8 = B$
10. $\frac{1}{2} \log_A 100 = 1$

Graph each of the following on rectangular graph paper.

11. $A = 4^B$
12. $A = (\frac{1}{4})^B$
13. $\log_4 A = B$
14. $\log X = Y$
15. On the same set of rectangular coordinate axes, plot the graph of $Y = e^X$ and $Y = \log_e X$.

Condense each statement into a single logarithmic statement.

16. $\log (X + 2) + \log (X - 3)$
17. $3 \ln X - \ln (X - 1)$
18. $\frac{1}{2}[2 \log (A + B) + \log (A - B) - \log 14]$
19. $\frac{1}{2} \log_2 (R + 14) + .15 \log_2 (R - 1) - \frac{1}{3} \log_2 R$
20. $\frac{1}{8}[\log (X - 1) + \log (X + 2)] + \log X - \frac{1}{4} \log (X + 10)$

Evaluate each of the following:

21. $\log 26.3$
22. $\log 8.94$
23. $\log .000706$
24. $\log 549.2$
25. $\ln 4.91$
26. $\ln 12.16$
27. $\ln .738$
28. $\ln 942.6$
29. Find N if $\log N = 4.301$.
30. Find N if $\log N = 2.8993$.
31. Find N if $\log N = 7.7543 - 10$.
32. Find N if $\log N = -3.52$.
33. Find N if $\ln N = 1.1939$.
34. Find N if $\ln N = 4.1744$.
35. Find N if $\ln N = .684$.
36. Find N if $\ln N = -.741$.

Evaluate each of the following, using logarithms:

37. $\dfrac{(62.7)(419)}{.359}$
38. $\dfrac{.0406}{(.0091)(83.5)}$
39. $54.1^{2.36}$

40. $7.9^{-.3}$

41. $\dfrac{(\sqrt{76})(8.07)}{14^3}$

42. $\dfrac{(16.2)^2}{(.025)(\sqrt[3]{94.7})}$

43. $\left[\dfrac{(\sqrt{249})42.8}{.913(\sqrt[3]{12.1})}\right]^2$

44. $\sqrt{\dfrac{(\sqrt[4]{86.2})6.71}{(\sqrt[3]{.0418})(10.8)^3}}$

Solve the following exponential equations:

45. $3^X = 14$
46. $2(4)^{1.4} = A$
47. $B^2 = 106$
48. $4^{(X+1)} = 3^X$
49. $.4^3 = X$
50. $4(A+1)^{.2} = 144$
51. $2.1^{(X+2)} = 15^{2X}$
52. $.1^{2X} = 1$
53. $400 = 800e^{-.140H}$ (use natural logs)
54. $I = 2.5e^{-1/2}$

Solve the following logarithmic equations:

55. $2.1 \log X = \log 5$
56. $\log(3 - X) - \log(X + 1) = 0$
57. $\log_8(X + 1) + \log_8 X = \log_8 6$
58. $\ln 2A + \ln A = \ln 16$
59. $\ln(B - 1) = 2.5$
60. $\log_2 Y - \log_2(Y + .5) = \log_2 14$

61. If the initial amount of bacteria is 400 and the rate of increase in the number of bacteria is .16 per hour, how long will it take to increase the number of bacteria to 10,000? The equation for this growth is $Y = Ie^{rt}$, where $I =$ the initial amount present, $r =$ the rate of growth, $t =$ the time, and $Y =$ the final amount present.

62. A formula for determining the amount of radioactive decay in a substance is $A = Ie^{rt}$, where A is the final amount present, $I =$ the initial amount present, $r =$ the rate of decay, and $t =$ the rate of time. If 100 grams decays to one-fourth that amount in two hours, what is the rate of decay?

63. Atmospheric pressure decreases according to the formula $p = 760e^{-.144h}$, where p is the pressure of mercury (millimeters) and h is the altitude (kilometers). At what altitude will the pressure be one-half the surface pressure of 760 millimeters?

64. If $S = P(1 + i)^n$ (see Section 12–6), find P if $S = \$2000$, $i = 1.5\%$, and $n = 12$.

65. The radioactive isotope (I) decreases with time (t). The following data was collected from an experiment.

t (sec)	0	20	40	60	80	100	120	140
I (counts/min.)	1460	1200	960	760	620	480	380	150

(a) Plot this data on rectangular coordinate graph paper.
(b) Plot this data on semilog graph paper.

66. When a charged capacitor is allowed to discharge through a resistor, the residual charge (R) can be determined as a function of time (t) by the formula $R = 45e^{-.0028t}$.
   (a) Make a table for $0 \leq t \leq 100$ seconds.
   (b) Plot this graph on rectangular coordinate graph paper.
   (c) Plot this graph on semilog graph paper.

67. Plot the following data on log-log graph paper and determine the equation relating P as a function of L.

L	.91	.61	.30	.15
P	1.53	1.26	.89	.63

68. The activity (I) of a radioactive isotope decreases with relation to time (t). Plot the following experimental data on semilog graph paper and determine the empirical equation. Use this equation to determine at what time the count will be 6.0 counts per second.

t (sec)	0	20	40	60	80	100
I (counts/sec)	12	10	8.4	7	5.8	4.9

chapter 13

Quadratic Equations

13–1. The Quadratic Equation

In Section 2–3 we defined a polynomial as an algebraic expression in which each term is of the type AX^N, where N is either zero or a positive integer. We also defined the degree of a polynomial to be the degree of the highest power term. A *polynomial equation*, then, will be an equation in which each term is of the type mentioned above. A polynomial equation of degree two is called a *quadratic equation*, and it is this type of equation that we will consider in this chapter. The general quadratic equation in one variable has the form

$$AX^2 + BX + C = 0$$

where A, B, and C are general constants and X could be replaced by any variable. Since a quadratic equation is defined to be of degree two, we should see that $A \neq 0$. However, B or C or both may be zero.

Example A:
1. $X^2 + 3X - 7 = 0$ is a quadratic equation with $A = 1, B = 3, C = -7$.
2. $5Y^2 - 8Y = 0$ is a quadratic equation with $A = 5, B = -8, C = 0$.

Sec. 13–1. The Quadratic Equation

3. $2T^2 + 9 = 0$ is a quadratic equation with $A = 2$, $B = 0$, $C = 9$.
4. $4R^2 = 6R - 1$ can be rewritten as $4R^2 - 6R + 1 = 0$ and is, therefore, a quadratic equation with $A = 4$, $B = -6$, $C = 1$.
5. $(P - 3)X^2 + (2P)X - 2$ is a quadratic equation with $A = P - 3$, $B = 2P$, $C = -2$.

Example B:
1. $6X - 7 = 0$ is not a quadratic equation due to the absence of a second degree term.
2. $8X^3 - 2X^2 + 9 = 0$ is not a quadratic equation due to the presence of a term with degree greater than two ($8X^3$).

As we have mentioned several times, a solution to an equation is any value that, when substituted for the involved variable, makes the equation a true statement. The complete solution for an equation is the set of all such individual solutions. These definitions also apply to quadratic equations.

Example C: Given the equation $X^2 - 5X + 6 = 0$,

3 is a solution to this equation since $3^2 - 5(3) + 6 = 0$.
2 is a solution to this equation since $2^2 - 5(2) + 6 = 0$.
5 is not a solution to this equation since $5^2 - 5(5) + 6 \neq 0$.

Example D: Given the equation $2Y^2 + 5Y - 3 = 0$,

$\frac{1}{2}$ is a solution to this equation since $2(\frac{1}{2})^2 + 5(\frac{1}{2}) - 3 = 0$.
-3 is a solution to this equation since $2(-3)^2 + 5(-3) - 3 = 0$.
1 is not a solution to this equation since $2(1)^2 + 5(1) - 3 \neq 0$.

We will state here and show later in this chapter that every quadratic equation has a complete solution consisting of two values (not necessarily distinct). Therefore, the complete solution for the equation in Example C ($X^2 - 5X + 6 = 0$) consists of the two numbers, 3 and 2. The complete solution for the equation in Example D ($2Y^2 + 5Y - 3 = 0$) consists of the two numbers, $\frac{1}{2}$ and -3.

The roots of a quadratic equation may be real, imaginary, or complex.

Example E: Is $(T = 1, -4)$ the complete solution of $T^2 + 2T - 3 = 0$?

$$(1)^2 + 2(1) - 3 = 0$$
$$(-4)^2 + 2(-4) - 3 \neq 0$$

1 is a solution of this equation but -4 is not. Therefore, $(T = 1, -4)$ is not the complete solution.

Example F: Is $(X = \pm 2j)$ the complete solution of $X^2 + 4 = 0$?

$$(2j)^2 + 4 = 4j^2 + 4 = -4 + 4 = 0$$
$$(-2j)^2 + 4 = 4j^2 + 4 = -4 + 4 = 0$$

Therefore, $(X = \pm 2j)$ is the complete solution.

Quadratic equations are important in a number of applied areas: an equation found in the study of motion is $s = vt + \frac{1}{2}at^2$, the formula for the surface area of a right circular cylinder is $A = 2\pi RH + 2\pi R^2$, and in the study of electric circuits, the equation $RI^2 + EI = K$ is encountered.

Exercises

State whether or not each of the following is a quadratic equation.

1. $X^2 - 3X = 8$
2. $R^2 - 3R = 2 - R^2$
3. $B^3 - 8B^2 = 2B + B^3$
4. $7Y - 3 = 2Y$
5. $TA^2 - (T + 1)A = 3A + 2$
6. $3X^2 - 7 = 3(X^2 - 2X)$
7. $\sqrt{2}S^2 - \sqrt{7}S = \sqrt{10}$
8. $5K^2 = 0$

For each of the following, determine whether each of the indicated values is a solution for the quadratic equation.

9. $X^2 - 2X - 15 = 0$ $(X = 3, 7, -2)$
10. $Y^2 + jY + 6 = 0$ $(Y = -1, 5, 2j)$
11. $2T^2 + 7T - 4 = 0$ $(T = -4, -\frac{1}{2}, 0)$
12. $4R^2 - 4R - 3 = 0$ $(R = \frac{1}{2}, -1, \frac{3}{2})$
13. $S^2 + 25 = 0$ $(S = 5, -5j, \frac{1}{5})$
14. $3P^2 + 4P = 0$ $(P = 2, 0, -\frac{4}{3})$

For each of the following, determine whether the sets of values correspond to the complete solution for the quadratic equation.

15. $Y^2 - 3Y - 4 = 0$ $(Y = -1, 4)$
16. $T^2 + 3jT + 18 = 0$ $(T = -6j, 3j)$
17. $2R^2 + 5R - 12 = 0$ $(R = 2, 5)$
18. $S^2 - 5S - 14 = 0$ $(S = -2, 7)$
19. $2X^2 + 3jX + 2 = 0$ $(X = -\frac{1}{2}j, -2j)$
20. $K^2 + 7K + 6 = 0$ $(K = -1, 6)$

13–2. Solving Quadratic Equations by Factoring

In this section we will consider quadratic equations whose quadratic expression $(AX^2 + BX + C)$ is factorable. In doing so, we will also make use of an idea that has been mentioned previously. That is, if the product of two quantities is equal to zero, then at least one of the quantities, possibly both, is equal to zero. To solve a quadratic equation (with zero as one member) by factoring then: factor the quadratic expression into a product of two linear factors, set each factor equal to zero, and solve. The solutions to the two factors will be the roots of the quadratic equation. (Refer to Section 2-4 for a complete discussion of factoring.)

Example A: Solve the equation $X^2 - 5X + 6 = 0$.

$$X^2 - 5X + 6 = 0$$
$$(X - 3)(X - 2) = 0$$
$$X - 3 = 0 \text{ and/or } X - 2 = 0$$

Therefore, $X = 3, 2$.

Check:
$$3^2 - 5(3) + 6 = 9 - 15 + 6 = 0$$
$$2^2 - 5(2) + 6 = 4 - 10 + 6 = 0$$

Example B: Solve the equation $2T^2 + 7T - 4 = 0$

$$2T^2 + 7T - 4 = 0$$
$$(2T - 1)(T + 4) = 0$$
$$2T - 1 = 0 \text{ and/or } T + 4 = 0$$

Therefore, $T = \frac{1}{2}, -4$.

Check:
$$2(\tfrac{1}{2})^2 + 7(\tfrac{1}{2}) - 4 = \tfrac{1}{2} + \tfrac{7}{2} - 4 = 0$$
$$2(-4)^2 + 7(-4) - 4 = 32 - 28 - 4 = 0$$

Example C: Solve the equation $A^2 - 49 = 0$.

$$A^2 - 49 = 0.$$
$$(A + 7)(A - 7) = 0$$
$$A + 7 = 0 \text{ and/or } A - 7 = 0$$

Therefore, $A = -7, 7$.

Check:
$$(-7)^2 - 49 = 49 - 49 = 0$$
$$(7)^2 - 49 = 49 - 49 = 0$$

Example D: Solve the equation $2Y^2 + 5Y = 0$.

$$2Y^2 + 5Y = 0$$
$$Y(2Y + 5) = 0$$
$$Y = 0 \text{ and/or } 2Y + 5 = 0$$

Therefore, $Y = 0, -\frac{5}{2}$.

Check:
$$2(0) + 5(0) = 0 + 0 = 0$$
$$2(-\tfrac{5}{2})^2 + 5(-\tfrac{5}{2}) = \tfrac{25}{2} - \tfrac{25}{2} = 0$$

Example E: Solve the equation $K^2 - 4K + 4 = 0$.

$$K^2 - 4K + 4 = 0$$
$$(K - 2)(K - 2) = 0$$
$$K - 2 = 0 \text{ and/or } K - 2 = 0$$

Therefore, $K = 2, 2$.

Check: $(2)^2 - 4(2) + 4 = 4 - 8 + 4 = 0$

NOTE: We can see from this example that it is possible for the two roots of a quadratic equation to be equal.

Example F: A certain rectangular field has an area of 1875 square feet. If the length of the field is 50 feet greater than the width, what are the dimensions of the field? Let the width of the field be W. Then the length of the field will be $W + 50$. Therefore,

$$W(W + 50) = 1875$$
$$W^2 + 50W = 1875$$
$$W^2 + 50W - 1875 = 0$$
$$(W + 75)(W - 25) = 0$$
$$W = -75, 25$$

Therefore, width is 25 feet and length is 75 feet.

NOTE: -75 is an inappropriate answer here and is disregarded.

Exercises

Solve each of the following quadratic equations by factoring.

1. $X^2 - 7X = 0$
2. $Y^2 = 8Y$
3. $T^2 - 25 = 0$
4. $B^2 - 121 = 0$

5. $A^2 + 2A - 8 = 0$
6. $K^2 - 2K - 15 = 0$
7. $M^2 - 4M - 12 = 0$
8. $V^2 + 3V - 10 = 0$
9. $R^2 + 2R + 1 = 0$
10. $P^2 - 6P + 9 = 0$
11. $2T^2 - 5T - 3 = 0$
12. $8B^2 - 14B - 15 = 0$
13. $3Y^2 - 10Y - 8 = 0$
14. $6K^2 + 13K - 15 = 0$
15. $4S^2 + 12S + 9 = 0$
16. $9X^2 - 6X + 1 = 0$
17. $6A^2 + 7A = 20$
18. $2A^2 + 5A = A^2 + 14$
19. $J^2 - 4 = 11J - 2J^2$
20. $20W^2 = 3 - 7W$
21. In a certain electric circuit, the relationship between current (I), voltage (E), and resistance (R) is given by $RI^2 + EI = 13{,}000$. If $R = 4$ ohms and $E = 60$ volts, find I.
22. When a projectile is fired into the air, its distance above the ground (s) in terms of time (t) is given by the equation $s = 128t - 16t^2$. How long after the projectile is fired will it hit the ground?
23. The sum of two numbers is -30, and their product is 176. What are the numbers?
24. A rectangular field has a perimeter of 300 meters and an area of 5000 square meters. What are the dimensions of that field?

13–3. Completing the Square

Solving quadratic equations by factoring is very convenient. However, this method cannot always be used since not all quadratic expressions are factorable. Therefore, we must develop other methods for solving quadratic equations that will work in all cases.

In this section we will look at the method of *completing the square* as applied to quadratic equations. This method is based on two rules that have been mentioned: the same quantity may be added to or subtracted from both members of an equation without changing the basic relationship (Section 2–6), and $\sqrt{A^2} = A$ (Section 2–1). Basically, we:

1. Arrange the quadratic equation in its general form $(AX^2 + BX + C = 0)$.
2. Move C to the other side of the equal sign.
3. Divide each term by A (if $A \neq 1$).
4. Add a number to both members that will make the quadratic expression a perfect square.
5. Take the square root of both sides.
6. Solve for X.

Example A: Solve the equation $X^2 + 6X - 3 = 0$ by completing the square.

$$X^2 + 6X - 3 = 0$$
$$X^2 + 6X = 3$$
$$X^2 + 6X + 9 = 3 + 9$$
$$(X + 3)^2 = 12$$
$$X + 3 = \pm\sqrt{12} = \pm 2\sqrt{3} \text{ or } \pm 3.46$$

Therefore, $X = -3 \pm 3.46 = .46$ or -6.46.

NOTE: When taking the square root we must not forget to use $\pm$, since $(+\sqrt{12})^2 = 12$ and $(-\sqrt{12})^2 = 12$.

In the previous example, we added 9 to both members of the equation to complete the square. In general, what should be added? We should recall here two important products:

$$(X + A)^2 = X^2 + 2AX + A^2$$
$$(X - A)^2 = X^2 - 2AX + A^2$$

Examining these products, we should see that A^2 is what must be added to both members of an equation to give us a perfect square. We should also see that $A^2 = [\frac{1}{2}(2A)]^2 = [\frac{1}{2}(-2A)]^2$. Therefore, to complete a square, we must take $\frac{1}{2}$ of the coefficient of the linear term and square it.

Example B: Solve the equation $Y^2 - 8Y + 7 = 0$ by completing the square.

$$Y^2 - 8Y + 7 = 0$$
$$Y^2 - 8Y = -7$$
$$Y^2 - 8Y + 16 = -7 + 16 \quad \{16 = [\tfrac{1}{2}(-8)]^2\}$$
$$(Y - 4)^2 = 9$$
$$Y - 4 = \pm 3$$
$$Y = 4 \pm 3$$

Therefore, $Y = 7$ or 1.

Example C: Solve the equation $2T^2 + 8T + 9 = 0$ by completing the square.

$$2T^2 + 8T + 9 = 0$$
$$2T^2 + 8T = -9$$
$$T^2 + 4T = -\tfrac{9}{2}$$

Sec. 13-3. Completing the Square

$$T^2 + 4T + 4 = -\tfrac{9}{2} + 4 \quad \{4 = [\tfrac{1}{2}(4)]^2\}$$
$$(T + 2)^2 = -\tfrac{1}{2}$$
$$T + 2 = \pm\sqrt{-.5} = \pm\sqrt{.5}j \text{ or } \pm.71j$$

Therefore, $T = -2 \pm .71j = -2 + .71j$ or $-2 - .71j$.

Example D: Solve the equation $5B^2 - 9B - 12 = 0$ by completing the square.

$$5B^2 - 9B - 12 = 0$$
$$5B^2 - 9B = 12$$
$$B^2 - \tfrac{9}{5}B = \tfrac{12}{5}$$
$$B^2 - \tfrac{9}{5}B + \tfrac{81}{100} = \tfrac{12}{5} + \tfrac{81}{100} \quad \left\{\tfrac{81}{100} = \left[\tfrac{1}{2}\left(-\tfrac{9}{5}\right)\right]^2\right\}$$
$$\left(B - \tfrac{9}{10}\right)^2 = \tfrac{12}{5} + \tfrac{81}{100} = \tfrac{240}{100} + \tfrac{81}{100} = \tfrac{321}{100}$$
$$B - \tfrac{9}{10} = \pm\tfrac{\sqrt{321}}{10} = \pm\tfrac{17.9}{10}$$

Therefore, $B = \tfrac{9}{10} \pm \tfrac{17.9}{10} = \tfrac{26.9}{10}$ or $\tfrac{-8.9}{10} = 2.69$ or $-.89$.

Example E: Solve the equation $3R^2 - 9R = 0$ by completing the square.

$$3R^2 - 9R = 0$$
$$R^2 - 3R = 0$$
$$R^2 - 3R + \tfrac{9}{4} = 0 + \tfrac{9}{4} \quad \{\tfrac{9}{4} = [\tfrac{1}{2}(-3)]^2\}$$
$$(R - \tfrac{3}{2})^2 = \tfrac{9}{4}$$
$$R - \tfrac{3}{2} = \pm\tfrac{3}{2}$$

Therefore, $R = \tfrac{3}{2} \pm \tfrac{3}{2} = 3$ or 0.

Example F: Solve the equation $5K^2 = 3K + 9$ by completing the square.

$$5K^2 = 3K + 9$$
$$5K^2 - 3K - 9 = 0$$
$$5K^2 - 3K = 9$$
$$K^2 - \tfrac{3}{5}K = \tfrac{9}{5}$$
$$K^2 - \tfrac{3}{5}K + \tfrac{9}{100} = \tfrac{9}{5} + \tfrac{9}{100} \quad \left\{\tfrac{9}{100} = \left[\tfrac{1}{2}\left(-\tfrac{3}{5}\right)\right]^2\right\}$$

$$\left(K - \frac{3}{10}\right)^2 = \frac{9}{5} + \frac{9}{100} = \frac{180}{100} + \frac{9}{100} = \frac{189}{100}$$

$$K - \frac{3}{10} = \pm\frac{\sqrt{189}}{10} = \pm\frac{13.7}{10}$$

Therefore, $K = \frac{3}{10} \pm \frac{13.7}{10} = \frac{16.7}{10}$ or $\frac{-10.7}{10} = 1.67$ or -1.07.

Example G: In an isosceles right triangle, the hypotenuse is 10 units longer than each of sides. What are the dimensions of the traingle? Since this is an isosceles triangle, the two sides will be equal. Let each one be represented by S. Therefore, the hypotenuse will be represented by $S + 10$. Using the Pythagorean Theorem, $S^2 + S^2 = (S + 10)^2$.

$$2S^2 = S^2 + 20S + 100$$
$$S^2 - 20S - 100 = 0$$
$$S^2 - 20S = 100$$
$$S^2 - 20S + 100 = 100 + 100 \quad \{100 = [\tfrac{1}{2}(-20)]^2\}$$
$$(S - 10)^2 = 200$$
$$S - 10 = \pm\sqrt{200} = \pm 14.1$$

Therefore, $S = 10 \pm 14.1 = 24.1$ (-4.1 is meaningless). Therefore, each of the sides is of length 24.1 and the hypotenuse is of length 34.1.

Exercises

Solve each of the following quadratic equations by the method of completing the square.

1. $X^2 + 18X = 0$
2. $Y^2 = 12Y$
3. $T^2 - 8T + 9 = 0$
4. $A^2 + 18A - 3 = 0$
5. $B^2 - 12B - 1 = 0$
6. $K^2 + 10K + 4 = 0$
7. $2V^2 - 8V + 3 = 0$
8. $3P^2 + 12P - 7 = 0$
9. $4Q^2 = 2Q - 11$
10. $5J = -6 - 2J^2$
11. $3S^2 + 1 = 8S - S^2$
12. $2(M^2 - 1) = 3M(M + 2)$
13. The formula for the surface area of a right circular cylinder is $A = 2\pi RH + 2\pi R^2$. If the height (H) of a certain cylinder is 16 centimeters and its surface area is 628 square centimeters, what is the radius (R) of the cylinder?
14. The sum of two resistances in a certain electric circuit is 11 ohms. The product of these two resistances is 28 ohms. What are the resistances?
15. A rectangular air duct is to be formed by bending a 68-inch wide piece of metal into a rectangle. The cross-sectional area of the duct has to be 2 square feet. What are the dimensions of the duct?

Sec. 13–4. The Quadratic Formula

16. When two resistors are connected in series, the resistance is $R = R_1 + R_2$. If they are connected in parallel, the resistance is

$$R = \frac{R_1 R_2}{R_1 + R_2}$$

What should the measure of each resistor be if the series resistance has to be 4 ohms and the parallel resistance has to be .5 ohms?

13–4. The Quadratic Formula

We will now apply the method of completing the square to the general quadratic equation ($AX^2 + BX + C = 0$) in order to develop a formula that may be used to solve any quadratic equation.

$$AX^2 + BX + C = 0$$
$$AX^2 + BX = -C$$
$$X^2 + \left(\frac{B}{A}\right)X = -\frac{C}{A}$$
$$X^2 + \left(\frac{B}{A}\right)X + \frac{B^2}{4A^2} = -\frac{C}{A} + \frac{B^2}{4A^2} \quad \left\{\frac{B^2}{4A^2} = \left[\frac{1}{2}\left(\frac{B}{A}\right)\right]^2\right\}$$
$$X^2 + \left(\frac{B}{A}\right)X + \frac{B^2}{4A^2} = \frac{B^2}{4A^2} - \frac{4AC}{4A^2}$$
$$\left(X + \frac{B}{2A}\right)^2 = \frac{B^2 - 4AC}{4A^2}$$
$$X + \frac{B}{2A} = \pm\frac{\sqrt{B^2 - 4AC}}{2A}$$
$$X = -\frac{B}{2A} \pm \frac{\sqrt{B^2 - 4AC}}{2A}$$
$$X = \frac{-B \pm \sqrt{B^2 - 4AC}}{2A}$$

This last expression is called the *quadratic formula* and may be used to solve any quadratic equation. Comparing this formula with the general equation with which we started, we can see that A is the coefficient of the squared term, B is the coefficient of the linear term, and C is the constant term. Thus, to use the formula, we determine these three values and substitute them in the appropriate positions.

Example A: Use the quadratic formula to solve the equation $X^2 - 3X - 7 = 0$. Here $A = 1$, $B = -3$, $C = -7$.

$$X = \frac{-B \pm \sqrt{B^2 - 4AC}}{2A} = \frac{-(-3) \pm \sqrt{(-3)^2 - 4(1)(-7)}}{2(1)}$$

$$= \frac{3 \pm \sqrt{9 + 28}}{2} = \frac{3 \pm \sqrt{37}}{2} = \frac{3 \pm 6.1}{2}$$

$$= \frac{9.1}{2} \text{ or } -\frac{3.1}{2}$$

$$= 4.55 \text{ or } -1.55$$

Example B: Use the quadratic formula to solve the equation $2Y^2 + 8Y - 12 = 0$. Here $A = 2$, $B = 8$, $C = -12$.

$$Y = \frac{-B \pm \sqrt{B^2 - 4AC}}{2A} = \frac{-8 \pm \sqrt{(8)^2 - 4(2)(-12)}}{2(2)}$$

$$= \frac{-8 \pm \sqrt{64 + 96}}{4} = \frac{-8 \pm \sqrt{160}}{4} = \frac{-8 \pm 12.6}{4}$$

$$= \frac{4.6}{4} \text{ or } -\frac{20.6}{4}$$

$$= 1.15 \text{ or } -5.15$$

Example C: Use the quadratic formula to solve the equation $2T^2 - 7T = 0$. Here $A = 2$, $B = -7$, $C = 0$.

$$T = \frac{-B \pm \sqrt{B^2 - 4AC}}{2A} = \frac{-(-7) \pm \sqrt{(-7)^2 - 4(2)(0)}}{2(2)}$$

$$= \frac{7 \pm \sqrt{49 - 0}}{4} = \frac{7 \pm 7}{4}$$

$$= \frac{14}{4} \text{ or } \frac{0}{4}$$

$$= 3.5 \text{ or } 0$$

Example D: Use the quadratic formula to solve the equation $3K^2 = 2K - 9$.

$$3K^2 - 2K + 9 = 0.$$

Here $A = 3$, $B = -2$, $C = 9$.

$$K = \frac{-B \pm \sqrt{B^2 - 4AC}}{2A} = \frac{-(-2) \pm \sqrt{(-2)^2 - 4(3)(9)}}{2(3)}$$

$$= \frac{2 \pm \sqrt{4 - 108}}{6} = \frac{2 \pm \sqrt{-104}}{6} = \frac{2 \pm \sqrt{104}j}{6} = \frac{2 \pm 10.2j}{6}$$

$$= .33 + 1.7j \text{ or } .33 - 1.7j$$

Sec. 13–4. The Quadratic Formula

Example E: The diagonal dimension of a square machine part is 8 millimeters longer than a side. If the area of the part is 420 square millimeters, what are the dimensions? Let S represent the side of the part.

Therefore, $S + 8$ represents the diagonal of the part. Using the Pythagorean Theorem, $S^2 + S^2 = (S + 8)^2$.

$$2S^2 = S^2 + 16S + 64$$
$$S^2 - 16S - 64 = 0$$

Here $A = 1$, $B = -16$, $C = -64$.

$$S = \frac{-B \pm \sqrt{B^2 - 4AC}}{2A} = \frac{-(-16) \pm \sqrt{(-16)^2 - 4(1)(-64)}}{2(1)}$$

$$= \frac{16 \pm \sqrt{256 + 256}}{2} = \frac{16 \pm \sqrt{512}}{2}$$

$$= \frac{16 \pm 22.6}{2} = \frac{38.6}{2} = 19.3 \quad \left(\frac{-6.6}{2} \text{ is meaningless}\right)$$

Therefore, the machine part is 19.3 millimeters by 19.3 millimeters.

Every quadratic equation can be written in the form $AX^2 + BX + C = 0$. From the development of the quadratic formula at the beginning of this section, we should see that the complete solution for an equation of this type is

$$X = \frac{-B + \sqrt{B^2 - 4AC}}{2A} \quad \text{or} \quad \frac{-B - \sqrt{B^2 - 4AC}}{2A}$$

If the two linear factors

$$\left[X - \left(\frac{-B + \sqrt{B^2 - 4AC}}{2A}\right)\right] \text{ and } \left[X - \left(\frac{-B - \sqrt{B^2 - 4AC}}{2A}\right)\right]$$

are multiplied together, we will obtain $AX^2 + BX + C$. Therefore, we have verified our statement in Section 13–1: *every quadratic equation* has a *complete solution* consisting of *exactly two values*. These values need not be distinct, and they may be real, imaginary, or complex.

Exercises Solve each of the following quadratic equations, using the quadratic formula.

1. $B^2 - 5B - 6 = 0$
2. $K^2 + 4K + 3 = 0$
3. $5T^2 - 14T - 3 = 0$
4. $6X^2 - 11X - 10 = 0$
5. $Y^2 - 7Y - 15 = 0$
6. $J^2 + 3J + 12 = 0$
7. $V^2 + 9V + 1 = 0$
8. $Z^2 - 4Z + 6 = 0$

9. $L^2 + 3L + 11 = 0$
10. $M^2 - 5M + 10 = 0$
11. $3P^2 - 2P - 1 = 0$
12. $4R^2 + 6R + 8 = 0$
13. $2X^2 - X - 7 = 0$
14. $5T^2 + T + 3 = 0$
15. $2K^2 - 8K + 12 = 0$
16. $3N^2 + 3N - 3 = 0$
17. $2B^2 = 5B - 9$
18. $3L = 8 - L^2$
19. $4V^2 - V = 12$
20. $Y^2 + 10Y = 20$

21. The distance (s) traveled by an object with an initial velocity (v) and acceleration (a) in terms of time (t) is given by $s = vt + \frac{1}{2}at^2$. If the object travels 763 feet with an initial velocity of 128 feet per second and an acceleration of 32 feet per second per second, for how many seconds has the object been traveling?

22. The sides of an isosceles right triangle are 2 units larger than the side of a certain square. The area of the square is 28 square units larger than the area of the triangle. What are the dimensions of the triangle and the square?

23. When two springs are connected in parallel, the equivalent spring factor is $K = K_1 + K_2$, and when they are connected in series, the equivalent spring factor is $K = \frac{K_1 K_2}{K_1 + K_2}$. What two springs will give a spring factor of 2 pounds per inch when connected in parallel and $\frac{1}{4}$ pound per inch when connected in series?

24. The distance above ground for an object propelled directly upward can be determined by $s = 1200t - 16t^2$, where s represents distance in meters and t represents time in seconds. How long will it take the object to reach a height of 4800 meters?

13-5. The Roots of a Quadratic Equation

It is often useful to know something about the roots of a quadratic equation without actually finding them. It is also helpful to know what kind of roots a quadratic equation will have, since if they do not fit the conditions of the probem (e.g., they are imaginary roots when real roots are desired), there is no need actually to go through the solution process.

The kind of roots that a quadratic equation will have can be determined by examining the expression $B^2 - 4AC$, which is found in the quadratic formula and is called the *discriminant*.

1. If $(B^2 - 4AC) > 0$, we will obtain two unequal real roots.

Example A: Consider the equation $T^2 + 3T - 6 = 0$.

Sec. 13–5. The Roots of a Quadratic Equation

$$B^2 - 4AC = (3)^2 - 4(1)(-6) = 9 + 24 = 33$$

$$T = \frac{-3 \pm \sqrt{33}}{2} = \frac{-3 \pm 5.74}{2} = \frac{2.74}{2} \text{ or } \frac{-8.74}{2} = 1.37 \text{ or } -4.37$$

1.37 and -4.37 are unequal real numbers.

2. If $(B^2 - 4AC) = 0$, we will obtain two equal real roots.

Example B: Consider the equation $4K^2 - 12K + 9 = 0$.

$$B^2 - 4AC = (-12)^2 - 4(4)(9) = 144 - 144 = 0$$

$$K = \frac{-(-12) \pm \sqrt{0}}{8} = \frac{12 \pm 0}{8} = \frac{12}{8} \text{ or } \frac{12}{8} = 1.5 \text{ or } 1.5$$

1.5 and 1.5 are equal real numbers.

3. If $(B^2 - 4AC) < 0$, we will obtain two unequal complex roots that are conjugates of one another.

Example C: Consider the equation $M^2 - 3M + 10 = 0$.

$$B^2 - 4AC = (-3)^2 - 4(1)(10) = 9 - 40 = -31$$

$$M = \frac{-(-3) \pm \sqrt{-31}}{2} = \frac{3 \pm \sqrt{31}j}{2} = \frac{3 \pm 5.57j}{2}$$

$$= 1.5 + 2.78j \text{ or } 1.5 - 2.78j.$$

$1.5 + 2.78j$ and $1.5 - 2.78j$ are unequal complex numbers and are also conjugates of one another.

Example D: In attempting to obtain the thickness (H) of a certain machine part, the equation $2H^2 + 5H + 9 = 0$ is obtained. Can this equation be used to find the thickness of the part?

$$B^2 - 4AC = (5)^2 - 4(2)(9) = 25 - 72 = -47$$

The roots will be complex (containing an imaginary part). Therefore, this equation cannot be used.

We should recall from the previous section that the two roots of a quadratic equation can be written as

$$\frac{-B + \sqrt{B^2 - 4AC}}{2A} \text{ and } \frac{-B - \sqrt{B^2 - 4AC}}{2A}$$

If we add these two quantities, we obtain:

$$\left(\frac{-B + \sqrt{B^2 - 4AC}}{2A}\right) + \left(\frac{-B - \sqrt{B^2 - 4AC}}{2A}\right)$$

$$= \frac{-B - B + \sqrt{B^2 - 4AC} - \sqrt{B^2 - 4AC}}{2A}$$

$$= \frac{-2B}{2A} = \frac{-B}{A}$$

Therefore, the *sum of the roots* of a quadratic equation is $\frac{-B}{A}$. If we multiply the same two quantities, we obtain:

$$\left(\frac{-B + \sqrt{B^2 - 4AC}}{2A}\right) \times \left(\frac{-B - \sqrt{B^2 - 4AC}}{2A}\right)$$

$$= \frac{(B^2) + (B\sqrt{B^2 - 4AC}) - (B\sqrt{B^2 - 4AC}) - (B^2 - 4AC)}{4A^2}$$

$$= \frac{B^2 - B^2 + 4AC}{4A^2} = \frac{4AC}{4A^2} = \frac{C}{A}$$

Therefore, the *product of the roots* of a quadratic equation is $\frac{C}{A}$.

Example E: Given the equation $X^2 - 5X - 6 = 0$, find the sum and product of the roots. Here $A = 1$, $B = -5$, $C = -6$.

$$\text{Sum of the roots} = \frac{-B}{A} = \frac{-(-5)}{1} = 5.$$

$$\text{Product of the roots} = \frac{C}{A} = \frac{-6}{1} = -6.$$

If this equation is solved we get:

$$X = \frac{-(-5) \pm \sqrt{(-5)^2 - 4(1)(-6)}}{2(1)} = \frac{5 \pm \sqrt{25 + 24}}{2} = \frac{5 \pm \sqrt{49}}{2}$$

$$= \frac{5 \pm 7}{2} = \frac{12}{2} \text{ or } -\frac{2}{2} = 6 \text{ or } -1$$

We can certainly see that $6 + (-1) = 5$ and $6 \times (-1) = -6$.

Example F: Given the equation $3P^2 - 8P + 6 = 0$, find the sum and product of the roots. Here $A = 3$, $B = -8$, $C = 6$.

$$\text{Sum of the roots} = \frac{-B}{A} = \frac{-(-8)}{3} = \frac{8}{3}.$$

Sec. 13–5. The Roots of a Quadratic Equation

$$\text{Product of the roots} = \frac{C}{A} = \frac{6}{3} = 2.$$

If this equation is solved we get:

$$P = \frac{-(-8) \pm \sqrt{(-8)^2 - 4(3)(6)}}{2(3)} = \frac{8 \pm \sqrt{64 - 72}}{6} = \frac{8 \pm \sqrt{-8}}{6}$$

$$= \frac{8 \pm \sqrt{8}\,j}{6} = \frac{8 + \sqrt{8}\,j}{6} \text{ or } \frac{8 - \sqrt{8}\,j}{6}$$

$$\left(\frac{8 + \sqrt{8}\,j}{6}\right) + \left(\frac{8 - \sqrt{8}\,j}{6}\right) = \frac{16}{6} = \frac{8}{3}$$

$$\left(\frac{8 + \sqrt{8}\,j}{6}\right) \times \left(\frac{8 - \sqrt{8}\,j}{6}\right) = \frac{64 - 8\sqrt{8}\,j + 8\sqrt{8}\,j - 8j^2}{36}$$

$$= \frac{64 + 8}{36} = \frac{72}{36} = 2$$

Once the roots of a quadratic equation are obtained, we may use the formulas for the sum and product of the roots as a check.

Exercises

For each of the following, use the discriminant to find the type of roots, and then verify this information by finding the solution.

1. $X^2 - 3X = 0$
2. $T^2 + 11T = 0$
3. $W^2 + 4 = 0$
4. $K^2 - 7 = 0$
5. $4B^2 - 20B + 25 = 0$
6. $9Y^2 - 24Y + 16 = 0$
7. $V^2 - 3V - 9 = 0$
8. $3M^2 + 2M - 10 = 0$
9. $5R^2 + R + 7 = 0$
10. $L^2 - 6L + 10 = 0$

For each of the following, find the sum and product of the roots, and then verify this information by finding the solution.

11. $T^2 + 7T = 0$
12. $S^2 - 12S = 0$
13. $X^2 - 8 = 0$
14. $R^2 + 10 = 0$
15. $4J^2 - 28J + 49 = 0$
16. $N^2 + 2\sqrt{2}\,N + 2 = 0$
17. $4Y^2 - 12Y + 1 = 0$
18. $2V^2 - V - 1 = 0$
19. $L^2 - 5L + 7 = 0$
20. $3W^2 + 6W + 4 = 0$

21. In attempting to obtain the dimensions of a rectangular field, the equation $3L^2 + 5L + 3 = 0$ is obtained, where L represents the length of the field. Can L be found from this equation?

22. Can the radius (R) of a right circular cylinder be found from the equation $2\pi R^2 + 6\pi R + 100 = 0$?

13–6. The Graph of a Quadratic Equation

Like many other equations, quadratic equations may be analyzed by means of their graphs. Also, certain points on their graphs and certain characteristics of their graphs will have important implications with regard to quadratic equations.

In order to construct the graph of a quadratic equation, it is necessary first to set the quadratic expression equal to a second variable. Then, the graph may be constructed and it will always be a *parabola*. When the quadratic equation is written in its general form ($Y = AX^2 + BX + C$), the parabola will open *upward if $A > 0$* and *downward if $A < 0$*. In Chapter 4 we pointed out that the roots of an equation correspond to the points at which the graph of the equation intersects the horizontal axis. This is certainly valid for quadratic equations. If the two roots are real and unequal ($B^2 - 4AC > 0$), the graph will intersect the horizontal axis in exactly two points. If the two roots are real and equal ($B^2 - 4AC = 0$), the graph will just touch the horizontal axis at a single point. If the two roots are complex conjugates of one another ($B^2 - 4AC < 0$), the graph will not intersect the horizontal axis.

Example A: Construct the graph of $X^2 - 4X + 3 = 0$. First, we let $Y = X^2 - 4X + 3$.

X	0	1	−1	2	−2	3	4
Y	3	0	8	−1	15	0	3

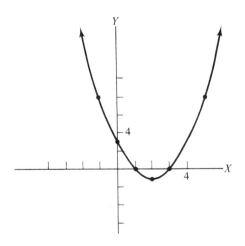

1. Since $A > 0(1)$, the parabola opens upward.
2. $X = \dfrac{-(-4) \pm \sqrt{(-4)^2 - 4(1)(3)}}{2(1)} = \dfrac{4 \pm \sqrt{4}}{2} = \dfrac{4 \pm 2}{2} = 3 \text{ or } 1.$

Sec. 13–6. The Graph of a Quadratic Equation

We can see that there are two real and unequal roots (3 and 1), and that the graph crosses the horizontal axis at exactly these two values.

Example B: Construct the graph of $-4T^2 + 4T - 1 = 0$. First, we let $V = -4T^2 + 4T - 1$.

T	0	1	−1	2	−2	½
V	−1	−1	−9	−9	−25	0

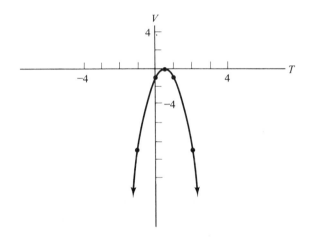

1. Since $A < 0(-4)$, the parabola opens downward.
2. $T = \dfrac{-4 \pm \sqrt{(4)^2 - 4(-4)(-1)}}{2(-4)} = \dfrac{-4 \pm \sqrt{0}}{-8} = \dfrac{-4 \pm 0}{-8} = \dfrac{1}{2}$ or $\dfrac{1}{2}$.

We can see that there are two real and equal roots (½ and ½), and that the graph just touches the horizontal axis at this value.

Example C: Construct the graph of $M^2 - 3M + 8 = 0$. First, we let $L = M^2 - 3M + 8$.

M	0	1	−1	2	−2	3/2	3
L	8	6	12	6	18	5¾	8

1. Since $A > 0(1)$, the parabola opens upward.
2. $M = \dfrac{-(-3) \pm \sqrt{(-3)^2 - 4(1)(8)}}{2(1)} = \dfrac{3 \pm \sqrt{-23}}{2} = \dfrac{3 \pm 4.8j}{2}$
   $= 1.5 + 2.4j$ or $1.5 - 2.4j$.

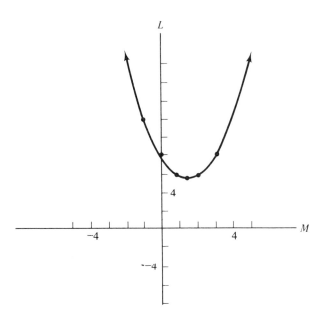

We can see that the two roots are complex conjugates of one another ($1.5 + 2.4j$ and $1.5 - 2.4j$), and that the graph does not cross the horizontal axis.

Since the graph of a quadratic equation is always a parabola, there will always be a maximum or minimum point on the graph (maximum point if the parabola opens downward and minimum point if the parabola opens upward). Also, since every quadratic equation has exactly two roots and a parabola is a symmetric curve (each half is a mirror image of the other), the roots will be in corresponding positions on the two halves of the parabola. If they were not in corresponding positions, either the graph would be tilted (a case that we are not considering) or the roots would not correspond to the horizontal intercepts (a contradiction). Figure 13-6.1 illustrates this. Therefore, the maximum or minimum point of the graph will be exactly halfway between the points corresponding to the roots. The abscissa of this point may be found by finding the

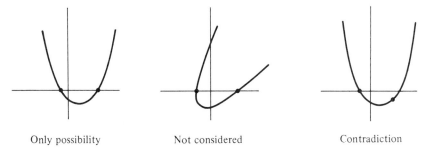

Only possibility   Not considered   Contradiction

FIGURE 13-6.1

Sec. 13–6. The Graph of a Quadratic Equation

sum of the two roots and dividing by 2. Since we have already seen that the sum of the roots is $\frac{-B}{A}$, the abscissa of the maximum or minimum point will be $\frac{-B}{2A}$. The ordinate may then be found by substituting this value into the quadratic equation. Also, the vertical line through this maximum or minimum point is called the *axis of symmetry* of the parabola. This means that the two halves of the parabola are mirror images of one another, with the axis of symmetry serving as the "mirror." The equation of this axis of symmetry will be $X = \frac{-B}{2A}$. (X could be replaced by any variable.)

Example D: Consider the quadratic equation and its graph from Example A.

$$X^2 - 4X + 3 = 0 \qquad Y = X^2 - 4X + 3$$

1. Mimimum point: $X = \frac{-B}{2A} = \frac{-(-4)}{2} = \frac{4}{2} = 2$
   $Y = (2)^2 - 3(2) + 3 = -1$
2. Axis of symmetry: $X = 2$

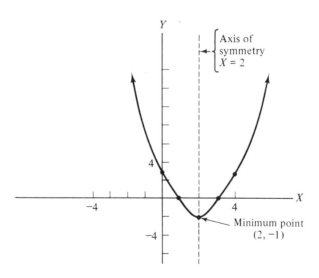

Minimum point (2, −1)

Example E: Consider the quadratic equation and its graph from Example B.

$$-4T^2 + 4T - 1 = 0 \qquad V = -4T^2 + 4T - 1$$

1. Maximum point: $T = \frac{-B}{2A} = \frac{-4}{2(-4)} = \frac{-4}{-8} = \frac{1}{2}$
   $V = -4\left(\frac{1}{2}\right)^2 + 4\left(\frac{1}{2}\right) - 1 = 0$

2. Axis of symmetry: $T = \frac{1}{2}$

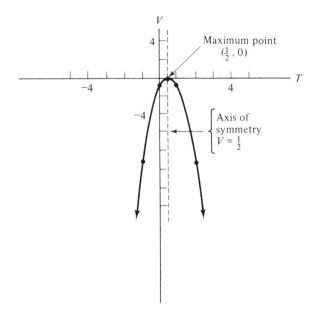

Example F: Consider the quadratic equation and its graph from Example C.

$$M^2 - 3M + 8 = 0 \qquad L = M^2 - 3M + 8$$

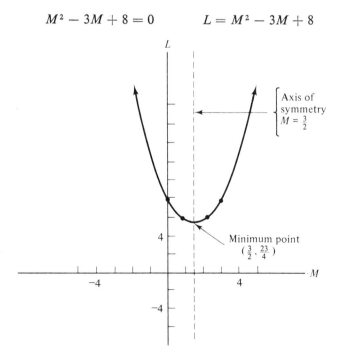

1. Minimum point: $M = \dfrac{-B}{2A} = \dfrac{-(-3)}{2(1)} = \dfrac{3}{2}$

$L = \left(\dfrac{3}{2}\right)^2 - 3\left(\dfrac{3}{2}\right) + 8 = \dfrac{23}{4}$

2. Axis of symmetry: $M = \dfrac{3}{2}$

Exercises

For each of the following quadratic equations, construct the graph and, from the graph, determine any real roots. If roots are complex simply state this fact. Also, determine the coordinates of the maximum or minimum point as well as the equation of the axis of symmetry.

1. $Y^2 + 4Y = 0$
2. $K^2 - 6K = 0$
3. $T^2 - 12 = 0$
4. $B^2 + 8B = 0$
5. $4L^2 - 12L + 9 = 0$
6. $9V^2 + 6V + 1 = 0$
7. $X^2 - 2X - 15 = 0$
8. $2M^2 + 7M - 4 = 0$
9. $W^2 + 2W - 3 = 0$
10. $6N^2 + 5N - 4 = 0$
11. $A^2 + 2A - 7 = 0$
12. $X^2 - 5X - 1 = 0$
13. $J^2 + 2J + 1 = 0$
14. $25P^2 + 20P + 4 = 0$
15. $S^2 - 4S + 8 = 0$
16. $2R^2 - R + 1 = 0$
17. $Q^2 - Q - 1 = 0$
18. $3Y^2 + 4Y + 2 = 0$
19. $U^2 - 5U = 8$
20. $2T^2 - 4T = 7 - T^2$

21. A ball is thrown directly upward from ground level. Its distance (s) above the ground is given by $s = 96t - 16t^2$. Determine from the graph of the function how long it will take for the ball to reach a height of (a) 100 feet, (b) 128 feet, (c) 144 feet.

22. A variable voltage (E) is related to time (t) by $E = 7t - 16t^2$. Determine from the graph of this function how long it will take the voltage to reach $\tfrac{1}{2}$ volt.

13-7. Quadratic Equations Involving Trigonometric Functions

Often, when equations involving trigonometric functions are encountered, they are quadratic in nature. In these situations, the equations may be solved using the methods described in this chapter. When a quadratic equation involves a trigonometric function, it is solved for the function. The trigonometric function is then solved for the angle, which we restrict to being between 0° (0) and 360° (2π) inclusive.

Ch. 13. Quadratic Equations

Example A:
1. The equation $\cos^2 X - 2 \cos X + 1 = 0$ would be solved for $\cos X$. $\cos X$ would then be solved for X.
2. The equation $2 \sin^2 Y + 3 \sin Y - 4 = 0$ would be solved for $\sin Y$. $\sin Y$ would then be solved for Y.

Example B: Solve the equation $\sin^2 X - 1 = 0$.

$$\sin^2 X - 1 = 0$$
$$(\sin X + 1)(\sin X - 1) = 0$$
$$\sin X + 1 = 0 \text{ and/or } \sin X - 1 = 0$$
$$\sin X = -1 \text{ or } \sin X = 1$$
$$X = 270°, 90°$$

Example C: Solve the equation $2 \cos^2 T - 5 \cos T - 3 = 0$.

$$2 \cos^2 T - 5 \cos T - 3 = 0$$
$$(2 \cos T + 1)(\cos T - 3) = 0$$
$$2 \cos T + 1 = 0 \text{ and/or } \cos T - 3 = 0$$
$$\cos T = -\tfrac{1}{2} \text{ or } \cos T = 3$$
$$T = 120°, 240°$$

NOTE: $\cos T = 3$ is an impossible situation, and thus, we get no solution from this factor.

Example D: Solve the equation $\tan^2 A - \tan A - 12 = 0$.

$$\tan^2 A - \tan A - 12 = 0$$
$$(\tan A + 3)(\tan A - 4) = 0$$
$$\tan A + 3 = 0 \text{ and/or } \tan A - 4 = 0$$
$$\tan A = -3 \text{ or } \tan A = 4$$
$$A = 108.4°, 288.4°, 75.9°, 255.9°$$

Example E: Solve the equation $\sin^2 B + 2 \sin B - 2 = 0$
This quadratic expression is not factorable.

$$\sin B = \frac{-2 \pm \sqrt{2^2 - 4(1)(-2)}}{2(1)} = \frac{-2 \pm \sqrt{12}}{2} = \frac{2 \pm 3.46}{2}$$
$$= \frac{5.46}{2} \text{ or } \frac{-1.46}{2} = 2.73 \text{ or } -.73$$

Sec. 13–7. Quadratic Equations Involving Trigonometric Functions

It is impossible for Sin B to equal 2.73.

$$\text{Sin } B = -.73$$
$$B = 226.9°, 313.1°$$

Example F: Solve the equation $2 \text{ Tan}^2 X - 3 \text{ Tan } X - 1 = 0$.

$2 \text{ Tan}^2 X - 3 \text{ Tan } X - 1 = 0$

$$\text{Tan } X = \frac{-(-3) \pm \sqrt{(-3)^2 - 4(2)(-1)}}{2(2)} = \frac{3 \pm \sqrt{17}}{4}$$

$$= \frac{3 \pm 4.1}{4} = \frac{7.1}{4} \text{ or } \frac{-1.1}{4} = 1.78 \text{ or } -.28$$

$$X = 60.7°, 240.7°, 164.3°, 344.3°$$

In some quadratic equations involving trigonometric functions, it may be necessary to make an appropriate substitution before solving.

Example G: Solve the equation $\text{Cos } 2T - \text{Cos } T - 1 = 0$. Here, we should recall that $\text{Cos } 2T = 2 \text{ Cos}^2 T - 1$.

$2 \text{ Cos}^2 T - \text{Cos } T - 2 = 0$

$$\text{Cos } T = \frac{-(-1) \pm \sqrt{(-1)^2 - 4(2)(-2)}}{2(2)} = \frac{1 \pm \sqrt{17}}{4}$$

$$= \frac{1 \pm 4.1}{4} = \frac{5.1}{4} \text{ or } \frac{-3.1}{4} = 1.28 \text{ or } -.78$$

$\text{Cos } T = 1.28$ is impossible

$\text{Cos } T = -.78$

$T = 141.3°, 218.7°$

Example H: Solve the equation $2 \text{ Cos}^2 X + \text{Sin } X + 1 = 0$. Here we should recall that $\text{Cos}^2 X = 1 - \text{Sin}^2 X$.

$2(1 - \text{Sin}^2 X) + \text{Sin } X + 1 = 0$
$2 - 2 \text{ Sin}^2 X + \text{Sin } X + 1 = 0$
$-2 \text{ Sin}^2 X + \text{Sin } X + 3 = 0$

$$\text{Sin } X = \frac{-1 \pm \sqrt{1^2 - 4(-2)(3)}}{2(-2)} = \frac{-1 \pm \sqrt{25}}{-4}$$

$$= \frac{-1 \pm 5}{-4} = \frac{4}{-4} \text{ or } \frac{-6}{-4} = -1 \text{ or } 1.5$$

$$\text{Sin } X = 1.5 \text{ is impossible}$$
$$\text{Sin } X = -1$$
$$X = 270°$$

Exercises Solve each of the following equations by an appropriate method.

1. $\text{Cos}^2 T - 1 = 0$
2. $\text{Cot}^2 B - 3 = 0$
3. $\text{Sin}^2 Y + 5 \text{ Sin } Y + 6 = 0$
4. $\text{Tan}^2 B - 4 \text{ Tan } B - 5 = 0$
5. $6 \text{ Cos}^2 X - \text{Cos } X - 1 = 0$
6. $12 \text{ Sin}^2 A + 11 \text{ Sin } A + 2 = 0$
7. $2 \text{ Cot}^2 T - 3 \text{ Cot } T - 9 = 0$
8. $\text{Sec}^2 K + 6 \text{ Sec } K + 8 = 0$
9. $2 \text{ Sin}^2 X - \text{Sin } X - 5 = 0$
10. $3 \text{ Cos}^2 M + 4 \text{ Cos } M - 2 = 0$
11. $\text{Tan}^2 V - 5 \text{ Tan } V + 1 = 0$
12. $\text{Csc}^2 P + 6 \text{ Csc } P + 8 = 0$
13. $\text{Cot}^2 X - 3 \text{ Cot } X - 5 = 0$
14. $4 \text{ Sin}^2 Y + \text{Sin } Y + 1 = 0$
15. $3 \text{ Sec}^2 B - 3 \text{ Sec } B - 3 = 0$
16. $5 \text{ Cos}^2 A + \text{Cos } A - 7 = 0$
17. $\text{Cos } 2T - \text{Sin } T - 4 = 0$
18. $\text{Sec}^2 \theta - 3 \text{ Tan } \theta + 1 = 0$
19. $\text{Sin}^2 R = 2 \text{ Cos } R$
20. $\text{Tan}^2 \theta = 3 - 2 \text{ Sec } \theta$
21. The displacement of a particle is given by $s = 2 \sin^2 t - \sin t - 1$, where s is the distance in meters and t is the time in seconds. Determine the time when $s = 0$.
22. The vertical displacement Y of an object at the end of a spring is given by $Y = 2 \text{ Cos}^2 t - \text{Cos } t$. Find the values of t (seconds) for which $Y = 0$.

13-8. Equations Solved by Quadratic Methods

There are many equations that are not quadratic, but can be converted into a quadratic equation by an appropriate mathematical procedure. They can then be solved by the methods described earlier in this chapter.

The first of this type of equation would be an equation that is in what is called *quadratic form*. That is, we have an equation consisting of a quantity, its square, and possibly a constant term. This type of equation can be made into a quadratic equation by substituting a linear variable for the basic quantity. The resulting equation is then solved. By referring back to the substitution, we may obtain the solution for the original equation. This solution *must be checked* since any time we perform an operation (in this case, substitution) involving the variable, we may introduce extraneous roots.

Example A: Solve the equation $X^4 - 5X^2 + 6 = 0$. Here we let $Y = X^2$. Therefore, the equation becomes $Y^2 - 5Y + 6 = 0$.

Sec. 13-8. Equations Solved by Quadratic Methods

$$(Y-3)(Y-2)=0$$

Therefore, $Y = 3, 2$. Since $Y = X^2$, $3 = X^2$ or $2 = X^2$. Therefore, $X = \pm\sqrt{3}$ or $\pm\sqrt{2}$.

Check:
$$(\sqrt{3})^4 - 5(\sqrt{3})^2 + 6 = 9 - 15 + 6 = 0$$
$$(-\sqrt{3})^4 - 5(-\sqrt{3})^2 + 6 = 9 - 15 + 6 = 0$$
$$(\sqrt{2})^4 - 5(\sqrt{2})^2 + 6 = 4 - 10 + 6 = 0$$
$$(-\sqrt{2})^4 - 5(-\sqrt{2})^2 + 6 = 4 - 10 + 6 = 0$$

Therefore, $X = \pm\sqrt{3}, \pm\sqrt{2}$.

Example B: Solve the equation $P - \sqrt{P} - 2 = 0$. Here we let $Q = \sqrt{P}$. Therefore, the equation becomes $Q^2 - Q - 2 = 0$.

$$(Q-2)(Q+1)=0$$

Therefore, $Q = 2, -1$. Since $Q = \sqrt{P}$, $2 = \sqrt{P}$ or $-1 = \sqrt{P}$. Therefore, $P = 4$. (By referring to our definition of square root in Section 2–1, we should realize that $\sqrt{P}$ could never be negative. Only $-\sqrt{P}$ can be negative).

Check: $\quad\quad 4 - \sqrt{4} - 2 = 4 - 2 - 2 = 0$

Therefore, $P = 4$.

Example C: Solve the equation $R^{-2} - 3R^{-1} - 4 = 0$. Here we let $S = R^{-1}$. Therefore, the equation becomes $S^2 - 3S - 4 = 0$.

$$(S-4)(S+1)=0$$

Therefore, $S = 4, -1$. Since $S = R^{-1}$, $4 = R^{-1}$ or $-1 = R^{-1}$. Therefore, $R = \frac{1}{4}$ or -1.

Check:
$$(\tfrac{1}{4})^{-2} - 3(\tfrac{1}{4})^{-1} - 4 = 16 - 12 - 4 = 0$$
$$(-1)^{-2} - 3(-1)^{-1} - 4 = 1 + 3 - 4 = 0$$

Therefore, $R = \frac{1}{4}, -1$.

Example D: Solve the equation $(T^2 - 2T)^2 - 5(T^2 - 2T) - 6 = 0$. Here we let $V = T^2 - 2T$. Therefore, the equation becomes $V^2 - 5V - 6 = 0$.

$$(V-6)(V+1)=0$$

Therefore, $V = 6, -1$. Since $V = T^2 - 2T$, $6 = T^2 - 2T$ or $-1 = T^2 - 2T$. Therefore, $T^2 - 2T - 6 = 0$ or $T^2 - 2T + 1 = 0$.

$$T = \frac{-(-2) \pm \sqrt{(-2)^2 - 4(1)(-6)}}{2(1)} \quad \bigg| \quad \begin{array}{l} (T-1)(T-1) = 0 \\ T = 1, 1 \end{array}$$

$$= \frac{2 \pm \sqrt{28}}{2} = \frac{2 \pm 5.3}{2}$$

$$= 3.65 \text{ or } -1.65$$

Check:
$$[(3.65)^2 - 2(3.65)]^2 - 5[(3.65)^2 - 2(3.65)] - 6 = 0$$
$$[(-1.65)^2 - 2(-1.65)]^2 - 5[(-1.65)^2 - 2(-1.65)] - 6 = 0$$
$$[(1)^2 - 2(1)]^2 - 5[(1)^2 - 2(1)] - 6 = 0$$

Therefore, $T = 3.65, -1.65, 1, 1$.

The second type of equation we will consider here is an equation involving *radicals*. By squaring both sides of such an equation, possibly more than once, we will obtain a quadratic equation. After solving, we must once again check our answers.

Example E: Solve the equation $(T - 3) = \sqrt{T - 1}$. Squaring both sides:

$$(T - 3)^2 = T - 1$$
$$T^2 - 6T + 9 = T - 1$$
$$T^2 - 7T + 10 = 0$$
$$(T - 5)(T - 2) = 0$$
$$T = 5, 2$$

Check: $(5 - 3) = \sqrt{5 - 1} = 2$
$(2 - 3) = -1$ and $\sqrt{2 - 1} = 1$ (2 does not check.)

Therefore, $T = 5$.

Example F: Solve the equation $\sqrt{3R} + \sqrt{2R + 1} = 11$.

$$\sqrt{2R + 1} = 11 - \sqrt{3R}$$

Squaring both sides

$$2R + 1 = 121 - 22\sqrt{3R} + 3R$$

Therefore, $-R - 120 = -22\sqrt{3R}$. Squaring both sides again

$$R^2 + 240R + 14{,}400 = 1452R$$
$$R^2 - 1212R + 14{,}400 = 0$$

Sec. 13-8. Equations Solved by Quadratic Methods

$$(R - 1200)(R - 12) = 0$$
$$R = 1200, 12$$

Check: $\sqrt{3600} + \sqrt{2401} = 60 + 49 \neq 11$ (1200 does not check.)
$\sqrt{36} + \sqrt{25} = 6 + 5 = 11$

Therefore, $R = 12$.

Certain logarithmic equations also yield quadratic equations when the properties of logarithms are applied.

Example G: Solve the equation $\log (2K + 3) + \log K = \log 20$.

$$\log (2K + 3)K = \log 20$$
$$\log (2K^2 + 3K) = \log 20$$

Therefore, $2K^2 + 3K = 20$.

$$2K^2 + 3K - 20 = 0$$
$$(2K - 5)(K + 4) = 0$$
$$K = \tfrac{5}{2}, -4$$

Since logarithms of negative numbers are not defined, $K = -4$ is not a solution. Therefore, $K = \tfrac{5}{2}$.

Example H: Solve the equation $\log (X - 3) + \log X = 1$.

$$\log (X - 3)X = 1$$
$$\log (X^2 - 3X) = 1$$
$$X^2 - 3X = 10$$
$$X^2 - 3X - 10 = 0$$
$$(X - 5)(X + 2) = 0$$
$$X = 5, -2$$

Since logarithms of negative numbers are not defined, $X = -2$ is not a solution. Therefore, $X = 5$.

Exercises   Solve each of the following equations by a method described in this section.

1. $T^4 - 7T^2 - 8 = 0$
2. $M^6 - M^3 - 2 = 0$
3. $K - 5\sqrt{K} - 36 = 0$
4. $Y^{1/2} - 6Y^{1/4} - 27 = 0$
5. $(X^2 - 4X)^2 + (X^2 - 4X) - 6 = 0$

6. $(V^2 - 2V)^2 - 6(V^2 - 2V) + 9 = 0$
7. $\sqrt{6 - X} = 3X - 4$
8. $V = \sqrt{3V}$
9. $P - 6 = \sqrt{2P - 12}$
10. $\sqrt{5 - P} = -2P - 5$
11. $\sqrt{3R + 1} - \sqrt{2R - 6} = 2$
12. $\sqrt{6Q} = 3 + \sqrt{Q + 3}$
13. $\log_2 (X - 2) + \log_2 (X) = 0$
14. $\log_7 (T^2 - 4) - \log_7 T = \log_7 3$
15. $\log_5 (L - 4) + \log_5 L = 1$
16. $\log (N + 3) + \log N = 1$
17. $\log_3 (2S - 5) + \log_3 S = \log_3 12$
18. $\log (X^2 + 3) - \log X = \log 6$
19. A certain rectangular area has a perimeter of 84 meters. If the width is equal to the square root of the length, what are the dimensions of the area?
20. An isosceles triangle has a perimeter of 22 units. If the two equal sides are of length $\sqrt{5X - 4}$, and the third side is of length $\sqrt{12X + 4}$, find the lengths of the three sides.

13-9. CHAPTER REVIEW

Solve each of the following quadratic equations by factoring.
1. $T^2 + 2T - 63 = 0$
2. $K^2 - 6K + 5 = 0$
3. $6R^2 + R - 2 = 0$
4. $5X^2 - 33X - 14 = 0$
5. $4M^2 - 11M + 6 = 0$
6. $3Y^2 - 7Y = 0$

Solve each of the following quadratic equations by completing the square.
7. $X^2 + 8X + 3 = 0$
8. $S^2 - 7S - 2 = 0$
9. $2P^2 - 6P + 12 = 0$
10. $3L^2 + 9L - 2 = 0$

Solve each of the following quadratic equations, using the quadratic formula.
11. $N^2 - 7N - 3 = 0$
12. $Y^2 + 5Y + 1 = 0$
13. $3J^2 - J + 1 = 0$
14. $5V^2 + 6V + 9 = 0$
15. $2X^2 = 4 - 11X$
16. $2W - 7 = 6W^2$

Solve each of the following by an appropriate method.
17. $\sin^2 T - \frac{1}{4} = 0$
18. $\tan^2 \theta - 2 \tan \theta - 3 = 0$
19. $6 \csc^2 X + 5 \csc X - 4 = 0$
20. $2 \tan^2 X + 4 \tan X + 1 = 0$
21. $\cos^2 Y - 2 \sin^2 Y = 0$
22. $\sin (2\theta) - \cos \theta = 0$
23. $X^4 - 7X^2 + 12 = 0$
24. $T^{1/2} + T^{1/4} - 20 = 0$
25. $B - \sqrt{B} - 2 = 0$

Sec. 13–9. Chapter Review

26. $(A^2 - 5A)^2 - 8(A^2 - 5A) + 12 = 0$
27. $\sqrt{3X} = X - 6$
28. $2X - 3 = \sqrt{6X + 1}$
29. $\sqrt{X} + \sqrt{2X + 7} = 8$
30. $\sqrt{3X + 4} - \sqrt{2X + 1} = 1$
31. $\log_9 (X - 3) + \log_9 X = 0$
32. $\log (X^2 - 5) - \log X = 1$
33. $\log_e (A - 4) + \log_e A = \log_e 5$
34. $\log (T - 3) + \log T = \log 18$

35. In a certain electric circuit, there are two resistances. The sum of these resistances is 30 ohms and their product is 216 ohms. Find the resistances.

36. A certain rectangular field has a perimeter of 280 meters. A diagonal of the field is 100 meters. Find the length and width of the field.

37. When a projectile is fired into the air, its distance above the ground (s) in terms of time (t) is given by the equation $s = 100t - 16t^2$. How long after the projectile is fired will it be 100 feet above the ground?

38. The surface area of a right circular cylinder is given by the formula $A = 2\pi RH + 2\pi R^2$. Find the radius (R) of a cylinder which has a height of 4 feet and a surface area of 48 square feet.

39. In a certain electric circuit, current (I), voltage (E), and resistance (R) are related by the equation $RI^2 + EI = 900$. If $R = 15$ ohms and $E = 60$ volts, find I.

40. A certain rectangle has a length of T centimeters and a width of $\sqrt{T}$ centimeters. If the perimeter is 40 centimeters, find the length and width.

41. A 3-centimeter square is cut out of each corner of a square sheet of metal and the sides are then turned up to form a rectangular container. If the volume of the container is 186 cubic centimeters, what should be the dimensions of the container?

chapter 14

Systems of Equations

14–1. Systems of Equations

Systems of two linear equations in two variables, where the graph of each equation represents a straight line, were solved in Chapter 6. This chapter will deal with the solution to systems of two equations in two variables, where the equations represent geometric figures other than straight lines.

A *system of two equations* refers to the situation in which two equations are related to each other in some way and a common solution to the equations is desired. The solution to the system of equations must satisfy each equation in the system.

Example A: Is $X = -1$ and $Y = -3$ a solution to the system of equations?

$$-2X^2 + X + 6 = Y$$
$$3X + 3 = Y$$

This question can be answered by checking the given values in each equation.

$$-2X^2 + X + 6 = Y$$
$$-2(-1)^2 + (-1) + 6 = -3$$

Sec. 14–1. Systems of Equations

$$-2 + (-1) + 6 = -3$$
$$3 \neq -3$$
$$3X + 3 = Y$$
$$3(-1) + 3 = -3$$
$$0 \neq -3$$

Neither equation is satisfied with the given values. Therefore, ($X = -1$ and $Y = -3$) is *not* a solution to this system of equations.

Example B: Is $X = 2$ and $Y = -3$ a solution to the system of equations?

$$X^2 + Y^2 = 13$$
$$X^3 - 11 = Y$$

or

$$X^2 + Y^2 = 13$$
$$(2)^2 + (-3)^2 = 13$$
$$4 + 9 = 13$$
$$13 = 13 \checkmark$$

The given values satisfy this equation.

$$X^3 - 11 = Y$$
$$(2)^3 - 11 = -3$$
$$8 - 11 = -3$$
$$-3 = -3 \checkmark$$

The given values satisfy this equation. Therefore, $X = 2$ and $Y = -3$ is a solution to this system of equations.

There are two techniques for finding the solution to a system of equations: one graphical and the other algebraic. The remainder of this chapter will deal with these techniques.

Exercises Determine whether the given values are a solution to the system of equations.

1. $2X - Y = 1$ $X = 0$ and $Y = -1$
   $3X^2 - Y^2 = 1$

2. $X^2 + Y^2 = 4$ $X = 1$ and $Y = 1$
   $Y = X^3$

3. $A = 2B^2 - 5B - 6$ $A = 0$ and $B = -\frac{1}{2}$
   $A^2 - 2B^2 = 6$

4. $R^2 + 3S^2 = 1$ $R = 0$ and $S = 1$
    $R = -3S + 1$

5. $2T^2 + 2V^2 = 8$ $T = 0$ and $V = 2$
    $T = 2V - 4$

6. $K^2 - 3L^2 = 6$ $K = 1$ and $L = 0$
    $2K^2 + 4L^2 = 9$

7. $3P + Q = 8$ $P = 2$ and $Q = 2$
    $4P^2 + Q = -6$

8. $5X^2 + 2Y = 10$ $X = 0$ and $Y = 5$
    $3X - 2Y^2 = 6$

9. $N = -3M^2 + 1$ (a) $N = -1$ and $M = 2$
    $N - 2M = 0$ (b) $N = \frac{1}{3}$ and $M = \frac{2}{3}$

10. $A^3 - 9A = B$ (a) $A = -3$ and $B = 0$
    $B = A^2 + 3A$ (b) $A = 0$ and $B = 0$
         (c) $A = 3$ and $B = 1$

14-2. Graphical Solution of Systems of Equations

In order to solve a system of equations graphically it is necessary to plot the graph of each equation in the rectangular coordinate system and determine the points of intersection, if any, of the graphs. In constructing the graphs, we should make use of all information we have discussed previously.

Example A: Solve the following system of equations graphically.

$$Y = 2X^2 - 1$$
$$X^2 + Y^2 = 4$$

X will be considered the independent variable and will be plotted horizontally. Graph the first equation by techniques learned in Chapter 13. The axis of symmetry is $X = 0$ with the vertex at $(0, -1)$. Additional points are determined by using a table of values.

X	1	−1	2	−2
Y	1	1	7	7

Graph the second equation by solving for Y, determining the X- and Y-intercepts and any excluded values, and constructing a table of values.

1. $Y = \pm\sqrt{4 - X^2}$

2. X-intercept $= \pm 2$ (found by letting $Y = 0$)
   Y-intercept $= \pm 2$ (found by letting $X = 0$)
3. $-2 \leq X \leq 2$ since other values of X will yield values of Y that are imaginary numbers.
4.

X	-1	1
Y	$\pm\sqrt{3}$	$\pm\sqrt{3}$

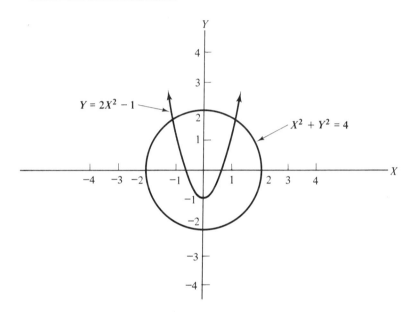

Read from the graph the coordinates of the points of intersection. A graphical solution should yield an answer precise to one decimal place. The coordinates of the points of intersection for this system of equations are $(1.2, 1.6)$ and $(-1.2, 1.6)$. If these ordered pairs are inserted into the original system of equations, they will not check because the ordered pairs are only approximate. Before deciding to solve a system of equations graphically, you must be willing to accept an answer that will be approximate. If an exact answer is desired, you will have to use an algebraic technique. The graphical technique will, at times, be the most appropriate method to use, since the algebraic technique can be very difficult to execute with certain types of equations.

Example B: Solve the following system of equations graphically.

$$9A^2 + 4B^2 = 25$$
$$B = 3^A$$

A will be considered the independent variable and will be plotted horizontally. Graph the first equation by solving for B, determining the A- and B-intercepts and any excluded values, and constructing a table of values.

1. $B = \pm\sqrt{\dfrac{25 - 9A^2}{4}}$

2. A-intercept $= \pm\tfrac{5}{3}$ (found by letting $B = 0$)
   B-intercept $= \pm\tfrac{5}{2}$ (found by letting $A = 0$)

3. $-\tfrac{5}{3} \le A \le \tfrac{5}{3}$ since other values of A will yield values of B that are imaginary numbers.

4.

A	1	−1
B	±2	±2

Graph the equation $B = 3^A$. Since this is an exponential function it is defined for all values of A.

A	−2	−1	0	1	2
B	$\tfrac{1}{9}$	$\tfrac{1}{3}$	1	3	9

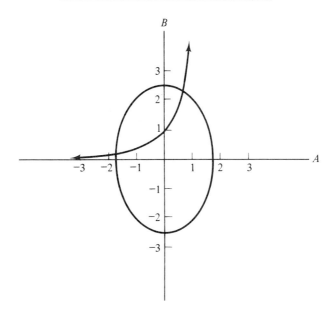

The coordinates of the points of intersection are $(-1.6, .4)$ and $(.5, 2.4)$.

Sec. 14–2. Graphical Solution of Systems of Equations

Example C: Solve the following system of equations graphically.

$$R = T^2 - 3T + 4$$
$$(T + 1)^2 + R^2 = 9$$

T will be considered the independent variable and will be plotted horizontally. Graph the first equation by determining the axis of symmetry, vertex, and a few other points.
  1. $T = \frac{3}{2}$ is the equation for the axis of symmetry.
  2. $(\frac{3}{2}, \frac{7}{4})$ is the vertex.
  3.

T	0	1	2	3
R	4	2	2	4

Graph the second equation by solving for R, finding the intercepts, excluded values, and a table of values.
  1. $R^2 = 9 - (T^2 + 2T + 1)$.
     $R = \pm\sqrt{8 - T^2 - 2T}$.
  2. The T-intercepts are found by letting $R = 0$.

$$T^2 + 2T - 8 = 0 \quad \text{(since 0 and } -0 \text{ are equivalent)}$$
$$(T + 4)(T - 2) = 0$$
$$T + 4 = 0 \qquad T - 2 = 0$$
$$T = -4 \qquad T = 2$$

The R-intercepts are found by letting $T = 0$.

$$(0 + 1)^2 + R^2 = 9$$
$$R^2 = 8$$
$$R = \pm 2\sqrt{2} = \pm 2.8$$

  3. $-4 \leq T \leq 2$ since any other values of T would yield imaginary values for R.
  4.

T	−3	−2	−1	1
R	±2.24	±2.8	±3	±2.24

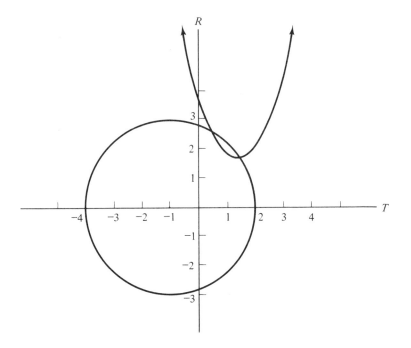

Read the coordinates of the points of intersection: they are (.8, 2.2) and (1.5, 1.8).

Example D: Solve the following system of equations graphically.

$$X^2 - Y^2 = 4$$
$$Y = \log_2 X$$

X will be considered the independent variable and will be plotted horizontally. Graph the first equation by solving for Y, finding the intercepts, excluded values, and a partial table of values.

1. $Y = \pm\sqrt{X^2 - 4}$
2. X-intercept: $X = \pm 2$ (found by letting $Y = 0$)
   Y-intercept: $Y = \pm\sqrt{-4}$ (found by letting $X = 0$)
   Since -4 is an imaginary number, there are no Y-intercepts.
3. $X \leq -2$ or $X \geq 2$ since any other values would yield imaginary numbers for Y.
4.

X	3	-3	4	-4
Y	±2.24	±2.24	±3.5	±3.5

Sec. 14–2. Graphical Solution of Systems of Equations

Graph the second equation by changing the equation to its equivalent exponential form and making a table of values.

1. $Y = \log_2 X \longrightarrow 2^Y = X$.
2.

X	$\frac{1}{8}$	$\frac{1}{4}$	$\frac{1}{2}$	1	2	4	8
Y	-3	-2	-1	0	1	2	3

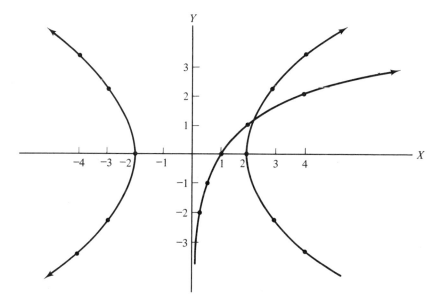

Read the coordinates of the point of intersection: they are (2.2, 1).

If there are *no points of intersection* of the two graphs, then there is *no real solution* to the system of equations.

Exercises

Solve the following systems of equations by graphing both equations in the rectangular coordinate system. Read the solutions to one decimal place.

1. $Y = -X + 2$
   $Y = X^2 + 2$

2. $X^2 + Y^2 = 9$
   $2X^2 + 5Y^2 = 14$

3. $B = 2A^2 - A + 1$
   $B^2 + A^2 = 4$

4. $Y = \dfrac{1}{X}$
   $X^2 + Y^2 = 1$

5. $R = S^3$
   $R = S$

6. $16A^2 - 9B^2 = 144$
   $A^2 + B^2 = 9$

7. $M = N^4 - 4N^2$
   $M = 3^N$

8. $A = \log_{10} B$
   $A = \log_{1/2} B$

9. $R = 3S^3 - 1$
   $RS = 4$

10. $Y = 2X^2 - 3X + 1$
    $Y = \operatorname{Sin} X$

11. $M = \operatorname{Cos} N$
    $M = 3^N$

12. $X^2 + Y^2 = 16$
    $X^2 - Y - 4 = 0$

13. $Y = \operatorname{Sin} X$
    $Y = \operatorname{Cos} X$

14. $M = 3N - 1$
    $M = \log_3 N$

For each of the following, set up a system of equations and solve graphically.

15. If the area of a circle equals πR^2 and the circumference of a circle equals $2\pi R$, for what value of R will the circumference be "numerically" equal to the area?

16. The result when the squares of two positive numbers are added is 65. The difference of these squares is 33. Find the two numbers.

17. A surveyor finds that a piece of property shaped like a right triangle has an area of 600 square meters. If the hypotenuse is 170 meters, find the dimensions of the legs of the triangle.

18. The pH value is defined by the relation pH $= -\log (\mathrm{H}^+)$, where H^+ is the hydrogen ion concentration. Construct the graph of this function and determine the range of values of H^+ when the pH is greater than 1 and the range of values of H^+ when the pH is less than 1.

14–3. Algebraic Solutions of Systems of Equations

The methods that will be employed to solve a system of equations algebraically are the same methods used in earlier chapters of this book. These are the elimination method, substitution method, and the methods explained in Chapter 13 for solving quadratic equations. At this time, we will not be able to solve algebraically all equations that were solved by graphical techniques. In this section we will consider algebraic solutions to combinations of linear and quadratic equations.

Example A: Solve the following system of equations algebraically.

$$X^2 + Y^2 = 16$$
$$Y = X^2 + 4$$

Since both equations have an X^2 term, the elimination method will be used.

Sec. 14–3. Algebraic Solutions of Systems of Equations

$$X^2 + Y^2 = 16$$
$$\underline{Y - X^2 = 4}$$
$$Y + Y^2 = 20$$

Since this new equation is a quadratic and can be factored, the solution is:

$$Y^2 + Y - 20 = 0$$
$$(Y + 5)(Y - 4) = 0$$
$$Y + 5 = 0 \qquad\qquad Y - 4 = 0$$
$$Y = -5 \qquad\qquad Y = 4$$

Each of these values must be used in either one of the original equations in order to find the corresponding values of X.

$$X^2 + Y^2 = 16 \qquad\qquad X^2 + Y^2 = 16$$
$$X^2 + (-5)^2 = 16 \qquad\qquad X^2 + (4)^2 = 16$$
$$X^2 = 16 - 25 \qquad\qquad X^2 = 0$$
$$X^2 = -9 \qquad\qquad X = 0$$
$$X = \pm\sqrt{-9}$$

Since substituting $Y = -5$ results in X being an imaginary number, $Y \neq -5$. Therefore, the solution to the system of equations is $X = 0$ and $Y = 4$. This solution would correspond to a single point of intersection of the graphs of the equations in this system.

Example B: Solve the following system of equations algebraically.

$$X^2 + 4Y^2 = 4$$
$$Y = \frac{X}{2} + 1$$

This system will be solved by substituting $\frac{X}{2} + 1$ in place of Y in the first equation.

$$X^2 + 4\left(\frac{X}{2} + 1\right)^2 = 4$$
$$X^2 + 4\left(\frac{X^2}{4} + X + 1\right) = 4$$
$$X^2 + X^2 + 4X + 4 = 4$$

Since this equation is quadratic and factorable it can be solved by the factoring technique.

$$2X^2 + 4X = 0$$
$$2X(X+2) = 0$$
$$2X = 0 \qquad X + 2 = 0$$
$$X = 0 \qquad X = -2$$

Substituting each of these values in either one of the equations of the system will yield corresponding values for Y.

$$Y = \frac{X}{2} + 1 \qquad Y = \frac{X}{2} + 1$$
$$= \frac{0}{2} + 1 \qquad = \frac{-2}{2} + 1$$
$$= 0 + 1 \qquad = -1 + 1$$
$$= 1 \qquad = 0$$

Therefore, $X = 0$ and $Y = 1$, and $X = -2$ and $Y = 0$ are the solutions to this system of equations. This solution would correspond to two points of intersection of the graphs of the equations in this system.

Example C: Solve the following system of equations algebraically.

$$A^2 - B^2 = 25$$
$$B = \frac{A}{3} - 3$$

Substituting for B in the first equation will yield a quadratic equation that must be solved by using the quadratic formula.

$$A^2 - \left(\frac{A}{3} - 3\right)^2 = 25$$
$$A^2 - \left(\frac{A^2}{9} - 2A + 9\right) = 25$$
$$A^2 - \frac{A^2}{9} + 2A - 9 = 25$$
$$9A^2 - A^2 + 18A - 81 = 225$$
$$8A^2 + 18A - 306 = 0$$
$$2(4A^2 + 9A - 153) = 0$$
$$4A^2 + 9A - 153 = 0$$

Sec. 14–3. *Algebraic Solutions of Systems of Equations*

$$A = \frac{-9 \pm \sqrt{81 + 2448}}{8}$$

$$= \frac{-9 \pm \sqrt{2529}}{8} = \frac{-9 \pm 50.29}{8}$$

$$= \frac{41.29}{8} \text{ and } \frac{-59.29}{8}$$

$$= 5.16 \text{ and } -7.41$$

$$B = \frac{5.16}{3} - 3 \text{ and } \frac{-7.41}{3} - 3$$

$$= -1.28 \text{ and } -5.47$$

Therefore, $A = 5.16$ and $B = -1.28$, $A = -7.41$ and $B = -5.47$ are the solutions to this system of equations. This solution would correspond to two points of intersection between the graphs of the equations in the system. It should also be noted that this solution is not exact since the roots determined from the quadratic formula were irrational numbers, and therefore, the roots are only approximate.

Example D: Solve the following system of equations algebraically. Use N as the independent variable.

$$4M^2 - 9N^2 = 36$$
$$9M^2 + 25N^2 = 225$$

This system is easily solved by eliminating M^2 and substituting a value for M into either one of the equations to find the corresponding values of N.

$$9(4M^2 - 9N^2) = (36)(9)$$
$$-4(9M^2 + 25N^2) = (225)(-4)$$
$$36M^2 - 81N^2 = 324$$
$$-36M^2 - 100N^2 = -900$$
$$\overline{-181N^2 = -576}$$

$$N^2 = \frac{-576}{-181} = 3.18$$

$$N = \pm 1.78$$

Substituting $N^2 = 3.18$ into either one of the original equations will yield a set of corresponding values for M.

$$4M^2 - 9(3.18) = 36$$
$$4M^2 = 36 + 28.62 = 64.62$$

$$M^2 = \frac{64.62}{4} = 16.16$$

$$M = \pm\sqrt{16.16} = \pm 4.02$$

The solution to the system of equations consists of four ordered pairs of numbers determined by associating each of the two values of M with each of the two values of N. Thus, the four solutions and four points of intersection, determined graphically, are (1.78, 4.02), (1.78, −4.02), (−1.78, 4.02), and (−1.78, −4.02). It should be noted that N and M are only approximate values.

Exercises Solve each system of equations by an appropriate algebraic method.

1. $Y = -X + 1$
    $X^2 + Y^2 = 4$

2. $2A^2 - 4 = B$
    $2A^2 + B^2 = 4$

3. $R = \frac{T}{2} + 3$
    $R = T^2 + \frac{1}{2}$

4. $M^2 - N^2 = 1$
    $N = 4M^2$

5. $S^2 + T^2 = 2$
    $2T^2 - S^2 = 4$

6. $A = \frac{1}{B}$
    $A = 2B + 1$

7. $Y = \sqrt{X}$
    $Y = X$

8. $A = B$
    $A = B^3$

9. $M^3 - 4M = N$
    $N = 2M$

10. $R = \frac{-1}{T}$
    $R = T + 1$

11. Ninety-six feet of fencing are required to enclose a rectangular area of 560 square feet. Find the dimensions of the area.

12. In the accompanying diagram, find A and B if it is known that $A^2 - B^2 = 119$.

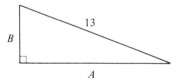

14-4. CHAPTER REVIEW

Solve the following systems of equations graphically.

1. $2Y - 3X + 4 = 0$
    $X^2 + Y^2 - 4 = 0$

2. $B = \frac{1}{2}A$
    $B = A^3$

Sec. 14–4. Chapter Review

3. $X = Y^2$
   $Y = X^2$

4. $R = T^2 - 4T + 5$
   $R = 2T^3 + T + 2$

5. $M^2 + N^2 = 16$
   $N = \frac{1}{2}M^2 - 2$

6. $4X^2 + 6Y^2 = 48$
   $X^2 + Y^2 = 4$

7. $A^2 - B^2 = 8$
   $B = A^2 - 1$

8. $Y^2 - 3X^2 = 16$
   $X^2 + Y^2 = 16$

9. $A = (\frac{1}{2})B$
   $A = \log_2 B$

10. $M^3 - 3M - N = 0$
    $M^2 + 9N^2 - 3 = 0$

Solve the following systems of equations algebraically.

11. $Y = X^2$
    $X^2 + Y^2 = 16$

12. $A = \sqrt{2}B$
    $A^2 - 3B^2 = -9$

13. $R = \frac{4}{S}$
    $R^2 + S^2 = 8$

14. $R = T^3 - 2T$
    $R = T$

15. $Y = X^4 - 4X^2$
    $Y = 2X^2$

16. $N = M^3$
    $N = M^3 + 5M^2 + 6M$

17. $3P^2 - Q^2 = 4$
    $P^2 + 4Q^2 = 10$

18. $4K^2 + 3L^2 = 12$
    $5K^2 - 4L^2 = 15$

19. $3A - 2B = -7$
    $AB = 5$

20. $T + 3V = 4$
    $4T^2 + 3T - 4 = V$

For each of the following, set up a system of equations and solve by an appropriate method.

21. A certain shipment of resistors costs $720. If the resistors cost $1.00 less each, an additional 36 resistors could be purchased. Find the current price of a resistor.

22. In a certain right triangle, the hypotenuse is 10 and one side is 3 times as long as the other. Find the lengths of the two sides.

chapter 15

Equations of Degree Greater than Two

15–1. The Remainder and Factor Theorems

In Section 2–3, we defined a polynomial as an algebraic expression in which each term is of the type AX^N, where X could be replaced by any variable and N must be zero or a positive integer. In Section 13–1, we defined a polynomial equation as an equation in which each term is of the type AX^N. The degree of a polynomial equation is the largest value of N that appears in the equation. Every polynomial equation of degree N will be a function and may be written in the form

$$f(X) = A_N X^N + A_{N-1} X^{N-1} + \ldots + A_2 X^2 + A_1 X + A_0$$

Example A:
1. $f(X) = 3X - 4$ is a polynomial equation of degree 1 in X.
2. $f(Y) = Y^2 - 2Y + 6$ is a polynomial equation of degree 2 in Y.
3. $f(T) = T^3 + 8T - 7$ is a polynomial equation of degree 3 in T.
4. $f(K) = K^5 + 5K^4 + 2K$ is a polynomial equation of degree 5 in K.

Polynomial equations of degree 1 are linear equations, which we have covered in detail in Chapter 6. Polynomial equations of degree 2 are quadratic

Sec. 15–1. The Remainder and Factor Theorems

equations, which we have talked about in Chapter 13. In this chapter we will discuss polynomial equations of degree greater than 2.

A polynomial of the type $f(X)$ may be divided by a factor of the type $(X - r)$ to obtain a quotient, $q(X)$, and a remainder (R). This division may be written as

$$\frac{f(X)}{X - r} = q(X) + \frac{R}{X - r} \quad \text{or} \quad f(X) = (X - r)q(X) + R$$

To perform the division, we employ the process of long division in the usual manner.

Example B: Divide $(2X^3 + 5X^2 - 7X + 1)$ by $(X - 2)$.

$$
\begin{array}{r}
2X^2 + 9X + 11 \\
X - 2 \overline{\smash{)}2X^3 + 5X^2 - 7X + 1} \\
\underline{2X^3 - 4X^2 } \\
9X^2 - 7X \\
\underline{9X^2 - 18X } \\
11X + 1 \\
\underline{11X - 22} \\
23
\end{array}
$$

Therefore, $(2X^3 + 5X^2 - 7X + 1) = (X - 2)(2X^2 + 9X + 11) + 23$.

Example C: Divide $(X^4 - 5X^2 + 6)$ by $(X + 1)$.

$$
\begin{array}{r}
X^3 - X^2 - 4X + 4 \\
X + 1 \overline{\smash{)}X^4 - 5X^2 + 6} \\
\underline{X^4 + X^3 } \\
-X^3 - 5X^2 \\
\underline{-X^3 - X^2 } \\
-4X^2 \\
\underline{-4X^2 - 4X } \\
4X + 6 \\
\underline{4X + 4} \\
2
\end{array}
$$

Therefore, $(X^4 - 5X^2 + 6) = (X + 1)(X^3 - X^2 - 4X + 4) + 2$.

If $X = r$, then the division, in general, can be written as $f(r) = (r - r)q(r) + R$ or $f(r) = R$. Thus, the remainder (R) is equal to the polynomial evaluated for r. This fact is extremely important and is called the *remainder theorem*. It states that *if a polynomial $f(x)$ is divided by a factor of the type $(X - r)$ until a constant remainder R is obtained, then $f(r) = R$.*

Example D:
1. Referring to Example B:
$$f(2) = 2(2)^3 + 5(2)^2 - 7(2) + 1 = 16 + 20 - 14 + 1 = 23.$$
2. Referring to Example C:
$$f(-1) = (-1)^4 - 5(-1)^2 + 6 = 1 - 5 + 6 = 2.$$
NOTE: Since $(X - r) = (X + 1)$, $r = -1$.

A special case of the remainder theorem is the *factor theorem*. This theorem states that *if $f(r) = R = 0$, then $(X - r)$ is a factor of $f(X)$ and r is a zero of $f(X)$.*

Example E: Is $(X - 2)$ a factor of $(X^3 - 4X^2 + X + 6)$?
$$f(2) = (2)^3 - 4(2)^2 + (2) + 6 = 8 - 16 + 2 + 6 = 0$$
Therefore, $(X - 2)$ is a factor of $(X^3 - 4X^2 + X + 6)$ and 2 is a zero of $(X^3 - 4X^2 + X + 6)$.

Example F: Is $(T + 3)$ a factor of $(3T^4 + 5T^3 - 3T)$?
$$f(-3) = 3(-3)^4 + 5(-3)^3 - 3(-3) = 243 - 135 + 9 = 117$$
Therefore, $(T + 3)$ is not a factor of $(3T^4 + 5T^3 - 3T)$ and -3 is not a zero of $(3T^4 + 5T^3 - 3T)$.

Example G: Is $(2K - 1)$ a factor of $(6K^3 - K^2 - 21K + 10)$? Here $(2K - 1)$ is not in the form $(X - r)$. Therefore, we must write $(2K - 1)$ as $2(K - \frac{1}{2})$. Thus, if $(K - \frac{1}{2}) = 0$, then $(2K - 1) = 0$.
$$f(\tfrac{1}{2}) = 6(\tfrac{1}{2})^3 - (\tfrac{1}{2})^2 - 21(\tfrac{1}{2}) + 10 = \tfrac{3}{4} - \tfrac{1}{4} - \tfrac{42}{4} + \tfrac{40}{4} = 0$$
Therefore, $(K - \frac{1}{2})$, and hence, $(2K - 1)$, are factors of $(6K^3 - K^2 - 21K + 10)$ and $\frac{1}{2}$ is a zero of $(6K^3 - K^2 - 21K + 10)$.

Exercises

For each of the following, find the remainder by long division and by the remainder theorem.
1. $(X^3 - 5X^2 - 10) \div (X - 2)$
2. $(7X^5 + 8X^2 - 3X) \div (X + 2)$

Sec. 15–2. Synthetic Division   427

3. $(Y^3 + 2Y^2 - 5Y - 6) \div (Y + 1)$
4. $(M^5 - 1) \div (M - 1)$
5. $(2T^3 - 3T^2 + 8T - 12) \div (T - 3)$
6. $(3P^4 + 10P^3 - 13P^2 - 20P) \div (3P - 5)$
7. $(R^4 + 4R^3 - 7R^2 - 22R + 24) \div (2R + 1)$
8. $(K^3 - 4K^2 + K + 6) \div (K - 3)$

For each of the following, use the factor theorem to determine whether or not the second expression is a factor of the first.

9. $(5Y^3 - 3Y^2 + 4), (Y + 2)$
10. $(R^4 - 2R^2 + 4R - 2), (2R + 1)$
11. $(T^2 - 2T - 3), (T - 3)$
12. $(4X^3 + X^2 - 16X - 4), (X - 2)$
13. $(P^5 + 1), (P + 1)$
14. $(2K^2 + 5K - 3), (2K - 1)$
15. $(3P^3 + 14P^2 + 7P - 4), (P + 4)$
16. $(S^3 + 4S^2 + 6S + 8), (3S - 4)$
17. $(Q^4 - 2Q^3 - 8), (Q + 2)$
18. $(M^3 + M^2 - M + 15), (M + 3)$

For each of the following, determine whether the given numbers are zeros of the polynomials.

19. $(2R^3 + 5R - 1), -1$
20. $(X^3 - 2X^2 - 9X + 18), 2$
21. $(R^5 - 4R^3 + 2R^2 - 6), 1$
22. $(2T^3 + 3T^2 - 8T - 12), \frac{1}{2}$
23. $(K^4 + 4K^3 - 7R^2 - 22R + 24), -4$
24. $(4Q^3 + 8Q^2 - 15Q - 9), -\frac{1}{2}$

15–2. Synthetic Division

In the last section we saw how the remainder theorem may be used to find the remainder when a polynomial, $f(X)$, is divided by a factor of the type $(X - r)$. If r is a large number or a fraction, or if the exponents in the polynomial are large, this procedure could get quite involved. Also, when testing roots of a polynomial equation, we must know the quotient as well as the remainder. Therefore, in this section we will look at another way of dividing $f(X)$ by $(X - r)$.

Consider $(2X^3 - 5X^2 + 6X - 7) \div (X - 1)$. The long division process follows.

$$
\begin{array}{r}
2X^2 - 3X + 3 \\
X - 1 \overline{)\, 2X^3 - 5X^2 + 6X - 7} \\
\underline{2X^3 - 2X^2 } \\
-3X^2 + 6X \\
\underline{-3X^2 + 3X } \\
3X - 7 \\
\underline{3X - 3} \\
-4
\end{array}
$$

In performing this division, we really worked only with the coefficients and, in the multiplication and subtraction operations, repeated many of them. Therefore, let us rewrite the division, omitting the variables and any identical terms.

$$
\begin{array}{r}
2 \;\; -3 \;\;\; 3 \\
-1 \overline{)\, 2 \;\; -5 \;\;\; 6 \;\; -7} \\
\underline{-2 } \\
-3 \\
\underline{ 3 } \\
3 \\
\underline{-3} \\
-4
\end{array}
$$

If we now move all the numbers up we obtain a more concise form.

$$
\begin{array}{r}
2 \;\; -3 \;\;\; 3 \\
-1 \overline{)\, 2 \;\; -5 \;\;\; 6 \;\; -7} \\
\underline{-2 \;\;\;\; 3 \;\; -3} \\
-3 \;\;\;\; 3 \;\; -4
\end{array}
$$

All of the coefficients of the quotient now appear in the bottom line except the 2. Therefore, we will repeat the 2 in the bottom line and omit the quotient above the dividend. Also, we will change the (-1) to 1, the actual value of r, and move it to the right. Our division problem is now written as follows.

$$
\begin{array}{r}
2 \;\; -5 \;\;\; 6 \;\; -7 \;\; \underline{|\,1} \\
\underline{-2 \;\;\;\; 3 \;\; -3 } \\
2 \;\; -3 \;\;\; 3 \;\; -4
\end{array}
$$

The numbers in the bottom line were originally found by subtraction. Since it is usually easier to add than subtract, let us change the signs of the numbers in the middle row and add. Our problem now becomes

Sec. 15–2. Synthetic Division

$$\begin{array}{rrrr|r}
2 & -5 & 6 & -7 & \underline{1} \\
  & 2 & -3 & 3 & \\
\hline
2 & -3 & 3 & -4 &
\end{array}$$

We should now note that the first three numbers in the bottom line, 2, −3, 3, are the coefficients of the quotient $(2X^2 - 3X + 3)$, and the last number, −4, is the remainder. If we look at the very last form of the division problem, we may state exactly what we have done. Write down the coefficients of $f(X)$ in descending powers of the variable, 2, −5, 6, −7 and write r (1) to the right. Rewrite the first coefficient, 2, in the bottom line. Multiply 1 times 2 and write the result, 2, in the middle row. Add 2 to −5 to obtain −3. Multiply 1 times −3 and write the result, −3, in the middle row. Add −3 to 6 to obtain 3. Multiply 1 times 3 and write the result, 3, in the middle row. Add 3 to −7 to obtain −4. This process is called *synthetic division* and the procedure may be formalized as follows.

To divide a polynomial $f(X)$ by a factor of the type $(X - r)$:

1. Write down the coefficients of $f(X)$ in descending powers of X. If any power of X is missing, use zero as a coefficient for that power. Then write r to the right.
2. Rewrite the first coefficient of $f(X)$ in the bottom row.
3. Multiply this first coefficient times r and write the result in the middle row under the second coefficient of $f(X)$.
4. Add these two numbers and write the result in the bottom row.
5. Now successively multiply the number in the bottom row times r, write the result in the middle row, and add it to the appropriate coefficient of $f(X)$ until there are as many numbers in the bottom row as the top row.
6. The last number on the right in the bottom row will be the remainder, while the other numbers read from left to right will be the coefficients, in descending powers of X, of the quotient.

Example A: Divide $(X^3 - 3X^2 - X + 2)$ by $(X - 2)$.

$$\begin{array}{rrrr|r}
1 & -3 & -1 & 2 & \underline{2} \\
  & 2 & -2 & -6 & \\
\hline
1 & -1 & -3 & -4 &
\end{array}$$

Therefore, the quotient is $(X^2 - X - 3)$ and the remainder is −4.

Example B: Divide $(T^4 - 5T^3 + T^2 - 2T + 6)$ by $(T + 4)$.

$$\begin{array}{rrrrr|r}
1 & -5 & 1 & -2 & 6 & \underline{-4} \\
  & -4 & 36 & -148 & 600 & \\
\hline
1 & -9 & 37 & -150 & 606 &
\end{array}$$

Therefore, the quotient is $(T^3 - 9T^2 + 37T - 150)$ and the remainder is 606.

Example C: Divide $(K^5 + 40)$ by $(K + 2)$.

$$\begin{array}{rrrrrr|r} 1 & 0 & 0 & 0 & 0 & 40 & -2 \\ & -2 & 4 & -8 & 16 & -32 & \\ \hline 1 & -2 & 4 & -8 & 16 & 8 & \end{array}$$

Therefore, the quotient is $(K^4 - 2K^3 + 4K^2 - 8K + 16)$ and the remainder is 8.

NOTE: Zeros represent the coefficients of the missing powers of K.

Example D: Divide $(4Y^3 + Y^2 - 16Y - 4)$ by $(Y - 2)$.

$$\begin{array}{rrrr|r} 4 & 1 & -16 & -4 & 2 \\ & 8 & 18 & 4 & \\ \hline 4 & 9 & 2 & 0 & \end{array}$$

Therefore, the quotient is $(4Y^2 + 9Y + 2)$ and the remainder is 0.

NOTE: Since the remainder is 0, $(Y - 2)$ is a factor of $(4Y^3 + Y^2 - 16Y - 4)$ and 2 is a zero of $(4Y^3 + Y^2 - 16Y - 4)$.

Example E: Divide $(3B^4 - 2B^3 + 6B^2 - 8)$ by $(2B - 1)$.

$$(2B - 1) = 2(B - \tfrac{1}{2})$$

Therefore, $r = \tfrac{1}{2}$.

$$\begin{array}{rrrrr|r} 3 & -2 & 6 & 0 & -8 & \tfrac{1}{2} \\ & 1.5 & -.25 & 2.875 & 1.4375 & \\ \hline 3 & -.5 & 5.75 & 2.875 & -6.5625 & \end{array}$$

The quotient is $(3B^3 - .5B^2 + 5.75B + 2.875)$ and the remainder is -6.5625.

Exercises

For each of the following, find the quotient and remainder using synthetic division.

1. $(2X^5 - X^4 + 3X^3 + X^2 - 6X - 9) \div (X - 1)$
2. $(5T^3 + 8T^2 - T - 10) \div (T + 2)$
3. $(Y^4 - 4Y^3 - Y^2 + Y - 100) \div (Y + 3)$
4. $(V^4 + V^3 - 2V^2 - 5V + 3) \div (V + 4)$
5. $(2A^4 - A^2 + 5A - 7) \div (A - 3)$
6. $(L^6 + 2L^3 + 3) \div (L + 1)$

Sec. 15–3. Number and Nature of the Roots of a Polynomial Equation

7. $(2Z^4 - Z^3 + 2Z^2 - 3Z + 1) \div (2Z - 1)$
8. $(6J^4 + 5J^3 - J^2 + 6J - 2) \div (3J - 1)$
9. $(K^5 + 1) \div (K + 1)$
10. $(7S^4 + 8S) \div (S + 3)$
11. $(B^3 + 6B^2 - 25B - 150) \div (B + 6)$
12. $(W^4 + 6W^3 + 5W^2 - 12W) \div (W + 3)$

For each of the following, use synthetic division to determine whether the second expression is a factor of the first.

13. $(T^3 - 1), (T - 1)$
14. $(3P^4 + 8P^2 - 9), (P + 1)$
15. $(M^5 \quad 32), (M + 2)$
16. $(Q^3 - 5Q^2 - 2Q + 24), (Q + 2)$

For each of the following, use synthetic division to determine whether the given numbers are zeros of the polynomials.

17. $(2R^4 - 6R^3 + R^2 - R + 7), 2$
18. $(3T^3 + 2T^2 - 4T + 1), \frac{1}{3}$
19. $(K^4 + 2K^3 - 15K^2 - 32K - 16), 4$
20. $(3Y^4 + 5Y + 6), -1$

15–3. Number and Nature of the Roots of a Polynomial Equation

Before we actually begin to find the roots of polynomial equations of degree greater than two, we will first develop some ideas that will make the root-finding procedure much easier.

First of all, we will state the *fundamental theorem of algebra: every polynomial equation has at least one root*. The root may be real or complex. This theorem can be proved in more advanced courses, but we will accept it here without proof.

Suppose we have a polynomial equation, $f(X) = 0$, and we are trying to find its roots. We know from the fundamental theorem that there is at least one root. Assume we have found this root by some means and call it r_1. Therefore, $f(X) = (X - r_1)f_1(X)$, where $f_1(X)$ is the polynomial quotient when $f(X)$ is divided by $(X - r_1)$. However, by the fundamental theorem, $f_1(X) = 0$ has at least one root which we will call r_2. Therefore, $f_1(X) = (X - r_2)f_2(X)$ and $f(X) = (X - r_1)(X - r_2)f_2(X)$. If we continue this process until a constant quotient, K, is obtained, we will have $f(X) = K(X - r_1)(X - r_2)\ldots(X - r_N)$. Each time a root is found, we obtain one linear factor and lower the degree of

Ch. 15. Equations of Degree Greater than Two

the quotient by one. Therefore, if $f(X)$ is of degree N, there will be N such factors. This observation leads to two theorems. First, *every polynomial of degree N can be factored into N linear factors.* Second, *every polynomial equation of degree N has exactly N roots.*

Example A: Consider the equation $f(X) = 3X^4 - X^3 - 39X^2 - 23X + 12 = 0$.

$$3X^4 - X^3 - 39X^2 - 23X + 12 = (X + 3)(3X^3 - 10X^2 - 9X + 4)$$
$$= (X + 3)(X + 1)(3X^2 - 13X + 4)$$
$$= (X + 3)(X + 1)(X - 4)(3X - 1)$$
$$= (X + 3)(X + 1)(X - 4)(X - \tfrac{1}{3})(3)$$

The degree of $f(X)$ is 4. There are four linear factors: $(X + 3), (X + 1), (X - 4)$, and $(X - \tfrac{1}{3})$. There are four roots of the equation: $-3, -1, 4$, and $\tfrac{1}{3}$.

Each time a root is found for a polynomial equation, the degree is lowered by one and we proceed with the quotient to find further roots. Once we reach the point where the quotient is of degree 2, we may always find the last two roots by using quadratic methods since a polynomial equation of degree 2 is a quadratic equation.

Example B: Solve the equation $3X^4 + 8X^3 - 35X^2 + 4X + 20 = 0$, given that 1 and 2 are roots.

```
      3    8   -35    4    20  | 1
                3    11  -24  -20
      ────────────────────────
      3   11   -24  -20    0   | 2
                6    34   20
      ────────────────────────
      3   17    10    0
```

Therefore,

$$(3X^4 + 8X^3 - 35X^2 + 4X + 20) = (X - 1)(X - 2)(3X^2 + 17X + 10) = 0$$

Therefore,

$$3X^2 + 17X + 10 = 0$$
$$(3X + 2)(X + 5) = 0$$
$$X = -\tfrac{2}{3}, -5$$

Therefore, the roots are $1, 2, -\tfrac{2}{3}$, and -5.

Example C: Solve the equation $T^4 - 21T^2 - 100 = 0$, given that 5 and -5 are roots.

Sec. 15–3. Number and Nature of the Roots of a Polynomial Equation

```
   1    0   -21    0   -100  |5
             5    25    20    100
   ─────────────────────────
   1    5     4    20     0  |-5
            -5     0   -20
   ─────────────────────────
   1    0     4     0
```

Therefore,

$$(T^4 - 21T^2 - 100) = (T - 5)(T + 5)(T^2 + 4) = 0$$

Therefore,

$$T^2 + 4 = 0$$
$$T^2 = -4$$
$$T = \pm\sqrt{-4}$$
$$T = \pm 2j$$

Therefore, the roots are 5, −5, 2*j*, and −2*j*.

Example D: Solve the equation $Y^4 - 2Y^3 - 7Y^2 - 12Y + 72 = 0$, given that 3 is a double root.

```
   1   -2   -7   -12    72  |3
         3    3   -12   -72
   ─────────────────────────
   1    1   -4   -24     0  |3
         3   12    24
   ─────────────────────────
   1    4    8     0
```

Therefore,

$$(Y^4 - 2Y^3 - 7Y^2 - 12Y + 72) = (Y - 3)(Y - 3)(Y^2 + 4Y + 8) = 0$$

Therefore,

$$Y^2 + 4Y + 8 = 0$$
$$Y = \frac{-4 \pm \sqrt{16 - 32}}{2}$$
$$= \frac{-4 \pm \sqrt{-16}}{2} = \frac{-4 \pm 4j}{2}$$
$$= -2 \pm 2j$$

Therefore, the roots are 3, 3, $-2 + 2j$, and $-2 - 2j$.

From these last two examples we should note that the last two roots obtained are complex and are conjugates of one another. In general, *any time* a complex number $(A + Bj)$ is a root of a polynomial equation, so is its conjugate $(A - Bj)$. This is true because any quadratic equation can be solved by the quadratic formula. When this formula is used, the two roots are of the form

$$\frac{-B + \sqrt{B^2 - 4AC}}{2A} \text{ and } \frac{-B - \sqrt{B^2 - 4AC}}{2A}$$

where the only difference between the roots is the sign preceding the radical. Therefore, *if there are any complex roots to a polynomial equation, there must be an even number of them and they must appear in conjugate pairs.*

Example E: Solve the equation $R^4 - 5R^3 - R^2 - 5R - 2 = 0$, given that j is a root.

$$\begin{array}{cccccc}
1 & -5 & -1 & -5 & -2 & \underline{|j} \\
 & j & -1-5j & 5-2j & 2 & \\
\hline
1 & -5+j & -2-5j & -2j & 0 & \underline{|-j} \\
 & -j & 5j & 2j & & \\
\hline
1 & -5 & -2 & 0 & &
\end{array}$$

Since j is a root, so is its conjugate $(-j)$.

Therefore, $R^4 - 5R^3 - R^2 - 5R - 2 = (R - j)(R + j)(R^2 - 5R - 2) = 0$.
Therefore, $R^2 - 5R - 2 = 0$.

$$R = \frac{5 \pm \sqrt{25 + 8}}{2}$$

$$= \frac{5 \pm \sqrt{33}}{2} = \frac{5 \pm 5.74}{2} = \frac{10.74}{2} \text{ or } \frac{-.74}{2} = 5.37 \text{ or } -.37$$

Therefore, the roots are j, $-j$, 5.37, and $-.37$.

Example F: Solve the equation $X^5 - 6X^4 + 22X^3 - 64X^2 + 117X - 90 = 0$, given that 2 and $2 + j$ are roots.

$$\begin{array}{cccccc}
1 & -6 & 22 & -64 & 117 & -90 \quad \underline{|2} \\
 & 2 & -8 & 28 & -72 & 90 \\
\hline
1 & -4 & 14 & -36 & 45 & 0 \quad \underline{|2+j} \\
 & 2+j & -5 & 18+9j & -45 & \\
\hline
1 & -2+j & 9 & -18+9j & 0 & \quad \underline{|2-j} \\
 & 2-j & 0 & 18-9j & & \\
\hline
1 & 0 & 9 & 0 & &
\end{array}$$

Since $(2 + j)$ is a root, so is its conjugate $(2 - j)$.

Therefore, $X^5 - 6X^4 + 22X^3 - 64X^2 + 117X - 90 = (X - 2)[X - (2 + j)][X - (2 - j)](X^2 + 9) = 0$.

Therefore,
$$X^2 + 9 = 0$$
$$X^2 = -9$$
$$X = \pm\sqrt{-9}$$
$$X = \pm 3j$$

Therefore, the roots are 2, $2 + j$, $2 - j$, $3j$, and $-3j$.

Exercises For each of the following, solve the equations given the roots indicated.
1. $X^3 + 2X^2 - 43X - 140 = 0$ (-5 is a root)
2. $T^3 - 12T^2 + 20T + 96 = 0$ (8 is a root)
3. $4Y^4 + 4Y^3 - 63Y^2 - 64Y - 16 = 0$ ($-\frac{1}{2}$ is a double root)
4. $6P^4 - 71P^3 + 146P^2 + 319P - 120 = 0$ ($\frac{1}{3}$ and 5 are roots)
5. $S^4 + 2S^3 - 4S^2 - 5S + 6 = 0$ (1 and -2 are roots)
6. $V^3 - 1 = 0$ (1 is a root)
7. $B^4 + 6B^3 - 19B^2 - 96B + 48 = 0$ (4 and -4 are roots)
8. $M^5 - M^4 - 2M^3 - M^2 + M + 2 = 0$ (1, -1, and 2 are roots)
9. $R^3 + 5R^2 + 9R + 5 = 0$ ($-2 + j$ is a root)
10. $A^5 + 11A^4 + 44A^3 + 176A^2 + 448A = 0$ ($4j$ and -7 are roots)
11. $Q^6 + 2Q^5 - 4Q^4 - 10Q^3 - 41Q^2 - 72Q - 36 = 0$ ($-1, -1,$ and $2j$ are roots)
12. $K^4 - 5K^3 + 32K^2 - 125K + 175 = 0$ ($-5j$ is a root)

15-4. Rational Roots of a Polynomial Equation

Suppose we have a polynomial equation $f(X) = A_N X^N + A_{N-1} X^{N-1} + \ldots + A_1 X + A_0 = 0$ with $A_N = 1$ and roots $r_1, r_2, \ldots r_N$. Then our equation may be written in factored form as $f(X) = (X - r_1)(X - r_2) \ldots (X - r_N)$. If these factors are multiplied together, the constant term of the result will be $(-r_1)(-r_2) \ldots (-r_N) = A_0$. Therefore, we can see that the roots, $r_1, r_2, \ldots r_N$, are factors of the constant term, A_0, in the original equation. Therefore, we may state that *when we have a polynomial equation of degree N and the coefficient of X is 1, then any integral roots of the equation must be factors of the constant term.*

Example A: Suppose $f(X) = X^3 - 3X^2 - 10X + 24 = 0$. Since the coefficient of X^3 is 1, any integral roots must be factors of 24. Therefore, the possible integral roots are $\pm 1, \pm 2, \pm 3, \pm 4, \pm 6, \pm 8, \pm 12$, and ± 24, since these are the possible factors of 24.

$$\begin{array}{rrrr|r} 1 & -3 & -10 & 24 & \underline{2} \\ & 2 & -2 & -24 & \\ \hline 1 & -1 & -12 & & \end{array}$$

$$X^2 - X - 12 = 0$$
$$(X - 4)(X + 3) = 0$$
$$X = 4, -3$$

Therefore, the roots are -3, 2, and 4, which are factors of 24.

$$f(X) = (X + 3)(X - 2)(X - 4) \text{ and } (3)(-2)(-4) = 24$$

Example B: Suppose $f(T) = T^4 - 5T^3 + 3T^2 - 45T - 54 = 0$. Since the coefficient of T^4 is 1, any integral roots must be factors of -54. Therefore, the possible integral roots are $\pm 1, \pm 2, \pm 3, \pm 6, \pm 9, \pm 18, \pm 27$, and ± 54.

$$\begin{array}{rrrrr|r} 1 & -5 & 3 & -45 & -54 & \underline{-1} \\ & -1 & 6 & -9 & 54 & \\ \hline 1 & -6 & 9 & -54 & & \underline{6} \\ & 6 & 0 & 54 & & \\ \hline 1 & 0 & 9 & & & \end{array}$$

$$T^2 + 9 = 0$$
$$T^2 = -9$$
$$T = \pm\sqrt{-9} = \pm 3j$$

Therefore, the roots are $-1, 6, -3j$, and $3j$. The only integral roots are -1 and 6 which are factors of -54.

$$f(T) = (T + 1)(T - 6)(T + 3j)(T - 3j) \quad \text{and} \quad (1)(-6)(3j)(-3j) = -54$$

Suppose we have a polynomial equation $f(X) = A_N X^N + A_{N-1} X^{N-1} + \ldots + A_1 X + A_0 = 0$ with $A_N \neq 1$ and roots $r_1, r_2, \ldots r_N$. We may rewrite our equation as

$$f(X) = A_N \left(X^N + \frac{A_{N-1} X^{N-1}}{A_N} + \ldots + \frac{A_1 X}{A_N} + \frac{A_0}{A_N} \right)$$

Then, in factored form the equation becomes $f(X) = K(X - r_1)(X - r_2) \ldots (X - r_N)$, where $K = A_N$. If these factors are then multiplied together, the constant term of the result will be

$$(-r_1)(-r_2) \ldots (-r_N) = \frac{A_0}{A_N}$$

Sec. 15–4. Rational Roots of a Polynomial Equation

Therefore, we can see that the roots $(r_1, r_2, \ldots r_N)$ are factors of the original constant divided by the highest power coefficient,

$$\frac{A_0}{A_N}$$

which is a fraction. In general then, we state that *when we have a polynomial equation of degree N, any rational roots must be factors of the constant term divided by factors of the highest power coefficient.*

Example C: Suppose $f(X) = 6X^3 - 5X^2 - 3X + 2 = 0$. Any rational roots must be factors of 2 ($\pm 1, \pm 2$), divided by factors of 6 ($\pm 1, \pm 2, \pm 3, \pm 6$). Therefore, the possible rational roots are $\pm 1, \pm \frac{1}{2}, \pm \frac{1}{3}, \pm \frac{1}{6}, \pm 2, \pm \frac{2}{3}$.

$$\begin{array}{rrrr|r}
6 & -5 & -3 & 2 & \underline{1} \\
  & 6 & 1 & -2 & \\
\hline
6 & 1 & -2 & &
\end{array}$$

$$6X^2 + X - 2 = 0$$
$$(3X + 2)(2X - 1) = 0$$
$$X = -\tfrac{2}{3}, \tfrac{1}{2}$$

Therefore, the roots are $-\frac{2}{3}, \frac{1}{2}$, and 1, which are factors of $\frac{2}{6}\left(\frac{A_0}{A_N}\right)$.

$$f(X) = 6(X^3 - \tfrac{5}{6}X^2 - \tfrac{3}{6}X + \tfrac{2}{6})$$
$$= 6(X + \tfrac{2}{3})(X - \tfrac{1}{2})(X - 1)$$

$$(\tfrac{2}{3})(-\tfrac{1}{2})(-1) = \tfrac{2}{6}$$

Example D: Suppose $f(B) = 16B^4 + 12B^3 - 48B^2 - 13B + 6 = 0$. Any rational roots must be factors of 6 ($\pm 1, \pm 2, \pm 3, \pm 6$), divided by factors of 16 ($\pm 1, \pm 2, \pm 4, \pm 8, \pm 16$). Therefore, the possible rational roots are $\pm 1, \pm \frac{1}{2}, \pm \frac{1}{4}, \pm \frac{1}{8}, \pm \frac{1}{16}, \pm 2, \pm 3, \pm \frac{3}{2}, \pm \frac{3}{4}, \pm \frac{3}{8}, \pm \frac{3}{16}$, and ± 6.

$$\begin{array}{rrrrr|r}
16 & 12 & -48 & -13 & 6 & \underline{-2} \\
 & -32 & 40 & 16 & -6 & \\
\hline
16 & -20 & -8 & 3 & & \underline{-\tfrac{1}{2}} \\
 & -8 & 14 & -3 & & \\
\hline
16 & -28 & 6 & & &
\end{array}$$

$$16B^2 - 28B + 6 = 0$$
$$2(8B^2 - 14B + 3) = 0$$
$$2(4B - 1)(2B - 3) = 0$$
$$B = \tfrac{1}{4}, \tfrac{3}{2}$$

Therefore, the roots are $-2, -\frac{1}{2}, \frac{1}{4},$ and $\frac{3}{2}$, which are factors of $\frac{6}{16} \left(\frac{A_0}{A_N}\right)$.

$$f(B) = 16(B^4 + \tfrac{12}{16}B^3 - \tfrac{48}{16}B^2 - \tfrac{13}{16}B + \tfrac{6}{16})$$
$$= 16(B+2)(B+\tfrac{1}{2})(B-\tfrac{1}{4})(B-\tfrac{3}{2})$$

$(2)(\tfrac{1}{2})(-\tfrac{1}{4})(-\tfrac{3}{2}) = \tfrac{3}{8} = \tfrac{6}{16}.$

As we can see from the previous two examples, there is often quite a large list of possible rational roots. With such a large list, it might take several attempts and a considerable amount of time to find the actual roots. Thus, we should make use of any help available to cut down on the time needed to use the "trial and error" method. One source of help in this area is *Descarte's rule of signs*, which states that *if we have a polynomial equation $f(X) = 0$, the maximum number of positive roots of $f(X)$ equals the number of sign changes in $f(X)$, and the maximum number of negative roots of $f(X)$ equals the number of sign changes in $f(-X)$.* $f(-X)$ is obtained by replacing X with $-X$ in the original function. We will accept this rule at this time without attempting to develop or prove it.

Example E: Determine the maximum number of positive and negative roots of the equation $X^3 + X^2 - 8X - 12 = 0$.

$$f(X) = X^3 + X^2 - 8X - 12 = 0$$
$$\underbrace{}_{\text{sign change}}$$

Therefore, there is at most one positive root.

$$f(-X) = (-X)^3 + (-X)^2 - 8(-X) - 12 = -X^3 + X^2 + 8X - 12 = 0$$
$$\underbrace{}_{\text{sign change}} \quad \underbrace{}_{\text{sign change}}$$

Therefore, there are at most two negative roots.

Example F: Determine the maximum number of positive and negative roots of the equation $Y^6 - Y^5 - 2Y^3 - 3Y^2 - Y - 2 = 0$.

$$f(Y) = Y^6 - Y^5 - 2Y^3 - 3Y^2 - Y - 2 = 0$$
$$\underbrace{}_{\text{sign change}}$$

Therefore, there is at most one positive root.

$$f(-Y) = (-Y)^6 - (-Y)^5 - 2(-Y)^3 - 3(-Y)^2 - (-Y) - 2$$
$$= Y^6 + Y^5 + 2Y^3 - 3Y^2 + Y - 2 = 0$$
$$\underbrace{}_{\text{sign change}} \underbrace{}_{\text{sign change}} \underbrace{}_{\text{sign change}}$$

Sec. 15-4. Rational Roots of a Polynomial Equation

Therefore, there are at most three negative roots. Since there are six roots altogether, this information tells us that there must be at least two complex roots.

Let us now summarize what we know about a polynomial equation $f(X) = 0$ of degree N:

1. There are exactly N roots.
2. The maximum number of positive roots equals the number of sign changes in $f(X)$, and the maximum number of negative roots equals the number of sign changes in $f(-X)$.
3. Any rational roots must be factors of the constant term divided by factors of the highest power coefficient.
4. Complex roots always occur in conjugate pairs.
5. The last two roots may always be found by quadratic methods.

Example G: Find the roots of $2X^3 - X^2 - 3X - 1 = 0$. There are three roots.

$$f(X) = 2X^3 - X^2 - 3X - 1 = 0$$

There is a maximum of one positive root.

$$f(-X) = -2X^3 - X^2 + 3X - 1 = 0$$

There are a maximum of two negative roots. Possible rational roots will be factors of -1 (± 1), divided by factors of 2 ($\pm 1, \pm 2$), or $\pm 1, \pm \frac{1}{2}$.

$$\begin{array}{rrrr|r}
2 & -1 & -3 & -1 & \underline{-\frac{1}{2}} \\
 & -1 & 1 & 1 & \\
\hline
2 & -2 & -2 & &
\end{array}$$

$$2X^2 - 2X - 2 = 0$$
$$2(X^2 - X - 1) = 0$$

$$X = \frac{-(-1) \pm \sqrt{1 - 4(-1)}}{4} = \frac{1 \pm \sqrt{5}}{4} = \frac{1 \pm 2.24}{4} = \frac{3.24}{4} \text{ or } \frac{-1.24}{4}$$

$$= .81 \quad \text{or} \quad -.31$$

Therefore, the roots are $-\frac{1}{2}$, $-.31$, and $.81$.

Example H: Find the roots of $R^5 + R^4 - 9R^3 - 5R^2 + 16R + 12 = 0$. There are five roots.

$$f(R) = R^5 + R^4 - 9R^3 - 5R^2 + 16R + 12 = 0$$

There are a maximum of two positive roots.

$$f(R) = -R^5 + R^4 + 9R^3 - 5R^2 - 16R + 12 = 0$$

There are a maximum of three negative roots. Possible rational roots will be factors of 12 ($\pm 1, \pm 2, \pm 3, \pm 4, \pm 6,$ and ± 12), divided by factors of 1 (± 1), or ($\pm 1, \pm 2, \pm 3, \pm 4, \pm 6,$ and ± 12).

$$\begin{array}{rrrrrr|r}
1 & 1 & -9 & -5 & 16 & 12 & \underline{2} \\
  & 2 & 6 & -6 & -22 & -12 & \\ \hline
1 & 3 & -3 & -11 & -6 & & \underline{-3} \\
  & -3 & 0 & 9 & 6 & & \\ \hline
1 & 0 & -3 & -2 & & & \underline{2} \\
  & 2 & 4 & 2 & & & \\ \hline
1 & 2 & 1 & & & &
\end{array}$$

$$R^2 + 2R + 1 = 0$$
$$(R + 1)(R + 1) = 0$$
$$R = -1, -1$$

Therefore, the roots are $-3, -1, -1, 2,$ and 2.

Exercises Find all of the roots for each of the following equations.

1. $X^3 + X^2 - 5X + 3 = 0$
2. $Y^3 + 2Y^2 - 19Y - 20 = 0$
3. $2S^3 - 5S^2 - 28S + 15 = 0$
4. $2Y^4 + 7Y^3 + 9Y^2 + 5Y + 1 = 0$
5. $P^4 + P^3 - 2P^2 - 4P - 8 = 0$
6. $9T^4 - 3T^3 + 34T^2 - 12T - 8 = 0$
7. $12K^4 + 44K^3 + 21K^2 - 11K - 6 = 0$
8. $B^4 - 11B^2 - 12B + 4 = 0$
9. $M^3 + 1 = 0$
10. $24R^3 + 26R^2 + 9R + 1 = 0$
11. $6V^4 + 7V^3 - 18V^2 - 4V + 3 = 0$
12. $A^5 + 3A^4 - 19A^3 - 33A^2 - 252A - 540 = 0$
13. From a rectangular piece of sheet metal, 30 centimeters by 24 centimeters, a rectangular container must be made by cutting a square from each corner and bending up the sides. How large a square must be cut from each corner if the volume of the container is to be 1296 cubic centimeters?

14. The path of a certain projectile is given by $Y = X^3 - 64X$ (distance in kilometers). Assuming that the terrain is level, how far from the starting position will the projectile hit the ground?

15–5. Irrational Roots of a Polynomial Equation

If none of the roots of a polynomial equation are known, we may find them by the method described in the previous section, provided that not more than two of the roots are irrational or complex. If there are more than two irrational or complex roots, the procedure just described will not work since we would be unable to reduce the equation to a quadratic. Therefore, other methods must be used to find irrational and complex roots when there are more than two of them.

Since irrational numbers are real numbers, any *irrational roots* may be found by the *graphical method*. That is, we could construct the graph of the equation and determine where the graph crosses the horizontal axis.

Example A: Solve the equation $X^4 - 7X^2 + 10 = 0$ graphically. Let $Y = X^4 - 7X^2 + 10$.

X	0	1	−1	2	−2	3	−3
Y	10	4	4	−2	−2	28	28

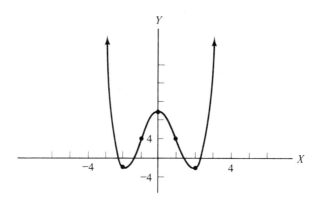

Therefore, the roots are -2.2, -1.4, 1.4, and 2.2.

The usual disadvantage of the graphical method is that it yields answers that are only approximate to one decimal place. This occurs here once again. Thus, if more precise answers are needed, other methods must be used.

One other method for finding irrational roots of a polynomial equation is by *linear interpolation*. This method is based upon the fact that *if two points of a graph are sufficiently close together, the actual curve between the two points may be approximated by a straight line*. Once this approximation is made, synthetic division is used to try different roots. In trying roots, we make use of the fact that the graph actually crosses the horizontal axis wherever there is a root. Therefore, with each approximation, we look for one value for which the remainder is positive and one value for which the remainder is negative. The root will be between these two values. Let us use this method to find the irrational root from Example A that lies between 1 and 2.

$$\left.\begin{array}{l} f(1) = 4 \\ f(2) = -2 \end{array}\right\}$$ Therefore, the root is between 1 and 2.

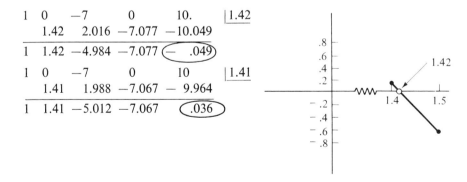

```
1   0    -7      0       10     | 1.6
    1.6   2.56  -7.10   -11.36
1   1.6  -4.44  -7.10  -  1.36

1   0    -7      0       10     | 1.5
    1.5   2.25  -7.12   -10.68
1   1.5  -4.75  -7.12  (- .68)

1   0    -7      0       10     | 1.4
    1.4   1.96  -7.06   - 9.88
1   1.4  -5.04  -7.06    (.12)
```

$$\left.\begin{array}{l} f(1.4) = .12 \\ f(1.5) = -.68 \end{array}\right\}$$ Therefore, the root is between 1.4 and 1.5.

```
1   0     -7      0        10.      | 1.42
    1.42   2.016 -7.077   -10.049
1   1.42  -4.984 -7.077  (- .049)

1   0     -7      0        10       | 1.41
    1.41   1.988 -7.067  -  9.964
1   1.41  -5.012 -7.067    (.036)
```

$$\left.\begin{array}{l} f(1.41) = .036 \\ f(1.42) = -.049 \end{array}\right\}$$ Therefore, the root is between 1.41 and 1.42.

Thus, the root to two decimal places is 1.41, since the absolute value of the

Sec. 15-5. Irrational Roots of a Polynomial Equation

remainder for 1.41 is smaller than the absolute value of the remainder for 1.42. This procedure could be repeated as many times as desired and the root obtained to whatever precision is needed.

Example B: For the equation $3T^3 + 13T^2 + 3T - 4 = 0$, there is an irrational root between -1 and 0. Find this root correct to two decimal places.

$$\left. \begin{array}{l} f(-1) = 3 \\ f(0) = -4 \end{array} \right\} \quad \text{Therefore, the root is between } -1 \text{ and } 0.$$

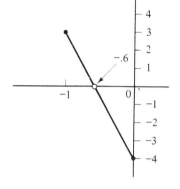

```
3    13      3      -4     |-.6
   - 1.8  -6.72   2.232
―――――――――――――――――――――――
3    11.2  -3.72  -1.768

3    13      3      -4     |-.7
   - 2.1   -7.63   3.241
―――――――――――――――――――――――
3    10.9  -4.63  (-.759)

3    13      3      -4     |-.8
   - 2.4   -8.48   4.384
―――――――――――――――――――――――
3    10.6  -5.48   (.384)
```

$$\left. \begin{array}{l} f(-.8) = .384 \\ f(-.7) = -.759 \end{array} \right\} \quad \text{Therefore, the root is between } -.8 \text{ and } -.7.$$

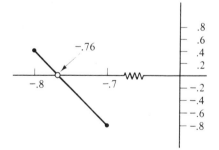

```
3    13      3      -4      |-.76
   - 2.28  -8.147   3.912
―――――――――――――――――――――――――
3    10.72 -5.147  (-.088)

3    13      3      -4      |-.77
   - 2.31  -8.23    4.027
―――――――――――――――――――――――――
3    10.69 -5.23    (.027)
```

Therefore, the root correct to two decimal places is $-.77$.

Exercises

Solve each of the following equations graphically.

1. $X^4 - 9X^2 + 14 = 0$
2. $T^4 + 7T^2 - 30 = 0$
3. $Y^5 - Y^4 - 6Y^3 + 5Y^2 + 5Y = 0$

4. $K^6 - 4K^5 + 2K^4 + 8K^3 - 8K^2 = 0$
5. $M^4 - M^3 - 6M^2 + 5M + 3 = 0$
6. $P^3 - 3P^2 - 8P - 4 = 0$

For each of the following equations, find the indicated irrational roots correct to two decimal places.

7. $A^3 - 3A + 1 = 0$ (root between 1 and 2)
8. $3R^3 + R^2 - 3 = 0$ (root between 0 and 1)
9. $Y^4 - 5Y^3 + 6Y^2 - 5Y + 5 = 0$ (root between 1 and 2)
10. $P^4 - 4P + 1 = 0$ (root between 0 and 1)
11. $2S^4 - S^3 + 6S^2 - 4S - 8 = 0$ (root between -1 and 0)
12. $T^3 + 5T^2 + T - 12 = 0$ (root between -3 and -2)

For each of the following, construct the graph and then determine any irrational roots correct to two decimal places by using linear interpolation.

13. $X^3 - 3X^2 - 10X + 30 = 0$
14. $V^4 - 10V^2 + 24 = 0$
15. $N^6 - 4 = 0$
16. $L^4 + L^3 - 10L^2 + 15L - 3 = 0$
17. An equation involving the thickness (X) of a certain machine part, in centimeters, is $X^3 + 3X^2 - X = 1$. Find the thickness correct to two decimal places.
18. A certain floatation device sinks in water to a depth (D), measured in inches, given by the equation $2D^3 + 9D^2 - 18D - 88 = 0$. Find the depth correct to two decimal places.

15-6. CHAPTER REVIEW

For each of the following, find the remainder, using the remainder theorem.

1. $(X^4 - 5X^3 + 3X^2 + 8X - 9) \div (X - 2)$
2. $(5T^4 + 7T^2 - 10) \div (T + 3)$
3. $(Y^5 - 18) \div (Y + 1)$
4. $(3K^4 - K^3 - K^2 + 9K + 2) \div (2K - 1)$

For each of the following, use the factor theorem to determine whether the second expression is a factor of the first.

5. $(M^3 - 5M + 8), (M - 1)$

Sec. 15–6. Chapter Review   445

6. $(3R^4 + R^3 - 2R + 9)$, $(3R + 1)$
7. $(2P^3 - 7P^2 - 17P + 10)$, $(P - 5)$
8. $(N^4 + N^3 + N^2 - 2N - 3)$, $(N + 1)$

For each of the following, find the quotient and the remainder, using synthetic division.

9. $(5Q^4 - 2Q^3 + 8Q - 1) \div (Q - 3)$
10. $(2L^5 - 46L^3 + L^2 - 9) \div (L - 5)$
11. $(8V^3 + 12V^2 + 6V + 1) \div (2V + 1)$
12. $(B^6 - 8B^3 + 12B^2 - B + 6) \div (B + 2)$

For each of the following, find all of the solutions, given the indicated roots.

13. $X^4 - 16 = 0$ ($2j$ is a root)
14. $Y^5 - 3Y^4 - 4Y^3 - Y^2 + 3Y + 4 = 0$ (-1, 1, and 4 are roots)
15. $24B^4 - 50B^3 - 65B^2 - 5B + 6 = 0$ ($-\frac{1}{2}$ and $\frac{1}{4}$ are roots)
16. $M^5 - 6M^4 + 4M^3 + 54M^2 - 117M = 0$ (-3 and $3 - 2j$ are roots)

Find all of the roots for the following equations.

17. $Y^3 - 8Y^2 + 20Y - 16 = 0$
18. $2V^4 + 3V^3 - 22V^2 + 7V + 60 = 0$
19. $2X^3 - X^2 - 8X - 5 = 0$
20. $2R^5 + 3R^4 + 9R^3 + 21R^2 - 35R = 0$
21. $2A^3 - 3A^2 - 11A + 6 = 0$
22. $2T^4 + T^3 + 3T^2 + 2T - 2 = 0$

Solve each of the following equations graphically.

23. $X^3 - 2X^2 - 2X - 7 = 0$
24. $2T^4 - T^3 + 6T^2 - 4T - 8 = 0$
25. $Y^4 - Y^3 - 2Y^2 - Y - 3 = 0$
26. $K^3 - 5K^2 + 3K = 0$

For each of the following find the indicated irrational roots correct to two decimal places, using linear interpolation.

27. $X^3 - 9X^2 - 2X + 80 = 0$ (root between 3 and 4)
28. $R^4 + 22R^2 - 75 = 0$ (root between -2 and -1)
29. $Z^5 + 5Z^4 + 7Z^3 + 45Z^2 - 18 = 0$ (root between 0 and 1)
30. $6U^4 - U^3 - 3U^2 - 4U - 2 = 0$ (root between -1 and 0)

31. A certain rectangular container has a volume of 15,680 cubic centimeters. If the length and width are equal, and the height is 8 centimeters less than the length and width, find the dimensions of the container.

32. In a certain situation, the work (W) in foot pounds, is related to time (t) in seconds, by the equation $W = 6t^2 - t^3$. After how many seconds is 25 foot pounds of work done?

33. Under certain conditions, the distance (s) that an object falls due to gravity is related to the time (t) by the equation $s = t^4 + 6t$. If s is measured in meters and t is measured in seconds, after how many seconds has the object fallen 15 meters? Express your answer correct to two decimal places.

34. The charge (q), in coulombs, of a certain electric circuit is related to time (t), in seconds, by the equation $q = 2t^3 + 4t$. After how many seconds will there be 2 coulombs of charge? Express your answer correct to two decimal places.

chapter 16

Inequalities

16–1. Definition and Properties of Inequalities

The importance of being able to set up and solve inequalities cannot be overemphasized. A large number of situations call for equations in order to express the proper relationship among sets of data. But many relationships among sets of data are inequalities. Thus, it is necessary for technically oriented people to be able to find solutions to inequalities.

An inequality is stated by using one of the following symbols: $<$ ("is less than"), $>$ ("is greater than"), or $\neq$ ("is not equal to"). These symbols were first introduced in Section 1–2.

If A, B are real numbers then,

1. $A < B$ if A is any number to the left of B on a number line.

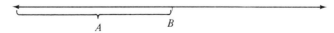

2. $A > B$ if A is any number to the right of B on a number line.

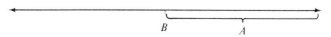

3. $A \neq B$ if A does not coincide with B on a number line.

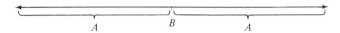

If $A < B$ then $A + C = B$, where C is a positive number. For example, the inequality $3 < 5$ is true because a positive number 2 must be added to 3 to equal 5.

An equivalent statement for $A < B$ is $B > A$. That is, the members of an inequality can be interchanged provided the order of the inequality is reversed. For example, if $6 < 10$, then $10 > 6$ expresses the same relationship.

There are three basic properties that are used to find the solution to a linear inequality.

1. If $A < B$, then $(A + C) < (B + C)$.
2. If $A < B$, then $A \cdot C < B \cdot C$ if $C > 0$.
3. If $A < B$, then $A \cdot C > B \cdot C$ if $C < 0$.

The first two properties allow for any number to be added to or subtracted from both members of an inequality, or for both members of an inequality to be multiplied or divided by any positive number. These properties are no different than the properties for solving an equation. The third property is different, stating that when both members of an inequality are multiplied or divided by a negative number the order of the inequality must be reversed.

To illustrate this last property, suppose a true inequality is multiplied by a negative number:

$-4 < 5$ a true inequality.

$(-1)(-4)$? $5(-1)$ multiply both members by a (-1).

$4 > -5$ the order of the inequality must be reversed in order to have a true inequality.

Example A: Solve each of the following inequalities using the basic properties of inequalities.

1. $\quad 2A - 6 < 4$
$\quad 2A - 6 + 6 < 4 + 6$
$\qquad\quad 2A < 10$
$\qquad (\tfrac{1}{2})2A < 10(\tfrac{1}{2})$
$\qquad\qquad A < 5$

2. $\quad 6B + 3 > 9$
$\quad 6B + 3 - 3 > 9 - 3$
$\qquad\quad 6B > 6$
$\qquad (\tfrac{1}{6})6B > 6(\tfrac{1}{6})$
$\qquad\qquad B > 1$

Sec. 16-2. Solutions of Basic Inequalities

3. $\quad -3X < 12$
$(-\tfrac{1}{3})(-3X) > 12(-\tfrac{1}{3})$
$X > -4$

Exercises Solve the following inequalities, using the basic properties of inequalities.

1. $2A + 5 < 3$
2. $-1 - 3B > 35$
3. $2X + 8 < 3X - 4$
4. $M + 3 - 3M > -2M + 23$
5. $7 - 2T < 32$
6. $4(12 - 2V) > 10(V + 4)$
7. $2 - (A + 3) < 4 - A$
8. $3[2 - (X - 2)] > 3 - 4(1 - X)$
9. $6 + 3(N + 2) > N + 16$
10. $8 - 2Y < 8 - (Y - 4)$
11. The sum of three times a number and 50 is greater than 47. What numbers will make this statement true?
12. If a student's marks on her first three tests were 60%, 75%, and 65%, what must be her range of marks on the fourth test in order to have an average of at least 70%?

16-2. Solution of Basic Inequalities

An inequality may be an *absolute inequality*, for which the solution is all real numbers, or a *conditional inequality*, for which the solution is a subset of the real numbers. This section will deal with the solutions of these two general types of inequalities and their graphs.

Example A: Find the solution for $X + 1 > X$.

$$X + 1 > X$$
$$X - X + 1 > X - X$$
$$1 > 0$$

This is an absolute inequality and it means that *all* real numbers will make the original inequality a true statement.

Example B: Find the solution to $2X + 3 > -4$.

$$2X + 3 - 3 > -4 - 3$$
$$2X > -7$$
$$(\tfrac{1}{2})2X > -7(\tfrac{1}{2})$$
$$X > -\tfrac{7}{2}$$

This last inequality states that all real numbers greater than $-\frac{7}{2}$ will make the original inequality a true statement and any other real number will make the original inequality a false statement. It is convenient to graph the set of real numbers, $X > -\frac{7}{2}$, on a number line. This graph represents the solution set.

There are two other types of conditional inequalities that use the connectors *and* and *or* in order to find the solution set. When the word *and* is used between two inequalities, both must be true. When the word *or* is used between two inequalities, one or the other must be true.

An inequality using the connector *and* would be any inequality with two simple inequalities in the same statement, such as $-5 < (X+4) < 10$. This inequality can be thought of as a quantity $(X+4)$ that is between -5 and 10. It means $-5 < (X+4)$ *and* $(X+4) < 10$. The basic properties explained in the previous section for inequalities with two members also apply to inequalities with three members.

$$-5 < (X + 4) < 10$$
$$-5 - 4 < X + 4 - 4 < 10 - 4$$
$$-9 < X < 6$$

This means that only the real numbers between -9 and 6 will satisfy the original statement because $-9 < X < 6$ means "all real numbers greater than -9 *and* all real numbers less than 6." A graph of this solution set is shown in Figure 16–2.1.

FIGURE 16–2.1

An inequality using the connector *or* would be $X + 5 \leq 9$. It means $X + 5 < 9$ *or* $X + 5 = 9$.

$$\begin{cases} X + 5 < 9 \text{ or } X + 5 = 9 \\ X < 4 \quad\quad \text{ or } X = 4 \\ \text{Thus, } X \leq 4 \end{cases} \quad \text{or} \quad \begin{cases} X + 5 \leq 9 \\ X + 5 - 5 \leq 9 - 5 \\ X \leq 4 \end{cases}$$

This means that only the real numbers less than 4 or equal to 4 will satisfy the original inequality. A graph of this solution set is shown in Figure 16–2.2.

Sec. 16-2. Solutions of Basic Inequalities

FIGURE 16-2.2

NOTE: Graphically, the solid circle at 4 means that this point is to be included.

Example C: Solve the inequality $-7 < (-\frac{1}{2}X + 3) < 13$.

$$-7 < -\tfrac{1}{2}X + 3 < 13$$
$$-10 < -\tfrac{1}{2}X < 10$$
$$20 > X > -20$$

The graph of this solution is

Example D: Solve the inequality $\frac{2}{3}X - 5 \geq 18$.

$$\tfrac{2}{3}X - 5 \geq 18$$
$$\tfrac{2}{3}X \geq 23$$
$$(\tfrac{3}{2})\tfrac{2}{3}X > 23(\tfrac{3}{2})$$
$$X \geq \tfrac{69}{2}$$

The graph of this solution is

Exercises Solve each of the following inequalities.

1. $2A + 8 \leq 3A - 4$
2. $5 - (B - 3) \geq 4 - B$
3. $6 + 2(N + 2) \geq N + 16$
4. $M + 3 - 4M \leq 2M + 23$
5. $A(1 - 2A) - 1 \leq 2[A(2 - A) - 1]$
6. $-(3 - X + X^2) \geq 9 + X(5 - X)$
7. $-4 < 2T - 6 < 4$
8. $0 \leq 5N + 6 \leq 1$
9. $-1 \leq -\tfrac{2}{3}T - 5 < 3$
10. $\tfrac{1}{2} < \tfrac{7}{8} - \tfrac{5}{3}Y \leq \tfrac{3}{4}$
11. $4(X - 1)^2 \geq 0$
12. $-4 \leq A + 4 \leq 4$

13. $-7 \leq 2A - 1 < 12$

14. $\dfrac{2A + 4}{3} \leq 12$

15. $-3 \leq \dfrac{3B - 6}{2} \leq 7$

16. Find a number such that the sum of three times the number and -4 is between -6 and 6.

17. The relationship between Fahrenheit and Celsius temperatures is given by the formula $C = \tfrac{5}{9}(F - 32)$. An experiment calls for the Celsius temperature to lie between $-10°$ and $15°$, inclusive. Within what range must we have the Fahrenheit temperature?

18. In order for a student to receive a B for a course he must have an average between 80% and 90%, including 80%. If his marks so far are 75%, 87%, and 67%, within what range must his fourth test lie?

16–3. Solution of Higher Degree Inequalities

Higher degree inequalities can be solved in a number of ways. One technique is to make one member of the inequality equal to zero and put the other member into factored form. The result is an inequality in which a product of factors is less than zero or negative, or an inequality in which a product of factors is greater than zero or positive. If the product is negative, the solution set can be found by finding the real numbers that make an odd number of the factors negative and the other factors positive. If the product is positive, the solution set can be found by finding the real numbers that make an even number of the factors negative or all factors positive.

Example A: Solve the following inequality: $(X - 2)(X + 3) < 0$. Make a sign graph for each linear factor by locating the critical value and show by a $+$ or $-$ the real numbers that will make that linear factor positive or negative. The critical value is found by setting each factor equal to zero and solving for the variable. The only real numbers that will satisfy this inequality are the real

```
              - - - - - - - - 0 + + + +
X - 2 ─────────────────────────┬──────────
                               2

              - - - 0 + + + + + + + +
X + 3 ──────────────┬─────────────────────
                   -3
Solution set  ──────●─────────●───────────
                   -3         2
```

numbers between -3 and 2. That is, $-3 < X < 2$. The reason for this is that the sign graph of $X - 2$ shows that the real numbers between -3 and 2 are

Sec. 16–3. Solution of Higher Degree Inequalities   453

negative and the sign graph of $X + 3$ shows that the real numbers between -3 and 2 are positive, and a negative number times a positive number yields a negative result.

Example B: Solve the following inequality: $(3X - 1)(X + 4) > 0$. Make a sign graph for each linear factor by locating the critical value and show by a $+$ or $-$ the real numbers that will make that factor positive or negative. Since both

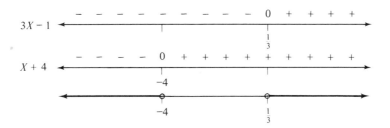

factors are negative, the product is positive for all real numbers less than -4. Both factors are positive; therefore, the product is positive for all real numbers greater than $\frac{1}{3}$ and the solution set is $X < -4$ or $X > \frac{1}{3}$.

Example C: Solve the following inequality $X^2 - 2X \leq 1$. Put the inequality into a form where one member is equal to zero. That is, $X^2 - 2X - 1 \leq 0$. Since this expression cannot be factored over the set of integers, the quadratic formula will give the roots for the quadratic equation and the linear factors can be found using these roots.

$$X = \frac{2 \pm \sqrt{4+4}}{2} = \frac{2 \pm 2\sqrt{2}}{2} = 1 + \sqrt{2}, 1 - \sqrt{2}$$

Using these values, the original inequality can be stated as a product of linear factors.

$$[X - (1 + \sqrt{2})][X - (1 - \sqrt{2})] \leq 0$$

$(X - 1 - \sqrt{2})(X - 1 + \sqrt{2}) \leq 0$

$X - 1 - \sqrt{2}$

$1 + \sqrt{2}$

$X - 1 + \sqrt{2}$

$1 - \sqrt{2}$

$1 - \sqrt{2}$ $1 + \sqrt{2}$

Thus, the solution set is $1 - \sqrt{2} \leq X \leq 1 + \sqrt{2}$.

Example D: Solve the following inequality: $(X - \frac{1}{2})(X + 2)(X - 4) < 0$. Since a product can be negative only when there is an *odd* number of negative factors, the solution to this inequality can be found by finding those real numbers that will yield an odd number of negative factors. In this case, that would be 1 or 3 negative factors. Thus, the solution is $X < -2$ or $\frac{1}{2} < X < 4$.

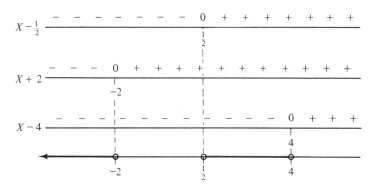

Example E: Solve the following inequality

$$\frac{(A + 3)(A - 1)}{5 - A} \leq 0$$

In this case, it is helpful to make the factor $5 - A$ equal to $A - 5$. This can be accomplished by multiplying each member of the inequality by -1.

$$\frac{1 \cdot}{-1 \cdot} \frac{(A + 3)(A - 1)}{(5 - A)} \geq 0(-1)$$

$$\frac{(A + 3)(A - 1)}{A - 5} \geq 0$$

The solution to this inequality will be all real numbers that will yield an *even* number of negative factors, since the quotient is positive. Thus, the solution set is $-3 \leq A \leq 1$ or $A > 5$. $A \neq 5$ because the denominator would be zero, which would make the inequality undefined.

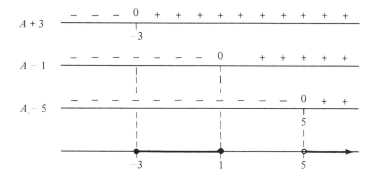

Sec. 16–3. Solution of Higher Degree Inequalities

Example F: Solve the inequality $X^3 + X^2 - 2X - 2 > 0$. This problem could be handled just like the previous ones except that it would be difficult to determine the linear factors.

Another technique for solving inequalities is to introduce another variable Y and find all real numbers X such that $Y > 0$. The graph of $Y = X^3 + X^2 - 2X - 2$ will give a curve in the rectangular coordinate system that will allow us to determine a solution in terms of X.

$$Y = X^3 + X^2 - 2X - 2$$

X	−2	−1	0	1	2	$-1\frac{1}{2}$	$\frac{1}{2}$	$1\frac{1}{2}$
Y	−2	0	−2	−2	6	$-\frac{1}{8}$	$-\frac{21}{8}$	$\frac{5}{8}$

The values of X for which $Y > 0$ would be $-1.4 < X < -1$ or $X > 1.4$.

Example G: Solve the inequality $X^4 - X^3 \leq 7X^2 - X - 6$ graphically.

$$X^4 - X^3 \leq 7X^2 - X - 6$$
$$X^4 - X^3 - 7X^2 + X + 6 \leq 0$$

Let

$$Y = X^4 - X^3 - 7X^2 + X + 6$$

X	0	1	−1	2	−2	3	−3	4	$-\frac{3}{2}$
Y	6	0	0	−12	0	0	48	90	$-\frac{45}{16}$

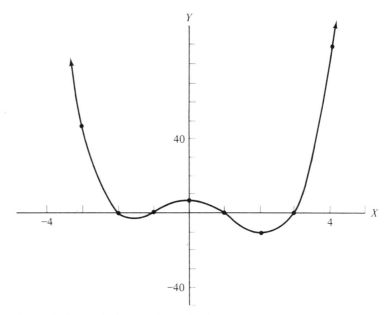

The values of X for which $Y \leq 0$ would be $-2 \leq X \leq -1$ or $1 \leq X \leq 3$.

Exercises

Solve each of the following inequalities.

1. $(X - 3)(X + 2) \leq 0$
2. $(A + 1)(A - 1) \geq 0$
3. $B^2 - 5B < 0$
4. $3T^2 + 9T \geq 0$
5. $N^2 - 4N - 21 < 0$
6. $A^2 + 11 \geq -12A$
7. $6X^2 - X - 2 < 0$
8. $Y^2 - Y + 1 \geq 0$
9. $T^2 - 2T - 4 \leq 0$
10. $X(X - 1)(X - 7) < 0$
11. $X(2X + 1)(X - 4) \geq 0$
12. $A^2(A - \frac{1}{2})(A + 3) \leq 0$
13. $T^3 - 4T < 0$
14. $B^3 - 2B^2 - B + 2 < 0$
15. $\dfrac{(Y - 2)(Y + 4)}{Y + 5} \leq 0$
16. $\dfrac{-3(T + 3)}{(T - 1)} \geq 0$
17. $\dfrac{(A + 2)(A - \frac{3}{2})}{2 - A} \geq 0$
18. $X^3 - 4X + 3 \leq 0$
19. $A^4 - 2A^2 \geq 0$
20. $\dfrac{1}{B - 2} \leq \dfrac{1}{B + 1}$

21. The deflection (D) of a beam is given by $D = S^3 - 243S + 1458$, where S is the length of the beam. For what values of S will the deflection be greater than zero?

22. The formula for the total surface area of a right circular cylinder is $S = 2\pi R^2 + 2\pi RH$. If the height (H) of a certain cylinder is 30 centimeters, what values of the radius will yield a surface area less than 400π square centimeters?

16–4. Systems of Inequalities

The solution to a system of inequalities will be a segment of a plane. We will graph systems of equations in this section to determine which part of the plane would make the system of inequalities true.

Example A: Solve the following system of inequalities.

$$Y \leq 2X + 1$$
$$Y > -\tfrac{1}{2}X - 1$$

The corresponding equations $Y = 2X + 1$ and $Y = -\tfrac{1}{2}X - 1$ will determine the boundary lines for the inequalities. Testing the coordinates of the origin (0, 0) with each inequality will tell whether the half plane containing the origin is in the solution set. Since $Y \leq 2X + 1$ is true when testing (0, 0), the half plane determined by the boundary line $Y = 2X + 1$ and containing the origin would be the solution to the first inequality. That is, all points in the half plane containing the origin up to and *including* the points on the boundary line are part of the solution set, ▨. Since $Y > -\tfrac{1}{2}X - 1$ is true when testing (0, 0), the half plane containing the origin and determined by the boundary line $Y = -\tfrac{1}{2}X - 1$ will be the solution set. That is, all points in the half plane containing the origin up to but *not including* the points on the boundary line are part of the solution set, ▨.

The segment of the plane that is the solution set for the system of inequalities is the intersection of the two half planes that satisfy the individual inequalities, ▨.

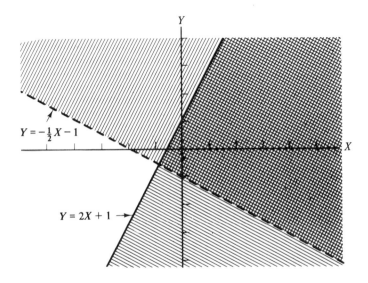

Example B: Solve the following system of inequalities.

$$Y \geq 2X^2 - 8X + 2$$
$$Y \leq X - 1$$

The corresponding equations will determine the boundary lines, and testing the coordinates of the origin will determine the segments of the plane that will satisfy each inequality. By using two different symbols for indicating individual solution sets, the intersection of the two will be the solution set for the system of inequalities.

$$Y = 2X^2 - 8X + 2$$

Axis of symmetry: $X = \frac{+8}{4} = 2$

Vertex = $(2, -6)$

Y-intercept = 2

$Y \geq 2X^2 - 8X + 2$

$Y = X - 1$

$Y \leq X - 1$

Solution set: .

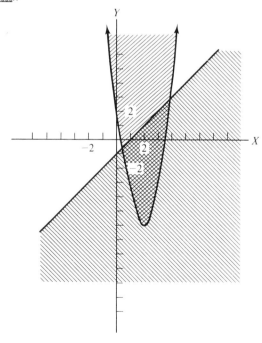

Sec. 16–5. Inequalities Involving Absolute Value   459

It should be noted that whenever the inequality symbols are $<$ or $>$, the boundary lines *will not* be part of the solution. When the inequality symbols are $\leq$ or $\geq$, the boundary lines *will* be part of the solution.

Exercises Solve each of the following systems of inequalities.

1. $Y \geq X^2 - 4$
   $Y \leq -1$

2. $A < B^2 - 2B$
   $A > 2B + 1$

3. $R^2 + S^2 \leq 16$
   $R > 3S^2 - 2$

4. $M^2 - N^2 > 12$
   $M^2 + 4N^2 < 16$

5. $Y \geq 3X^3 + 6X$
   $Y \geq 3X + 2$

6. $-3R + 2T \leq 6$
   $4R - T \geq 8$

7. $Y \leq 10^x$
   $Y \leq 1 - X^2$

8. $Y \leq \log_{10} X$
   $Y > X - 1$

9. $X^2 + Y^2 \leq 36$
   $X \geq 3$
   $Y \geq 3$

10. $A > 3$
    $B > -1$
    $A + B > 2$

11. The area of a circle is given by $A = \pi R^2$ and the circumference of a circle is given by $C = 2\pi R$. What value of R will make the area greater than 25π square feet and the circumference less than 14π feet?

12. The area of a square is given by $A = S^2$ and the perimeter of a square is given by $P = 4S$. What value of S will make the area less than 62 square centimeters and the perimeter greater than 28 centimeters?

16–5. Inequalities Involving Absolute Value

The basic idea of an absolute value inequality was discussed in Section 1–3. This section will extend the basic concept to additional absolute value problems.

In general absolute value inequalities are defined as follows:

1. If $|X| < A$, then $-A < X < A$. The solution set is a line segment from $-A$ to A, not including its end points.

2. If $|X| > A$, then $X < -A$ or $X > A$. The solution set is all real numbers outside of A and $-A$. That is, the solution set consists of two half lines, one to the left of $-A$ and one to the right of A.

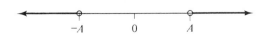

These two types of inequalities can be thought of as all real numbers that are less than A units from a reference point, zero ($|X| < A$) or all real numbers greater than A units from a reference point, zero ($|X| > A$).

An extension of this idea is to consider $|X - B| < A$, which is similar to $|X| < A$ except the reference point is no longer zero but some other real number, B.

1. $|X - B| < A$
Graphically:

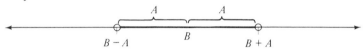

A represents a distance of A units from the reference point B. $B + A$ and $B - A$ represent the coordinates of the end points. The open line segment between $B - A$ and $B + A$ represents all real numbers whose distance is less than A units from B.

Algebraically: This solution set is stated as

$$-A < X - B < A$$

or

$$B - A < X < B + A$$

2. $|X - B| > A$
Graphically:

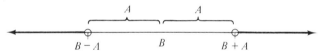

A represents a distance of A units from B. $B + A$ and $B - A$ represent the coordinates of the end points of two half lines, where the distance of each real number on these half lines is greater than A units from B.

Algebraically: This solution set is stated as

$$X - B > A \quad \text{or} \quad X - B < -A$$
$$X > B + A \quad \text{or} \quad X < B - A$$

Example A: Solve the absolute value inequality $|X - 4| < 3$. A graphical solution to this problem can easily be found. The reference point is 4 and the solution set will be all real numbers less than a distance of three units from 4.

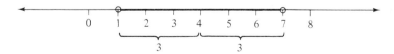

That is,

$$-3 < (X - 4) < 3 \quad \text{or} \quad 1 < X < 7$$

Sec. 16–5. Inequalities Involving Absolute Value

Example B: Solve the absolute value inequality $|2X - 6| < 5$. This problem could also be solved graphically, but it may be easier to determine the critical values by using the algebraic technique.

$$|2X - 6| < 5$$
$$-5 < (2X - 6) < 5$$
$$1 < 2X < 11$$
$$\tfrac{1}{2} < X < \tfrac{11}{2}$$

Example C: Solve the absolute value inequality $|-3X - 4| > 6$. Since this inequality uses the greater-than sign, the connector *or* will have to be used and the solution set will be all real numbers outside of two critical values. These critical values are best determined by the following algebraic technique.

$$|-3X - 4| > 6$$

$-3X - 4 > 6$	or	$-3X - 4 < -6$
$-3X > 10$		$-3X < -2$
$X < -\tfrac{10}{3}$	or	$X > \tfrac{2}{3}$

Example D: Solve the absolute value inequality $|X + 1| < 2$. This absolute value inequality is a little different from the previous examples because it states the sum of two real numbers instead of the difference. The expression $X + 1$ can be expressed as a difference by stating $X - (-1)$, which means the reference value is -1. A graphical solution follows.

$$|X - (-1)| < 2$$

The algebraic technique is no different for this type of absolute value inequality than it was for the previous examples.

Example E: Solve the absolute value inequality $|A^2 + 3A - 2| < 2$. This statement is equivalent to

$$A^2 + 3A - 2 < 2 \quad \text{and} \quad A^2 + 3A - 2 > -2$$

Since each statement is quadratic, it can be solved by making one member of the inequality equal to zero and putting the other member into factored form.

$$A^2 + 3A - 4 < 0 \quad \text{and} \quad A^2 + 3A > 0$$
$$(A + 4)(A - 1) < 0 \qquad A(A + 3) > 0$$

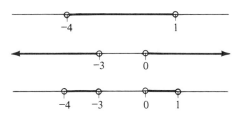

Because the connector *and* is used, the solution set will be the intersection of the solution set of $A^2 + 3A - 4 < 0$ and $A^2 + 3A > 0$. Thus, the solution set is:

Therefore,
$$-4 < A < -3 \quad \text{or} \quad 0 < A < 1$$

Exercises

Determine the solution set of each of the following using the definition of absolute value inequalities.

1. $|X - 2| > 3$
2. $|A + 5| < 8$
3. $|T - 4| \leq 5$
4. $|N + 1| > 10$

Solve the following inequalities algebraically and represent the solution set on a line graph.

5. $|3X + 5| \leq 2$
6. $|2X + 7| \geq 1$
7. $|2 + \frac{3}{2}A| < \frac{1}{2}$
8. $\left|\dfrac{4A - 3}{4}\right| > 6$

Sec. 16–6. Chapter Review

9. $|4X - 5| \leq 4$

10. $\left|\dfrac{2A + 1}{3}\right| < 3$

11. $|-2T + 4| \geq -5$

12. $|2A - 1| \leq 0$

Write an absolute value inequality for each of the following.

13. The distance from A to 3 is greater than 4 units.
14. The distance from X to -2 is greater than 1 unit.
15. The distance from N to 5 is less than 3 units.
16. The distance from Y to -3 is less than 6 units.

Solve each of the following algebraically.

17. $|T^2 - 3T + 1| < 1$

18. $|Y^2 - Y + 4| > 2$

16–6. CHAPTER REVIEW

Solve each inequality algebraically and represent the solution set on a number line.

1. $X + 6 \geq 8$
2. $A - 5 \leq 7$
3. $4A - 2 > -1 + 3A$
4. $5B + 3 \leq B - 1$
5. $\dfrac{2T + 3}{2} \leq 5$
6. $\dfrac{3N - 4}{2} > 12$
7. $A(A + 2) \leq 4$
8. $T(T - 3) \geq 0$
9. $(A + 1)(A - 3) \leq 0$
10. $(T + 2)(T + 5) \geq 0$
11. $X^2 - 3X > 4$
12. $T^2 - 5T \leq 6$
13. $A^2 < 4$
14. $T^2 > 5$
15. $4N^2 + 1 < 0$
16. $\dfrac{3}{A - 2} > 0$
17. $\dfrac{2}{X} \leq 4$
18. $\dfrac{X}{X + 2} < 0$
19. $\dfrac{-1}{3 - A} \geq 0$
20. $\dfrac{2}{B - 2} \geq \dfrac{3}{B}$
21. $A(A - 2)(A + 3) \leq 0$
22. $\dfrac{A^2(A + 1)}{A - 3} \geq 0$
23. $\dfrac{R^2}{2} < \dfrac{4}{R}$
24. $T^3 - 9T \leq 0$
25. $\dfrac{1}{4A + 1} > \dfrac{2}{A - 5}$
26. $\dfrac{T}{T + 2} < 4$
27. $-5 < -2A < 9$
28. $-1 \leq 2B - 1 \leq 3$

29. $-2 \leq \frac{2X+3}{3} \leq 4$
30. $0 \leq 5 - T \leq 3$
31. $1 \leq 1 - \frac{1}{2}A < 5$
32. $-2 \leq \frac{2}{3}T + 3 \leq -1$
33. $|A - 1| < 3$
34. $|\frac{1}{2}T + 5| > 4$
35. $|6 - B| < 1$
36. $\left|\frac{2A - 3}{2}\right| > 3$
37. $|X + \frac{1}{2}| \leq 0$
38. $|Y + \frac{1}{3}| < 0$
39. $|T + 6| \leq -1$
40. $|2N - 1| \geq -2$
41. $|A^2 - 2A + 3| > 3$
42. $|R^2 - 5R| < 6$

Solve each of the following systems of inequalities.

43. $Y - 2X < 3$
    $Y + 3X > 6$
44. $Y \leq 2X^2 + 7$
    $Y \geq 9 - 3X^2$
45. $A - 3B < 2$
    $A \leq B^2 - B + 1$
46. $R^2 + S^2 \leq 16$
    $2R + 5S \geq 1$

47. Write an inequality that states that B is within 4 units of 3.
48. Write an inequality that states that $2A$ is within 3 units of -1.
49. Write an inequality that states the distance from $4T$ to -3 is greater than 1.
50. If a cyclist increases her speed by 5 kilometers per hour, she could travel in 3 hours a distance at least as great as that which now takes her 4 hours. What is the greatest speed she achieves before increasing her speed?
51. The equation $H = 112t - 16t^2$ gives the height (H), in feet, of an object in terms of time (t), in seconds. For what values of t is H greater than 160 feet?
52. In a right triangle, one leg must be twice the length of the other and the hypotenuse must be less than 10. What must be the lengths of the sides?
53. An environmental technician finds that one water purification system can treat 25,000 liters less than twice as much as another system in the same plant each day. If these systems together must purify at least 170,000 liters per day, what is the least number of liters that each of them must be able to purify?
54. A chemical lab technician finds that after washing a rectangular piece of fabric, with a length 4 times its width, in a certain chemical solution, its width was unchanged but its length was 2 centimeters shorter than before the wash. If the piece of material decreased at most 24 square centimeters in area, find the maximum values of its original dimensions.

chapter 17

Variation

17-1. Variation

A technician collects and studies sets of data in order to find a precise way to express the relationship between the sets of data.

A spring made from a particular type of metal is measured each time a weight is attached to it. We want to develop a precise description of the relationship between the amount of stretch (L) in the spring and the force (F) applied to the spring. The following data was collected by testing the spring.

F (kg)	0	10	20	30	40
L (cm)	0	1	2	3	4

From this table it is easy to see that the length of the spring increases 1 centimeter each time an additional 10 kilograms of force is applied to the spring. A verbal description of this relationship is "the length of the stretch of the spring is directly related to or varies directly with the amount of force applied." A precise empirical equation that expresses the same relationship and that can be used in determining a quantitative analysis is $L = .1F$, where L is the length of the stretch in centimeters and F represents the force in kilograms.

If a technician has an understanding of the different types of variation it will enable that person to give a precise description of the relationship between sets of data and to write an empirical equation relating the sets of data.

In this chapter we will consider direct, inverse, and joint variation, and the characteristics of each type.

17-2. Direct Variation

Direct variation means that two quantities always exist in the same proportion to each other. Different ways of expressing this type of variation are:

1. Y varies as X.
2. Y varies directly as X.
3. Y is directly proportional to X.

An equation for a direct variation is $Y = KX$, where K is known as the *constant of proportionality*, and Y and X are the variables. Any convenient variables could replace Y and X. The graph of a direct variation may be increasing (if K is positive) or decreasing (if K is negative), and if the variation is linear it will be a straight line with a slope equal to K.

A general procedure for handling problems that are direct variation problems is to use one set of values for the two variables in order to determine the value of K, substitute that value for K in the general equation $Y = KX$, and use the new equation to find any additional values that are required. Substitute dimensional quantities in order to find K as a dimensional quantity, although K may or may not be a dimensional quantity.

Example A: Write an equation stating the relationship between I and t, and determine the value of I after 1.8 seconds. The constant of proportionality in

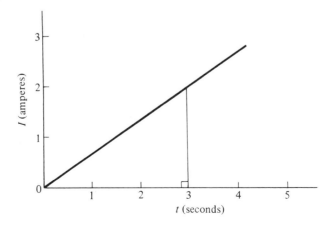

Sec. 17-2. Direct Variation

the equation of a direct proportion is the slope of the line. Thus, the value of K in this case is $\frac{2}{3}$. t is the independent variable and I is the dependent variable. Therefore, the equation would be $I = \frac{2}{3}t$.

Substitute 1.8 in place of t.

$$I = \tfrac{2}{3}(1.8)$$
$$= 1.2 \text{ amp}$$

Example B: Write an equation stating the relationship between P (the power developed by an engine in horsepower) and Ap (the piston area in square centimeters). Determine P when $Ap = 53$ cm².

P (hp)	75	150	225	300
Ap (cm²)	20	40	60	80

The table of values shows that there is a *constant* rate of change in area that corresponds to a *constant* rate of change in horsepower. Therefore, horsepower varies directly with piston area, and the variation is linear.

1. A general equation is $P = K(Ap)$.
2. Substituting 75 hp for P and 20 cm² for Ap then

$$K = \frac{75 \text{ hp}}{20 \text{ cm}^2} = 3.75 \text{ hp/cm}^2$$

3. The equation is $P = 3.75\, Ap$. Use the equation to determine P when $Ap = 53$ cm².
4. $P = \left(\dfrac{3.75 \text{ hp}}{\text{cm}^2}\right)\left(\dfrac{53 \text{ cm}^2}{1}\right)$
   $= 198.75$ hp

NOTE: In this case the dimensional units for K were substituted into the equation as a check on the dimensional soundness of the answer. From the table of values, it is obvious that the units for P must be expressed in terms of horsepower.

Example C: Y varies directly with X. If $Y = 2$ when $X = \frac{1}{4}$, find Y when $X = 16$.

1. Find the value of K.

$$Y = KX \quad \text{or} \quad K = \frac{Y}{X} = \frac{2}{\frac{1}{4}} = 8$$

2. Determine Y when $X = 16$.

$$Y = 8X$$

$$Y = 8(16)$$
$$Y = 128$$

Example D: The force needed to stretch a certain spring is proportional to the amount the spring is to be stretched. If it takes 5 kilograms of force to stretch a certain spring 12 centimeters, what force will be necessary to stretch the spring 18 centimeters?

$$F = KL$$
$$5 \text{ kg} = K(12 \text{ cm})$$
$$K = \frac{5 \text{ kg}}{12 \text{ cm}}$$
$$F = \frac{5 \text{ kg}}{12 \text{ cm}} L = \left(\frac{5 \text{ kg}}{12 \text{ cm}}\right)\left(\frac{18 \text{ cm}}{1}\right)$$
$$= \frac{90}{12} \text{kg} = 7.5 \text{ kg}$$

Example E: The area of a circle is directly proportional to the square of the radius of the circle. If the area is 25π square centimeters when the radius is 5 centimeters, find the area when the radius is 5.4 centimeters.

$$A = KR^2$$
$$25\pi = K(25)$$
$$\pi = K$$
$$A = \pi R^2$$
$$= \pi(5.4)^2 = 29.16\pi \text{ cm}^2$$

Exercises

From each graph determine an appropriate equation using the variables listed on the axes.

1.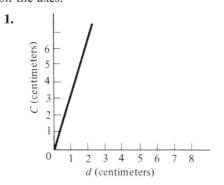

2.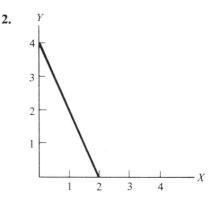

Sec. 17-2. Direct Variation

3.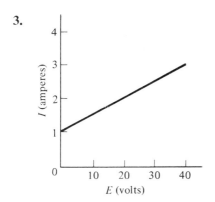

4.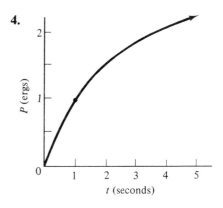

From each table of values write an equation that relates the sets of data.

5.

t (sec)	0	1	2	3
d (ft)	0	88	176	264

6.

T (°C)	0	5	10	15	20
L (mm)	0	2.25	4.50	6.75	9

7.

t (sec)	0	1	2	3	4	5
L (cm)	0	2	8	18	32	50

Write an appropriate equation for each of the following statements.

8. The distance that a ball rolls along an inclined plane is proportional to the time squared.

9. Property tax varies directly as the assessed valuation of a piece of real estate.

10. The scaled distance on a topographical map is proportional to the actual distance. ($K = 2 \times 10^{-3}$)

11. The volume of a gas kept at a constant pressure varies directly as the absolute temperature.

In each of the following write an appropriate equation and use this equation to answer each question.

12. A particular rocket uses alcohol as fuel and liquid oxygen as oxidizer. The amount of alcohol consumed is directly proportional to the consumption of liquid oxygen. If the rocket consumes 41 kilograms of alcohol and during the same time consumes 253 kilograms of liquid oxygen, how much alcohol would be consumed when 726 kilograms of liquid oxygen are consumed?

13. When water freezes, it expands 9% of its volume. What volume of water would come from 437 cubic centimeters of ice?

14. The amount of hydrogen released from water is directly related to the amount of sodium added to the water. If 1.2 grams of hydrogen is released when 27.6 grams of sodium are added to water, how much hydrogen is produced when 150 grams of sodium are added to the water?

15. The amount of hemoglobin is proportional to the volume of blood: $K = \frac{.12 \text{ gm}}{\text{cm}^3}$. How many grams of hemoglobin would 138 cm³ of blood contain?

16. The surface area of a sphere varies as the square of the radius. The area is 16π square centimeters when the radius is 2 centimeters. Determine the formula for the surface area of a sphere, and find the area when the radius is 9 centimeters.

17. The distance an object falls when under the influence of gravity is directly proportional to the square of the time of the fall. If an object falls 36 feet in 1.5 seconds, how far will it fall in 3 seconds?

17–3. Inverse Variation

Inverse variation means that the product of two quantities is constant. Different ways of expressing this type of variation are:

1. Y is inversely proportional to X.
2. Y varies indirectly as X.

The equation for an inverse variation is

$$Y \cdot X = K \quad \text{or} \quad Y = \frac{K}{X}$$

where K is known as the constant of proportionality and Y and X are the variables.

Sec. 17-3. Inverse Variation

A graph that represents this type of relationship may be either increasing (if K is negative) or decreasing (if K is positive).

Example A: The following data express the relationship between pressure and volume at a constant temperature.

V (m³)	5	10	15	20	30	35
P (gm/cm²)	140	70	46.6	35	23.3	20

1. Write an equation from this data and use the equation to determine the pressure at a volume of 23.2 cubic meters.
2. Graph this relationship.

Solution:
1. Since the data shows the product of the pressure and volume to be constant, the relationship is an inverse variation and the general equation is $P = \dfrac{K}{V}$.

    The value of K can be determined by substituting any of the pairs of P and V from the table. Thus,

    $$140 = \frac{K}{5}$$
    $$K = (140)(5)$$
    $$K = 700$$

    Therefore, the equation is

    $$P = \frac{700}{V}$$

    In order to determine the pressure at 23.2 cubic meters, substitute this value into the equation.

    $$P = \frac{700}{23.2}$$
    $$P = 30.172 \text{ gm/cm}^2$$

    NOTE: Since the dimension of P is already known, it is not necessary to include the dimension of K when solving.

2. The graph of this relationship would be in quadrant I since both the volume and pressure are positive.

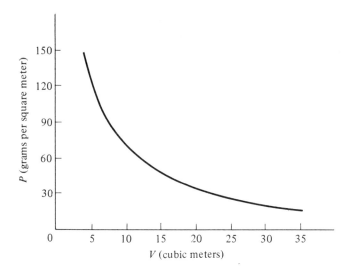

Example B: The altitude of a triangle varies inversely as the length of its base for a constant area. If the altitude is 100 centimeters when the base is 10 centimeters, write the equation for this relationship, and use this equation to determine the altitude of the triangle when the base is 21 centimeters. Since the relationship is an inverse variation relationship, the general equation $Y = \dfrac{K}{X}$ would apply. Adapting this equation for altitude (H) and base (B), the equation with the proper variables would be

$$H = \frac{K}{B}$$

Substituting 100 for H and 10 for B, the constant of proportionality would be

$$K = HB = (100 \text{ cm})(10 \text{ cm}) = 1000 \text{ cm}^2$$

Thus, the precise equation is $H = \dfrac{1000}{B}$.

The altitude of the triangle when the base is 21 can be determined from this equation

$$H = \frac{1000}{21}$$

$$H = 47.619 \text{ cm}$$

The units of dimension of K are square centimeters, which is correct since in the original problem it was stated for a constant area. It really isn't necessary to include the units of dimension to determine the unit of dimension for the altitude since the proper unit for altitude in this problem would be centimeters.

Example C: The strength (S) of a horizontal wooden beam with a rectangular cross section varies inversely with the distance (L) between supports. If a beam

Sec. 17–3. Inverse Variation

12 feet long and supported at both ends has a safe load capacity of 840 pounds, what is the safe load capacity of a similar beam 15 feet long?

$$S = \frac{K}{L}$$

$$840 \text{ lb} = \frac{K}{12 \text{ ft}}$$

$$K = 840 \text{ lb } (12 \text{ ft}) = 10{,}080 \text{ ft lb}$$

$$S = \frac{10{,}080 \text{ ft lb}}{L}$$

$$= \frac{10{,}080 \text{ ft lb}}{15 \text{ ft}}$$

$$= 672 \text{ lb}$$

Exercises Determine the equation from the graph for each of the following:

1.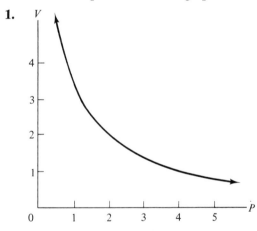

2.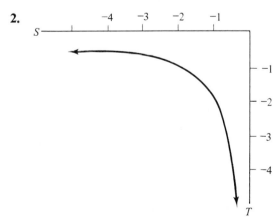

Determine the equation from the table of data for each of the following:

3.

Y	2	4	8	16
X	1	$\frac{1}{2}$	$\frac{1}{4}$	$\frac{1}{8}$

4.

R	4	$3\frac{1}{3}$	2	1.74
Y	1	1.2	2	2.3

5.

I		10^3	10^2	1
D		10	10^2	10^4

Write an appropriate equation for each statement.

6. The force of attraction between two particles varies inversely as the square of the distance between their centers.

7. The frequency (F) of a radio wave is inversely proportional to the wave length (L).

8. The intensity of illumination (I) varies inversely as the square of its distance (d) from the source of light.

9. The pressure of a gas (P) at a constant temperature is inversely proportional to its volume (V).

In each of the following write an appropriate equation and use this equation to answer each question.

10. The exposure time for photographing an object is inversely proportional to the square of the lens diameter. If the exposure time is $\frac{1}{200}$ second when the lens diameter is 1.0 centimeter, find the exposure time when the diameter is doubled.

11. Construct a graph of P vs. V in problem (9) above, if $P = 5$ gm/cm³ when $V = 120$ cm³.

12. The mass of an object is inversely proportional to the square of its distance from the center of the earth. If the mass of an object on earth is 200 kilograms and the radius of the earth is 4000 miles, what would the mass of the object be at 150 kilometers above the earth?

13. The intensity of a light source is inversely proportional to the square of the distance from the source. If the intensity is 30 candle power at a distance of 200 centimeters, what is the intensity at a distance of 140 centimeters?

14. For a given length of wire, the resistance varies inversely as the cross-sectional area. If the resistance of a certain length of wire is .2 ohm when the cross-sectional area is .05 square inch, what is the resistance when the cross-sectional area is .08 square inch?

17–4. Joint Variation

When a quantity is dependent upon more than one other quantity, the relationship is a *joint variation*. Joint variation is a combination of direct and/or inverse variation.

The equation for a joint variation involves a product of individual variations.

$$Y = KXZ \quad \text{or} \quad Y = \frac{KX}{Z} \quad \text{or} \quad Y = \frac{K}{XZ}$$

Example A: When an object is rotating, its centripetal force varies *directly* with its mass and the square of its speed and *inversely* with the radius of rotation.

1. An equation for centripetal force would be $F = \frac{KMS^2}{R}$, where $F =$ force, $M =$ mass, $S =$ speed, $R =$ radius of rotation, and $K =$ the constant of proportionality.
2. Determine the value of K for the given set of values of the remaining variables: $M = 50$ g, $S = \frac{120 \text{ cm}}{\text{sec}}$, $R = 60$ cm, and $F = 1200$ dynes.

$$F = \frac{KMS^2}{R}$$

$$1200 \text{ dynes} = \frac{K(50 \text{ g}) \frac{14{,}400 \text{ cm}^2}{\text{sec}^2}}{60 \text{ cm}}$$

$$\left(\frac{60 \text{ cm}}{1}\right)\left(\frac{1200 \text{ dynes}}{50 \text{ g}}\right)\left(\frac{\text{sec}^2}{14{,}400 \text{ cm}}\right) = K$$

$$.1 = K$$

3. The equation now becomes

$$F = \frac{.1MS^2}{R}$$

This equation can be used to determine other values of centripetal force.

4. Determine the centripetal force for the same object as in part (2) but when the radius of rotation has been increased to 120 cm (or doubled).

$$F = \frac{.1(50 \text{ g})\left(\frac{14{,}400 \text{ cm}}{\text{sec}^2}\right)}{120 \text{ cm}}$$

$$F = 600 \text{ dynes}$$

Example B: The volume of a cylindrical container is proportional to the square of the radius and its height.
1. What is the constant of proportionality?
2. What happens to the volume when the height is doubled?
3. What happens to the volume when the radius is doubled?

Solution: The formula for the volume of a cylinder is

$$V = \pi R^2 H$$

1. Thus, the constant of proportionality is π.
2. When the height is doubled it can be represented by $2H$. Substituting this value into the formula yields $V = 2\pi R^2 H$; thus, the volume is doubled.
3. When the radius is doubled it can be represented by $2R$. Substituting this value into the formula yields $V = \pi(2R)^2 H = 4\pi R^2 H$; thus, the volume is increased 4 times.

Example C: The electrical resistance of a wire varies directly as its length and inversely as its cross-sectional area. Write an equation for this relationship and determine the resistance when the length is 100 meters and the cross-sectional area is .02 square centimeter. The constant of proportionality for the wire is 2.0×10^{-4}. The equation is $R = \dfrac{(2.0 \times 10^{-4})L}{A}$, where $L =$ the length, $A =$ the cross-sectional area, and $R =$ the resistance. The resistance in ohms for a wire length of 100 meters with a cross-sectional area of .02 square centimeter is

$$R = \frac{2.0 \times 10^{-4} \, (10^4 \, \cancel{cm})}{(.02) \, \cancel{cm^2}} = 100 \text{ ohms/cm}$$
$$\phantom{R = \frac{2.0}{x}} \text{cm}$$

NOTE: $L = 100$ meters $= 10{,}000$ cm.

It should be noted that it is very impractical (or impossible) to graph a joint variation because of the number of variables involved.

Exercises

1. If A varies jointly as B and C^2, and A is 48 when B is 8 and C is 3, what is the value of A when B is 12 and C is 2?

2. The volume of a cone varies jointly as the altitude (H) and the square of the radius (R) of the base.
   (a) Write a general equation if the volume is 113 cubic centimeters when the altitude is 12 centimeters and the radius is 3 centimeters.
   (b) Using this equation find the volume of the cone whose altitude is 13 centimeters and whose radius is 12 centimeters.

3. The energy released each second by a black body varies jointly with the area of the body and the fourth power of its Kelvin temperature. When the temperature is 100° K and the area is 10 square meters, the energy released is 5.0 joules per second. Find the energy released when the area is 18 square meters.

4. When an object moves in a circular path, the centripetal force varies directly as the square of the velocity and inversely as the radius of the circle. If the force is 25,000 dynes for a mass of 100 grams rotating at 100 centimeters per second in a circle of radius 50 centimeters, find the force if the radius of the circle is increased to 100 centimeters.

5. The heat developed in a wire carrying a current for a certain time varies jointly as the resistance and the square of the current. If a current of 150 milliamperes produces 100 calories of heat in a wire with a resistance of 75 ohms, how many calories are produced by a current of 300 milliamperes and 120 ohms?

6. The volume (V) of a gas varies directly as its temperature (T) and inversely as its pressure (P). A gas occupies 30 cubic centimeters at a temperature of 200° A (absolute) and a pressure of 25 kilograms per square centimeter. What will the volume be if the temperature is increased to 300° A and the pressure decreased to 20 kilograms per square centimeter?

7. If Y varies jointly as X and Z, what is the resulting change in Y when X is doubled and Z is halved?

8. If S varies directly as T^2 and inversely as R, what is the resulting change in S when T is doubled?

17-5. CHAPTER REVIEW

Each of the following tables represents a direct or inverse variation. Write an equation for each set of data.

1.

T (newtons)	0	4	9	16	25
V (m/sec)	0	20	30	40	50

2.

I (amp)	0	2	4	6	8
H (cal)	0	12	48	108	192

3.

t (sec)	0	1	2	3	4	5
s (m)	0	1	4	9	16	25

4.

V (cm³)	10	20	30	40	50
P (atm)	60	30	20	15	12

5.

t (sec)	0	1	2	3	4	5
s (m)	0	1.5	8	13.5	24	37.5

Write an appropriate equation for each of the following statements.

6. The current in a conductor is directly proportional to the voltage across the conductor.

7. The kinetic energy of a body varies directly with the square of the speed of that body.

8. The velocity of transverse waves in a cord is proportional to the square root of the tension in the cord.

9. The electrical resistance of a wire varies directly with its length and inversely as its cross-sectional area.

Solve each of the following variation problems.

10. A liquid will exert a pressure on an object that is directly proportional to the depth of the object beneath the surface of the liquid. If a liquid exerts a pressure of 25 grams per square centimeter at a depth of 19 centimeters, what would the pressure be at 60 centimeters?

11. The distance an object falls is directly proportional to the square of the length of time it falls. If the object falls 16 meters in 2 seconds, how far will the same object fall in 10 seconds?

12. The volume of a cone varies jointly as the altitude and the square of the radius of the base. If the volume of the cone is 20π cubic centimeters when the radius is 4 centimeters and the altitude is 3.75 centimeters, find the volume of the cone whose altitude is 10 centimeters and whose radius is 15 centimeters.

13. The force between two electrical charges is inversely proportional to the square of the distance between the charges. Two charges exert a force of

200 dynes on each other when they are 5 centimeters apart. What force will be exerted by the charges when they are 2 centimeters apart?

14. The effort is inversely proportional to its distance from the fulcrum. If the fulcrum is 2 meters from an effort of 50 kilograms, how far is the fulcrum from an effort of 125 kilograms?

15. According to Ohm's law, the voltage across a given resistor is proportional to the current in the resistor. If there are 27 volts across a certain resistor in which a current of 12 amperes is flowing, how much current is flowing when the voltage is 22.5 volts?

16. The kinetic energy of a moving object varies jointly as the mass of the object and the square of its velocity. If the kinetic energy of a 20-gram object moving at 15 centimeters per second is 2250 ergs, what will be the kinetic energy of the same object moving at 12 centimeters per second?

17. An object falling under the influence of gravity falls a distance which varies jointly as the acceleration of the object due to gravity and the square of the time of fall. If a certain object falls 64 feet in 2 seconds, how long will it take the object to fall 400 feet? The acceleration due to gravity is 32 feet per second per second.

18. The volume of a right circular cylinder varies jointly as the square of the radius and the height of the cylinder. The volume of a certain cylinder with a radius of 4 centimeters and a height of 8 centimeters is 128π cubic centimeters. What is the radius of a cylinder that is 7 centimeters high and has a volume of 63π cubic centimeters?

chapter 18

Plane Analytic Geometry

18–1. Basic Definitions and Terminology

Previously, we have discussed in detail various types of equations including linear, quadratic, higher degree, exponential, logarithmic, and trigonometric equations. In each case we constructed tables of values and drew graphs for numerous examples. In this chapter, we will look further at the relationships that exist between certain types of equations and their graphs. Specifically, we will consider linear and quadratic equations. Our main objectives will be to *analyze* certain *geometric* figures and curves by examining the equations that represent them, and to construct the graphs of certain equations without writing tables of values. These graphs will be constructed in the rectangular coordinate system (i.e., in a plane); hence, this endeavor is called *plane analytic geometry*.

As indicated, we will consider *linear equations* in one or two variables for which the graph will be a *straight line* (refer to Chapter 6) and *quadratic equations* in two variables for which the graph will be a *parabola*, *circle*, *ellipse*, or *hyperbola*. These last four curves are usually called the *conic sections* since they may be obtained by intersecting a cone with a plane (see Figure 18–1.1).

Sec. 18-2. The Distance Formula

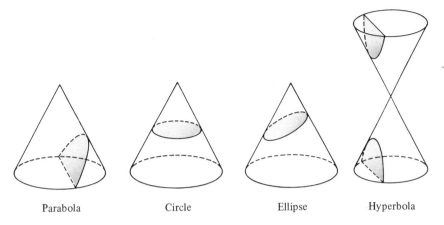

FIGURE 18-1.1

Exercises

Refer to the definition of a linear equation in Section 6-1 and the definition of a quadratic equation in Section 13-1, and state whether each of the following is a linear equation, quadratic equation, or neither.

1. $2X - 3Y = 8$
2. $X^2 - 3X + 2Y = 1$
3. $2A = 3B$
4. $4R + \dfrac{5}{S} = 20$
5. $3\sqrt{T} - 4T^2 = 0$
6. $P^2 + Q^2 = 9$
7. $5R + 2S - 3T = 0$
8. $\dfrac{2}{X} = 8$
9. $4A + 3AB + 6 = 0$
10. $P^3 + P^2 + P + 1 = 0$
11. $5M^2 + 4N^2 = 20$
12. $K^2 + L^2 + 2K - 6L + 6 = 0$
13. $Y = X^3$
14. $A = 5B^2 + 2B - 7$
15. $R^2 - S^2 = 1$
16. $X^2 = Y^2 + 3Y - 4$
17. $T^2 + 5\sqrt{T} = 0$
18. $P^{-1} + P^{-2} = 6$
19. $XY = 18$
20. $K^4 + 3K^2 - 9 = 0$

18-2. The Distance Formula

One concept of extreme importance in our study of plane analytic geometry is the *distance between any two points* in the coordinate plane.

If two points lie on a line parallel to the horizontal axis, the distance between them is the absolute value of the difference of their abscissas (see Figure 18-2.1).

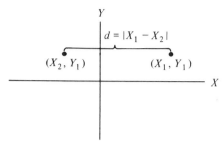

FIGURE 18–2.1

Example A: Find the distance between the points (6, 3) and (−5, 3).

$$d = |6 - (-5)| = |11| = 11$$

If two points lie on a line parallel to the vertical axis, the distance between them is the absolute value of the difference of their ordinates (see Figure 18–2.2).

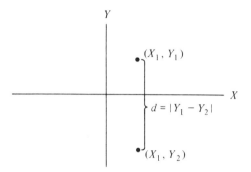

FIGURE 18–2.2

Example B: Find the distance between the points (4, 7) and (4, −2).

$$d = |7 - (-2)| = |9| = 9$$

If two points are on a line that is not parallel to either axis, we find the distance between them by making use of the Pythagorean Theorem (see Figure 18–2.3).

$$d = \sqrt{|X_2 - X_1|^2 + |Y_2 - Y_1|^2} = \textbf{Distance Formula}$$

Example C: Find the distance between the points (2, 5) and (−4, 3).

Sec. 18–2. The Distance Formula

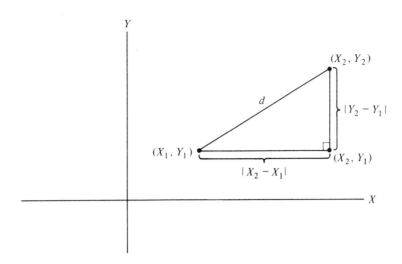

FIGURE 18–2.3

$d = \sqrt{|2-(-4)|^2 + |5-3|^2}$
$ = \sqrt{36 + 4}$
$ = \sqrt{40}$
$ = 6.32$

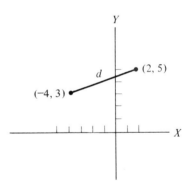

Example D: Find the distance between the points $(-3, 4)$ and $(2, -4)$.

$d = \sqrt{|-3-2|^2 + |4-(-4)|^2}$
$ = \sqrt{25 + 64}$
$ = \sqrt{89}$
$ = 9.43$

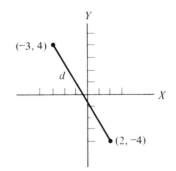

Another important concept is the *midpoint* of a line segment. This is the point exactly halfway between two given points and on the line segment joining the two points (see Figure 18-2.4).

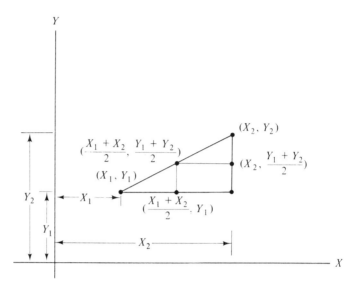

FIGURE 18-2.4

$$C = \left(\frac{X_1 + X_2}{2}, \frac{Y_1 + Y_2}{2}\right) = \text{Midpoint Formula}$$

Example E: Find the midpoint of the line segment joining the points (2, 3) and (6, 7).

$$C = \left(\frac{6+2}{2}, \frac{7+3}{2}\right)$$
$$= (4, 5)$$

Example F: Find the midpoint of the line segment joining the points $(-2, 6)$ and $(4, -1)$.

Sec. 18-3. The Straight Line

$$C = \left(\frac{-2+4}{2}, \frac{6+(-1)}{2}\right)$$
$$= (1, \tfrac{5}{2})$$

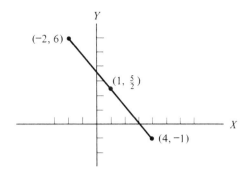

Exercises

Find the distance between each of the following pairs of points and also find the midpoint of the line segment joining them.

1. $(2, 5)$ and $(7, 11)$
2. $(-2, 6)$ and $(-2, -4)$
3. $(-3, -5)$ and $(4, -5)$
4. $(6, 8)$ and $(3, 4)$
5. $(5, -9)$ and $(-7, 3)$
6. $(1, -3)$ and $(0, -2)$
7. $(-4, 2)$ and $(5, -1)$
8. $(-1, 0)$ and $(-7, 5)$
9. $(\tfrac{1}{2}, 1)$ and $(3, -\tfrac{1}{2})$
10. $(2, 6)$ and $(\tfrac{1}{2}, -\tfrac{1}{4})$
11. $(0, 0)$ and $(\tfrac{2}{3}, -7)$
12. $(.7, -.5)$ and $(-.3, -.6)$
13. Determine whether the points $(2, 1)$, $(5, 1)$, and $(5, 5)$ are the vertices of a right triangle.
14. Determine whether the points $(-3, 1)$, $(-5, 8)$, and $(-1, 8)$ are the vertices of an isosceles triangle.
15. Determine whether the points $(-2, 1)$, $(1, 4)$, $(1, -3)$, and $(4, 0)$ are the vertices of a parallelogram.
16. Determine whether the points $(-3, 3)$, $(-5, 2)$, $(2, -1)$ and $(0, -2)$ are the vertices of a parallelogram.

18-3. The Straight Line

In Chapter 6, we introduced the concept of *slope* as a measure of the *direction* of a line. Since a straight line does not change direction (i.e., its direction is constant), we will define a *straight line* as a "curve" with *constant slope*. We also pointed out in Chapter 6 that if (X_1, Y_1) is a fixed point on a straight line and (X, Y) is any other point on the same line, then the slope (M) of the straight line is given by

$$M = \frac{Y - Y_1}{X - X_1}$$

(i.e., the vertical change between the two points divided by the horizontal change between the same two points). See Figure 18–3.1.

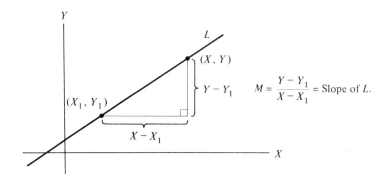

FIGURE 18–3.1

Example A: Find the slope of the straight line which passes through the points $(-2, -1)$ and $(4, 2)$.

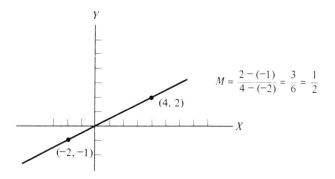

NOTE: It is extremely important that in finding the differences $(Y - Y_1)$ and $(X - X_1)$, we subtract in the same direction for both of them.

If we multiply both sides of the equation

$$M = \frac{Y - Y_1}{X - X_1}$$

by $(X - X_1)$, we may write the result as $(Y - Y_1) = M(X - X_1)$. This is called *the point-slope form* of the equation of a straight line. In other words, if we know

Sec. 18-3. The Straight Line

the slope of a straight line and one point on the line (or if we can find them), we can write the equation of the line.

Example B: Find the equation of the straight line with a slope of 3 and passing through the point $(2, -3)$.

$$M = 3 \text{ and } (X_1, Y_1) = (2, -3)$$

Therefore,

$$(Y - Y_1) = M(X - X_1)$$
$$[Y - (-3)] = 3(X - 2)$$
$$Y + 3 = 3X - 6$$
$$Y - 3X + 9 = 0$$

Example C: Find the equation of the straight line passing through the points $(-3, 2)$ and $(4, -1)$.

$$M = \frac{-1 - 2}{4 - (-3)} = \frac{-3}{7}$$

(X_1, Y_1) could be either of the given points. We will use $(4, -1)$. Therefore,

$$(Y - Y_1) = M(X - X_1)$$
$$[Y - (-1)] = \frac{-3}{7}(X - 4)$$
$$Y + 1 = \frac{-3X}{7} + \frac{12}{7}$$
$$7Y + 7 = -3X + 12$$
$$7Y + 3X - 5 = 0$$

If the point (X_1, Y_1) is the *vertical intercept* of the straight line, it will have coordinates $(0, B)$ and our equation will become $(Y - B) = M(X - 0)$ or $Y = MX + B$. This is called the *slope-intercept* form of the equation of a straight line.

Example D: Find the equation of the straight line with a slope of $\frac{1}{2}$ and a vertical intercept of $(0, -3)$.

$$M = \frac{1}{2} \text{ and } B = -3$$

Therefore,

$$Y = MX + B$$
$$Y = \frac{1}{2}X - 3$$

$$2Y = X - 6$$
$$2Y - X + 6 = 0$$

In each of the last three examples, we should note that our final result consisted of an X term, a Y term, and a constant. Also, we have collected all of these terms on one side of the equal sign with zero on the other side. Therefore, we define the *general form* of the equation of a straight line as

$$AX + BY + C = 0 \quad (X \text{ and } Y \text{ could be replaced by any variables})$$

We will now examine a few special types of straight lines and their slopes. Previously, we have shown (in Chapter 6) that the equation of a horizontal line is $Y = B$. Since there is no vertical change in a horizontal line, such a line has a slope of

$$M = \frac{0}{X - X_1} = 0$$

(see Figure 18–3.2). A vertical line has an equation of the form $X = A$ and,

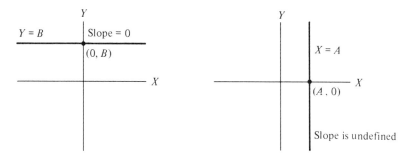

FIGURE 18–3.2

since there is no horizontal change in such a line, the slope of this type of line is undefined,

$$M = \frac{Y - Y_1}{0}$$

(see Figure 18–3.2).

Another way of describing the slope of a straight line is that it equals the tangent of the positive angle measured from the horizontal axis to the line. This angle is called *the angle of inclination* of the straight line (see Figure 18–3.3).

If a straight line is directed upward to the right, its angle of inclination will be between 0° and 90°. Since the tangent of such an angle is positive, any straight line directed *upward to the right* will have a *positive slope*. A straight

Sec. 18–3. The Straight Line

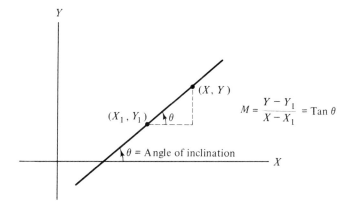

FIGURE 18-3.3

line directed downward to the right will have a *negative slope* since such a line would have an angle of inclination between 90° and 180° and the tangent of such an angle is negative (see Figure 18–3.4).

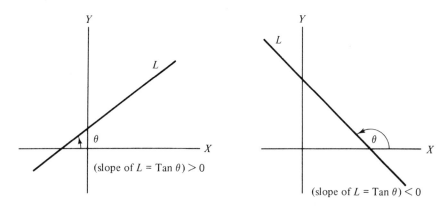

FIGURE 18-3.4

From this definition of slope, we can derive a number of important ideas. Since parallel lines by definition do not intersect, they must have equal angles of inclination. Therefore, the *slopes of parallel lines are equal* (see Figure 18–3.5).

If two lines are perpendicular, their angles of inclination differ by 90°. That is, $\theta_2 = \theta_1 + 90°$. Therefore,

$$-\theta_1 = 90° - \theta_2$$
$$\text{Tan}(-\theta_1) = \text{Tan}(90° - \theta_2)$$
$$-\text{Tan}\,\theta_1 = \text{Cot}\,\theta_2$$

$$-\text{Tan } \theta_1 = \frac{1}{\text{Tan } \theta_2}$$

$$-\text{Tan } \theta_1 \text{ Tan } \theta_2 = 1$$

$$\text{Tan } \theta_2 = -\frac{1}{\text{Tan } \theta_1}$$

$$M_2 = -\frac{1}{M_1}$$

Thus, the *slopes of perpendicular lines* are *negative reciprocals* of one another (see Figure 18–3.6).

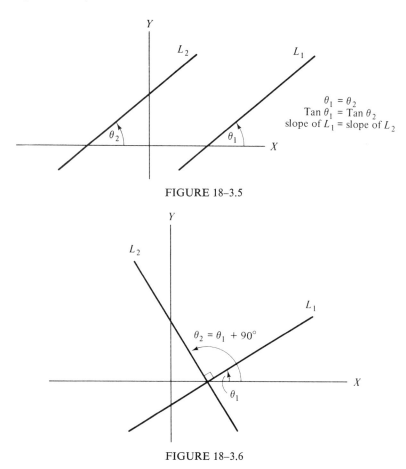

FIGURE 18–3.5

FIGURE 18–3.6

Example E: Find the slopes of the straight lines joining the pairs of points $[(-2, 3)$ and $(4, 0)]$ and $[(-3, -2)$ and $(-1, 2)]$, and determine whether the lines are perpendicular.

Sec. 18–3. The Straight Line

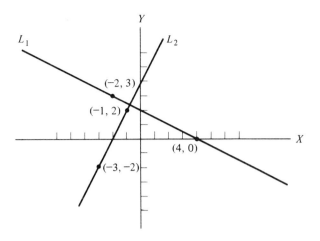

$$\text{Slope of } L_1 = M_1 = \frac{3 - 0}{-2 - 4} = \frac{3}{-6} = -\frac{1}{2}$$

$$\text{Slope of } L_2 = M_2 = \frac{2 - (-2)}{-1 - (-3)} = \frac{4}{2} = 2$$

Therefore, L_1 and L_2 are perpendicular lines since their slopes are negative reciprocals.

Exercises

For each of the following, find the equation of the straight lines with the given properties. Write each equation in the general form. In each of the problems (1)–(22), consider X the independent variable and Y the dependent variable.

1. Has a slope of -2 and passes through the point $(5, -2)$.
2. Has a slope of $\frac{2}{3}$ and passes through the point $(-4, -3)$.
3. Passes through the points $(-1, -3)$ and $(-2, 4)$.
4. Passes through the points $(-3, 7)$ and $(8, -3)$.
5. Has an inclination of $135°$ and a vertical intercept of -5.
6. Has an inclination of $63.5°$ and a horizontal intercept of 4.
7. Is perpendicular to the horizontal axis and passes through the point $(\frac{2}{3}, -\frac{1}{2})$.
8. Is parallel to the horizontal axis and passes through the point $(-6, -\frac{5}{3})$.
9. Passes through the point $(2, 5)$ and is parallel to the line $Y - 2X + 1 = 0$.
10. Is perpendicular to the line $2X + 6Y - 4 = 0$ and has a vertical intercept of -3.
11. Has a horizontal intercept of $(-2, 0)$ and a vertical intercept of $(0, -\frac{4}{5})$.
12. Passes through the origin and has an inclination of $45°$.

Write each of the following in slope-intercept form, list the slope and vertical intercept, and draw the graph.

13. $3X - Y + 8 = 0$
14. $X + 5Y + 6 = 0$
15. $2X - 4Y + 7 = 0$
16. $4X - Y - 2 = 0$
17. $5X - 4Y - 8 = 0$
18. $X + 3Y = 0$

Determine whether the following pairs of lines are parallel, perpendicular, or neither.

19. $(2X - Y - 7 = 0)$ and $(4X - 2Y + 8 = 0)$
20. $(3X + 4Y - 1 = 0)$ and $(X + 3Y + 2 = 0)$
21. $(Y - 5X + 4 = 0)$ and $(Y + 5X - 7 = 0)$
22. $(3Y - X + 6 = 0)$ and $(3X + Y - 5 = 0)$
23. The perimeter of a rectangular area is given in terms of its length and width by the equation $P = 2L + 2W$. If the width is fixed at 20 yards, graph the perimeter as a function of length and determine from the graph the perimeter when the length is 15 yards.
24. The relationship between Celsius and Fahrenheit temperatures is linear. If 0°C corresponds to 32°F and 100°C corresponds to 212°F, determine the equation which relates these temperatures, graph F as a function of C, and from the graph determine the Fahrenheit temperature corresponding to 30°C.
25. The resistance of a certain resistor increases by .01 ohm for every increase of 1°C. If the resistance is 3.5 ohms at 0°C, determine the equation relating resistance and temperature, graph resistance as a function of temperature, and from the graph determine the resistance when the temperature is 25°C.
26. For an object propelled vertically into the air with an initial velocity v_0 and subject to constant acceleration, the relationship between velocity and time is linear. If $v_0 = 144$ feet per second and the constant acceleration is -32 feet per second per second, determine the relationship between velocity and time, graph velocity as a function of time, and from the graph determine when the object reaches its maximum height (i.e., when the velocity is 0).

18–4. The Parabola

In the next four sections we will define the conic sections as mentioned in Section 18–1. The first of these curves to be considered is the *parabola*. We have encountered this type of curve previously. In Chapter 4, several of the graphs we constructed were parabolas, and in Chapter 13, we saw that the graph of a quadratic equation in one variable (when set equal to a second variable) is

Sec. 18–4. The Parabola

always a parabola. We now *define* a parabola as the locus, or path, of a point that moves so that it is always equidistant from a given point and a given line. The given point is called the *focus*, the given line is called the *directrix*, the line through the focus and perpendicular to the directrix is called the *axis* (discussed in Chapter 13 and called the axis of symmetry), and the point of the parabola that is on the axis is called the *vertex* (see Figure 18–4.1). We will only consider here parabolas opening up, down, right, or left.

Let us now consider a parabola with a vertex at some point (H, K) and an axis which is a vertical line (see Figure 18–4.2). Using the definition of a parabola with the distance formula, we have:

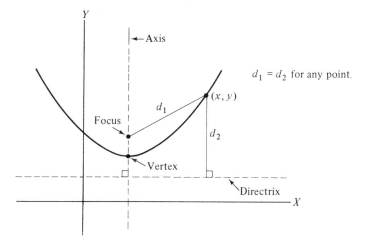

FIGURE 18–4.1

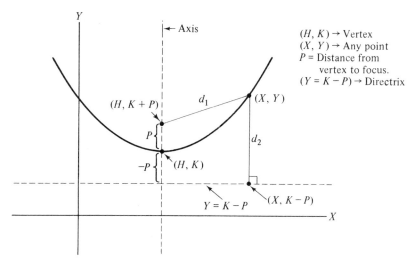

FIGURE 18–4.2

$$d_1 = d_2$$
$$\sqrt{(X-H)^2 + (Y-K-P)^2} = \sqrt{(X-X)^2 + (Y-K+P)^2}$$
$$(X-H)^2 + (Y-K-P)^2 = (X-X)^2 + (Y-K+P)^2$$
$$X^2 - 2XH + H^2 + \cancel{Y^2} + \cancel{K^2} + \cancel{P^2} - \cancel{2YK} - 2YP + 2KP$$
$$= \cancel{Y^2} + \cancel{K^2} + \cancel{P^2} - \cancel{2YK} + 2YP - 2KP$$
$$X^2 - 2XH + H^2 = 4YP - 4KP$$
$$(X-H)^2 = 4P(Y-K)$$

Therefore, the equation of a parabola with vertex at (H, K), focus P units from the vertex, and opening up or down is $(X-H)^2 = 4P(Y-K)$. If P is positive, the parabola opens *upward*. If P is negative, the parabola opens *downward*. We should note that if $(H, K) = (0, 0)$, then the vertex will be at the origin and the equation becomes $X^2 = 4PY$. Also, X and Y may be replaced by any variables or interchanged.

Example A: For the parabola represented by the equation $X^2 - 6X - 12Y - 15 = 0$, find the vertex, focus, and directrix, and sketch the graph. (Consider X the independent variable.)

$$X^2 - 6X - 12Y - 15 = 0$$
$$X^2 - 6X = 12Y + 15$$
$$X^2 - 6X + 9 = 12Y + 15 + 9 \quad \text{9 is added to both sides in order}$$
$$(X-3)^2 = 12Y + 24 \quad \text{to yield a perfect square on the left.}$$
$$(X-3)^2 = 12(Y+2)$$

Therefore, the vertex is at $(3, -2)$ and, since $4P = 12$, $P = 3$, meaning that we have a parabola opening upward.

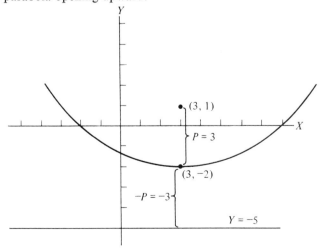

Sec. 18–4. The Parabola

Vertex ⟶ (3, −2)

Focus ⟶ (3, 1)

Directrix ⟶ $Y = -5$

Example B: For the parabola represented by the equation $X^2 + 4Y - 8 = 0$, find the vertex, focus, and directrix, and sketch the graph. (Consider X the independent variable.)

$$X^2 + 4Y - 8 = 0$$
$$X^2 = -4Y + 8$$
$$(X - 0)^2 = -4(Y - 2)$$

Therefore, the vertex is at (0, 2) and, since $4P = -4$, $P = -1$, meaning that we have a parabola opening downward.

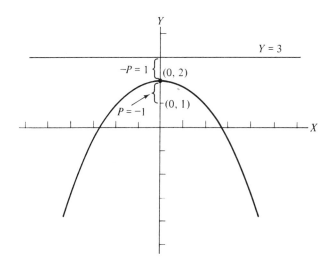

Vertex ⟶ (0, 2)

Focus ⟶ (0, 1)

Directrix ⟶ $Y = 3$

By going through a similar procedure, it can be shown that the equation of a parabola with vertex at (H, K) and an axis which is horizontal is $(Y - K)^2 = 4P(X - H)$, where (X, Y) is any point on the parabola. If P is *positive*, the parabola opens to the *right*. If P is *negative*, the parabola opens to the *left*. If $(H, K) = (0, 0)$, then we have the vertex at the origin and the equation becomes $Y^2 = 4PX$.

Example C: For the parabola represented by the equation $Y^2 + 2Y - 8X - 7 = 0$, find the vertex, focus, and directrix, and sketch the graph. (Consider X the independent variable.)

$$Y^2 + 2Y - 8X - 7 = 0$$
$$Y^2 + 2Y = 8X + 7$$
$$Y^2 + 2Y + 1 = 8X + 7 + 1$$
$$(Y + 1)^2 = 8X + 8$$
$$(Y + 1)^2 = 8(X + 1)$$

Therefore, the vertex is at $(-1, -1)$ and, since $4P = 8$, $P = 2$, meaning that we have a parabola opening to the right.

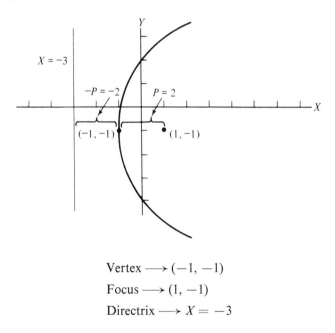

Vertex $\longrightarrow (-1, -1)$
Focus $\longrightarrow (1, -1)$
Directrix $\longrightarrow X = -3$

Example D: Find the equation of the parabola with vertex at $(2, -5)$ and focus at $(0, -5)$. Sketch the graph. (Use X and Y and consider X the independent variable.)

$$(H, K) = (2, -5)$$
$$P = -2$$

By plotting the vertex and focus, we should see that this parabola will open to the left.

Sec. 18–4. The Parabola

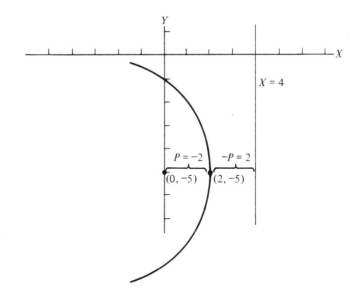

Therefore,
$$(Y - K)^2 = 4P(X - H)$$
$$(Y + 5)^2 = -8(X - 2)$$
$$Y^2 + 10Y + 25 = -8X + 16$$
$$Y^2 + 10Y + 8X + 9 = 0$$

From the previous examples we should note that any parabola opening up, down, right, or left will have an equation that can be written as $AX^2 + BX + CY + D = 0$ or $AY^2 + BY + CX + D = 0$. In each case, only one of the variables (X or Y) is squared, and these equations are called the *general forms* of the equation of a parabola. Since X and Y may be replaced by any letters, we may generalize as follows. If an equation contains two variables with exactly one of them squared, the equation represents a parabola. If the independent variable is squared, the parabola opens up or down. If the dependent variable is squared, the parabola opens right or left.

Exercises For each of the following, determine the vertex, focus, and directrix, and sketch the graph.

1. $X^2 - 10Y = 0$ X is the independent variable.
2. $B^2 + 12A = 0$ A is the independent variable.
3. $T^2 - 10V + 40 = 0$ T is the independent variable.
4. $K^2 + 9L + 18 = 0$ L is the independent variable.

5. $X^2 - 8X + 6Y + 34 = 0$ X is the independent variable.
6. $R^2 + 4R - 2S + 14 = 0$ S is the independent variable.
7. $P^2 + 6P - 14Q - 5 = 0$ Q is the independent variable.
8. $2A^2 - 4A + 14B - 26 = 0$ B is the independent variable.
9. $X^2 + 6X - 12Y + 15 = 0$ X is the independent variable.
10. $T^2 - 10T + 8V + 31 = 0$ T is the independent variable.

For each of the following, find the equation of the parabola with the given conditions and sketch the graph. Write each equation in the general form.

11. Vertex at $(0, 0)$, focus at $(0, 3)$. Use X and Y with X the independent variable.
12. Vertex at $(0, 0)$, directrix $X = -2$. Use X and Y with X the independent variable.
13. Vertex at $(-2, 0)$, directrix $A = 4$. Use A and B with B the independent variable.
14. Vertex at $(0, -1)$, focus at $(-3, -1)$. Use T and V with T the independent variable.
15. Focus at $(2, 3)$, directrix $X = -4$. Use X and Y with X the independent variable.
16. Focus at $(-4, 1)$, directrix $K = -3$. Use K and L with L the independent variable.
17. Vertex at $(3, \frac{1}{2})$, focus at $(-1, \frac{1}{2})$. Use M and N with M the independent variable.
18. Vertex at $(-\frac{1}{4}, 2)$, directrix $R = -\frac{5}{4}$. Use R and S with R the independent variable.
19. Focus at $(-4, 3)$, directrix $Q = 0$. Use P and Q with P the independent variable.
20. Vertex at $(-2, -6)$, directrix $V = 0$. Use T and V with V the independent variable.
21. In optics, any light ray emanating from the focus of a parabolic reflecting surface and directed at the surface will be reflected off parallel to the axis of the parabola. If a light ray from the focus of the parabola $Y^2 - 9X = 0$ strikes the parabola at the point $(4, 6)$, what is the equation of the reflected light ray? (Consider X the independent variable.)
22. The arch of a certain bridge is parabolic in shape. If the span of the arch, at the points where it meets the ground, is 100 feet and the maximum height above this span is 25 feet, find the equation of the arch of the bridge. Choose a convenient point as the origin of the coordinate system.

Sec. 18–5. The Circle

23. If the area of a circle is given by the equation $A = \pi R^2$, sketch the graph of this equation using R as the independent variable and determine the vertex, focus, and directrix.

24. The volume of a right circular cone is given by the equation $V = \frac{1}{3}\pi R^2 H$. Sketch the graph of this equation for a cone with a height of 12 meters, using R as the independent variable, and then determine the vertex, focus, and directrix.

18–5. The Circle

The next curve we will look at in detail is the *circle*. A circle is the locus, or path, of a point that moves so that it is always a fixed distance from a fixed point. The fixed point is called the *center* and the fixed distance is called the *radius* (see Figure 18–5.1).

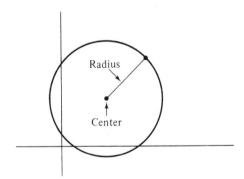

FIGURE 18–5.1

Let us now consider a circle with center at some point (H, K) and a radius of R (see Figure 18–5.2). Using the distance formula we have

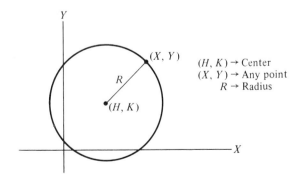

FIGURE 18–5.2

$$\sqrt{(X-H)^2+(Y-K)^2}=R$$
$$(X-H)^2+(Y-K)^2=R^2$$

Therefore, the equation of a circle with a *center at* (H, K) and a *radius of* R is

$$(X-H)^2+(Y-K)^2=R^2.$$

We should note that if $(H, K) = (0, 0)$, the center is at the origin and the equation becomes $X^2 + Y^2 = R^2$.

Example A: For the circle represented by the equation $X^2 + Y^2 - 6X - 4Y - 12 = 0$, determine the center and radius and sketch the graph. (Consider X the independent variable.)

$$X^2 + Y^2 - 6X - 4Y - 12 = 0$$
$$X^2 - 6X + Y^2 - 4Y = 12$$
$$X^2 - 6X + 9 + Y^2 - 4Y + 4 = 12 + 9 + 4$$
$$(X-3)^2 + (Y-2)^2 = 25$$

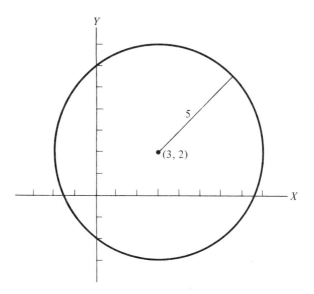

Therefore, center $= (3, 2)$ and radius $= 5$. Every point on the circle is exactly 5 units from the point $(3, 2)$.

Example B: For the circle represented by the equation $T^2 + V^2 + 2T + 4V - 4 = 0$, determine the center and radius, and sketch the graph. (Use T as the independent variable.)

Sec. 18–5. The Circle

$$T^2 + V^2 + 2T + 4V - 4 = 0$$
$$T^2 + 2T + V^2 + 4V = 4$$
$$T^2 + 2T + 1 + V^2 + 4V + 4 = 4 + 1 + 4$$
$$(T + 1)^2 + (V + 2)^2 = 9$$

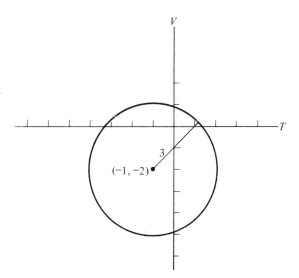

Therefore, center $= (-1, -2)$ and radius $= 3$. Every point on the circle is exactly 3 units from the point $(-1, -2)$.

Example C: Find the equation of the circle with center at $(2, -4)$ and a radius of $\sqrt{3}$. Sketch the graph. (Use X and Y, and consider X the independent variable.)

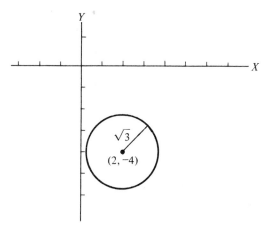

Therefore,
$$(X - H)^2 + (Y - K)^2 = R^2$$

$$(X-2)^2 + (Y+4)^2 = (\sqrt{3})^2$$
$$X^2 - 4X + 4 + Y^2 + 8Y + 16 = 3$$
$$X^2 + Y^2 - 4X + 8Y + 17 = 0$$

From the previous examples we should note that any circle will have an equation that can be written as

$$X^2 + Y^2 + AX + BY + C = 0$$

This is called the *general form* of the *equation of a circle*. We should see that the basic characteristic of this equation is that both variables are squared and the squared terms have coefficients of 1. If these coefficients are the same but are not equal to 1, we should divide each term of the equation by their common value.

Example D: For the circle represented by the equation $2X^2 + 2Y^2 - 4Y - 6 = 0$, determine the center and radius, and sketch the graph. (Consider X the independent variable.)

$$2X^2 + 2Y^2 - 4Y - 6 = 0$$
$$X^2 + Y^2 - 2Y - 3 = 0$$
$$X^2 + Y^2 - 2Y = 3$$
$$X^2 + Y^2 - 2Y + 1 = 3 + 1$$
$$X^2 + (Y-1)^2 = 4$$

Therefore, center = (0, 1) and radius = 2.

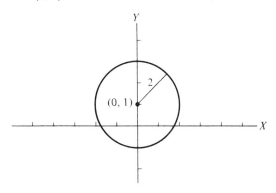

Exercises

For each of the following, determine the center and radius, and sketch the graph.

1. $X^2 + Y^2 - 9 = 0$ X is the independent variable.
2. $M^2 + N^2 - 6 = 0$ M is the independent variable.

Sec. 18–5. The Circle

3. $K^2 + L^2 + 4L - 21 = 0$ K is the independent variable.
4. $A^2 + B^2 - 2A - 15 = 0$ A is the independent variable.
5. $P^2 + Q^2 + 2P - 8Q + 16 = 0$ P is the independent variable.
6. $X^2 + Y^2 - 10X + 2Y + 25 = 0$ X is the independent variable.
7. $R^2 + S^2 + 6R + 10S + 26 = 0$ R is the independent variable.
8. $T^2 + V^2 - 8T - 14V + 60 = 0$ T is the independent variable.
9. $2X^2 + 2Y^2 - 8X + 10Y - 24 = 0$ X is the independent variable.
10. $3M^2 + 3N^2 + 18M - 24N - 31 = 0$ M is the independent variable.

For each of the following, find the equation of the circle with the given conditions and sketch the graph. Write each equation in the general form.

11. Center at $(-4, 0)$ and radius of 2. Use X and Y with X the independent variable.
12. Center at $(0, 0)$ and radius of $\sqrt{5}$. Use T and V with T the independent variable.
13. Center at $(2, -3)$ and radius of $\sqrt{11}$. Use M and N with M the independent variable.
14. Center at $(4, \frac{1}{2})$ and radius of 1. Use K and L with K the independent variable.
15. Center at $(0, -1)$ and passes through $(4, 1)$. Use R and S with R the independent variable.
16. Center at $(-2, 4)$ and passes through $(1, -1)$. Use A and B with A the independent variable.
17. Center at $(-1, 3)$ and tangent to the line $X = 3$. Use X and Y with X the independent variable.
18. Center at $(2, -2)$ and tangent to the line $V = -4$. Use T and V with T the independent variable.
19. The points $(-2, -3)$ and $(2, 5)$ are the ends of a diameter. Use X and Y with X the independent variable.
20. The points $(-5, -3)$ and $(3, 1)$ are the ends of a diameter. Use P and Q with P the independent variable.
21. The Pythagorean Theorem for a right triangle states: $A^2 + B^2 = C^2$, where C is the hypotenuse. If the hypotenuse of a certain right triangle must be 12, draw the graph that represents the possible values for the sides A and B.
22. Two circular machine parts have radii R_1 and R_2. The sum of the surface areas of the two parts must be 62.8 square centimeters. Draw the graph representing the possible values for the radii of the two parts.

23. Find, graphically, any zeros of the relationship $X^2 + Y^2 + 6X - 2Y + 6 = 0$. (Use X as the independent variable.)

24. Find, graphically, any zeros of the relationship $K^2 + L^2 + 4K + 10L + 21 = 0$. (Use K as the independent variable.)

18–6. The Ellipse

We will now consider a curve known as an *ellipse*. An ellipse is the locus, or path, of a point that moves so that the sum of its distances from two fixed points is constant. The two fixed points are called the *foci* of the ellipse. Figure 18–6.1

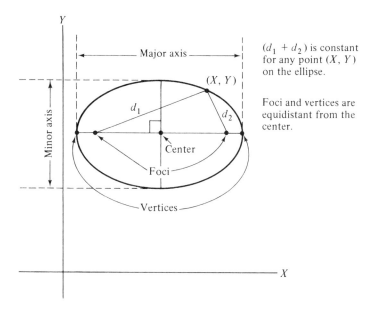

FIGURE 18–6.1

shows all of the important concepts associated with the ellipse. We will only consider here ellipses with axes that are horizontal and vertical.

Let us now consider an ellipse with a center at some point (H, K) and a major axis which is horizontal. Also, let $2A$ represent the length of the major axis, let $2B$ represent the length of the minor axis, and let $2C$ represent the distance from one focus to the other (see Figure 18–6.2). Using point P(one of the vertices) as a point of the ellipse, we should see that the sum of the distances from this point to the foci is $2A$. This sum must then be obtained for the distances from any other point on the ellipse. Therefore, using this definition of the ellipse with the distance formula we have

$$d_1 + d_2 = 2A$$

Sec. 18–6. The Ellipse

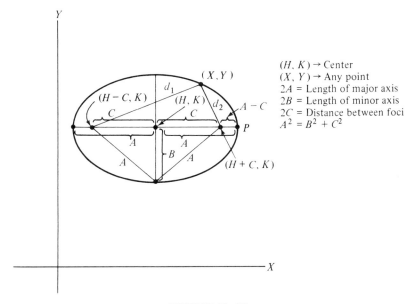

FIGURE 18-6.2

$\sqrt{(X-H+C)^2+(Y-K)^2}$
$\qquad + \sqrt{(X-H-C)^2+(Y-K)^2} = 2A$
$\qquad\qquad \sqrt{(X-H+C)^2+(Y-K)^2} = 2A - \sqrt{(X-H-C)^2+(Y-K)^2}$
$\qquad\qquad \cancel{X^2}+\cancel{H^2}+\cancel{C^2}-\cancel{2XH}+2XC$
$\qquad\qquad\qquad -2HC+\cancel{Y^2}+\cancel{K^2}-\cancel{2YK} = 4A^2 - 4A\sqrt{(X-H-C)^2+(Y-K)^2}$
$\qquad\qquad\qquad\qquad\qquad +\cancel{X^2}+\cancel{H^2}+\cancel{C^2}-\cancel{2XH}-2XC$
$\qquad\qquad\qquad\qquad\qquad +2HC+\cancel{Y^2}+\cancel{K^2}-\cancel{2YK}$
$\qquad\qquad 4XC - 4HC - 4A^2 = -4A\sqrt{(X-H-C)^2+(Y-K)^2}$
$\qquad\qquad XC - HC - A^2 = -A\sqrt{(X-H-C)^2+(Y-K)^2}$
$X^2C^2 + H^2C^2 + A^4$
$\qquad -2XHC^2 - 2XCA^2 + 2HCA^2 = A^2[(X^2+H^2+C^2-2XH-2XC$
$\qquad\qquad\qquad\qquad\qquad\qquad\qquad +2HC)+(Y-K)^2]$
$X^2C^2 + H^2C^2 + A^4$
$\qquad -2XHC^2 - \cancel{2XCA^2} + \cancel{2HCA^2} = A^2X^2 + A^2H^2 + A^2C^2 - 2XHA^2$
$\qquad\qquad\qquad\qquad\qquad\qquad - \cancel{2XCA^2} + \cancel{2HCA^2} + A^2(Y-K)^2$
$C^2(X^2 - 2XH + H^2) - A^2$
$\qquad (X^2 - 2XH + H^2) - A^2(Y-K)^2 = A^2C^2 - A^4$
$A^2(X^2 - 2XH + H^2) - C^2$
$\qquad (X^2 - 2XH + H^2) + A^2(Y-K)^2 = A^4 - A^2C^2$

$$A^2(X-H)^2$$
$$-C^2(X-H)^2 + A^2(Y-K)^2 = A^2(A^2-C^2)$$
$$(A^2-C^2)(X-H)^2 + A^2(Y-K)^2 = A^2(A^2-C^2)$$

But, $\quad (A^2 - C^2) = B^2 \quad$ (refer back to the graph)

Therefore,

$$B^2(X-H)^2 + A^2(Y-K)^2 = A^2B^2$$

$$\frac{(X-H)^2}{A^2} + \frac{(Y-K)^2}{B^2} = 1$$

Thus, the equation of an ellipse with center at (H, K), a horizontal major axis of $2A$, and a vertical minor axis of $2B$ is

$$\frac{(X-H)^2}{A^2} + \frac{(Y-K)^2}{B^2} = 1$$

By using the same procedure it can be shown that if the major axis is vertical, the equation becomes

$$\frac{(X-H)^2}{B^2} + \frac{(Y-K)^2}{A^2} = 1$$

In other words, since $A > B$, if the major axis is horizontal, the independent variable will be over the larger number, and if the major axis is vertical, the dependent variable will be over the larger number.

Example A: If X is the independent variable and Y is the dependent variable, the ellipse represented by

$$\frac{(X-3)^2}{8} + \frac{(Y-2)^2}{6} = 1$$

would have a horizontal major axis, while the ellipse represented by

$$\frac{(X+1)^2}{9} + \frac{(Y-4)^2}{12} = 1$$

would have a vertical major axis. Also, if $(H, K) = (0, 0)$, the center is at the origin and the equation becomes

$$\frac{X^2}{A^2} + \frac{Y^2}{B^2} = 1 \quad \text{or} \quad \frac{X^2}{B^2} + \frac{Y^2}{A^2} = 1$$

Example B: For the ellipse represented by the equation $16X^2 + 25Y^2 - 64X - 50Y - 311 = 0$, determine the center, vertices, foci, and major and minor axes, and sketch the graph. (Consider X the independent variable.)

Sec. 18-6. The Ellipse

$$16X^2 + 25Y^2 - 64X - 50Y - 311 = 0$$
$$16X^2 - 64X + 25Y^2 - 50Y = 311$$
$$16(X^2 - 4X) + 25(Y^2 - 2Y) = 311$$
$$16(X^2 - 4X + 4) + 25(Y^2 - 2Y + 1) = 311 + 64 + 25$$
$$16(X - 2)^2 + 25(Y - 1)^2 = 400$$
$$\frac{(X - 2)^2}{25} + \frac{(Y - 1)^2}{16} = 1$$

Therefore, the center = (2, 1), major axis = $2\sqrt{25} = 10$, and minor axis = $2\sqrt{16} = 8$. Also, major axis is horizontal.

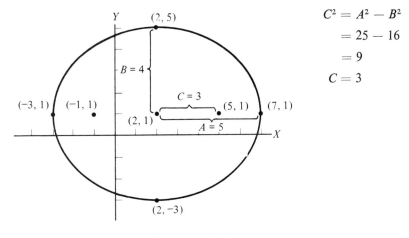

$$C^2 = A^2 - B^2$$
$$= 25 - 16$$
$$= 9$$
$$C = 3$$

Center = (2, 1)
Vertices = (7, 1) and (−3, 1)
Foci = (5, 1) and (−1, 1)
Major Axis = 10
Minor Axis = 8

Example C: For the ellipse represented by the equation $12T^2 + 4V^2 + 16V - 32 = 0$, determine the center, vertices, foci, and major and minor axes, and sketch the graph. (Use T as the independent variable.)

$$12T^2 + 4V^2 + 16V - 32 = 0$$
$$12T^2 + 4(V^2 + 4V) = 32$$
$$12T^2 + 4(V^2 + 4V + 4) = 32 + 16$$
$$12T^2 + 4(V + 2)^2 = 48$$
$$\frac{T^2}{4} + \frac{(V + 2)^2}{12} = 1$$

Therefore, the center = (0, −2), major axis = $2\sqrt{12} = 6.93$, and minor axis = $2\sqrt{4} = 4$. Also, major axis is vertical.

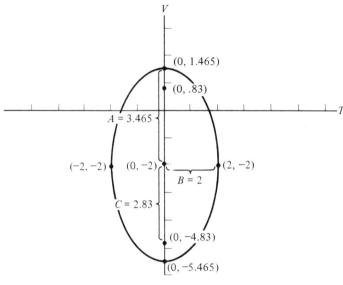

$$C^2 = A^2 - B^2$$
$$= 12 - 4$$
$$= 8$$
$$C = 2.83$$

Center $= (0, -2)$
Vertices $= (0, 1.465)$ and $(0, -5.465)$
Foci $= (0, .83)$ and $(0, -4.83)$
Major Axis $= 6.93$
Minor Axis $= 4$

Example D: Determine the equation of the ellipse with center at $(-1, -2)$, minor axis $= 6$, and a vertex at $(-1, 3)$. (Use M and N with M the independent variable.)

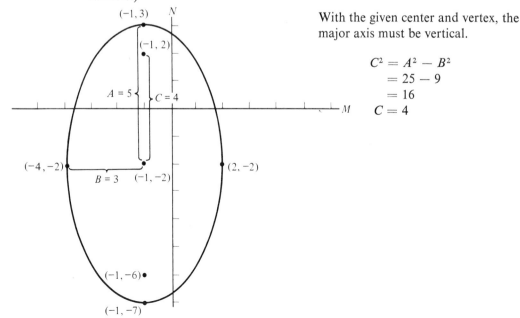

With the given center and vertex, the major axis must be vertical.

$$C^2 = A^2 - B^2$$
$$= 25 - 9$$
$$= 16$$
$$C = 4$$

Sec. 18–6. The Ellipse

$$\frac{(M-H)^2}{B^2} + \frac{(N-K)^2}{A^2} = 1$$

$$\frac{(M+1)^2}{9} + \frac{(N+2)^2}{25} = 1$$

$$25(M^2 + 2M + 1) + 9(N^2 + 4N + 4) = 225$$
$$25M^2 + 50M + 25 + 9N^2 + 36N + 36 = 225$$
$$25M^2 + 9N^2 + 50M + 36N - 164 = 0$$

From the previous examples, we should note that any ellipse with axes that are horizontal and vertical will have an equation that may be written in the form $AX^2 + BY^2 + CX + DY + F = 0$. This is called the *general form of the equation of an ellipse*. The basic characteristic of this equation is that both variables are squared, with the coefficients of the squared terms positive and unequal.

Exercises

For each of the following, determine the center, vertices, foci, and major and minor axes, and sketch the graph.

1. $4X^2 + 9Y^2 - 36 = 0$ X is the independent variable.
2. $T^2 + 16V^2 - 16 = 0$ V is the independent variable.
3. $3M^2 + 5N^2 - 10N - 25 = 0$ M is the independent variable.
4. $2K^2 + 3L^2 + 8K - 10 = 0$ L is the independent variable.
5. $16A^2 + 25B^2 - 32A + 50B - 359 = 0$ A is the independent variable.
6. $4P^2 + 5Q^2 + 24P - 20Q + 36 = 0$ Q is the independent variable.
7. $2X^2 + 7Y^2 + 8X + 56Y + 106 = 0$ X is the independent variable.
8. $2R^2 + 3S^2 - 16R + 6S + 29 = 0$ R is the independent variable.
9. $A^2 + 3B^2 - 10A + 18B + 37 = 0$ B is the independent variable.
10. $3T^2 + 2V^2 + 24T - 20V + 74 = 0$ T is the independent variable.

For each of the following, find the equation of the ellipse with the given conditions and sketch the graph. Write each equation in the general form.

11. Center at (0, 0), vertex at (8, 0), and focus at (6, 0). Use X and Y with X the independent variable.
12. Center at (0, 0) focus at (0, 3), and a minor axis of 8. Use T and V with T the independent variable.
13. Foci at (1, 5) and (1, −1), and a major axis of 8. Use K and L with K the independent variable.
14. Foci at (1, 3) and (−1, 3), and a minor axis of 2. Use P and Q with P the independent variable.

15. Center at $(-2, 4)$, vertex at $(-2, 9)$, and a minor axis of 6. Use M and N with M the independent variable.

16. Vertices at $(3, 8)$ and $(3, -2)$, and a focus at $(3, 0)$. Use R and S with R the independent variable.

17. Find graphically any zeros of the equation $9X^2 + 16Y^2 - 72X - 32Y - 416 = 0$. Consider X the independent variable.

18. A square and a circle are cut from several different pieces of sheet metal. If the sum of the area of the square and the circle must always be 4π square units, sketch the graph showing the relationship between the radius of the circle (R) and the side of the square (S). Use R as the independent variable.

18–7. The Hyperbola

The last curve we will consider in this chapter is the *hyperbola*. A hyperbola is the locus, or path, of a point that moves so that the difference of its distances from two fixed points is constant. The two fixed points are called the *foci* of the hyperbola. Figure 18–7.1 shows all of the important concepts associated with

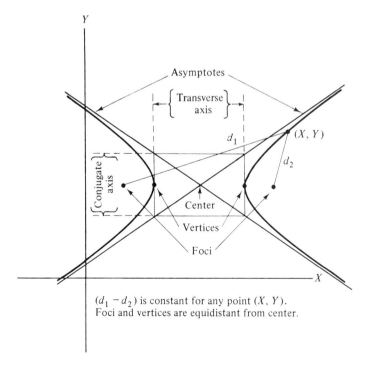

$(d_1 - d_2)$ is constant for any point (X, Y).
Foci and vertices are equidistant from center.

FIGURE 18–7.1

Sec. 18–7. The Hyperbola

the hyperbola. We will only consider here hyperbolas opening right and left or up and down.

Let us now consider a hyperbola with a center at some point (H, K) and a transverse axis that is horizontal (i.e., a hyperbola opening right and left). Also, let $2A$ represent the length of the transverse axis, let $2B$ represent the length of the conjugate axis, and let $2C$ represent the distance from one focus to the other (see Figure 18–7.2). Using point P (one of the vertices) as a point of the hyper-

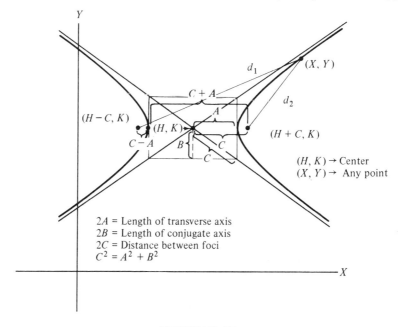

FIGURE 18–7.2

bola, we should see that the difference of the distances from this point to the foci is $2A$. The same difference must then be obtained for the distances from any other point on the hyperbola. Therefore, using this definition of the hyperbola with the distance formula, we have:

$$d_1 - d_2 = 2A$$
$$\sqrt{(X - H + C)^2 + (Y - K)^2}$$
$$- \sqrt{(X - H - C)^2 + (Y - K)^2} = 2A$$
$$\sqrt{(X - H + C)^2 + (Y - K)^2} = 2A + \sqrt{(X - H - C)^2 + (Y - K)^2}$$
$$\cancel{X^2} + \cancel{H^2} + \cancel{C^2} - \cancel{2XH} + 2XC$$
$$- 2HC + \cancel{Y^2} + \cancel{K^2} - \cancel{2YK} = 4A^2 + 4A\sqrt{(X - H - C)^2 + (Y - K)^2}$$
$$+ \cancel{X^2} + \cancel{H^2} + \cancel{C^2} - \cancel{2XH} - 2XC$$
$$+ 2HC + \cancel{Y^2} + \cancel{K^2} - \cancel{2YK}$$

$$4XC - 4HC - 4A^2 = 4A\sqrt{(X-H-C)^2 + (Y-K)^2}$$
$$XC - HC - A^2 = A\sqrt{(X-H-C)^2 + (Y-K)^2}$$
$$X^2C^2 + H^2C^2 + A^4 - 2XHC^2$$
$$- 2XCA^2 + 2HCA^2 = A^2[(X^2 + H^2 + C^2 - 2XH$$
$$- 2XC + 2HC) + (Y-K)^2]$$
$$X^2C^2 + H^2C^2 + A^4 - 2XHC^2$$
$$-\cancel{2XCA^2} + \cancel{2HCA^2} = A^2X^2 + A^2H^2 + A^2C^2 - 2XHA^2$$
$$-\cancel{2XCA^2} + \cancel{2HCA^2}$$
$$+ A^2(Y-K)^2$$

$$C^2(X^2 - 2XH + H^2) - A^2$$
$$(X^2 - 2XH + H^2) - A^2(Y-K)^2 = A^2C^2 - A^4$$
$$C^2(X-H)^2 - A^2(X-H)^2$$
$$- A^2(Y-K)^2 = A^2(C^2 - A^2)$$
$$(C^2 - A^2)(X-H)^2$$
$$- A^2(Y-K)^2 = A^2(C^2 - A^2)$$

But, $\qquad (C^2 - A^2) = B^2$ (refer back to the graph).

Therefore, $B^2(X-H)^2 - A^2(Y-K)^2 = A^2B^2$

$$\frac{(X-H)^2}{A^2} - \frac{(Y-K)^2}{B^2} = 1$$

Thus, the equation of a hyperbola with center at (H, K), a horizontal transverse axis of $2A$, and a vertical conjugate axis of $2B$ is

$$\frac{(X-H)^2}{A^2} - \frac{(Y-K)^2}{B^2} = 1$$

By using the same procedure, it can be shown that if the transverse axis is vertical, the equation becomes

$$\frac{(Y-K)^2}{A^2} - \frac{(X-H)^2}{B^2} = 1$$

In other words, if the transverse axis is horizontal, the independent variable is associated with the plus sign, and if the transverse axis is vertical, the dependent variable is associated with the plus sign.

Example A: If X is the independent variable and Y is the dependent variable, the hyperbola represented by

$$\frac{(X+2)^2}{7} - \frac{(Y-3)^2}{9} = 1$$

would have a horizontal transverse axis, while the hyperbola represented by

$$\frac{(Y-1)^2}{4} - \frac{(X+5)^2}{2} = 1$$

would have a vertical transverse axis.

NOTE: For a hyperbola, A or B may be larger, whereas with the ellipse, A is always greater than B.

From the graph we should also note that the asymptotes of the hyperbola both pass through the point (H, K) and have slopes of

$$\frac{B}{A} \quad \text{and} \quad \frac{-B}{A}$$

Therefore, the equations of these asymptotes, using the point-slope form for the equation of a straight line, are

$$(Y - K) = \frac{B}{A}(X - H) \quad \text{and} \quad (Y - K) = \frac{-B}{A}(X - H)$$

$\left[(Y - K) = \pm\frac{A}{B}(X - H) \text{ if the transverse axis is vertical.}\right]$ Also, if $(H, K) = (0, 0)$, the center will be at the origin and the equation becomes

$$\frac{X^2}{A^2} - \frac{Y^2}{B^2} = 1 \quad \text{or} \quad \frac{Y^2}{A^2} - \frac{X^2}{B^2} = 1$$

Example B: For the hyperbola represented by $9X^2 - 4Y^2 - 36X + 8Y - 4 = 0$, determine the center, vertices, foci, asymptotes, and transverse and conjugate axes, and sketch the graph. (Consider X the independent variable.)

$$9X^2 - 4Y^2 - 36X + 8Y - 4 = 0$$
$$9X^2 - 36X - 4Y^2 + 8Y = 4$$
$$9(X^2 - 4X) - 4(Y^2 - 2Y) = 4$$

$$9(X^2 - 4X + 4) - 4(Y^2 - 2Y + 1) = 4 + 36 - 4$$
$$9(X - 2)^2 - 4(Y - 1)^2 = 36$$
$$\frac{(X - 2)^2}{4} - \frac{(Y - 1)^2}{9} = 1$$

Therefore, the center $= (2, 1)$, transverse axis $= 2\sqrt{4} = 4$, and conjugate axis $= 2\sqrt{9} = 6$.

Also, transverse axis is horizontal.

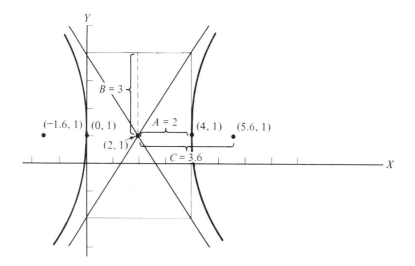

$$C^2 = A^2 + B^2$$
$$= 4 + 9$$
$$= 13$$
$$C = 3.6$$

Center $= (2, 1)$
Vertices $= (0, 1)$ and $(4, 1)$
Foci $= (-1.6, 1)$ and $(5.6, 1)$
Transverse axis $= 4$
Conjugate axis $= 6$
Asymptotes: $(Y - 1) = \frac{3}{2}(X - 2)$ or $2Y - 3X + 4 = 0$
$(Y - 1) = \frac{-3}{2}(X - 2)$ or $2Y + 3X - 8 = 0$

Example C: For the hyperbola represented by $N^2 - M^2 + 2N + 6M - 10 = 0$, determine the center, vertices, foci, asymptotes, and transverse and conjugate axes, and sketch the graph. (Consider M the independent variable.)

$$N^2 - M^2 + 2N + 6M - 10 = 0$$
$$N^2 + 2N - M^2 + 6M = 10$$

Sec. 18–7. The Hyperbola 515

$$(N^2 + 2N) - (M^2 - 6M) = 10$$
$$(N^2 + 2N + 1) - (M^2 - 6M + 9) = 10 + 1 - 9$$
$$(N + 1)^2 - (M - 3)^2 = 2$$
$$\frac{(N + 1)^2}{2} - \frac{(M - 3)^2}{2} = 1$$

Therefore, the center = $(-1, 3)$, transverse axis = $2\sqrt{2} = 2.8$, and conjugate axis = $2\sqrt{2} = 2.8$.

Also, transverse axis is vertical.

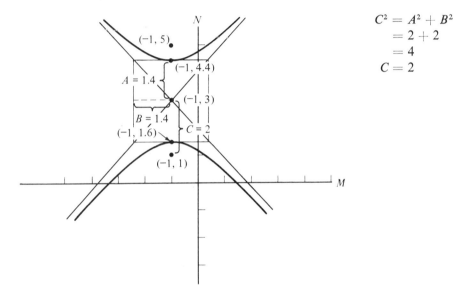

$$C^2 = A^2 + B^2$$
$$= 2 + 2$$
$$= 4$$
$$C = 2$$

Center = $(-1, 3)$
Vertices = $(-1, 1.6)$ and $(-1, 4.4)$
Foci = $(-1, 1)$ and $(-1, 5)$
Transverse axis = 2.8
Conjugate axis = 2.8
Asymptotes: $(N - 3) = 1(M + 1)$ or $N - M - 4 = 0$
$(N - 3) = -1(M + 1)$ or $N + M - 2 = 0$

Example D: Find the equation of the hyperbola with center at $(2, -3)$, a conjugate axis of 8, and a focus at $(-3, -3)$. (Use R and S with R the independent variable.) With the given center and focus, the transverse axis must be horizontal.

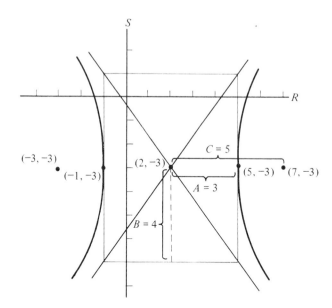

$$C^2 = A^2 + B^2$$
$$C^2 - B^2 = A^2$$
$$25 - 16 = 9 = A^2$$
$$3 = A$$

$$\frac{(R-H)^2}{A^2} - \frac{(S-K)^2}{B^2} = 1$$

$$\frac{(R-2)^2}{9} - \frac{(S+3)^2}{16} = 1$$

$$16(R^2 - 4R + 4) - 9(S^2 + 6S + 9) = 144$$

$$16R^2 - 64R + 64 - 9S^2 - 54S - 81 = 144$$

$$16R^2 - 9S^2 - 64R - 54S - 161 = 0$$

From the previous examples, we should note that any hyperbola with axes that are horizontal and vertical will have an equation that may be written in the form

$$AX^2 - BY^2 + CX + DY + E = 0 \quad \text{or} \quad AY^2 - BX^2 + CX + DY + E = 0$$

These are called the *general forms of the equation of a hyperbola*. The basic characteristic of these equations is that both variables are squared, with the coefficients of the squared terms of opposite signs. If the coefficient of the square of the independent variable is positive, the hyperbola opens right and left. If the coefficient of the square of the dependent variable is positive, the hyperbola opens up and down.

Another type of equation for which the graph is a hyperbola is

$$XY = K$$

Sec. 18–7. The Hyperbola

If $K > 0$, the hyperbola will be in the first and third quadrants. If $K < 0$, the hyperbola will be in the second and fourth quadrants. In both cases, the center will be at the origin, the coordinate axes will be the asymptotes, and the vertices and foci will be on the line $Y = X$ or $Y = -X$.

Example E: Sketch the graph of $XY = 6$. (Use X as the independent variable.)

X	1	−1	2	−2	3	−3	6	−6
Y	6	−6	3	−3	2	−2	1	−1

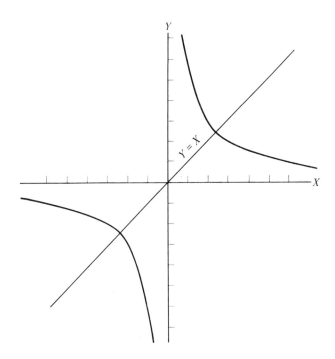

Example F: Sketch the graph of $MN = -4$. (Use M as the independent variable.)

M	1	−1	2	−2	4	−4
N	−4	4	−2	2	−1	1

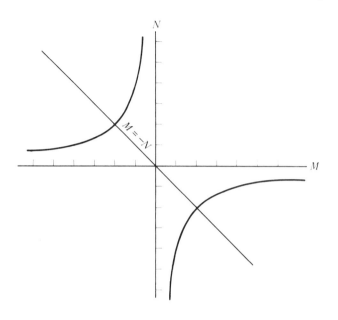

Exercises

For each of the following, determine the center, vertices, foci, asymptotes, and transverse and conjugate axes, and sketch the graph.

1. $4X^2 - 9Y^2 - 36 = 0$ X is the independent variable.
2. $3V^2 - 2T^2 - 6 = 0$ T is the independent variable.
3. $5S^2 - 2R^2 - 20S + 10 = 0$ R is the independent variable.
4. $4M^2 - 3N^2 + 24M + 24 = 0$ M is the independent variable.
5. $2A^2 - B^2 - 12A + 2B + 11 = 0$ A is the independent variable.
6. $Q^2 - P^2 + 8Q - 4P + 8 = 0$ P is the independent variable.
7. $X^2 - Y^2 - 10X - 2Y + 23 = 0$ X is the independent variable.
8. $3L^2 - 3K^2 + 12L - 12K - 9 = 0$ K is the independent variable.
9. $3T^2 - 8V^2 - 12T - 64V - 140 = 0$ T is the independent variable.
10. $2S^2 - 5R^2 + 8S - 60R - 182 = 0$ R is the independent variable.

For each of the following, find the equation of the hyperbola with the given conditions and sketch the graph. Write each equation in the general form.

11. Center at $(0, 0)$, focus at $(0, 5)$, and vertex at $(0, -3)$. Use X and Y with X the independent variable.
12. Center at $(0, 0)$, focus at $(4, 0)$, and a conjugate axis of 6. Use R and S with R the independent variable.

13. Foci at (2, 3) and (2, −5), and a transverse axis of 4. Use M and N with M the independent variable.
14. Foci at (−6, −1) and (0, −1), and a conjugate axis of 5. Use K and L with K the independent variable.
15. Vertices at (−5, 2) and (1, 2), and a conjugate axis of 6. Use P and Q with P the independent variable.
16. Center at (4, −3), vertex at (4, 0), and focus at (4, −7). Use T and V with T the independent variable.

Sketch the graphs of each of the following.

17. $XY = 8$ Use X as the independent variable.
18. $RS = -10$ Use R as the independent variable.
19. Find graphically any zeros of the equation $5T^2 - 4V^2 - 40T - 24V + 24 = 0$. Consider V the independent variable.
20. One version of Boyle's law is that for a perfect gas, the product of the pressure (P) and volume (V) is constant. If temperature is held constant, one corresponding set of values for a certain gas is $P = 4$ atmospheres and $V = 8$ liters. Sketch the graph showing the relationship between pressure and volume. Use V as the independent variable.
21. The formula for the area of a rectangle is $A = LW$. If the area (A) of a certain rectangular field must be 1600 square meters, sketch the graph showing the possible values for the length (L) and the width (W). Use L as the independent variable.
22. Refer to problem (18) at the end of the previous section. If the area of the square must be greater than the area of the circle by 2π square units, sketch the graph showing the relationship between the radius of the circle (R) and the side of the square (S). Use S as the independent variable.

18–8. The General Second Degree Equation

In this section we will summarize the general characteristics of the equations of a straight line, a parabola, a circle, an ellipse, and an hyperbola. Each of these curves can be represented by an equation of the form

$$AX^2 + BY^2 + CXY + DX + EY + F = 0$$

This is called the *general second degree equation in two variables*. In this equation, A, B, C, D, E, and F represent constants and their values determine which kind of curve is represented by the equation.

1. If $A = B = C = 0$, the equation represents a straight line.

Example A:
$$2X + 3Y - 8 = 0$$
$$5X - 7Y + 2 = 0$$

2. If $A = 0$ or $B = 0$ (but not both) and $C = 0$, the equation represents a parabola.

Example B:
$$3X^2 + X - 2Y + 7 = 0$$
$$Y^2 - X + 4Y + 6 = 0$$

3. If $A = B$ and $C = 0$, the equation represents a circle.

Example C:
$$X^2 + Y^2 + 3X - 2Y + 8 = 0$$
$$2X^2 + 2Y^2 - 7Y + 2 = 0$$

4. If $A \neq B$ (but they have the same sign) and $C = 0$, the equation represents an ellipse.

Example D:
$$2X^2 + 5Y^2 - X + 3Y = 0$$
$$3X^2 + Y^2 + 6X - Y + 7 = 0$$

5. (a) If A and B have different signs and $C = 0$
   (b) If $A = B = D = E = 0$ and $C \neq 0$
   } The equation represents an hyperbola.

Example E:
$$X^2 - Y^2 + 4X - 6Y - 12 = 0$$
$$3X^2 - 4Y^2 - 5X + Y + 7 = 0$$
$$XY - 5 = 0$$
$$2XY + 8 = 0$$

The accompanying table gives a complete summary of the equations of each of the curves we have discussed. Assume X to be the independent variable in each of the equations.

	Straight Line	Parabola	Circle	Ellipse	Hyperbola
Basic Equation	$(Y - Y_1) = M(X - X_1)$ $Y = MX + B$	$(X - H)^2 = 4P(Y - K)$ $(Y - K)^2 = 4P(X - H)$	$(X - H)^2 + (Y - K)^2 = R^2$	$\dfrac{(X - H)^2}{A^2} + \dfrac{(Y - K)^2}{B^2} = 1$ $\dfrac{(Y - K)^2}{A^2} + \dfrac{(X - H)^2}{B^2} = 1$	$\dfrac{(X - H)^2}{A^2} - \dfrac{(Y - K)^2}{B^2} = 1$ $\dfrac{(Y - K)^2}{A^2} - \dfrac{(X - H)^2}{B^2} = 1$ $XY = K$
Significant Information and Relationships	M = Slope (X_1, Y_1) = any point B = Vertical Intercept	(H, K) = Vertex $P = \begin{cases} \text{distance from} \\ \text{vertex to focus} \end{cases}$ $-P = \begin{cases} \text{distance from} \\ \text{vertex to} \\ \text{directrix} \end{cases}$	(H, K) = Center R = Radius	(H, K) = Center $A = \begin{cases} \text{distance from} \\ \text{center to vertex} \end{cases}$ $B = \begin{cases} \text{distance from} \\ \text{center to ends} \\ \text{of minor axis} \end{cases}$ $C = \begin{cases} \text{distance from} \\ \text{center to focus} \end{cases}$ Major Axis = $2A$ Minor Axis = $2B$ distance between foci = $2C$ $A^2 = B^2 + C^2$	(H, K) = Center $A = \begin{cases} \text{distance from} \\ \text{center to vertex} \end{cases}$ $B = \begin{cases} \text{distance from} \\ \text{center to ends} \\ \text{of conjugate axis} \end{cases}$ $C = \begin{cases} \text{distance from} \\ \text{center to focus} \end{cases}$ Transverse Axis = $2A$ Conjugate Axis = $2B$ distance between foci = $2C$ $C^2 = A^2 + B^2$ Asymptotes: $(Y - K) = \pm \dfrac{B}{A}(X - H)$
General Equation	$DX + EY + F = 0$	$AX^2 + DX + EY + F = 0$ $BY^2 + DX + EY + F = 0$	$X^2 + Y^2 + DX + EY + F = 0$	$AX^2 + BY^2 + DX + EY + F = 0$	$AX^2 - BY^2 + DX + EY + F = 0$ $BY^2 - AX^2 + DX + EY + F = 0$ $XY = K$

Exercises State whether each of the following equations represents a straight line, parabola, circle, ellipse, hyperbola, or none of these curves.

1. $Y = \dfrac{6}{X}$
2. $X^2 = 4 - Y^2$
3. $2M = 3N$
4. $R^2 - S^2 = 9$
5. $8P - 5Q + 6 = 0$
6. $K^2 + 3K - L + 4 = 0$
7. $Y = 1$
8. $T^3 + 3T - 1 = 0$
9. $R = 9 - S^2$
10. $4M^2 + 4N^2 + 8N = 11$
11. $T - 4V = 9 - V^2$
12. $K^2 - 2KL = 2L(L - K)$
13. $6P(P - 3) = 4Q(1 - Q)$
14. $R = -5S + 6$
15. $Y = 2\sqrt{X} - 3$
16. $12PQ = 24$
17. $T^2(T - 1) = 3V(4 - V) + T^3$
18. $\frac{2}{3} - 7K = \frac{5}{2} - 4L$
19. $X(Y - 4X) = X^2 + XY + Y^2 + 1$
20. $M(M + N) = M(N - 3) - (2 + N^2)$

18–9. CHAPTER REVIEW

For each of the following, determine all pertinent information that applies in each case (slope, center, vertices, directrix, radius, foci, axes) and sketch the graph.

1. $5X - 4Y + 3 = 0$ X is the independent variable.
2. $3M + 5N - 4 = 0$ M is the independent variable.
3. $R^2 - 4R - 6S + 28 = 0$ R is the independent variable.
4. $V^2 + 6V + 10T - 1 = 0$ T is the independent variable.
5. $K^2 + L^2 - 4K - 6L - 3 = 0$ K is the independent variable.
6. $2P^2 + 2Q^2 + 4P + 12Q - 30 = 0$ P is the independent variable.
7. $A^2 + 2B^2 - 2A - 8B + 1 = 0$ A is the independent variable.
8. $4X^2 + Y^2 + 16X + 4Y + 8 = 0$ X is the independent variable.
9. $3M^2 - 2N^2 - 4N - 8 = 0$ M is the independent variable.
10. $2R^2 - 3S^2 - 8R + 18S - 31 = 0$ R is the independent variable.

For each of the following, find the equation of the curve with the given conditions and sketch the graph. Write each equation in general form.

11. Straight line, passes through the points $(-2, -5)$ and $(4, -1)$. Use T and V with T the independent variable.

12. Straight line, an inclination of 125° and a horizontal intercept of −3. Use K and L with K the independent variable.

13. Parabola, vertex at (−4, 2) and directrix $P = −6$. Use P and Q with P the independent variable.

14. Parabola, focus at (1, 0) and directrix $N = −4$. Use M and N with M the independent variable.

15. Circle, center at (−3, 2) and radius $\sqrt{10}$. Use R and S with R the independent variable.

16. Circle, center at (2, −4) and passes through (0, 1). Use X and Y with X the independent variable.

17. Ellipse, center at (0, 2), focus at (3, 2), and a minor axis of 8. Use A and B with A the independent variable.

18. Ellipse, vertices at (1, 5) and (1, −3), and a minor axis of 6. Use X and Y with X the independent variable.

19. Hyperbola, center at (−1, 1), vertex at (−1, −1), and a focus at (−1, 4). Use T and V with T the independent variable.

20. Hyperbola, vertices at (3, 2) and (−1, 2), and a conjugate axis of 5. Use M and N with M the independent variable.

Solve each of the following systems of equations graphically.

21. $16X^2 + 25Y^2 − 96X − 50Y − 231 = 0$
    $X^2 − Y^2 − 6X + 2Y − 1 = 0$
    Use X as the independent variable.

22. $X^2 − Y^2 + 4Y − 8 = 0$
    $16Y^2 − X^2 − 64Y + 48 = 0$
    Use X as the independent variable.

23. Sketch the graph which shows the circumference of a circle as a function of its radius.

24. The relationship between the volume (V) and base (B) of a pyramid is linear if the height is held constant. If the volume of a certain pyramid is 4 cubic feet when the base is 12 square feet, and if the volume increases 1 cubic foot for every increase of 3 square feet in the base, determine the equation relating V and B, and sketch the graph of V as a function of B.

25. The distance (s) above the ground of a projectile fired vertically into the air is given in terms of time (t) by the equation $s = 9t − 2t^2$. Sketch the graph of s as a function of t and determine after how many seconds the projectile will hit the ground.

26. The total surface area (S) of a right circular cylinder as a function of the radius (R) of the base and the height (H) is given by the equation $S = 2\pi R^2 + 2\pi RH$. If the height is fixed at 15 centimeters for a certain cylinder, sketch the graph of S as a function of R.

27. Sketch the graph showing the possible lengths for the rafters (*X* and *Y*) of the roof truss represented by the accompanying diagram. Use *X* as the independent variable.

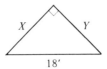

28. Two pulleys have radii of 2 feet and 3 feet. The centers are on the same vertical line and the edges of the pulleys must be exactly 6.5 feet apart. If the smaller pulley must be on the bottom, use the origin as its center and sketch the pulleys.

29. A designer is making a model for an elliptical machine part. Two pins are inserted in a drawing board, a piece of string is looped over both pins, and an outline is made by keeping the string taut. If the part is to measure 10 centimeters by 8 centimeters, how long should the loop of string be and how far apart should the pins be placed?

30. The formula for the area of a triangle is $A = \tfrac{1}{2}BH$. If the area is to be held constant at 24 square inches, sketch the graph showing the relationship between *B* and *H*. Use *B* as the independent variable.

appendix A

Geometric Figures and Formulas

1. Lines:

Parallel Lines

Intersecting Lines

Perpendicular Lines

2. Angles:

Acute Angle: between 0° and 90°.

Obtuse Angle: between 90° and 180°.

Right Angle: equals 90°.

Straight Angle: equals 180°.

Adjacent Angles: common side and vertex.

Vertical Angles: formed by intersecting lines. They are equal.

Alternate Interior Angles: formed by a line intersecting a pair of parallel lines. They are equal.

Complementary Angles: two acute angles whose sum is 90°.

Supplementary Angles: two positive angles whose sum is 180°.

3. Square:

Perimeter: $P = 4S$
Area: $A = S^2$

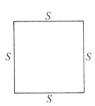

4. Rectangle:

Perimeter: $P = 2L + 2w$
Area: $A = LW$

5. Triangle:

Equilateral Triangle: Three equal angles and three equal sides.

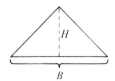

App. A. Geometric Figures and Formulas   **527**

Isosceles Triangle: Two angles are equal and the sides opposite these angles are equal.

Right Triangle: One angle equals 90°. If a triangle is not a right triangle it is called *oblique*.

Pythagorean Theorem:
$\left.\begin{array}{l} C^2 = A^2 + B^2 \\ C = \sqrt{A^2 + B^2} \end{array}\right\}$ Equivalent

Scalene Triangles: No two angles or sides are equal.

Similar Triangles: Corresponding angles are equal and corresponding sides are proportional.

Congruent Triangles: Corresponding angles and sides are equal.

For all triangles: Sum of three angles is 180°. Area = $\frac{1}{2}BH$.

6. Circle:

Circumference: $C = 2\pi R$ or $C = \pi D$
Area: $A = \pi R^2$ or $A = \dfrac{\pi D^2}{4}$

7. Rhombus:

Perimeter: $P = 4S$
Area: $A = SH$

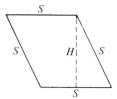

8. Parallelogram:

Perimeter: $P = 2S + 2B$
Area: $A = BH$

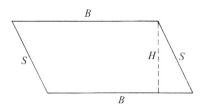

9. Isosceles Trapezoid:

Perimeter: $P = A + B + 2S$
Area: $A = \frac{1}{2}(A + B)H$

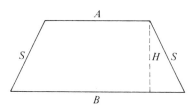

10. General Trapezoid:

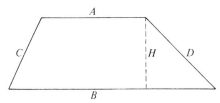

Perimeter: $P = A + B + C + D$
Area: $A = \frac{1}{2}(A + B)H$

11. Cube:

Surface Area: $A = 6E^2$
Volume: $V = E^3$

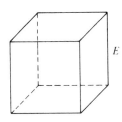

App. A. Geometric Figures and Formulas   529

12. Rectangular Solid:

Surface Area:
$$A = 2(LW + LH + WH)$$
Volume: $V = LWH$

13. Right Circular Cylinder:

Surface Area: $A = 2\pi RH + 2\pi R^2$
Volume: $V = \pi R^2 H$

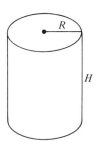

14. Cylinder or Prism with Parallel Bases:

Volume: $V = BH$

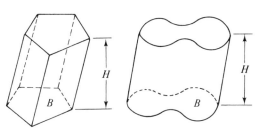

15. Right Circular Cone:

Surface Area: $A = \pi RL + \pi R^2$
Volume: $V = \tfrac{1}{3}\pi R^2 H$

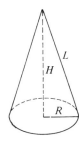

16. Cone or Pyramid:

Volume: $V = \frac{1}{3}BH$

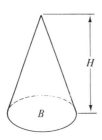
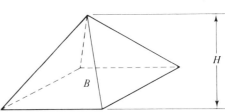

17. Sphere:

Surface Area: $A = 4\pi R^2$
Volume: $V = \frac{4}{3}\pi R^3$

appendix B

Operations with Hand Calculators

B-1. Exact and Approximate Numbers

Before actually looking at various operations that may be performed on a hand calculator, we must first discuss the precision and accuracy of different types of numbers. In general, when combining numbers in some manner to obtain a result, we should realize that the result cannot be more precise or accurate than the original numbers with which we started. Therefore, even though most calculators give us eight- or ten-digit results, we must round these results to correspond to the information given.

Basically, there are two types of numbers: exact and approximate. An *exact number* is obtained by definition or a counting process. An *approximate number* is obtained by some type of measurement.

Example A:
1. If we say there are 16 ounces in 1 pound or there are 1000 meters in 1 kilometer, the numbers 16, 1, 1000, and 1 are exact since they are arrived at by definition.
2. If we count the number of dollars in our wallet and obtain a figure of 23, this number is exact.

3. If we measure a person's weight to be 163 pounds, this number is approximate.
4. If we measure the distance between two points to be 63.5 meters, this number is approximate.

The same number may be exact in one instance and approximate in another. If we say there are 60 seconds in 1 minute, the number 60 is exact since it is a definition. If we are watching the second hand on our watch and then state, after what we determine to be 1 revolution, that 60 seconds have passed, the number 60 is approximate since it involved a measurement on our part. So anytime that a measurement of any type is involved, no matter how precise or accurate the measuring instrument used, the values obtained are approximate.

In working with numbers, it is often necessary to determine the number of *significant digits*. All digits are significant with the exception of zeros used solely for the purpose of placing the decimal point properly.

Example B:
1. The number 45.68 has four significant digits.
2. The number 670 has two significant digits since the zero is present only to insure the decimal point being properly placed.
3. The number 7090 has three significant digits (7, 0, 9) since the zero at the end is used to place the decimal point but not the zero between 7 and 9.
4. The number .183 has three significant digits.
5. The number .0035 has two significant digits since the zeros at the beginning of this number are used to place the decimal point properly.
6. The number .08060 has four significant digits (8, 0, 6, 0) since the zero at the beginning is used to place the decimal point but the other zeros are not.
7. The number 60.007 has five significant digits since none of these zeros is used to place the decimal point.

From the examples above we should see that, with the exception of zeros appearing at the end of a whole number (40500) or at the beginning of a number less than 1 (.0005090), all digits are considered significant.

When we use numbers in calculations, we must be concerned with two things: the number of decimal places and the number of significant digits. The *number of decimal places* determines the precision of a number, whereas the *number of significant digits* determines its accuracy.

Example C:
1. The number 48.79 has two-place precision and four-digit accuracy.
2. The number .0005 has four-place precision and one-digit accuracy.
3. The number .26 has two-place precision and two-digit accuracy.

Example D:
1. Given the two numbers, 65.9 and .36, the first is more accurate (three significant digits) and the second is more precise (two decimal places).
2. Given the numbers, 2.0007 and 492,000, the first is more accurate (five significant digits) and more precise (four decimal places).

In most problems, we usually *round off numbers* either during or after calculations. To round a number to N digits or N decimal places, look at the $(N + 1)$ digit or the digit in the $(N + 1)$ decimal place. If this digit is 5 or greater, increase the previous digit by 1. If the digit is less than 5, simply omit it and all digits that follow.

Example E:
1. Round 93,260 to three significant digits. We look at the fourth significant digit, 6. Since it is greater than 5, we increase the previous digit by 1. Therefore, $93,260 = 93,300$.
2. Round 48.592 to four significant digits. We look at the fifth significant digit, 2. Since it is less than 5, we omit it and anything that follows. Therefore, $48.592 = 48.59$.
3. Round 3.1416 to two decimal places. We look at the digit in the third decimal place, 1. Since it is less than 5, we omit it and anything that follows. Therefore, $3.1416 = 3.14$.
4. Round 65.273 to one decimal place. We look at the digit in the second decimal place, 7. Since it is greater than 5, we increase the previous digit by 1. Therefore, $65.273 = 65.3$.

When performing operations involving approximate numbers, our results certainly cannot be more precise or accurate than any of the numbers with which we began. In general, the following rules should be observed.

1. When approximate numbers are added or subtracted, the result should be expressed with the precision of the least precise number with which we started.
2. When approximate numbers are multiplied or divided, the result should be expressed with the accuracy of the least accurate number with which we started.
3. When finding a root of an approximate number, the root should be expressed with the same accuracy as the number.

Example F: Two consecutive sections of roadway are measured to be 14.7 kilometers and 9.346 kilometers. The overall length will be 14.7 km + 9.846 km = 24.546 km. This answer should be rounded to 24.5 kilometers to correspond with 14.7 kilometers, the least precise number with which we started.

Example G: The length and width of a certain rectangular piece of land are measured to be 165.47 feet and 73.6 feet. The area of this land will be 165.47 ft × 73.6 ft = 12,178.592 ft². This should be rounded to 12,200 square feet to correspond with 73.6 feet, the least accurate number with which we started.

Exercises

In each of the following, determine whether the numbers are exact or approximate.
1. There are 2000 pounds in 1 ton.
2. The book costs $14.25.
3. The specific gravity of alcohol is .81.
4. The velocity of sound is 1088 feet per second.
5. The two cities are 127 kilometers apart.
6. A certain person is 5 feet, 11 inches tall.
7. The trip took $4\frac{1}{2}$ hours.
8. The tax rate in our town is 18%.
9. There are 2.2 pounds in 1 kilogram.
10. Rounded to three significant digits, the answer is 19,800.

Determine the precision and accuracy of each of the following.
11. 56.81
12. 2.108
13. 170.6
14. 98,300
15. 407.060
16. .00305

Round each of the following to three and to two significant figures.
17. 41,825
18. 1942
19. 16.356
20. .009407
21. 2.006
22. 119.98

Round each of the following to three and to two decimal places.
23. 18.9208
24. 2.7154
25. .000825
26. .070204
27. 126.4538
28. 35.8997

For each of the following, assume all numbers to be approximate. Perform the indicated operations and then round the answer appropriately.
29. (94.73) + (.0084) + (7.516)
30. (126.91) − (19.408) − (.735)
31. (14.8) × (165.2)
32. (.7045) × (.00083)
33. (24,600) ÷ (.7)
34. (73.59) ÷ (27,942)
35. (4.73)²
36. (126.8)³
37. $\sqrt{784.3}$
38. $\sqrt[3]{.0502}$
39. (249.8 × 12.1) − (93.7)
40. (35.92 + 7.086) ÷ (260)

App. B-2. Basic Operations

41. Three consecutive distances are measured to be 93.5 meters, 128 meters, and 56.47 meters. What is the total distance?
42. If voltage = current × resistance, what is the voltage in a certain circuit with a current of .25 amp and a resistance of 16.7 ohms?
43. If one car travels 55 miles per hour and another travels 98 kilometers per hour, which car is traveling faster and by how many kilometers per hour?
44. Four separate weights are measured to be 240.3 pounds, 206.75 pounds, 227.33 pounds, and 218.125 pounds. What is the sum of these weights?
45. If the two legs of a right triangle are 16.8 and 7.3, find the hypotenuse.
46. If the area of a circle is given by the formula $A = \pi R^2$, find the radius of a circle with an area of 192.25 square feet.

B-2. Basic Operations

While the procedure to be followed in performing different types of calculations will vary slightly with the different brands and styles of hand calculators, the methods described in this appendix will be applicable in the majority of cases. Also, this appendix is not intended to be a manual for any particular hand calculator, but rather an outline of some of the things that may be done by using one.

Basic operations are performed on a hand calculator exactly as they are written. If numbers are approximate, results should be rounded in accordance with the rules stated in the previous section.

Example A: Find 73 + 156
Enter 73
Press +
Enter 156
Press =
Read answer → 229

Example B: Find 94.7 + 473.27 + 286.54
Enter 94.7
Press +
Enter 473.27
Press +
Enter 286.54
Press =
Read answer → 854.51
= 854.5

Example C: Find 19.52 − 7.96
Enter 19.52
Press −
Enter 7.96
Press =
Read answer → 11.56

Example D: Find 792.78 − 912.89
Enter 792.78
Press −
Enter 912.89
Press =
Read answer →
−120.11

Example E: Find 57 × 147
Enter 57
Press ×
Enter 147
Press =
Read answer → 8379
        = 8400

Example F: Find 28.6 × 6.28
                    × 42.53
Enter 28.6
Press ×
Enter 6.28
Press ×
Enter 42.53
Press =
Read answer →
    7638.7282 = 7640

Example G: Find 483 ÷ 57
Enter 483
Press ÷
Enter 57
Press =
Read answer →
    8.4736842 = 8.5

Example H: Find 5526 ÷ 48.3
                    ÷ 5.62
Enter 5526
Press ÷
Enter 48.3
Press ÷
Enter 5.62
Press =
Read answer →
    20.357638 = 20.4

Reciprocals may be found in one of two ways. If the calculator has a reciprocal key, $\frac{1}{X}$, they may be found directly. Otherwise, we find the reciprocal of a number by dividing it into 1.

Example I: Find the reciprocal of 25.
1. Enter 25
   Press $\frac{1}{X}$
   Read answer → .04

2. Enter 1
   Press ÷
   Enter 25
   Press =
   Read answer → .04

Example J: Find the reciprocal of 6.28.
1. Enter 6.28
   Press $\frac{1}{X}$
   Read answer → .1592356
             = .159

2. Enter 1
   Press ÷
   Enter 6.28
   Press =
   Read answer → .1592356
             = .159

If a calculator has a key marked +/−, the sign of any number may be changed after it is entered.

App. B–3. Scientific Notation

Example K: Find $73 \times (-56)$
Enter 73
Press $\times$
Enter 56
Press $+/-$
Press $=$
Read answer $\rightarrow -4088$
$\phantom{\text{Read answer}} = -4100$

Example L: Find $(-86.3) \div 142.9$
Enter 86.3
Press $+/-$
Press $\div$
Enter 142.9
Press $=$
Read answer $\rightarrow$
$\phantom{\text{Read answer}} -.6039188 = -.604$

Exercises In each of the following, perform the indicated calculations and round the results according to the rules stated in the previous section.

1. $49.63 + 194.7$
2. $2406.91 + 1937.66$
3. $17.6 + 32.9 + 54.1$
4. $154.2 + 98.35 + 241.06$
5. $12.8 + 26.72 + 9.3 + 45.38$
6. $426.9 + 558.3 + 391.4 + 683.7$
7. $296.3 - 83.81$
8. $75.62 - 28.7$
9. $18.36 - 34.75$
10. $652.49 - 831.57$
11. $41.76 - 19.2 - 23.8$
12. $87.2 - 41.6 - 35.71$
13. 45.9×58.6
14. 1220×94.8
15. $.035 \times .645$
16. $.743 \times .129$
17. $83.6 \times 91.4 \times 123.8$
18. $16.3 \times 107.4 \times 37.8$
19. $592.7 \div 65.9$
20. $346.5 \div 47.2$
21. $.591 \div 12.6$
22. $11.72 \div .037$
23. $46.8 \div 2.54 \div 8.06$
24. $2483 \div 305 \div 26.5$
25. Find the reciprocal of 16.
26. Find the reciprocal of 3.14.
27. Find the reciprocal of .68.
28. Find the reciprocal of .052.
29. Find the reciprocal of 280.
30. Find the reciprocal of 4120.
31. $(-426) \times .294$
32. $1430 \times (-.083)$
33. $(-.00836) \times (-5260)$
34. $8140 \div (-65,200)$
35. $(-.308) \div 0703$
36. $(-7009) \div (-274)$

B–3. Scientific Notation

Some calculators are designed so that we may make use of scientific notation. Those with this capacity will have a key marked EE or EEX. By making use of this feature, we may write or work with numbers that would otherwise be too large or too small to be handled in the normal manner. A complete discussion of scientific notation may be found in Section 2–2 of this text.

Example A: Express 93,000,000 in scientific notation on the calculator.
$$93,000,000 = 9.3 \times 10^7$$
Enter 9.3
Press EE
Enter 7
Read answer $\rightarrow$ 9.3__7

Example B: Express .0000592 in scientific notation on the calculator.
$$.0000592 = 5.92 \times 10^{-5}$$
Enter 5.92
Press EE
Enter 5
Press $+/-$
Read answer $\rightarrow$
$\qquad$ 5.92__-5

Example C: Find $68,000,000 \times 4,000,000$, using scientific notation.
$68,000,000 \times 4,000,000 = (6.8 \times 10^7) \times (4.0 \times 10^6)$
Enter 6.8
Press EE
Enter 7
Press $\times$
Enter 4.0
Press EE
Enter 6
Press $=$
Read answer $\rightarrow$ 2.72__14
$= 2.72 \times 10^{14} = 3 \times 10^{14}$

Example D: Find $.0000492 \times .000261 \times 25,600$, using scientific notation.
$.0000492 \times .000261 \times 25,600 = (4.92 \times 10^{-5}) \times (2.61 \times 10^{-4}) \times (2.56 \times 10^4)$
Enter 4.92
Press EE
Enter 5
Press $+/-$
Press $\times$
Enter 2.61
Press EE
Enter 4
Press $+/-$
Press $\times$
Enter 2.56
Press EE
Enter 4
Press $=$
Read answer $\rightarrow$
3.2873472__$-4 = 3.29 \times 10^{-4}$

Example E: Find $52,700 \div 36,800,000$, using scientific notation.

$52,700 \div 36,800,000$
$= (5.27 \times 10^4) \div (3.68 \times 10^7)$
Enter 5.27
Press EE
Enter 4
Press $\div$
Enter 3.68
Press EE

Example F: Find $163,000,000 \div 52,900,000 \div .000841$, using scientific notation.
$163,000,000 \div 52,900,000 \div .000841 = (1.63 \times 10^8) \div (5.29 \times 10^7) \div (8.41 \times 10^{-4})$
Enter 1.63
Press EE
Enter 8
Press $\div$

App. B–4. Powers and Roots

           Enter 7   Enter 5.29
           Press =   Press EE
           Read answer →   Enter 7
1.4320652__$-3 = 1.43 \times 10^{-3}$ Press ÷
                                                                    Enter 8.41
                                                                    Press EE
                                                                    Enter 4
                                                                    Press +/−
                                                                    Press =
                                                                    Read answer →
                                                  3.6638351__$3 = 3.66 \times 10^{3}$

Exercises   Express each of the following in scientific notation using the calculator.

    1. 3.14   **2.** 186,000   **3.** .00946

    4. .000000527   **5.** 48,700,000   **6.** 6.28

Perform each of the following calculations on the calculator using scientific notation.

    7. 28,700 × .0000749   **8.** 2,900,000 × 5,500

    9. .000843 × .0000358   **10.** .00681 × 145,000

    11. 490,000 × 1,463,000 × 9,200   **12.** .000463 × .00278 × 51,800

    13. 672,000 × .00003806 × 25,620   **14.** 726,000 ÷ 59,000,000

    15. .000947 ÷ .00706   **16.** .0000628 ÷ 3728

    17. 275,000 ÷ .09408   **18.** 38,000,000 ÷ 162 ÷ 8,249

    19. .00297 ÷ .00006092 ÷ .00783   **20.** 597,000 ÷ 74,100 ÷ .000683

B–4. Powers and Roots

Powers and *roots* are found by employing different methods depending on the type of calculator being used. With a calculator having only the four basic functions with no special capacity for finding powers of numbers, we may treat the process of *raising numbers to powers* the same as *successive multiplication*.

Example A: Find 6^4 *Example B:* Find 45^3
                  $6^4 = 6 \times 6 \times 6 \times 6$ $45^3 = 45 \times 45 \times 45$
                  Enter 6   Enter 45
                  Press ×   Press ×
                  Enter 6   Enter 45
                  Press ×   Press ×

Enter 6
Press ×
Enter 6
Press =
Read answer → 1296

Enter 45
Press =
Read answer → 91125

Example C: Find 7^{-3}

$$7^{-3} = \frac{1}{7^3} = \frac{1}{7 \times 7 \times 7}$$

Enter 7
Press ×
Enter 7
Press ×
Enter 7
Press =
Read result → 343
Clear   or Press $\frac{1}{X}$
Enter 1
Press ÷   Read answer → .00292
Enter 343
Press =
Read answer → .00292

With a calculator that has a key marked X^2 for *finding squares* of numbers, the successive multiplication process is somewhat refined.

Example D: Find 37^2
Enter 37
Press X^2
Read answer → 1369

Example E: Find 25^{-2}

$$25^{-2} = \frac{1}{25^2} = \frac{1}{25 \times 25}$$

Enter 25
Press X^2
Read result → 625
Clear   or Press $\frac{1}{X}$
Enter 1
Press ÷   Read answer → .0016
Enter 625
Press =
Read answer → .0016

App. B-4. Powers and Roots

For integral powers other than 2 or -2, the following rules may be used with a calculator that has a key for squaring numbers, X^2.

$$A^3 = A^2 \times A$$
$$A^4 = (A^2)^2$$
$$A^5 = (A^2)^2 \times A$$
$$A^6 = (A^2 \times A)^2$$
$$A^7 = (A^2 \times A)^2 \times A$$
$$A^8 = [(A^2)^2]^2$$
$$A^9 = [(A^2)^2]^2 \times A$$
$$A^{10} = [(A^2)^2 \times A]^2$$

Example F: Find 16^5
$16^5 = (16^2)^2 \times 16$
Enter 16
Press X^2
Press X^2
Press $\times$
Enter 16
Press $=$
Read answer $\rightarrow$
1,048,576

Example G: Find 9^8
$9^8 = [(9^2)^2]^2$
Enter 9
Press X^2
Press X^2
Press X^2
Read answer $\rightarrow$
43,046,721

With a calculator that has a key marked X^Y, any integral power may be found directly.

Example H: Find 7^5
Enter 7
Press X^Y
Enter 5
Press $=$
Read answer $\rightarrow$ 16807

Example I: Find 4^{-6}
Enter 4
Press X^Y
Enter 6   or Press $=$
Press $+/-$ Press $\frac{1}{X}$
Press $=$
Read answer $\rightarrow$ .0002441   Read answer $\rightarrow$ .0002441

With calculators that do not have the capacity for finding roots directly, we must use an *iterative* process. We will describe here such processes for finding square and cube roots.

To find the square root of N ($\sqrt{N}$):

1. Select an approximation, A.
2. Divide N by A, $\left(\dfrac{N}{A}\right)$.
3. Add A to the result, $\left(\dfrac{N}{A} + A\right)$.
4. Multiply this result by .5, $\left[.5\left(\dfrac{N}{A} + A\right)\right]$.
5. Compare this result to original approximation and then use it as the next approximation.

Example J: Find $\sqrt{10}$
1. Select an approximation, 3.1.

$$10 \div 3.1 = 3.2258$$
$$3.2258 + 3.1 = 6.3258$$
$$.5 \times .63258 = 3.1629$$

2. Select a new approximation, 3.16.

$$10 \div 3.16 = 3.1646$$
$$3.1646 + 3.16 = 6.3246$$
$$.5 \times 6.3246 = 3.1623$$

Therefore, $\sqrt{10} = 3.16$ to two decimal places.

Example K: Find $\sqrt{500}$
1. Select an approximation, 22.

$$500 \div 22 = 22.727$$
$$22.727 + 22 = 44.727$$
$$.5 \times 44.727 = 22.364$$

2. Select a new approximation, 22.36.

$$500 \div 22.36 = 22.361$$
$$22.361 + 22.36 = 44.721$$
$$.5 \times 44.721 = 22.361$$

Therefore, $\sqrt{500} = 22.36$ to two decimal places.

App. B–4. Powers and Roots

To find the cube root of $N(\sqrt[3]{N})$:

1. Select an approximation, A.
2. Cube A, (A^3).
3. Divide N by A^3, $\left(\dfrac{N}{A^3}\right)$.
4. Add 2 to this result, $\left(\dfrac{N}{A^3} + 2\right)$.
5. Multiply this result by A, $\left[A\left(\dfrac{N}{A^3} + 2\right)\right]$.
6. Divide this result by 3, $\left[\dfrac{A\left(\dfrac{N}{A^3} + 2\right)}{3}\right]$.
7. Compare this result to the original approximation and then use it as the new approximation.

Example L: Find $\sqrt[3]{12}$

1. Select an approximation, 2.4.

$$2.4^3 = 13.824$$
$$12 \div 13.824 = .8681$$
$$.8681 + 2 = 2.8681$$
$$2.4 \times 2.8681 = 6.88344$$
$$6.88344 \div 3 = 2.294$$

2. Select a new approximation (2.29).

$$2.29^3 = 12.009$$
$$12 \div 12.009 = .9993$$
$$.9993 + 2 = 2.993$$
$$2.29 \times 2.993 = 6.8684$$
$$6.8684 \div 3 = 2.2895$$

Therefore, $\sqrt[3]{12} = 2.29$ to 2 decimal places.

If a calculator has a square root key, $\sqrt{X}$, *square roots* may be found directly.

Example M: Find $\sqrt{73}$
  Enter 73
  Press $\sqrt{X}$
  Read answer → 8.544 = 8.5

If a calculator has a general root key ($\sqrt[x]{X}$), *any root* may be found directly.

Example N: Find $\sqrt[3]{61}$
Enter 61
Press $\sqrt[x]{X}$
Enter 3
Press $=$
Read answer $\rightarrow$ 3.9367
$= 3.9$

Example O: Find $\sqrt[4]{168}$
Enter 168
Press $\sqrt[x]{X}$
Enter 4
Press $=$
Read answer $\rightarrow$ 3.6

NOTE: Fourth roots may also be handled as double square roots, since $\sqrt[4]{N} = \sqrt{\sqrt{N}}$.

Exercises Evaluate each of the following.

1. 5^4
2. 12^3
3. 18^2
4. 71^3
5. 11^5
6. 7.6^3
7. 41.5^2
8. 34.2^4
9. 6.3^7
10. 2.5^8
11. 123^3
12. $.9^4$
13. 10^{-3}
14. 67^{-2}
15. 108^{-2}
16. $.74^{-3}$
17. $.83^6$
18. 2.1^{10}
19. $.084^3$
20. 52.3^3

Find each of the following roots.

21. $\sqrt{15}$
22. $\sqrt{943}$
23. $\sqrt{.76}$
24. $\sqrt{.254}$
25. $\sqrt{6540}$
26. $\sqrt{42{,}000}$
27. $\sqrt{.038}$
28. $\sqrt{93}$
29. $\sqrt[3]{24}$
30. $\sqrt[3]{109}$
31. $\sqrt[3]{.41}$
32. $\sqrt[3]{.092}$
33. $\sqrt[3]{.726}$
34. $\sqrt[3]{186{,}000}$
35. $\sqrt[3]{58{,}300}$
36. $\sqrt[3]{382}$
37. $\sqrt[4]{38}$
38. $\sqrt[4]{.65}$
39. $\sqrt[4]{2{,}900{,}000}$
40. $\sqrt[4]{1285}$

B–5. Trigonometric Functions

Although there exist formulas that may be used to find trigonometric functions with any hand calculator, we will discuss here only those calculators that do this directly.

With calculators that have a key for each of the *trigonometric functions* (SIN, COS, TAN, CSC, SEC, COT), we simply enter the angle and press the appropriate key. If the calculator has a *degree-radian* switch, we must make sure it is in the right position.

App. B–5. Trigonometric Functions

Example A: Find Sin 58°
Make sure the degree-radian
switch is in the degree position.
    Enter 58
    Press SIN
    Read answer $\rightarrow$ .8480

Example B: Find Csc 296°
Make sure the degree-radian switch
is in the degree position.
    Enter 296
    Press CSC
    Read answer $\rightarrow$ -1.113

Example C: Find Tan 2.1. Make sure the degree-radian switch is in the radian position. If there is no such switch, we multiply 2.1 times 57.3° to obtain 120.33°.
Enter 2.1 (radian position)   or Enter 120.33 (degree position)
Press TAN   Press TAN
Read answer $\rightarrow$ -1.71 Read answer $\rightarrow$ -1.71

With calculators that have a key for only the three basic trigonometric functions (SIN, COS, TAN), we make use of the reciprocal relationships, since $\operatorname{Csc} \theta = \frac{1}{\operatorname{Sin} \theta}$, $\operatorname{Sec} \theta = \frac{1}{\operatorname{Cos} \theta}$, and $\operatorname{Cot} \theta = \frac{1}{\operatorname{Tan} \theta}$.

Example D: Find Sec 65°
Make sure the degree-radian switch
is in the degree position.
    Enter 65
    Press COS
    Press $\frac{1}{X}$
    Read answer $\rightarrow$ 2.366

Example E: Find Cot 312°
Make sure the degree-radian switch is
in the degree position.
    Enter 312
    Press TAN
    Press $\frac{1}{X}$
    Read answer $\rightarrow$ $-.9004$

If a calculator has an ARC key or keys marked SIN^{-1}, COS^{-1}, TAN^{-1}, CSC^{-1}, SEC^{-1}, COT^{-1}, we may find an angle when a functional value is given. The position of the degree-radian switch will determine whether the answer will be expressed in degrees or radians.

Example F: If Cos θ = .8572, find θ in degrees.
Make sure the degree-radian switch is in the degree position.
Enter .8572   or Enter .8572
Press ARC   Press SIN^{-1}
Press SIN   Read answer $\rightarrow$ 31
Read answer $\rightarrow$ 31
Therefore, Cos 31° = .8572. Also, Cos 329° = .8572, since 360° $-$ 31° = 329°.

Example G: If Tan θ = 1.376, find θ in radians.
  1. Make sure the degree-radian switch is in the radian position.
      Enter 1.376   or Enter 1.376

Press ARC
Press TAN
Read answer → .943

2. If there is no radian-degree switch:
Enter 1.376
Press ARC⎫
Press TAN⎭   or Press TAN⁻¹
Read answer → 54

Press TAN⁻¹
Read answer → .943

Divide 54° by 57.3° to obtain .943. Therefore, Tan .943 = 1.376. Also, Tan 2.197 = 1.376, since 3.14 − .943 = 2.197.

With keys for just the three basic trigonometric functions or for just SIN⁻¹, COS⁻¹, and TAN⁻¹, we again make use of the reciprocal relationships.

Example H: If Csc θ = 2.924, find θ in degrees. Make sure the degree-radian switch is in the degree position.

Enter 2.924
Press $\frac{1}{X}$
Press ARC
Press SIN
Read answer → 20°

or Enter 2.924
Press $\frac{1}{X}$
Press SIN⁻¹
Read answer → 20°

Therefore, Csc 20° = 2.924. Also, Csc 160° = 2.924, since 180° − 20° = 160°.

Example I: If Cot θ = −.5774, find θ in degrees. Make sure the degree-radian switch is in the degree position.

Enter −.5774
Press $\frac{1}{X}$
Press ARC
Press TAN
Read answer → −60° = 300°

or Enter −.5774
Press $\frac{1}{X}$
Press TAN⁻¹
Read answer → −60° = 300°

Therefore, Cot 300° = −.5774. Also, Cot 120° = −.5774 since 180° − 60° = 120°.

From Examples F, G, H, and I, we should note that whenever a *functional value* is given, there will always be *two possible answers* since each of the trigonometric functions is positive in two quadrants and negative in the other two, and since every angle has a first quadrant reference angle. A complete discussion of these ideas is presented in Section 7-7.

Exercises

Find the indicated trigonometric function for each of the following:

1. Sin 36°
2. Cos 95°
3. Tan 54.2°
4. Csc .7°
5. Sec 162°
6. Cot 108°
7. Sin 245°
8. Sec 342°
9. Cos 265°
10. Tan 300°
11. Cos (−58°)
12. Cot (−125°)
13. Tan 3
14. Sin .8
15. Cos .92
16. Sec 4.2
17. Cot 1.5
18. Csc (−2)
19. Sin (−.6)
20. Tan .5

For each of the following, find θ in degrees. List both answers.

21. Cos θ = .3987
22. Tan θ = .4986
23. Sin θ = .0175
24. Csc θ = 3.236
25. Cot θ = .0875
26. Sec θ = 1.743
27. Tan θ = −.4557
28. Sin θ = −.8704
29. Cot θ = −1.327
30. Csc θ = −1.325
31. Sec θ = −1.022
32. Cos θ = −.1045

For each of the following, find θ in radians. List both answers.

33. Sin θ = .3090
34. Cos θ = .1736
35. Tan θ = .2217
36. Csc θ = −1.051
37. Sec θ = −2
38. Cot θ = 11.43

B–6. Logarithms

We will discuss here the finding of common and natural logarithms and antilogarithms with calculators with which we may do so directly. A complete explanation and discussion of logarithms may be found in Chapter 12.

With calculators that have keys marked LOG and LN, we may find the *common (LOG) and/or natural (LN) logarithm* of any number. We should recall that common logarithms are logarithms to the base ten and natural logarithms are logarithms to the base e.

Example A: Find log 16
Enter 16
Press LOG
Read answer → 1.2041

Example B: Find log 48,200
Enter 48,200
Press LOG
Read answer → 4.6830

Example C: Find log .00738
  Enter .00738
  Press LOG
  Read answer ⟶ −2.1319

Example D: Find ln 1.8
  Enter 1.8
  Press LN
  Read answer ⟶ .5878

Example E: Find ln 3720
  Enter 3720
  Press LN
  Read answer ⟶ 8.2215

Example F: Find ln .000815
  Enter .000815
  Press LN
  Read answer ⟶ −7.1123

If a calculator has a key marked X^Y, *common antilogarithms* may be found directly by making use of the basic definition of a logarithm. That is, if log $N = T$, then $N = 10^T$.

Example G: If log $N = 4.301$, find N.
  $N = 10^{4.301}$
  Enter 10
  Press X^Y
  Enter 4.301
  Press =
  Read answer ⟶ 20,000

Example H: If log $N = 1.7193$, find N.
  $N = 10^{1.7193}$
  Enter 10
  Press X^Y
  Enter 1.7193
  Press =
  Read answer ⟶ 52.4

Example I: If log $N = -2.1451$, find N.
  $N = 10^{-2.1451}$
  Enter 10
  Press X^Y
  Enter 2.1451
  Press +/−
  Press =
  Read answer ⟶ .00716

Example J: If log $N = 9.8871 - 10$, find N.
  $9.8871 - 10 = -.1129$
  $N = 10^{-.1129}$
  Enter 10
  Press X^Y
  Enter .1129
  Press +/−
  Press =
  Read answer ⟶ .771

If a calculator has a key marked e^X, *natural antilogarithms* may be found directly since, if ln $N = T$, then $N = e^T$.

Example K: If ln $N = 1.3083$, find N.
  $N = e^{1.3083}$
  Enter 1.3083
  Press e^X
  Read answer ⟶ 3.7

Example L: If ln $N = 9.72$, find N.
  $N = e^{9.72}$
  Enter 9.72
  Press e^X
  Read answer ⟶
          16,647.24

App. B-7. Combined Operations

Example M: If $\ln N = -2.51$, find N.
$N = e^{-2.51}$
Enter 2.51
Press $+/-$
Press e^x
Read answer $\to$ .08127

Exercises Find each of the following logarithms.

1. log 52	2. log 427	3. log .38
4. log 36,000	5. log .091	6. log .00072
7. log 8.05	8. log 278,400	9. log .314
10. log .00186	11. ln 8.3	12. ln 126
13. ln .057	14. ln .249	15. ln 8750
16. ln 27.3	17. ln 681	18. ln .00054
19. ln .00794	20. ln 83.2	

Find the common antilogarithm of each of the following numbers.

21. 5.408	22. 18.635	23. 2.093	24. .617
25. .0704	26. 4.169	27. $-.947$	28. -2.802
29. -3.562	30. -1.845		

Find the natural antilogarithm of each of the following numbers.

31. .9163	32. 4.4998	33. 1.4816	34. 2.2083
35. 1.6094	36. 12.802	37. .0736	38. $-.914$
39. -2.0541	40. -5.9120		

B-7. Combined Operations

We will now look at some problems involving more than one operation. The problems included here can easily be solved with a calculator that has keys for the *basic operations* $(+, -, \times, \div)$, *changing signs* $(+/-)$, *reciprocals* $\left(\frac{1}{X}\right)$, *squares* (X^2), and *square roots* $(\sqrt{X})$. More involved problems may be solved with a calculator that has keys for memory and other functions, but we will not consider them here.

In general, problems with more than one operation are solved on a calculator just as they appear.

Example A: Find $(23.6 + 18.7) \times 12.4$
  Enter 23.6
  Press $+$
  Enter 18.7
  Press $\times$
  Enter 12.4
  Press $=$
  Read answer $\rightarrow$ 524.52

Example B: Find $(3.14 \times 36) \div 4$
  Enter 3.14
  Press $\times$
  Enter 36
  Press $\div$
  Enter 4
  Press $=$
  Read answer $\rightarrow$ 28.26

Example C: Find $2.54 \times 26.8 \times 19.7$
  Enter 2.54
  Press $\times$
  Enter 26.8
  Press $\times$
  Enter 19.7
  Press $=$
  Read answer $\rightarrow$ 1341.02

Example D: Find $\dfrac{1}{85 \times .046 \times 12.1}$
  Enter 85
  Press $\times$
  Enter .046
  Press $\times$
  Enter 12.1
  Press $=$
  Press $\dfrac{1}{X}$
  Read answer $\rightarrow$ .0211

Example E: Find $\dfrac{2200 \times 6.28 \times 60}{5280 \times .5}$
  Enter 2200
  Press $\div$
  Enter 5280
  Press $\times$
  Enter 6.28
  Press $\div$
  Enter .5
  Press $\times$
  Enter 60
  Press $=$
  Read answer $\rightarrow$ 313.9$\bar{9}$
  $= 314$

Example F: Find $(.78 + 17.65)^2$
  Enter .78
  Press $+$
  Enter 17.65
  Press X^2
  Read answer $\rightarrow$ 339.66

Example G: Find $\sqrt{.074 \times 93.6}$
  Enter .074
  Press $\times$
  Enter 93.6
  Press $\sqrt{X}$
  Read answer $\rightarrow$ 2.63

Example H: Find $(\sqrt{10} + .76)^2$
  Enter 10
  Press $\sqrt{X}$
  Press $+$
  Enter .76
  Press X^2
  Read answer $\rightarrow$ 15.38

App. B–8. Formulas

Exercises   Perform the indicated operations for each of the following.

1. $129.3 + 48.6 - 95.3$
2. $841.2 - 648.7 + 123.9$
3. $.94 \times .735 \times 54.8$
4. $86.2 \times 48.7 \times .348$
5. $(.76 \times 198.3) + 45.4$
6. $(2486 - 1593) \times .397$
7. $(4852 \div 7520) + .61$
8. $(522.8 \div 654.7) - .83$
9. $\dfrac{48.2 \times 19.7 \times .073}{.83 \times 124.1}$
10. $\dfrac{2400 \times 6.28 \times 60}{5280 \times .25}$
11. $\dfrac{1}{14.3 \times 186.7 \times .072}$
12. $\dfrac{1}{(92.7 + 17.3) \times .64}$
13. $(.72 + 9.6 - 1.53)^2$
14. $(.031 \times 48.7)^2$
15. $\sqrt{19.4 \times 107.2}$
16. $\sqrt{68.3 + 129.4}$
17. $\left(\dfrac{1}{9 \times 12.6}\right)^2$
18. $\left[\dfrac{1}{(16 + 11.8) \times .72}\right]^2$
19. $\sqrt{6.3^2 + 24.6}$
20. $(\sqrt{19} - 2.7)^2$

B–8. Formulas

Most mathematical formulas involve combined operations similar to those mentioned in the previous section. Although there are numerous formulas that we could consider, we will look at two that are extremely important in scientific and technological areas: the *Pythagorean Theorem* and the *quadratic formula*.

The Pythagorean Theorem states that in any right triangle, the hypotenuse is equal to the square root of the sum of the squares of the other two sides (see Figure B–8.1). We may rewrite this equation so that we may use it with a basic calculator.

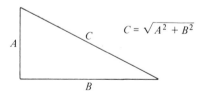

FIGURE B–8.1

$$C = \sqrt{\dfrac{B^2}{B^2}(A^2 + B^2)} = \sqrt{B^2\left(\dfrac{A^2}{B^2} + 1\right)} = \left[\sqrt{\left(\dfrac{A}{B}\right)^2 + 1}\right]B$$

Therefore, with a calculator:

Enter A
Press $\div$
Enter B

Press X^2
Press $+$
Enter 1
Press $\sqrt{X}$
Press $\times$
Enter B
Press $=$
Read answer

Example A: Find $\sqrt{3^2 + 4^2}$

$$\sqrt{3^2 + 4^2} = \left[\sqrt{\left(\frac{3}{4}\right)^2 + 1}\right] \times 4$$

Enter 3
Press $\div$
Enter 4
Press X^2
Press $+$
Enter 1
Press $\sqrt{X}$
Press $\times$
Enter 4
Press $=$
Read answer $\rightarrow$ 5

Example B: Find $\sqrt{12.6^2 + 23.1^2}$

$$\sqrt{12.6^2 + 23.1^2} = \left[\sqrt{\left(\frac{12.6}{23.1}\right)^2 + 1}\right] \times 23.1$$

Enter 12.6
Press $\div$
Enter 23.1
Press X^2
Press $+$
Enter 1
Press $\sqrt{X}$
Press $\times$
Enter 23.1
Press $=$
Read answer $\rightarrow$ 26.31

If we have a quadratic equation in the form $AX^2 + BX + C = 0$, the roots of such an equation may be found by the quadratic formula, which is

$$X = \frac{-B \pm \sqrt{B^2 - 4AC}}{2A}$$

(A complete discussion of quadratic equations may be found in Chapter 13.) For use with a hand calculator this formula would be rewritten as follows:

$$X = \frac{\pm\sqrt{-4AC + B^2} - B}{2A} = \frac{\pm\sqrt{\frac{B^2}{B^2}(-4AC + B^2)} - B}{2A}$$

$$= \frac{\left[\pm\left(\sqrt{\frac{-4AC}{B^2} + 1}\right)B\right] - B}{2A}$$

The quantity

$$\left[\pm\left(\sqrt{\frac{-4AC}{B^2} + 1}\right)B\right]$$

App. B-8. Formulas

will be the intermediate result. Also, only *real roots* may be found using a hand calculator. *Therefore, with a calculator:*

    Enter B
    Press X^2
    Press $\frac{1}{X}$
    Press $\times$
    Enter 4
    Press $+/-$
    Press $\times$
    Enter A
    Press $\times$
    Enter C
    Press $+$
    Enter 1
    Press $\sqrt{X}$
    Press $\times$
    Enter B
    Press $=$
    Read intermediate result $\rightarrow I$

Enter I	Enter $-I$
Press $-$	Press $-$
Enter B	Enter B
Press $\div$	Press $\div$
Enter 2	Enter 2
Press $\div$	Press $\div$
Enter A	Enter A
Press $=$	Press $=$
Read first root	Read second root

Example C: Solve the equation $X^2 - 7X + 12 = 0$.

$$X = \frac{\pm\left[\left(\sqrt{\frac{-4(1)(12)}{(-7)^2} + 1}\right)(-7)\right] - (-7)}{2(1)}$$

    Enter 7
    Press $+/-$
    Press X^2
    Press $\frac{1}{X}$
    Press $\times$
    Enter 4
    Press $+/-$
    Press $\times$
    Enter 1

Press ×
Enter 12
Press +
Enter 1
Press √X
Press ×
Enter 7
Press +/−
Press =
Read intermediate result → −1

Enter 1   Enter 1
Press +/−   Press −
Press −   Enter 7
Enter 7   Press +/−
Press +/−   Press ÷
Press ÷   Enter 2
Enter 2   Press ÷
Press ÷   Enter 1
Enter 1   Press =
Press =   Read second root → 4
Read first root → 3

Example D: Solve the equation $2T^2 + 8T + 3 = 0$.

$$T = \frac{\pm\left[\left(\sqrt{\frac{-4(2)(3)}{8^2} + 1}\right)8\right] - 8}{2(2)}$$

Enter 8
Press X²
Press $\frac{1}{X}$
Press ×
Enter 4
Press +/−
Press ×
Enter 2
Press ×
Enter 3
Press +
Enter 1
Press √X
Press ×
Enter 8
Press =
Read intermediate result → 6.32

App. B–8. Formulas

Enter 6.32	Enter 6.32
Press −	Press +/−
Enter 8	Press −
Press ÷	Enter 8
Enter 2	Press ÷
Press ÷	Enter 2
Enter 2	Press ÷
Press =	Enter 2
Read first root → −.42	Press =
	Read second root → −3.58

Exercises Evaluate each of the following.

1. $\sqrt{13^2 + 21^2}$
2. $\sqrt{45^2 + 62^2}$
3. $\sqrt{.72^2 + .86^2}$
4. $\sqrt{.23^2 + .14^2}$
5. $\sqrt{62.8^2 + 37.9^2}$
6. $\sqrt{(-15)^2 + 83^2}$
7. $\sqrt{126^2 + (-91)^2}$
8. $\sqrt{41.7^2 + 52.8^2}$
9. $\sqrt{(-17)^2 + (-23)^2}$
10. $\sqrt{4150^2 + 3820^2}$

Solve each of the following quadratic equations.

11. $X^2 - 5X + 6 = 0$
12. $Y^2 - 2Y - 8 = 0$
13. $T^2 - 9 = 0$
14. $M^2 + 5M = 0$
15. $2C^2 + 5C - 12 = 0$
16. $3V^2 + 5V + 2 = 0$
17. $6K^2 - 11K + 4 = 0$
18. $8P^2 + 10P - 3 = 0$
19. $2S^2 - S - 9 = 0$
20. $3Q^2 + 2Q - 7 = 0$
21. $L^2 + 8L + 10 = 0$
22. $2R^2 - 7R + 4 = 0$

Tables

TABLE 1. POWERS AND ROOTS

Although Table 1 only lists squares, square roots, cubes, and cube roots of numbers from 1 to 100, we can find these quantities for other numbers by making use of scientific notation.

Number	Square	Square Root	Cube	Cube Root	Number	Square	Square Root	Cube	Cube Root
1	1	1.000	1	1.000	51	2,601	7.141	132,651	3.708
2	4	1.414	8	1.260	52	2,704	7.211	140,608	3.733
3	9	1.732	27	1.442	53	2,809	7.280	148,877	3.756
4	16	2.000	64	1.587	54	2,916	7.348	157,464	3.780
5	25	2.236	125	1.710	55	3,025	7.416	166,375	3.803
6	36	2.449	216	1.817	56	3,136	7.483	175,616	3.826
7	49	2.646	343	1.913	57	3,249	7.550	185,193	3.849
8	64	2.828	512	2.000	58	3,364	7.616	195,112	3.871
9	81	3.000	729	2.080	59	3,481	7.681	205,379	3.893
10	100	3.162	1,000	2.154	60	3,600	7.746	216,000	3.915
11	121	3.317	1,331	2.224	61	3,721	7.810	226,981	3.936
12	144	3.464	1,728	2.289	62	3,844	7.874	238,328	3.958
13	169	3.606	2,197	2.351	63	3,969	7.937	250,047	3.979
14	196	3.742	2,744	2.410	64	4,096	8.000	262,144	4.000
15	225	3.873	3,375	2.466	65	4,225	8.062	274,625	4.021
16	256	4.000	4,096	2.520	66	4,356	8.124	287,496	4.041
17	289	4.123	4,913	2.571	67	4,489	8.185	300,763	4.062
18	324	4.243	5,832	2.621	68	4,624	8.246	314,432	4.082
19	361	4.359	6,859	2.668	69	4,761	8.307	328,509	4.102
20	400	4.472	8,000	2.714	70	4,900	8.367	343,000	4.121
21	441	4.583	9,261	2.759	71	5,041	8.426	357,911	4.141
22	484	4.690	10,648	2.802	72	5,184	8.485	373,248	4.160
23	529	4.796	12,167	2.844	73	5,329	8.544	389,017	4.179
24	576	4.899	13,824	2.884	74	5,476	8.602	405,224	4.198
25	625	5.000	15,625	2.924	75	5,625	8.660	421,875	4.217
26	676	5.099	17,576	2.962	76	5,776	8.718	438,976	4.236
27	729	5.196	19,683	3.000	77	5,929	8.775	456,533	4.254
28	784	5.292	21,952	3.037	78	6,084	8.832	474,552	4.273
29	841	5.385	24,389	3.072	79	6,241	8.888	493,039	4.291
30	900	5.477	27,000	3.107	80	6,400	8.944	512,000	4.309
31	961	5.568	29,791	3.141	81	6,561	9.000	531,441	4.327
32	1,024	5.657	32,768	3.175	82	6,724	9.055	551,368	4.344
33	1,089	5.745	35,937	3.208	83	6,889	9.110	571,787	4.362
34	1,156	5.831	39,304	3.240	84	7,056	9.165	592,704	4.380
35	1,225	5.916	42,875	3.271	85	7,225	9.220	614,125	4.397
36	1,296	6.000	46,656	3.302	86	7,396	9.274	636,056	4.414
37	1,369	6.083	50,653	3.332	87	7,569	9.327	658,503	4.431
38	1,444	6.164	54,872	3.362	88	7,744	9.381	681,472	4.448
39	1,521	6.245	59,319	3.391	89	7,921	9.434	704,969	4.465
40	1,600	6.325	64,000	3.420	90	8,100	9.487	729,000	4.481
41	1,681	6.403	68,921	3.448	91	8,281	9.539	753,571	4.498
42	1,764	6.481	74,088	3.476	92	8,464	9.592	778,688	4.514
43	1,849	6.557	79,507	3.503	93	8,649	9.644	804,357	4.531
44	1,936	6.633	85,184	3.530	94	8,836	9.695	830,584	4.547
45	2,025	6.708	91,125	3.557	95	9,025	9.747	857,375	4.563
46	2,116	6.782	97,336	3.583	96	9,216	9.798	884,736	4.579
47	2,209	6.856	103,823	3.609	97	9,409	9.849	912,673	4.595
48	2,304	6.928	110,592	3.634	98	9,604	9.899	941,192	4.610
49	2,401	7.000	117,649	3.659	99	9,801	9.950	970,299	4.626
50	2,500	7.071	125,000	3.684	100	10,000	10.000	1,000,000	4.642

TABLE 2. TRIGONOMETRIC FUNCTIONS

Degrees	Radians	Sin	Cos	Tan	Cot	Sec	Csc		
0°00′	.0000	.0000	1.0000	.0000	—	1.000	—	1.5708	90°00′
10	029	029	000	029	343.8	000	343.8	679	50
20	058	058	000	058	171.9	000	171.9	650	40
30	.0087	.0087	1.0000	.0087	114.6	1.000	114.6	1.5621	30
40	116	116	.999	116	85.94	000	85.95	592	20
50	145	145	999	145	68.75	000	68.76	563	10
1°00′	.0175	.0175	.9998	.0175	57.29	1.000	57.30	1.5533	89°00′
10	204	204	998	204	49.10	000	49.11	504	50
20	233	233	997	233	42.96	000	42.98	475	40
30	.0262	.0262	.9997	.0262	38.19	1.000	38.20	1.5446	30
40	291	291	996	291	34.37	000	34.38	417	20
50	320	320	995	320	31.24	001	31.26	388	10
2°00′	.0349	.0349	.9994	.0349	28.64	1.001	28.65	1.5359	88°00′
10	378	378	993	378	26.43	001	26.45	330	50
20	407	407	992	407	24.54	001	24.56	301	40
30	.0436	.0436	.9990	.0437	22.90	1.001	22.93	1.5272	30
40	465	465	989	466	21.47	001	21.49	243	20
50	495	494	988	495	20.21	001	20.23	213	10
3°00′	.0524	.0523	.9986	.0524	19.08	1.001	19.11	1.5184	87°00′
10	553	552	985	553	18.07	002	18.10	155	50
20	582	581	983	582	17.17	002	17.20	126	40
30	.0611	.0610	.9981	.0612	16.35	1.002	16.38	1.5097	30
40	640	640	980	641	15.60	002	15.64	068	20
50	669	669	978	670	14.92	002	14.96	039	10
4°00′	.0698	.0698	.9976	.0699	14.30	1.002	14.34	1.5010	86°00′
10	727	727	974	729	13.73	003	13.76	981	50
20	756	756	971	758	13.20	003	13.23	952	40
30	.0785	.0785	.9969	.0787	12.71	1.003	12.75	1.4923	30
40	814	814	967	816	12.25	003	12.29	893	20
50	844	843	964	846	11.83	004	11.87	864	10
5°00′	.0873	.0872	.9962	.0875	11.43	1.004	11.47	1.4835	85°00′
10	902	901	959	904	11.06	004	11.10	806	50
20	931	929	957	934	10.71	004	10.76	777	40
30	.0960	.0958	.9954	.0963	10.39	1.005	10.43	1.4748	30
40	989	987	951	992	10.08	005	10.13	719	20
50	.1018	.1016	948	.1022	9.788	005	9.839	690	10
6°00′	.1047	.1045	.9945	.1051	9.514	1.006	9.567	1.4661	84°00′
10	076	074	942	080	9.255	006	9.309	632	50
20	105	103	939	110	9.010	006	9.065	603	40
30	.1134	.1132	.9936	.1139	8.777	1.006	8.834	1.4573	30
40	164	161	932	169	8.556	007	8.614	544	20
50	193	190	929	198	8.345	007	8.405	515	10
7°00′	.1222	.1219	.9925	.1228	8.144	1.008	8.206	1.4486	83°00′
10	251	248	922	257	7.953	008	8.016	457	50
20	280	276	918	287	7.770	008	7.834	428	40
30	.1309	.1305	.9914	.1317	7.596	1.009	7.661	1.4399	30
40	338	334	911	346	7.429	009	7.496	370	20
50	367	363	907	376	7.269	009	7.337	341	10
8°00′	.1396	.1392	.9903	.1405	7.115	1.010	7.185	1.4312	82°00′
10	425	421	899	435	6.968	010	7.040	283	50
20	454	449	894	465	6.827	011	6.900	254	40
30	.1484	.1478	.9890	.1495	6.691	1.011	6.765	1.4224	30
40	513	507	886	524	6.561	012	6.636	195	20
50	542	536	881	554	6.435	012	6.512	166	10
9°00′	.1571	.1564	.9877	.1584	6.314	1.012	6.392	1.4137	81°00′
		Cos	Sin	Cot	Tan	Csc	Sec	Radians	Degrees

TABLE 2. TRIGONOMETRIC FUNCTIONS (*Cont.*)

Degrees	Radians	Sin	Cos	Tan	Cot	Sec	Csc		
9°00′	.1571	.1564	.9877	.1584	6.314	1.012	6.392	1.4137	81°00′
10	600	593	872	614	197	013	277	108	50
20	629	622	868	644	084	013	166	079	40
30	.1658	.1650	.9863	.1673	5.976	1.014	6.059	1.4050	30
40	687	679	858	703	871	014	955	021	20
50	716	708	853	733	769	015	855	992	10
10°00′	.1745	.1736	.9848	.1763	5.671	1.015	5.759	1.3963	80°00′
10	774	765	843	793	576	016	665	934	50
20	804	794	838	823	485	016	575	904	40
30	.1833	.1822	.9833	.1853	5.396	1.017	5.487	1.3875	30
40	862	851	827	883	309	018	403	846	20
50	891	880	822	914	226	018	320	817	10
11°00′	.1920	.1908	.9816	.1944	5.145	1.019	5.241	1.3788	79°00′
10	949	937	811	974	066	019	164	759	50
20	978	965	805	004	989	020	089	730	40
30	.2007	.1994	.9799	.2035	4.915	1.020	5.016	1.3701	30
40	036	.022	793	065	843	021	945	672	20
50	065	051	787	095	773	022	876	643	10
12°00′	.2094	.2079	.9781	.2126	4.705	1.022	4.810	1.3614	78°00′
10	123	108	775	156	638	023	745	584	50
20	153	136	769	186	574	024	682	555	40
30	.2182	.2164	.9763	.2217	4.511	1.024	4.620	1.3526	30
40	211	193	757	247	449	025	560	497	20
50	240	221	750	278	390	026	502	468	10
13°00′	.2269	.2250	.9744	.2309	4.331	1.026	4.445	1.3439	77°00′
10	298	278	737	339	275	027	390	410	50
20	327	306	730	370	219	028	336	381	40
30	.2356	.2334	.9724	.2401	4.165	1.028	4.284	1.3352	30
40	385	363	717	432	113	029	232	323	20
50	414	391	710	462	061	030	182	294	10
14°00′	.2443	.2419	.9703	.2493	4.011	1.031	4.134	1.3265	76°00′
10	473	447	696	524	962	031	088	235	50
20	502	476	689	555	914	032	039	206	40
30	.2531	.2504	.9681	.2586	3.867	1.033	3.994	1.3177	30
40	560	532	674	617	821	034	950	148	20
50	589	560	667	648	776	034	906	119	10
15°00′	.2618	.2588	.9659	.2679	3.732	1.035	3.864	1.3090	75°00′
10	647	616	652	711	689	036	822	061	50
20	676	644	644	742	647	037	782	032	40
30	.2705	.2672	.9636	.2773	3.606	1.038	3.742	1.3003	30
40	734	700	628	805	566	039	703	974	20
50	763	728	621	836	526	039	665	945	10
16°00′	.2793	.2756	.9613	.2867	3.487	1.040	3.628	1.2915	74°00′
10	822	784	605	899	450	041	592	886	50
20	851	812	596	931	412	042	556	857	40
30	.2880	.2840	.9588	.2962	3.376	1.043	3.521	1.2828	30
40	909	868	580	994	340	044	487	799	20
50	938	896	572	.3026	305	045	453	770	10
17°00′	.2967	.2924	.9563	.3057	3.271	1.046	3.420	1.2741	73°00′
10	996	952	555	089	237	047	388	712	50
20	.3025	979	546	121	204	048	356	683	40
30	054	.3007	.9537	.3153	3.172	1.049	3.326	1.2654	30
40	083	035	528	185	140	049	295	625	20
50	113	062	520	217	108	050	265	595	10
18°00′	.3142	.3090	.9511	.3249	3.078	1.051	3.236	1.2566	72°00′
		Cos	Sin	Cot	Tan	Csc	Sec	Radians	Degrees

TABLE 2. TRIGONOMETRIC FUNCTIONS (*Cont.*)

Degrees	Radians	Sin	Cos	Tan	Cot	Sec	Csc		
18°00′	.3142	.3090	.9511	.3249	3.078	1.051	3.236	1.2566	72°00′
10	171	118	502	281	047	052	207	537	50
20	200	145	492	314	018	053	179	508	40
30	.3229	.3173	.9483	.3346	2.989	1.054	3.152	1.2479	30
40	258	201	474	378	960	056	124	450	20
50	287	228	465	411	932	057	098	421	10
19°00′	.3316	.3256	.9455	.3443	2.904	1.058	3.072	1.2392	71°00′
10	345	283	446	476	877	059	046	363	50
20	374	311	436	508	850	060	021	334	40
30	.3403	.3338	.9426	.3541	2.824	1.061	2.996	1.2305	30
40	432	365	417	574	798	062	971	275	20
50	462	393	407	607	773	063	947	246	10
20°00′	.3491	.3420	.9397	.3640	2.747	1.064	2.924	1.2217	70°00′
10	520	448	387	673	723	065	901	188	50
20	549	475	377	706	699	066	878	159	40
30	.3578	.3502	.9367	.3739	2.675	1.068	2.855	1.2130	30
40	607	529	356	772	651	069	833	101	20
50	636	557	346	805	628	070	812	072	10
21°00′	.3665	.3584	.9336	.3839	2.605	1.071	2.790	1.2043	69°00′
10	694	611	325	872	583	072	769	014	50
20	723	638	315	906	560	074	749	985	40
30	.3752	.3665	.9304	.3939	2.539	1.075	2.729	1.1956	30
40	782	692	293	973	517	076	709	926	20
50	811	719	283	006	496	077	689	897	10
22°00′	.3840	.3746	.9272	.4040	2.475	1.079	2.669	1.1868	68°00′
10	869	773	261	074	455	080	650	839	50
20	898	800	250	108	434	081	632	810	40
30	.3927	.3827	.9239	.4142	2.414	1.082	2.613	1.1781	30
40	956	854	228	176	394	084	595	752	20
50	985	881	216	210	375	085	577	723	10
23°00′	.4014	.3907	.9205	.4245	2.356	1.086	2.559	1.1694	67°00′
10	043	934	194	279	337	088	542	665	50
20	072	961	182	314	318	089	525	636	40
30	.4102	.3987	.9171	.4348	2.300	1.090	2.508	1.1606	30
40	131	014	159	383	282	092	491	577	20
50	160	041	147	417	264	093	475	548	10
24°00′	.4189	.4067	.9135	.4452	2.246	1.095	2.459	1.1519	66°00′
10	218	094	124	487	229	096	443	490	50
20	247	120	112	522	211	097	427	461	40
30	.4276	.4147	.9100	.4557	2.194	1.099	2.411	1.1432	30
40	305	173	088	592	177	100	396	403	20
50	334	200	075	628	161	102	381	374	10
25°00′	.4363	.4226	.9063	.4663	2.145	1.103	2.366	1.1345	65°00′
10	392	253	051	699	128	105	352	316	50
20	422	279	038	734	112	106	337	286	40
30	.4451	.4305	.9026	.4770	2.097	1.108	2.323	1.1257	30
40	480	331	013	806	081	109	309	228	20
50	509	358	001	841	066	111	295	199	10
26°00′	.4538	.4384	.8988	.4877	2.050	1.113	2.281	1.1170	64°00′
10	567	410	975	913	035	114	268	141	50
20	596	436	962	950	020	116	254	112	40
30	.4625	.4462	.8949	.4986	2.006	1.117	2.241	1.1083	30
40	654	488	936	022	991	119	228	054	20
50	683	514	923	059	977	121	215	025	10
27°00′	.4712	.4540	.8910	.5095	1.963	1.122	2.203	1.0996	63°00′
		Cos	Sin	Cot	Tan	Csc	Sec	Radians	Degrees

TABLE 2. TRIGONOMETRIC FUNCTIONS (*Cont.*)

Degrees	Radians	Sin	Cos	Tan	Cot	Sec	Csc	
27°00'	.4712	.4540	.8910	.5095	1.963	1.122	2.203	1.0996	63°00'
10	741	566	897	132	949	124	190	966	50
20	771	592	884	169	935	126	178	937	40
30	.4800	.4617	.8870	.5206	1.921	1.127	2.166	1.0908	30
40	829	643	857	243	907	129	154	879	20
50	858	669	843	280	894	131	142	850	10
28°00'	.4887	.4695	.8829	.5317	1.881	1.133	2.130	1.0821	62°00'
10	916	720	816	354	868	134	118	792	50
20	945	746	802	392	855	136	107	763	40
30	.4974	.4772	.8788	.5430	1.842	1.138	2.096	1.0734	30
40	003	797	774	467	829	140	085	705	20
50	032	823	760	505	816	142	074	676	10
29°00'	.5061	.4848	.8746	.5543	1.804	1.143	2.063	1.0647	61°00'
10	091	874	732	581	792	145	052	617	50
20	120	899	718	619	780	147	041	588	40
30	.5149	.4924	.8704	.5658	1.767	1.149	2.031	1.0559	30
40	178	950	689	696	756	151	020	530	20
50	207	975	675	735	744	153	010	501	10
30°00'	.5236	.5000	.8660	.5774	1.732	1.155	2.000	1.0472	60°00'
10	265	025	646	812	720	157	990	443	50
20	294	050	631	851	709	159	980	414	40
30	.5323	.5075	.8616	.5890	1.698	1.161	1.970	1.0385	30
40	352	100	601	930	686	163	961	356	20
50	381	125	587	969	675	165	951	327	10
31°00'	.5411	.5150	.8572	.6009	1.664	1.167	1.942	1.0297	59°00'
10	440	175	557	048	653	169	932	268	50
20	469	200	542	088	643	171	923	239	40
30	.5498	.5225	.8526	.6128	1.632	1.173	1.914	1.0210	30
40	527	250	511	168	621	175	905	181	20
50	556	275	496	208	611	177	896	152	10
32°00'	.5585	.5299	.8480	.6249	1.600	1.179	1.887	1.0123	58°00'
10	614	324	465	289	590	181	878	094	50
20	643	348	450	330	580	184	870	065	40
30	.5672	.5373	.8434	.6371	1.570	1.186	1.861	1.0036	30
40	701	398	418	412	560	188	853	0007	20
50	730	422	403	453	550	190	844	977	10
33°00'	.5760	.5446	.8387	.6494	1.540	1.192	1.836	.9948	57°00'
10	789	471	371	536	530	195	828	919	50
20	818	495	355	577	520	197	820	890	40
30	.5847	.5519	.8339	.6619	1.511	1.199	1.812	.9861	30
40	876	544	323	661	501	202	804	832	20
50	905	568	307	703	492	204	796	803	10
34°00'	.5934	.5592	.8290	.6745	1.483	1.206	1.788	.9774	56°00'
10	963	616	274	787	473	209	781	745	50
20	992	640	258	830	464	211	773	715	40
30	.6021	.5664	.8241	.6873	1.455	1.213	1.766	.9687	30
40	050	688	225	916	446	216	758	657	20
50	080	712	208	959	437	218	751	628	10
35°00'	.6109	.5736	.8192	.7002	1.428	1.221	1.743	.9599	55°00'
10	138	760	175	046	419	223	736	570	50
20	167	783	158	089	411	226	729	541	40
30	.6196	.5807	.8141	.7133	1.402	1.228	1.722	.9512	30
40	225	831	124	177	393	231	715	483	20
50	254	854	107	221	385	233	708	454	10
36°00'	.6283	.5878	.8090	.7265	1.376	1.236	1.701	.9425	54°00'
		Cos	Sin	Cot	Tan	Csc	Sec	Radians	Degrees

TABLE 2. TRIGONOMETRIC FUNCTIONS (*Cont.*)

Degrees	Radians	Sin	Cos	Tan	Cot	Sec	Csc		
36°00′	.6283	.5878	.8090	.7265	1.376	1.236	1.701	.9425	54°00′
10	312	901	073	310	368	239	695	396	50
20	341	925	056	355	360	241	688	367	40
30	.6370	.5948	.8039	.7400	1.351	1.244	1.681	.9338	30
40	400	972	021	445	343	247	675	308	20
50	429	995	004	490	335	249	668	279	10
37°00′	.6458	.6018	.7986	.7536	1.327	1.252	1.662	.9250	53°00′
10	487	041	969	581	319	255	655	221	50
20	516	065	951	627	311	258	649	192	40
30	.6545	.6088	.7934	.7673	.1303	1.260	1.643	.9163	30
40	574	111	916	720	295	263	636	134	20
50	603	134	898	766	288	266	630	105	10
38°00′	.6632	.6157	.7880	.7813	1.280	1.269	1.624	.9076	52°00′
10	661	180	862	860	272	272	618	047	50
20	690	202	844	907	265	275	612	018	40
30	.6720	.6225	.7826	.7954	1.257	1.278	1.606	.8988	30
40	749	248	808	002	250	281	601	959	20
50	778	271	790	050	242	284	595	930	10
39°00′	.6807	.6293	.7771	.8098	1.235	1.287	1.589	.8901	51°00′
10	836	316	753	146	228	290	583	872	50
20	865	338	735	195	220	293	578	843	40
30	.6894	.6361	.7716	.8243	1.213	1.296	1.572	.8814	30
40	923	383	698	292	206	299	567	785	20
50	952	406	679	342	199	302	561	756	10
40°00′	.6981	.6428	.7660	.8391	1.192	1.305	1.556	.8727	50°00′
10	010	450	642	441	185	309	550	698	50
20	039	472	623	491	178	312	545	668	40
30	.7069	.6494	.7604	.8541	1.171	1.315	1.540	.8639	30
40	098	517	585	591	164	318	535	610	20
50	127	539	566	642	157	322	529	581	10
41°00′	.7156	.6561	.7547	.8693	1.150	1.325	1.524	.8552	49°00′
10	185	583	528	744	144	328	519	523	50
20	214	604	509	796	137	332	514	494	40
30	.7243	.6626	.7490	.8847	1.130	1.335	1.509	.8465	30
40	272	648	470	899	124	339	504	436	20
50	301	670	451	952	117	342	499	407	10
42°00′	.7330	.6691	.7431	.9004	1.111	1.346	1.494	.8378	48°00′
10	359	713	412	057	104	349	490	348	50
20	389	734	392	110	098	353	485	319	40
30	.7418	.6756	.7373	.9163	1.091	1.356	1.480	.8290	30
40	447	777	353	217	085	360	476	261	20
50	476	799	333	271	079	364	471	232	10
43°00′	.7505	.6820	.7314	.9325	1.072	1.367	1.466	.8203	47°00′
10	534	841	294	380	066	371	462	174	50
20	563	862	274	435	060	375	457	145	40
30	.7592	.6884	.7254	.9490	1.054	1.379	1.453	.8116	30
40	621	905	234	545	048	382	448	087	20
50	650	926	214	601	042	386	444	058	10
44°00′	.7679	.6947	.7193	.9657	1.036	1.390	1.440	.8029	46°00′
10	709	967	173	713	030	394	435	999	50
20	738	988	153	770	024	398	431	970	40
30	.7767	.7009	.7133	.9827	1.018	1.402	1.427	.7941	30
40	796	030	112	884	012	406	423	912	20
50	825	050	092	942	006	410	418	883	10
45°00′	.7854	.7071	.7071	1.000	1.000	1.414	1.414	.7854	45°00′
		Cos	Sin	Cot	Tan	Csc	Sec	Radians	Degrees

TABLE 3. COMMON LOGARITHMS OF NUMBERS

N	0	1	2	3	4	5	6	7	8	9
10	0000	0043	0086	0128	0170	0212	0253	0294	0334	0374
11	0414	0453	0492	0531	0569	0607	0645	0682	0719	0755
12	0792	0828	0864	0899	0934	0969	1004	1038	1072	1106
13	1139	1173	1206	1239	1271	1303	1335	1367	1399	1430
14	1461	1492	1523	1553	1584	1614	1644	1673	1703	1732
15	1761	1790	1818	1847	1875	1903	1931	1959	1987	2014
16	2041	2068	2095	2122	2148	2175	2201	2227	2253	2279
17	2304	2330	2355	2380	2405	2430	2455	2480	2504	2529
18	2553	2577	2601	2625	2648	2672	2695	2718	2742	2765
19	2788	2810	2833	2856	2878	2900	2923	2945	2967	2989
20	3010	3032	3054	3075	3096	3118	3139	3160	3181	3201
21	3222	3243	3263	3284	3304	3324	3345	3365	3385	3404
22	3424	3444	3464	3483	3502	3522	3541	3560	3579	3598
23	3617	3636	3655	3674	3692	3711	3729	3747	3766	3784
24	3802	3820	3838	3856	3874	3892	3909	3927	3945	3962
25	3979	3997	4014	4031	4048	4065	4082	4099	4116	4133
26	4150	4166	4183	4200	4216	4232	4249	4265	4281	4298
27	4314	4330	4346	4362	4378	4393	4409	4425	4440	4456
28	4472	4487	4502	4518	4533	4548	4564	4579	4594	4609
29	4624	4639	4654	4669	4683	4698	4713	4728	4742	4757
30	4771	4786	4800	4814	4829	4843	4857	4871	4886	4900
31	4914	4928	4942	4955	4969	4983	4997	5011	5024	5038
32	5051	5065	5079	5092	5105	5119	5132	5145	5159	5172
33	5185	5198	5211	5224	5237	5250	5263	5276	5289	5302
34	5315	5328	5340	5353	5366	5378	5391	5403	5416	5428
35	5441	5453	5465	5478	5490	5502	5514	5527	5539	5551
36	5563	5575	5587	5599	5611	5623	5635	5647	5658	5670
37	5682	5694	5705	5717	5729	5740	5752	5763	5775	5786
38	5798	5809	5821	5832	5843	5855	5866	5877	5888	5899
39	5911	5922	5933	5944	5955	5966	5977	5988	5999	6010
40	6021	6031	6042	6053	6064	6075	6085	6096	6107	6117
41	6128	6138	6149	6160	6170	6180	6191	6201	6212	6222
42	6232	6243	6253	6263	6274	6284	6294	6304	6314	6325
43	6335	6345	6355	6365	6375	6385	6395	6405	6415	6425
44	6435	6444	6454	6464	6474	6484	6493	6503	6513	6522
45	6532	6542	6551	6561	6571	6580	6590	6599	6609	6618
46	6628	6637	6646	6656	6665	6675	6684	6693	6702	6712
47	6721	6730	6739	6749	6758	6767	6776	6785	6794	6803
48	6812	6821	6830	6839	6848	6857	6866	6875	6884	6893
49	6902	6911	6920	6928	6937	6946	6955	6964	6972	6981
50	6990	6998	7007	7016	7024	7033	7042	7050	7059	7067
51	7076	7084	7093	7101	7110	7118	7126	7135	7143	7152
52	7160	7168	7177	7185	7193	7202	7210	7218	7226	7235
53	7243	7251	7259	7267	7275	7284	7292	7300	7308	7316
54	7324	7332	7340	7348	7356	7364	7372	7380	7388	7396

TABLE 3. COMMON LOGARITHMS OF NUMBERS (*Cont.*)

N	0	1	2	3	4	5	6	7	8	9
55	7404	7412	7419	7427	7435	7443	7451	7459	7466	7474
56	7482	7490	7497	7505	7513	7520	7528	7536	7543	7551
57	7559	7566	7574	7582	7589	7597	7604	7612	7619	7627
58	7634	7642	7649	7657	7664	7672	7679	7686	7694	7701
59	7709	7716	7723	7731	7738	7745	7752	7760	7767	7774
60	7782	7789	7796	7803	7810	7818	7825	7832	7839	7846
61	7853	7860	7868	7875	7882	7889	7896	7903	7910	7917
62	7924	7931	7938	7945	7952	7959	7966	7973	7980	7987
63	7993	8000	8007	8014	8021	8028	8035	8041	8048	8055
64	8062	8069	8075	8082	8089	8096	8102	8109	8116	8122
65	8129	8136	8142	8149	8156	8162	8169	8176	8182	8189
66	8195	8202	8209	8215	8222	8228	8235	8241	8248	8254
67	8261	8267	8274	8280	8287	8293	8299	8306	8312	8319
68	8325	8331	8338	8344	8351	8357	8363	8370	8376	8382
69	8388	8395	8401	8407	8414	8420	8426	8432	8439	8445
70	8451	8457	8463	8470	8476	8482	8488	8494	8500	8506
71	8513	8519	8525	8531	8537	8543	8549	8555	8561	8567
72	8573	8579	8585	8591	8597	8603	8609	8615	8621	8627
73	8633	8639	8645	8651	8657	8663	8669	8675	8681	8686
74	8692	8698	8704	8710	8716	8722	8727	8733	8739	8745
75	8751	8756	8762	8768	8774	8779	8785	8791	8797	8802
76	8808	8814	8820	8825	8831	8837	8842	8848	8854	8859
77	8865	8871	8876	8882	8887	8893	8899	8904	8910	8915
78	8921	8927	8932	8938	8943	8949	8954	8960	8965	8971
79	8976	8982	8987	8993	8998	9004	9009	9015	9020	9025
80	9031	9036	9042	9047	9053	9058	9063	9069	9074	9079
81	9085	9090	9096	9101	9106	9112	9117	9122	9128	9133
82	9138	9143	9149	9154	9159	9165	9170	9175	9180	9186
83	9191	9196	9201	9206	9212	9217	9222	9227	9232	9238
84	9243	9248	9253	9258	9263	9269	9274	9279	9284	9289
85	9294	9299	9304	9309	9315	9320	9325	9330	9335	9340
86	9345	9350	9355	9360	9365	9370	9375	9380	9385	9390
87	9395	9400	9405	9410	9415	9420	9425	9430	9435	9440
88	9445	9450	9455	9460	9465	9469	9474	9479	9484	9489
89	9494	9499	9504	9509	9513	9518	9523	9528	9533	9538
90	9542	9547	9552	9557	9562	9566	9571	9576	9581	9586
91	9590	9595	9600	9605	9609	9614	9619	9624	9628	9633
92	9638	9643	9647	9652	9657	9661	9666	9671	9675	9680
93	9685	9689	9694	9699	9703	9708	9713	9717	9722	9727
94	9731	9736	9741	9745	9750	9754	9759	9763	9768	9773
95	9777	9782	9786	9791	9795	9800	9805	9809	9814	9818
96	9823	9827	9832	9836	9841	9845	9850	9854	9859	9863
97	9868	9872	9877	9881	9886	9890	9894	9899	9903	9908
98	9912	9917	9921	9926	9930	9934	9939	9943	9948	9952
99	9956	9961	9965	9969	9974	9978	9983	9987	9991	9996

TABLE 4. NATURAL LOGARITHMS OF NUMBERS

n	$\log_e n$	n	$\log_e n$	n	$\log_e n$
0.0		4.5	1.5041	9.0	2.1972
0.1	7.6974*	4.6	1.5261	9.1	2.2083
0.2	8.3906*	4.7	1.5476	9.2	2.2192
0.3	8.7960*	4.8	1.5686	9.3	2.2300
0.4	9.0837*	4.9	1.5892	9.4	2.2407
0.5	9.3069*	5.0	1.6094	9.5	2.2513
0.6	9.4892*	5.1	1.6292	9.6	2.2618
0.7	9.6433*	5.2	1.6487	9.7	2.2721
0.8	9.7769*	5.3	1.6677	9.8	2.2824
0.9	9.8946*	5.4	1.6864	9.9	2.2925
1.0	0.0000	5.5	1.7047	10	2.3026
1.1	0.0953	5.6	1.7228	11	2.3979
1.2	0.1823	5.7	1.7405	12	2.4849
1.3	0.2624	5.8	1.7579	13	2.5649
1.4	0.3365	5.9	1.7750	14	2.6391
1.5	0.4055	6.0	1.7918	15	2.7081
1.6	0.4700	6.1	1.8083	16	2.7726
1.7	0.5306	6.2	1.8245	17	2.8332
1.8	0.5878	6.3	1.8405	18	2.8904
1.9	0.6419	6.4	1.8563	19	2.9444
2.0	0.6931	6.5	1.8718	20	2.9957
2.1	0.7419	6.6	1.8871	25	3.2189
2.2	0.7885	6.7	1.9021	30	3.4012
2.3	0.8329	6.8	1.9169	35	3.5553
2.4	0.8755	6.9	1.9315	40	3.6889
2.5	0.9163	7.0	1.9459	45	3.8067
2.6	0.9555	7.1	1.9601	50	3.9120
2.7	0.9933	7.2	1.9741	55	4.0073
2.8	1.0296	7.3	1.9879	60	4.0943
2.9	1.0647	7.4	2.0015	65	4.1744
3.0	1.0986	7.5	2.0149	70	4.2485
3.1	1.1314	7.6	2.0281	75	4.3175
3.2	1.1632	7.7	2.0412	80	4.3820
3.3	1.1939	7.8	2.0541	85	4.4427
3.4	1.2238	7.9	2.0669	90	4.4998
3.5	1.2528	8.0	2.0794	95	4.5539
3.6	1.2809	8.1	2.0919	100	4.6052
3.7	1.3083	8.2	2.1041		
3.8	1.3350	8.3	2.1163		
3.9	1.3610	8.4	2.1282		
4.0	1.3863	8.5	2.1401		
4.1	1.4110	8.6	2.1518		
4.2	1.4351	8.7	2.1633		
4.3	1.4586	8.8	2.1748		
4.4	1.4816	8.9	2.1861		

*Attach -10 to these logarithms.

TABLE 5. VALUES OF e^x AND e^{-x}

x	e^x	e^{-x}	x	e^x	e^{-x}
0.00	1.0000	1.0000	2.5	12.182	0.0821
0.05	1.0513	0.9512	2.6	13.464	0.0743
0.10	1.1052	0.9048	2.7	14.880	0.0672
0.15	1.1618	0.8607	2.8	16.445	0.0608
0.20	1.2214	0.8187	2.9	18.174	0.0550
0.25	1.2840	0.7788	3.0	20.086	0.0498
0.30	1.3499	0.7408	3.1	22.198	0.0450
0.35	1.4191	0.7047	3.2	24.533	0.0408
0.40	1.4918	0.6703	3.3	27.113	0.0369
0.45	1.5683	0.6376	3.4	29.964	0.0334
0.50	1.6487	0.6065	3.5	33.115	0.0302
0.55	1.7333	0.5769	3.6	36.598	0.0273
0.60	1.8221	0.5488	3.7	40.447	0.0247
0.65	1.9155	0.5220	3.8	44.701	0.0224
0.70	2.0138	0.4966	3.9	49.402	0.0202
0.75	2.1170	0.4724	4.0	54.598	0.0183
0.80	2.2255	0.4493	4.1	60.340	0.0166
0.85	2.3369	0.4274	4.2	66.686	0.0150
0.90	2.4596	0.4066	4.3	73.700	0.0136
0.95	2.5857	0.3867	4.4	81.451	0.0123
1.0	2.7183	0.3679	4.5	90.017	0.0111
1.1	3.0042	0.3329	4.6	99.484	0.0101
1.2	3.3201	0.3012	4.7	109.95	0.0091
1.3	3.6693	0.2725	4.8	121.51	0.0082
1.4	4.0552	0.2466	4.9	134.29	0.0074
1.5	4.4817	0.2231	5	148.41	0.0067
1.6	4.9530	0.2019	6	403.43	0.0025
1.7	5.4739	0.1827	7	1096.6	0.0009
1.8	6.0496	0.1653	8	2981.0	0.0003
1.9	6.6859	0.1496	9	8103.1	0.0001
2.0	7.3891	0.1353	10	22026	0.00005
2.1	8.1662	0.1225			
2.2	9.0250	0.1108			
2.3	9.9742	0.1003			
2.4	11.023	0.0907			

TABLE 6. TRIGONOMETRIC FORMULAS

$\text{Sin } A = \dfrac{1}{\text{Csc } A}$ $\qquad$ $\text{Csc } A = \dfrac{1}{\text{Sin } A}$ $\qquad$ $(\text{Sin } A)(\text{Csc } A) = 1$

$\text{Cos } A = \dfrac{1}{\text{Sec } A}$ $\qquad$ $\text{Sec } A = \dfrac{1}{\text{Cos } A}$ $\qquad$ $(\text{Cos } A)(\text{Sec } A) = 1$

$\text{Tan } A = \dfrac{1}{\text{Cot } A}$ $\qquad$ $\text{Cot } A = \dfrac{1}{\text{Tan } A}$ $\qquad$ $(\text{Tan } A)(\text{Cot } A) = 1$

$\text{Tan } A = \dfrac{\text{Sin } A}{\text{Cos } A}$ $\qquad$ $\text{Cot } A = \dfrac{\text{Cos } A}{\text{Sin } A}$

$\text{Sin}^2 A + \text{Cos}^2 A = 1$ $\qquad$ $\text{Sec}^2 A - \text{Tan}^2 A = 1$ $\qquad$ $\text{Csc}^2 A - \text{Cot}^2 A = 1$

$\text{Sin}^2 A = 1 - \text{Cos}^2 A$ $\qquad$ $\text{Sec}^2 A = 1 + \text{Tan}^2 A$ $\qquad$ $\text{Csc}^2 A = 1 + \text{Cot}^2 A$

$\text{Cos}^2 A = 1 - \text{Sin}^2 A$ $\qquad$ $\text{Tan}^2 A = \text{Sec}^2 A - 1$ $\qquad$ $\text{Cot}^2 A = \text{Csc}^2 A - 1$

$\text{Sin } A = \pm\sqrt{1 - \text{Cos}^2 A}$ $\qquad$ $\text{Sec } A = \pm\sqrt{1 + \text{Tan}^2 A}$ $\qquad$ $\text{Csc } A = \pm\sqrt{1 + \text{Cot}^2 A}$

$\text{Cos } A = \pm\sqrt{1 - \text{Sin}^2 A}$ $\qquad$ $\text{Tan } A = \pm\sqrt{\text{Sec}^2 A - 1}$ $\qquad$ $\text{Cot } A = \pm\sqrt{\text{Csc}^2 A - 1}$

$\text{Sin } (A + B) = (\text{Sin } A)(\text{Cos } B) + (\text{Cos } A)(\text{Sin } B)$

$\text{Cos } (A + B) = (\text{Cos } A)(\text{Cos } B) - (\text{Sin } A)(\text{Sin } B)$

$\text{Tan } (A + B) = \dfrac{\text{Tan } A + \text{Tan } B}{1 - (\text{Tan } A)(\text{Tan } B)}$

$\text{Sin } (A - B) = (\text{Sin } A)(\text{Cos } B) - (\text{Cos } A)(\text{Sin } B)$

$\text{Cos } (A - B) = (\text{Cos } A)(\text{Cos } B) + (\text{Sin } A)(\text{Sin } B)$

$\text{Tan } (A - B) = \dfrac{\text{Tan } A - \text{Tan } B}{1 + (\text{Tan } A)(\text{Tan } B)}$

$\text{Sin } 2A = 2(\text{Sin } A)(\text{Cos } A)$

$\text{Cos } 2A = \text{Cos}^2 A - \text{Sin}^2 A = 2\text{Cos}^2 A - 1 = 1 - 2\text{Sin}^2 A$

$\text{Tan } 2A = \dfrac{2 \text{Tan } A}{1 - \text{Tan}^2 A}$

$\text{Cot } 2A = \dfrac{\text{Cot}^2 A - 1}{2 \text{Cot } A}$

$\text{Sin } \dfrac{A}{2} = \pm\sqrt{\dfrac{1 - \text{Cos } A}{2}}$

$\text{Cos } \dfrac{A}{2} = \pm\sqrt{\dfrac{1 + \text{Cos } A}{2}}$

$\text{Tan } \dfrac{A}{2} = \pm\sqrt{\dfrac{1 - \text{Cos } A}{1 + \text{Cos } A}} = \dfrac{1 - \text{Cos } A}{\text{Sin } A} = \dfrac{\text{Sin } A}{1 + \text{Cos } A}$

$\text{Sin } A + \text{Sin } B = 2 \text{Sin}\left(\dfrac{A + B}{2}\right) \text{Cos}\left(\dfrac{A - B}{2}\right)$

$\text{Sin } A - \text{Sin } B = 2 \text{Sin}\left(\dfrac{A - B}{2}\right) \text{Cos}\left(\dfrac{A + B}{2}\right)$

$\text{Cos } A + \text{Cos } B = 2 \text{Cos}\left(\dfrac{A + B}{2}\right) \text{Cos}\left(\dfrac{A - B}{2}\right)$

$\text{Cos } A - \text{Cos } B = -2 \text{Sin}\left(\dfrac{A + B}{2}\right) \text{Sin}\left(\dfrac{A - B}{2}\right)$

TABLE 6. TRIGONOMETRIC FORMULAS (*Cont.*)

$$\operatorname{Tan} A + \operatorname{Tan} B = \frac{\operatorname{Sin}\ (A+B)}{(\operatorname{Cos} A)(\operatorname{Cos} B)}$$

$$\operatorname{Tan} A - \operatorname{Tan} B = \frac{\operatorname{Sin}(A-B)}{(\operatorname{Cos} A)(\operatorname{Cos} B)}$$

$$\operatorname{Cot} A + \operatorname{Cot} B = \frac{\operatorname{Sin}(A+B)}{(\operatorname{Sin} A)(\operatorname{Sin} B)}$$

$$\operatorname{Cot} A - \operatorname{Cot} B = \frac{-\operatorname{Sin}(A-B)}{(\operatorname{Sin} A)(\operatorname{Sin} B)}$$

$$\operatorname{Sin}^2 A - \operatorname{Sin}^2 B = \operatorname{Sin}(A+B)\operatorname{Sin}(A-B)$$

$$\operatorname{Cos}^2 A - \operatorname{Sin}^2 B = \operatorname{Cos}(A+B)\operatorname{Cos}(A-B)$$

Answers to Odd-Numbered Problems

Chapter 1

Section 1-1

1. Composite
3. Prime
5. Prime
7. Composite
9. Prime
11. Composite
13. Rational
15. Rational
17. Rational
19. Rational
21. Irrational
23. Irrational
25. Imaginary
27. Real
29. Real
31. Real
33. Imaginary
35. Real

37.

39.

41.

43.

570 Answers to Odd-Numbered Problems

45. $-\sqrt{11}$

47. $-\frac{9}{4}$

Section 1–2

1. $5 > -5$
3. $\frac{3}{2} < 1.6$
5. $-7 > -10$
7. $0 > -\pi$
9. $-12 < -7$
11. $\frac{7}{9} < \frac{\pi}{2}$

13. -5

15. -4.1

17. $X < 6$

19. $B < \frac{1}{2}$

21. $K > 3$

23. $M > 0$

Section 1–3

1. 9
3. $\frac{4}{3}$
5. 2.7
7. $2\pi = 6.28$
9. $3\sqrt{10} = 9.487$

11. $A = \pm 6$

13. $P = \pm\sqrt{11}$

15. $X > 3$ or $X < -3$

17. $-\frac{14}{5} < T < \frac{14}{5}$

19. $S \neq 0$

Section 1–4

1. -23
3. $-\frac{2}{3}$
5. $-\sqrt{3}$
7. $\frac{1}{8}$
9. 7
11. $-\frac{1}{\sqrt{2}}$
13. $4M + 24$
15. $-3X - 21$
17. 13

Answers to Odd-Numbered Problems   571

19. −9.52   **21.** 2   **23.** −$\frac{2}{21}$ or −.0952   **25.** 15

Section 1–5

1. −4   **3.** −9   **5.** 0   **7.** 0   **9.** 0   **11.** Undefined

Section 1–6 Chapter Review

1. Real-Rational   **3.** Real-Rational   **5.** Real-Irrational
7. Real-Irrational   **9.** Real-Rational   **11.** Real-Rational

13.

15.

17.

19. $K = \pm 11$

21. $-7 < B < 7$

23. $L > 1.5$ or $L < -1.5$

25. $-\frac{13}{4}$ or -3.25 **27.** 1   **29.** 1

Chapter 2

Section 2–1

1. 64   **3.** 1   **5.** 14   **7.** $125A^3$ **9.** $\frac{1}{25}$ **11.** $\frac{3}{X^2}$
13. T^7 **15.** $2K^7$ **17.** X^4 **19.** $\frac{1}{Y^3}$ **21.** K^{12}
23. $\frac{125}{X^3}$ **25.** A^3B^4 **27.** $\frac{2R}{S^2}$ **29.** 5   **31.** 2
33. −3   **35.** 6   **37.** −6   **39.** 2   **41.** 216   **43.** $\frac{1}{4}$
45. $\frac{1}{27}$ **47.** X **49.** Y^2 **51.** T **53.** $2\sqrt{5}$ or 4.472
55. 3   **57.** $\sqrt{5}$ or 2.236   **59.** $\sqrt[8]{2}$ or 1.09   **61.** $2\sqrt{3}$ or 3.464
63. $4\sqrt{2}$ or 5.657   **65.** $\sqrt{2}$ or 1.414

Section 2–2

1. 5.608×10^4 **3.** 1.86×10^5 **5.** 4.91×10^0
7. 9.11×10^{-28} **9.** 2.51×10^{-1} **11.** 78,100,000
13. 4.05   **15.** .0983   **17.** 8.001   **19.** .445
21. 3×10^{14} **23.** 1.05   **25.** 10,000   **27.** 8×10^{24}

29. 8.987×10^{-5}
31. .0000000000000000000000000166035
33. 299,790,200
35. 100,000
37. 3,000,000,000,000,000
39. 1.013×10^6

Section 2-3

1. 0 **3.** $-4R$ **5.** $X^3 + 10Y^3$ **7.** $X - Y$
9. $-10A - 9B$ **11.** $6\sqrt{T} - 2\sqrt{V}$ **13.** $-2\sqrt{2}$ or -2.828
15. $35A^8$ **17.** $15X^2$ **19.** $\sqrt{6}\, X^7 Y$ or $2.449 X^7 Y$
21. $2A^2 - 13AB + 15B^2$ **23.** $8R^3S - 12R^2S^2 + 4R^4S$
25. $2X^3 + 3X^2Y - 4XY^2 + Y^3$ **27.** $A^2 + 6AB + 9B^2$
29. $R^3 - R^2S - RS^2 + S^3$ **31.** $X^3 + \sqrt{5}\, X^2 - \sqrt{5}\, X - 5$
33. $-16 + \sqrt{21}$ or -11.417 **35.** $3A^2$ **37.** $2RS^4$
39. $\dfrac{7K^2}{L} - \dfrac{1}{L} + \dfrac{L}{5}$

Section 2-4

1. $7(M - 4N)$ **3.** $7Z(3Z - 1)$ **5.** $3(5A + B - 5C)$
7. $(X - Y)(A - B)$ **9.** $X^{34}(X^2 + 1)$ **11.** $(X + 2)(X - 2)$
13. $(I + 4)(I - 4)$ **15.** $(2V + 5)(2V - 5)$ **17.** $(11 + Z)(11 - Z)$
19. $(A + B + Y)(A + B - Y)$ **21.** $(A + 2)^2$ **23.** $(2V - 3)^2$
25. $(2M + 5N)^2$ **27.** $5(6R - 5)^2$ **29.** $(2I + 3)^2$
31. $(A + 5)(A + 1)$ **33.** $(U + 4)(U + 2)$
35. $(B + 6)(B - 2)$ **37.** $(V - 8)(V + 2)$
39. $3(X + 8Y)(X + 6Y)$ **41.** $(2R + 5S)(2R + S)$
43. $(3Y + 1)(2Y + 3)$ **45.** Not factorable
47. $3(5T + 3)(T - 2)$ **49.** $(3S - 2)^2$

Section 2-5

1. $\dfrac{3}{11}$ **3.** $\dfrac{A^2}{2X^3}$ **5.** $\dfrac{T-5}{T+5}$ **7.** $\dfrac{29}{24}$
9. $\dfrac{-7A^2 + 2A + 5}{A^3}$ **11.** $\dfrac{TV - T^2}{V(V+1)}$ **13.** $\dfrac{-K^3 + 5K^2 + 2K - 6}{K(K+2)(K-2)}$
15. $\dfrac{3}{28}$ **17.** $\dfrac{A^2}{15B^3}$ **19.** $\dfrac{R+1}{R^2(R-4)}$ **21.** $\dfrac{T-2}{(T+2)(3T-1)}$
23. $\dfrac{4}{3}$ **25.** $\dfrac{9B^3}{4A}$ **27.** $\dfrac{2}{M^3(M+4)}$ **29.** $\dfrac{-3T^3 + 3T^2 + 20}{4T^2}$
31. $\dfrac{2}{Y(Y-1)}$ **33.** $\dfrac{2X - 5X^2}{7X^2 + 6}$ **35.** $4:1$
37. $4:3$ or $1\tfrac{1}{3}:1$ **39.** $36:5$ or $7\tfrac{1}{5}:1$ **41.** $90\,\text{lb}:150\,\text{lb}$
43. $900\,\text{ft}^2:1500\,\text{ft}^2:2100\,\text{ft}^2$

Section 2-6

1. $T = 10$ **3.** $R = 2$ **5.** $N = -\tfrac{15}{4}$ **7.** $V = 1$
9. $X = \tfrac{5}{4}$ **11.** $B = \tfrac{1}{9}$ **13.** No solution **15.** $V = \tfrac{3}{4}$
17. $P = 9$ **19.** No solution **21.** No solution
23. $P = \pm 12$ **25.** $T = -2, 6$ **27.** $N = \pm 4$
29. No solution **31.** $R = \dfrac{V}{I}$ **33.** $W = \dfrac{FH}{L}$

Answers to Odd-Numbered Problems   573

35. $A = \dfrac{M + PX - RX}{P}$

37. 8459.7 gal

39. 30 ohms, 120 ohms

41. 3.1425 mi

Section 2-7 Chapter Review

1. -3 **3.** -2 **5.** 6   **7.** 32   **9.** $\tfrac{1}{2}$ **11.** $4XY^2$
13. $7K - 2M$ **15.** $6XY - X^2 - 3Y^2$ **17.** $4T - 7V$
19. $15X^2 - 29XY - 14Y^2$ **21.** $62 - 8\sqrt{21}$ or 25.339
23. $R^4 + 6R^2S^2 + 9S^4$ **25.** $4A^6 - 27$
27. $2T^3 + 5T^2 - 22T + 15$ **29.** $\dfrac{3X^4}{A^2}$
31. $\dfrac{4M}{3N} - \dfrac{2}{3MN} - \dfrac{N}{15}$ **33.** $2MN(MN - 4M + 6)$
35. $(X - 7)^2$ **37.** $(2L - 5)(L + 6)$ **39.** $3(4T + 3)(2T - 5)$
41. $\dfrac{K + 8}{2K}$ **43.** $\dfrac{6XY - 5XY^2 + 24X}{8Y^3}$ **45.** $\dfrac{-TV^2 - 2TV + 20V}{T(T - 4)}$
47. $3B$ **49.** $\dfrac{10R^2}{3}$ **51.** $\dfrac{6X^2 + 35}{12XY}$ **53.** $T = 3$
55. $T = -\tfrac{2}{5}$ **57.** $T = 7$ **59.** $T = 16$ **61.** $T = \dfrac{5 - 4S}{2}$
63. $T = \dfrac{5}{R^2 - RS}$ **65.** $T = \pm 4$ **67.** $T = 15$ **69.** $T = 5$
71. $T = -2, 3$ **73.** $\dfrac{R_2 R_3 + R_1 R_3 + R_1 R_2}{R_1 R_2 R_3}$ **75.** $H = \dfrac{3V}{\pi R^2}$
77. $D = \dfrac{2WIH}{E}$ **79.** $R_1 = \dfrac{R_2 R}{R_2 - R}$ **81.** \$600, \$1800
83. 17.75 in.

Chapter 3

Section 3-1

1. 207.6 in.   **3.** 152,460 ft²   **5.** 4800 lb
7. 18,832 yd   **9.** 3.5 qt   **11.** 915,200 yd
13. 14.88 acres   **15.** 1.875 mi   **17.** 12.91 hp
19. 90.91 rods

Section 3-2

1. 31,600,000 micrometers   **3.** 16,800,000 m²   **5.** 3,700,000 l
7. 30 cal   **9.** 1.436 grams   **11.** .194 kl   **13.** 34.036 cm
15. 30°C   **17.** 298.1 mi   **19.** 29.385 qt

Section 3-3

1. 3000 cal   **3.** 9 Btu   **5.** 14 ft   **7.** 156 m²
9. 2 ft/sec   **11.** 12 ft   **13.** 5418 ft²   **15.** 48 in.³
17. 3000 kg/m   **19.** 23 ft/sec   **21.** 7 ft   **23.** 13.825 min
25. 2.128 ft/m

Section 3-4

1. 16,896 ft
3. 25.8 km
5. 68.18 kg
7. 70.4 ft/sec
9. 42.328 oz
11. 1.098 km
13. 1.254×10^9 ergs
15. 4.21
17. 6.38 km/l
19. 16,108.5 in.^{3}

Section 3-5

1. 144 ft
3. 13.85 cm^2
5. 17 ft
7. 110.7 kg
9. lb/sec

Section 3-6 Chapter Review

1. 293.7 ft
3. 1.543 yd^3
5. 2016 hr
7. .738 m
9. 27,400 cc
11. .0063 km^2
13. 48.4 mi
15. 275.877 qt
17. 8.45 kg/cm^2
19. .951 qt
21. 351.2 ft^3
23. 66 ft/sec
25. 17.22 m/sec
27. 6.012 ft
29. 32 ft/sec^2
31. 231.3 lb

Chapter 4

Section 4-1

1. Function
3. Relation
5. Function
7. Relation
9. Function
11. Yes, yes
13. Yes, yes
15. Yes, $X = \dfrac{Y-1}{4}$, yes
17. No, $V = T^2 - 1$, yes
19. $A = f(S) = S^2$
21. $p = f(L) = 2L + 16$
23. $A = f(C) = \dfrac{C^2}{4\pi}$

Section 4-2

1. X is independent.
   Y is dependent.
   Domain: $X =$ all real numbers
   Range: $Y \geq 3$
   Function
3. X is independent.
   M is dependent.
   Domain: $X \geq 0$
   Range: $M \geq 0$
   Function
5. W is independent.
   Z is dependent.
   Domain: $-5 \leq W \leq 5$
   Range: $-5 \leq Z \leq 5$
   Not a function
7. B is independent.
   A is dependent.
   Domain: $B \neq 0$
   Range: $A \neq 0$
   Function
9. Y is independent.
   X is dependent.
   Domain: $Y =$ all real numbers
   Range: $X =$ all real numbers
   Function
11. 4, 8
13. $-5, -2$
15. Undefined, $\frac{9}{4}$
17. $4V^2 - 2V + 3, V^2 + V + 3$

Answers to Odd-Numbered Problems   575

Section 4-3

1. $(1, 3), (-3, 0), (-1, -1), (3, -2\frac{1}{2})$

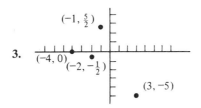

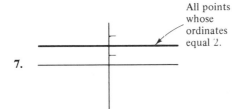

9. Fourth   **11.** 0   **13.** First and third

17. Yes   **19.** X, it is independent.

Section 4-4

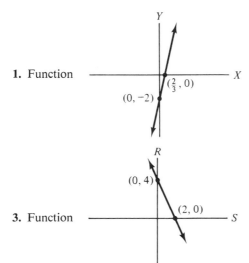

1. Function

3. Function

576 *Answers to Odd-Numbered Problems*

5. Function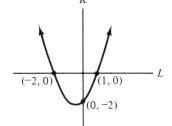

7. Function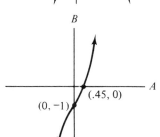

9. Function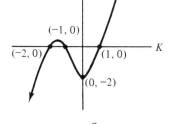

11. Function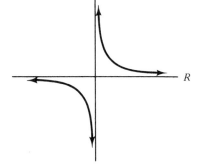

13. Function

Answers to Odd-Numbered Problems

15. Function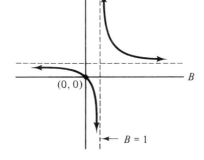

17. Function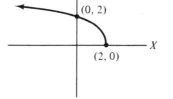

19. Not a function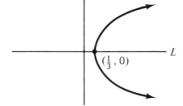

21. Function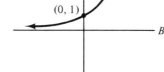

Section 4–5

1.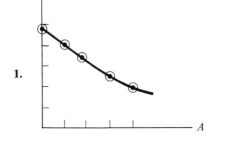

3.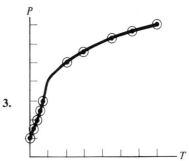

5.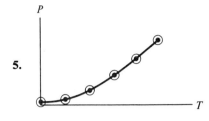

7.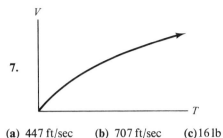

(a) 447 ft/sec (b) 707 ft/sec (c) 16 lb

9. (a)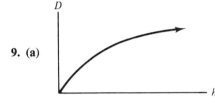

(b) 22.5 km

11.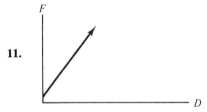

$F = 3D + 1$

Section 4-6

1. $A = \frac{5}{3}$
3. $T = -\frac{2}{3}, \frac{1}{2}$
5. $S = -4, 0$
7. $A = \frac{8}{5}$
9. $X = -2.9, .9$
11. $M = -2.2, 3.2$
13. $V = -2.1$
15. No real solution
17. $T = -11$
19. $V = -\frac{5}{2}, 1$
21. 4 sec, 1.06 or 2.94 sec

Section 4-7 Chapter Review

1. Function
   Domain: B = all real numbers
   Range: A = all real numbers
3. Function
   Domain: S = all real numbers
   Range: R = all real numbers
5. Function
   Domain: $X \neq 1$
   Range: $Y \neq 1$
7. Relation
   Domain: Q = all real numbers
   Range, $P \geq 1$ or $P \leq -1$
9. $1, -19$
11. $-5, \frac{25}{9}$
13. Undefined, $\frac{2}{3}$
15. $0, 4X^2 - 6X - 4$

17.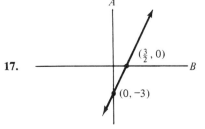

Answers to Odd-Numbered Problems **579**

19. 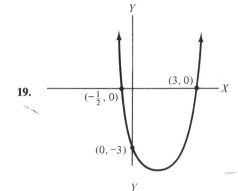   **21.**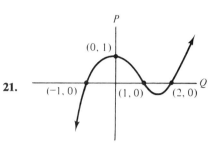

23. 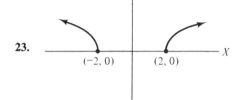   **25.**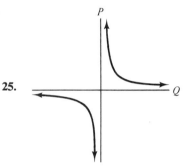

27. 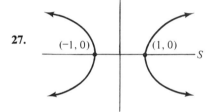   **29.** $X = -\frac{7}{2}$ **31.** $R = -2.5, .67$

33. $T = 0, 3$ **35.** $K = 0$ **37.** $A = \pm 3$ **39.**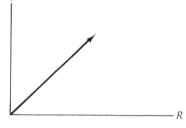

41.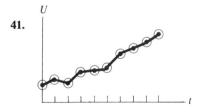

43.    **45.**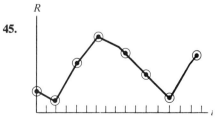

47. 6 ft   **49.** $t = 2.5$ sec or 3 sec   **51.**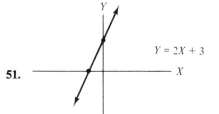

$Y = 2X + 3$

53.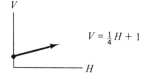

$V = \frac{1}{4}H + 1$

Chapter 5

Section 5–1

1. Mean = 13
Median = 14
Mode = 14

3. Mean = 63
Median = 62
Mode = 56 and 72

5. Mean = 34
Median = 32
Mode = 32

7. Mean = 53°
Median = 54°
Mode = 47°, 55°, 58°, 63°

9. Mean = 37.8 in.
Median = 38.15 in.
Mode = 38.5 in.

Section 5–2

1. 7.48   **3.** 1.56   **5.** 8.08, 54%

Section 5–3

1. Mean = 146.9 kg
Median = 146.8 kg
Modal Class = 146 kg–154 kg

3. Mean = 180.8 mi
Median = 181 mi
Modal Class = 175 mi–178 mi
179 mi–182 mi

Section 5–4

1. 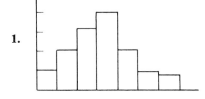   **3.**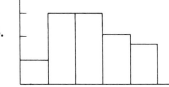

Answers to Odd-Numbered Problems

5.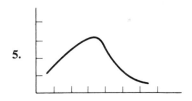
7.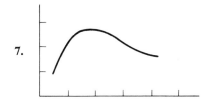

Section 5-5 Chapter Review

1. Mean = 157
   Median = 164
   Mode = 172
   Standard Deviation = 22.54

3. Mean = 127.7 mm
   Median = 128.5 mm
   Mode = 130 mm
   Standard Deviation = 13.79 mm
   Within one standard deviation → 75%

5. Mean = 1337.07 lb
   Median = 1388 lb
   Mode = 1350 lb, 1445 lb, 1450 lb
   Standard Deviation = 149.5 lb
   Within one standard deviation → 70%

7. Mean = $31,113.79
   Median = $31,000
   Modal Class = $34,000 − $35,999

9.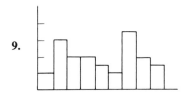
11.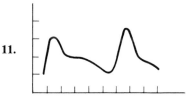

Chapter 6

Section 6-1

1. Yes   3. No   5. Yes   7. No   9. Yes
11. Yes 13. No   15. Yes 17. No   19. Yes

21.
23.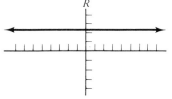

25.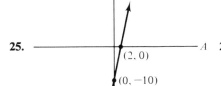
27.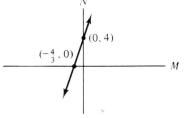

29.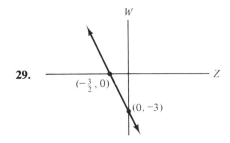

Section 6-2

1. $M = 2$ and $B = 2$ **3.** $M = 0$ and $B = 1$ **5.** $M = \frac{2}{3}$ and $B = -2$

7.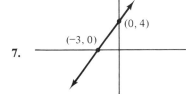

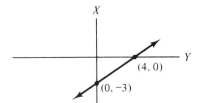

9. 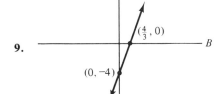 **11.**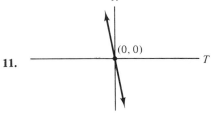

13. $S = \frac{5}{4}T + 10$ **15.** $C = \frac{5}{4}D - \frac{1}{4}$ **17.** $V = 3T + 4$

19. 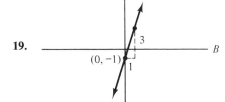 **21.**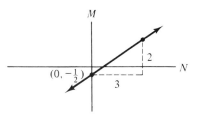

23. 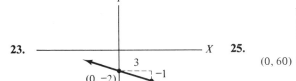 **25.**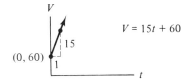

Answers to Odd-Numbered Problems 583

27. (a) $V_0 \geq 0$ (b) $a =$ any real number (c)

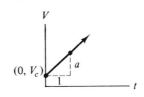

Section 6-3

1. No   3. Yes   5. No   7. Yes   9. No

Section 6-4

1. $X = 6, Y = -2$
3. $S - \frac{1}{5}, T = \frac{2}{3}$
5. $X = 1, Y = 1$
7. $V_1 = 7, V_2 = -5$
9. $S = \frac{1}{3}, T = 0$
11. 10 mph, 2 mph
13. 30 acres @ $300, 50 acres @ $400

Section 6-5

1. $P = -3, Q = 3$
3. $T = 8, V = 7$
5. $X = 2, Y = 3$
7. $A = 3, B = -2$
9. $X = -2, Y = 1$
11. $X = 5, Y = 3$
13. $R = 2, S = 2$
15. $M = 10, N = 10$
17. $P = 2, Q = \frac{3}{5}$
19. $X = -\frac{4}{7}, Y = \frac{19}{7}$
21. 50 mph, 40 mph
23. 60 gal of 55%, 40 gal of 30%
25. $L = 12$ m, $W = 8$ m

Section 6-6

1. -10 3. 78   5. 21   7. 0   9. 0
11. $X = \frac{23}{14}, Y = -\frac{22}{14} = -\frac{11}{7}$
13. $W = \frac{10}{43}, Z = \frac{40}{43}$
15. $P = \frac{47}{6}, Q = \frac{7}{2}$
17. $A = \frac{23}{7}, B = \frac{5}{7}$
19. No solution
21. $W = \frac{59}{37}, Z = -\frac{3}{37}$
23. 12 mph, 3 mph
25. 12 ft, 4 ft

Section 6-7

1. $A = 4, B = -3, C = -5$
3. $L = -48, M = -10, N = -28$
5. $U = \frac{4}{5}, V = \frac{32}{5}, W = \frac{16}{5}$
7. Dependent
9. $P = 13, Q = 22, R = -1$
11. $L = 3, M = -4, N = \frac{3}{2}$
13. $S = 74\%, N = 8\%, C = 18\%$

Section 6-8

1. -158 3. -94 5. 276
7. $A = -\frac{4}{11}, B = \frac{12}{11}, C = -\frac{3}{11}$
9. $R = -4, S = 2, T = -5$
11. $P = 1, Q = 2, R = 0$
13. $A = 1, B = 2, C = 1$
15. $R = 3, S = -2, T = 4$
17. $P = 1, Q = 2, R = 3$
19. $5000 @ 5.5%, $7000 @ 6%, $3000 @ 6.5%

Section 6-9   Chapter Review

1. $X = 3, Y = -2$
3. $R = 1, S = 3$
5. $V = 1, W = -2$
7. $K = -1, L = -1$
9. $A = 2, B = 3$
11. $R = 3, S = -4$
13. $V - 2, W = 1$
15. $A = -1, B = 1$
17. -14

584   Answers to Odd-Numbered Problems

19. 22   **21.** 0   **23.** $X = 6, Y = 3$ **25.** $R = 2, S = 1$
27. $T = 3, U = 4$ **29.** $A = 2, B = 3$ **31.** $A = -\frac{1}{3}, B = -\frac{1}{14}$
33. $A = 4, B = 1, C = 2$ **35.** $R = 4, S = -2, T = -3$
37. $U = \frac{1}{2}, V = -2, W = 1$ **39.** -6 **41.** 108   **43.** -6
45. $X = 1, Y = 2, Z = 3$ **47.** $R = 2, S = -1, T = -\frac{1}{3}$
49. $I = 1, J = -5, K = 4$ **51.** 15 mph, 3 mph
53. $\frac{76}{11}$ in., $\frac{171}{11}$ in.   **55.** \$3000 @ 5%, \$1000 @ 6%
57. $X = 1, Y = 1, Z = 1$ **59.** 3.25 ohms, 8.75 ohms, 10 ohms

Chapter 7

Section 7–1

1. 19.2′   **3.** 36″   **5.** 48′   **7.** .233°   **9.** .45′
11. .4°   **13.** 45.84°   **15.** 212.01°   **17.** 302.54°
19. 108°   **21.** 36°   **23.** 829.8°   **25.** 1.267
27. 2.293   **29.** 5.379   **31.** $\frac{5\pi}{4}$ or 1.25π
33. $\frac{5\pi}{12}$ or $.417\pi$ **35.** $.229\pi$ **37.** 434.5°, $-285.5°$
39. 458°20′, $-261°40′$ **41.** 11.11, -1.45 **43.** $2.5\pi, -1.5\pi$
45. (a) $A, D, F, G, J, L, N, C, Q, T$ (b) $B, C, E, H, I, K, M, P, R, S$
       (c) $A \& B, C \& D, E \& F, G \& H, A \& H, B \& G, C \& F, D \& E, I \& J,$
           $J \& K, K \& L, L \& I, M \& N, N \& S, S \& T, T \& M, O \& P, P \& Q,$
           $Q \& R, R \& O$
       (d) $A \& G, B \& H, C \& E, D \& F, I \& K, J \& L, M \& S, N \& T, O \& Q,$
           $P \& R$
       (e) $M \& E, N \& F, H \& P, G \& O$
       (f) $A \& B, B \& G, G \& H, H \& A, C \& D, D \& E, E \& F, F \& C, I \& J,$
           $J \& K, K \& L, L \& I, M \& N, N \& S, S \& T, T \& M, O \& P, P \& Q,$
           $Q \& R, R \& O$

Section 7–2

1. $X = 40°, B = 16$ **3.** All angles = 60°   **5.** $C = 5, X = 36.9°$
                                     All sides = 7
7. 21 square ft   **9.** 17.5 square in.
11. $25\sqrt{3}$ or 43.3 square ft

Section 7–3

1. 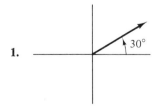   **3.**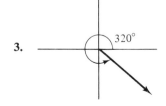

Answers to Odd-Numbered Problems

5.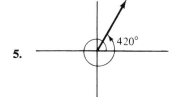

7.

9.

11.

13.

15.

17.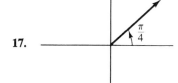

19.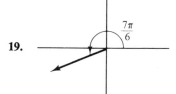

21. $135°, \dfrac{3\pi}{4}, 2.356$ 23. $-270°, -\dfrac{3\pi}{2}, -4.71$ 25. $240°, \dfrac{4\pi}{3}, 4.188$

Section 7–4

1. $\cos\theta = \dfrac{4}{5} = \dfrac{8}{10} = .8$
   $\tan\theta = \dfrac{3}{4} = \dfrac{6}{8} = .75$

3. $\tan\theta = \dfrac{9}{6} = \dfrac{13.5}{9} = 1.5$
   $\csc\theta = \dfrac{10.82}{9} = \dfrac{16.22}{13.5} = 1.2$

5. $\sin\theta = \tfrac{3}{5} = .6$ $\csc\theta = \tfrac{5}{3} = 1.67$
   $\cos\theta = \tfrac{4}{5} = .8$ $\sec\theta = \tfrac{5}{4} = 1.25$
   $\tan\theta = \tfrac{3}{4} = .75$ $\cot\theta = \tfrac{4}{3} = 1.33$

7. $\sin\theta = .27$ $\csc\theta = 3.64$
   $\cos\theta = .96$ $\sec\theta = 1.04$
   $\tan\theta = .29$ $\cot\theta = 3.5$

9. $\sin\theta = .6$ $\csc\theta = 1.67$
   $\cos\theta = -.8$ $\sec\theta = -1.25$
   $\tan\theta = -.75$ $\cot\theta = -1.33$

11. $\sin\theta = -.89$ $\csc\theta = -1.12$
    $\cos\theta = -.45$ $\sec\theta = -2.24$
    $\tan\theta = 2$ $\cot\theta = .5$

13. $\sin\theta = .71$ $\csc\theta = 1.41$
    $\cos\theta = -.71$ $\sec\theta = -1.41$
    $\tan\theta = -1$ $\cot\theta = -1$

15. $\sec\theta = 1.12$ 17. $\csc\theta = 1.02$ 19. $\sin\theta = .81$

Section 7–5

1. Sin 30° = .5 Csc 30° = 2
   Cos 30° = .866 Sec 30° = 1.155
   Tan 30° = .5774 Cot 30° = 1.732
3. Sin 65° = .9063 Csc 65° = 1.103
   Cos 65° = .4226 Sec 65° = 2.366
   Tan 65° = 2.145 Cot 65° = .4663
5. .2756
7. 2.154
9. 7.596
11. 9.255
13. .6626
15. .4806
17. .6398
19. 1.019
21. .9184
23. 1.464
25. 19°
27. 17°50′ or 17.83°
29. 68°50′ or 68.83°
31. 27°50′ or 27.83°
33. 39°30′ or 39.5°
35. 6°10′ or 6.17°
37. 49°34′ or 49.57°
39. 43°14′ or 43.23°
41. 8°44′ or 8.73°
43. 73°36′ or 73.6°
45. $\theta = 67°29′$ or 67.38°

Section 7–6

1. $A = 67.4°$ $a = 6$
   $B = 22.6°$ $b = 2.5$
   $C = 90°$ $c = 6.5$
3. $A = .73$ $a = 100.7$
   $B = .84$ $b = 112.2$
   $C = 1.57$ $c = 151$
5. $A = 33°12′$ $a = 2.55$
   $B = 56°48′$ $b = 3.89$
   $C = 90°$ $c = 4.65$
7. No solution
9. $\alpha = 65.93°$ $\beta = 24.07°$ $a_3 = 9.4$ ft/sec^2
11. 1.192 mi
13. 805 m, 3220 m/hr
15. Sides = 16.6 cm
    Altitude = 13.13 cm

Section 7–7

	Sin	Cos	Tan	Csc	Sec	Cot
1.	.7071	−.7071	−1	1.414	−1.414	−1
3.	.9092	−.4163	−2.184	1.10	−2.402	−.4578
5.	−.3420	−.9397	.3640	−2.924	−1.064	2.747
7.	.8631	−.5050	−1.709	1.159	−1.980	−.5851
9.	−.0523	−.9986	.0524	−19.11	−1.001	19.08
11.	−.9257	.3944	−2.330	−1.080	2.535	−.4292
13.	.1736	.9848	.1763	5.759	1.015	5.671
15.	.6157	−.7880	−.7813	1.624	−1.269	−1.280
17.	−.8660	−.5	1.732	−1.155	−2	.5774
19.	−.2588	.9659	−.2679	−3.864	1.035	−3.732
21.	.8192	−.5736	−1.428	1.221	−1.743	−.7002
23.	.9905	.1374	7.207	1.009	7.276	.1388

25. $\theta = 135°, 315°$
27. No solution
29. $\theta = 24.5°, 155.5°$
31. $\theta = 1.05, 2.09$
33. $\theta = 0, 3.14$
35. $\theta = .2, 6.08$

Section 7–8

1. $S = 11.78$ in.
3. $\theta = 2.3$
5. $R = 6.92$ ft
7. $A = 1341.8$ cm^2
9. $\theta = 3.55$
11. $R = 6$ in.
13. $v = 14$ m/sec
15. $\omega = 211{,}200$ rad/hr or 58.67 rad/sec
17. $R = 1.168$ ft
19. .419 ft
21. 56.55 ft^2

Answers to Odd-Numbered Problems   587

23. 53,856 mi/day   25. −2.346 amps

Section 7–9 Chapter Review

1. 143.9°   3. 35.5°   5. .178π   7. 2.215
9. 12 ft²   11. 16 in.²
13. $\sin\theta = -\frac{4}{5} = -.8$ $\csc\theta = -\frac{5}{4} = -1.25$
    $\cos\theta = \frac{3}{5} = .6$ $\sec\theta = \frac{5}{3} = 1.67$
    $\tan\theta = -\frac{4}{3} = -1.33$ $\cot\theta = -\frac{3}{4} = -.75$
15. $\sin\theta = .9864$ $\csc\theta = 1.0138$
    $\cos\theta = .1644$ $\sec\theta = 6.0828$
    $\tan\theta = 6$ $\cot\theta = .1667$
17. .5150   19. .6743   21. .9321   23. 1.253
25. −.8966   27. .2309   29. −1.231   31. −.8746
33. −.1848   35. .9397   37. 49.7°   39. 68.5°
41. 50.3°   43. .1047   45. .666   47. 1.446
49. $A = 33.3°, 146.7°$ 51. $A = 136.4°, 223.6°$
53. $A = 1.8, 4.95$ 55. $A = 3.51, 5.91$
57. $B = 52°, b = 84.2, c = 106.9$ 59. $B = .47, a = 106.9, b = 54.5$
61. 71.8 mi   63. 94.05 ft   65. 16.75 km   67. 392.7 mi²
69. 9.97 rev/sec   71. 178 v

Chapter 8

Section 8–1

1. Scalar   3. Scalar   5. Vector   7. Scalar   9. Vector

11. 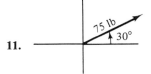   13.

15. 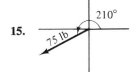   17.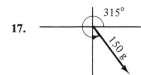

19.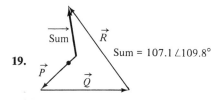
Sum = 107.1 ∠109.8°

21.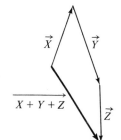

23.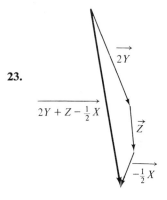

25. $\vec{F_H} = 11.6$, $\vec{F_V} = 12.4$
27. $\vec{A_H} = -.0823$, $\vec{A_V} = .0251$
29. $\vec{P_H} = 0$, $\vec{P_V} = 12$
31. $\vec{Z_H} = 34$, $\vec{Z_V} = 0$
33. $\vec{R} = 23/145.6°$
35. $\vec{R} = .857/212.5°$
37. $\vec{R} = 14.04/163.95°$

Section 8–2

1. $V_w = 38.18$ mph
   $V_s = 38.18$ mph
3. 1500.67 mph $/1.7°$ S of E
5. 135.2 kg
7. Cable = 473.3 lb, Brace = 428.95 lb
9. 62.5 lb
11. 12.49 mph $/39.2°$ E of N
13. .2 tons or 400 lb
15. Yes

Section 8–3

1.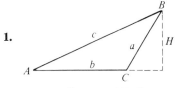

$\text{Sin } A = \dfrac{h}{c} \longrightarrow h = c \text{ Sin } A$

$\text{Sin } C = \text{Sin } (180° - C) = \dfrac{h}{a} \longrightarrow h = a \text{ Sin } C$

$c \text{ Sin } A = a \text{ Sin } C$

$\dfrac{\text{Sin } A}{a} = \dfrac{\text{Sin } C}{c} = \dfrac{\text{Sin } B}{b}$

3. $A_1 = 55.4°$, $C_1 = 103.6°$, $c_1 = 103.7$; $A_2 = 124.6°$, $C_2 = 34.4°$, $c_2 = 59.9$
5. $A = 126°$, $a = .7815$, $c = .0674$
7. $L = 9.54$ m
9. 564 kg, 693 kg
11. 32.3 ft

Section 8–4

1.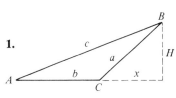

$\text{Sin } A = \dfrac{h}{c} \longrightarrow h = c \text{ Sin } A$

$\text{Cos } A = \dfrac{b + x}{c} \longrightarrow x = c \text{ Cos } A - b$

$a^2 = h^2 + x^2$
$= c^2 \text{ Sin}^2 A + c^2 \text{ Cos}^2 A + b^2 - 2bc \text{ Cos } A$
$= b^2 + c^2 (\text{Sin}^2 A + \text{Cos}^2 A) - 2bc \text{ Cos } A$
$= b^2 + c^2 - 2bc \text{ Cos } A$

3. $B = 28.84°$, $C = 20.16°$, $a = 10.95$
5. $A = 77.2°$, $B = 25.6°$, $C = 77.2°$
7. $\vec{R} = 70$ kg $/231.8°$
9. $\alpha = 33.2°$

Section 8–5 Chapter Review

1. $\vec{A_H} = -1.25$, $\vec{A_V} = -2.17$
3. $\vec{H} = 10$ mph, $\vec{V} = -17.3$ mph
5. $14.1/-.45°$
7. 882 km/hr $/2.07°$ N of E
9. 13.14 km
11. 74.96 km
13. 99 km
15. 2582 m

588

Answers to Odd-Numbered Problems 589

17. 35.8 kg $\begin{cases} 29° \text{ with a } 16.8\text{-kg force} \\ 21.2° \text{ with a } 22.6\text{-kg force} \end{cases}$

19. $\vec{H} = 63.04$ kg, $\vec{V} = 49.3$ kg

21. 230 kg /125.1°; 230 kg /305.1°

23. 70.7 kg

Chapter 9

Section 9–1

1.

X	$-\pi$	$-\dfrac{3\pi}{4}$	$-\dfrac{\pi}{2}$	$-\dfrac{\pi}{4}$	0	$\dfrac{\pi}{4}$	$\dfrac{\pi}{2}$	$\dfrac{3\pi}{4}$	π
Sin X	0	−.707	−1	−.707	0	.707	1	.707	0
Cos X	−1	−.707	0	.707	1	.707	0	−.707	−1
Tan X	0	1	und	−1	0	1	und	−1	0

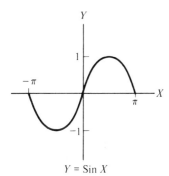

Y = Sin X

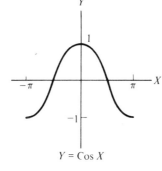

Y = Cos X

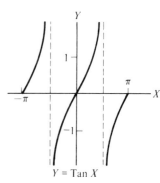

Y = Tan X

X	$-\pi$	$-\dfrac{5\pi}{6}$	$-\dfrac{2\pi}{3}$	$-\dfrac{\pi}{2}$	$-\dfrac{\pi}{3}$	$-\dfrac{\pi}{6}$	0	$\dfrac{\pi}{6}$	$\dfrac{\pi}{3}$	$\dfrac{\pi}{2}$
3. Sin X	0	−.5	−.866	−1	−.866	−.5	0	.5	.866	1
5. Tan X	0	.577	1.732	und	−1.732	−.577	0	.577	1.732	und
7. Sec X	−1	−1.155	−2	und	2	1.155	1	1.155	2	und

X	$\dfrac{2\pi}{3}$	$\dfrac{5\pi}{6}$	π	$\dfrac{7\pi}{6}$	$\dfrac{4\pi}{3}$	$\dfrac{3\pi}{2}$	$\dfrac{5\pi}{3}$	$\dfrac{11\pi}{6}$	2π
Sin X	.866	.5	0	−.5	−.866	−1	−.866	−.5	0
Tan X	−1.732	−.577	0	.577	1.732	und	−1.732	−.577	0
Sec X	−2	−1.155	−1	−1.155	−2	und	2	1.155	1

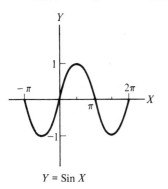

Y = Sin X

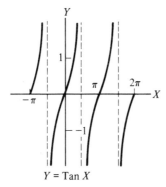

Y = Tan X

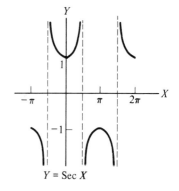

Y = Sec X

Section 9–2

1. Amp. = 2, Per. = $\frac{\pi}{4}$, Dis. = 0   **3.** Amp. = 6, Per. = 2π, Dis. = $\frac{\pi}{2}$ right

5. Amp. = 1, Per. = 1, Dis. = $\frac{1}{2}$ right   **7.** Amp. = 5, Per. = π, Dis. = 3 right

9. Amp. = 1, Per. = $\frac{\pi}{4}$, Dis. = $\frac{\pi}{8}$ left

11.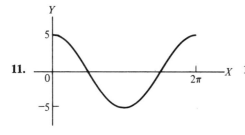

13.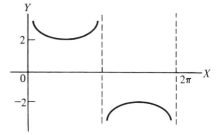

15.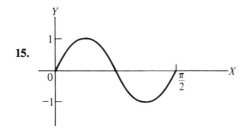

17.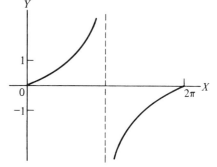

19.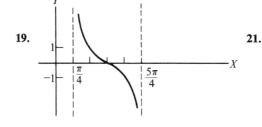

21.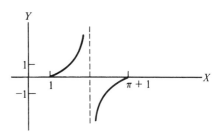

23.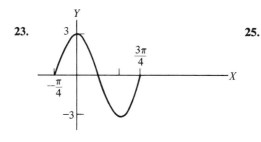

25.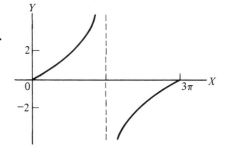

Answers to Odd-Numbered Problems 591

27.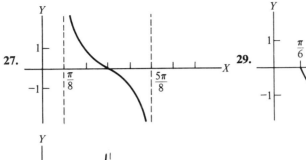

29.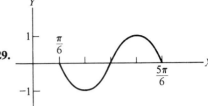

31.

33.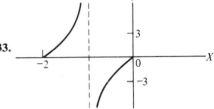

Section 9–3

1.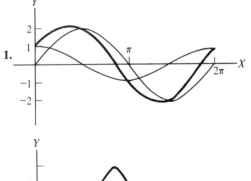

3.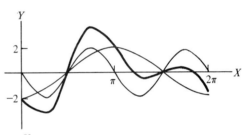

5.

7.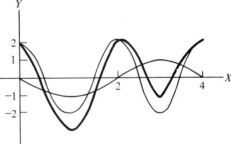

9.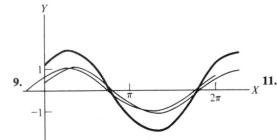

11.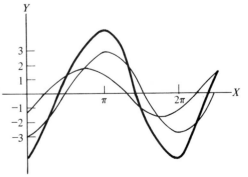

Answers to Odd-Numbered Problems

13.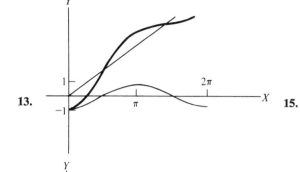

15.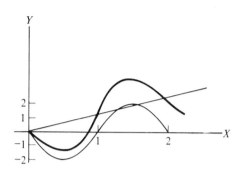

17.

19.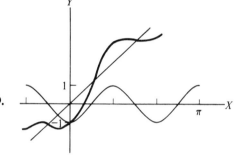

21.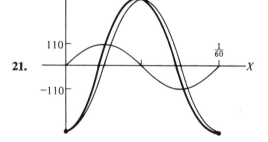

Section 9–4

1.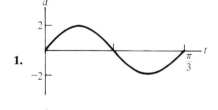

3.

5.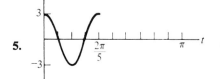

7.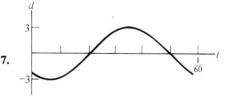

Answers to Odd-Numbered Problems

9.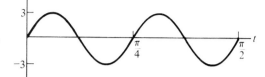

11.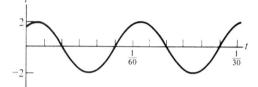

13.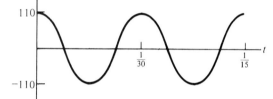

15.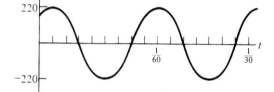

17.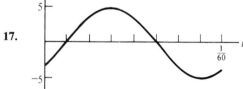

(a) 3.5 amps (b) 0 amps
(c) .004, .009, .012, .017 sec (d) .003, .010, .011, .018 sec

19.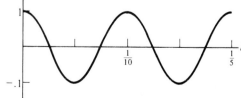

Section 9–5 Chapter Review

1. Amp. $= 2$, Per. $= \frac{2\pi}{3}$, Dis. $= 0$

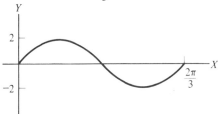

3. Amp. $= 3$, Per. $= \frac{\pi}{2}$, Dis. $= 0$

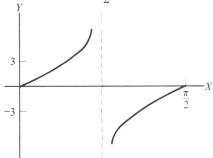

5. Amp. $= 3$, Per. $= 2\pi$, Dis. $= \pi$ right

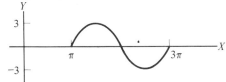

7. Amp. $= 2$, Per. $= 2\pi$, Dis. $= \frac{1}{2}$ left

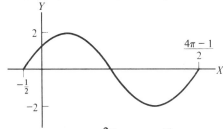

9. Amp. $= 1$, Per. $= \frac{2\pi}{3}$, Dis. $= \frac{\pi}{3}$ right

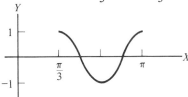

11. Amp. = 1, Per. = 2, Dis. = 1 right

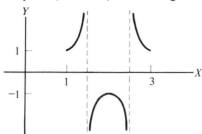

13. Amp. = 2, Per. = 2π, Dis. = $\frac{\pi}{2}$ right

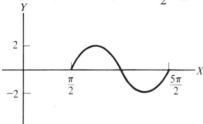

15. Amp. = 1, Per. = 2, Dis. = 1 left

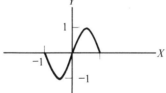

17. Amp. = 4, Per. = $\frac{\pi}{3}$, Dis. = $\frac{\pi}{12}$ left

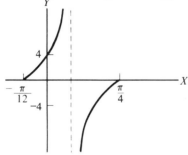

19. Amp. = 3, Per. = 4, Dis. = 1 right

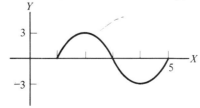

596 *Answers to Odd-Numbered Problems*

21.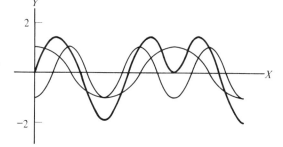

23.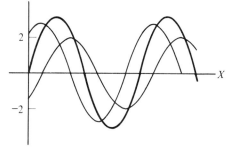

25.

27.

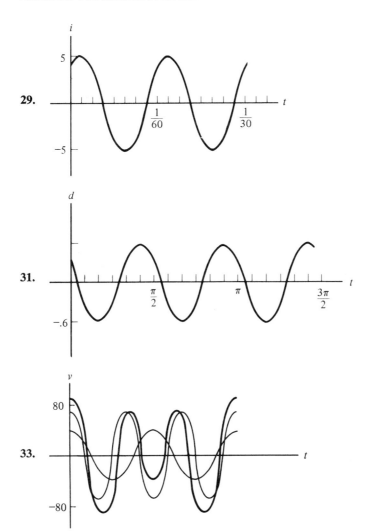

Chapter 10

Section 10–1

1. $\cos 52° = \dfrac{1}{\sec 52°} = .6167$ 3. $\sin^2 2 + \cos^2 2 = .8267 + .1733 = 1$

5. $\sin(27° + 79°) = \sin 106° = .9613$
   $\sin 27° \cos 79° + \cos 27° \sin 79° = (.4540)(.1908) + (.8910)(.9816)$
   $= .0866 + .8746 = .9613$

7. $\sin 70° = .9397$
   $2 \sin 35° \cos 35° = 2(.5736)(.8192)$
   $= .9397$

9. $\sin 105° = \sin (60° + 45°)$
$= \sin 60° \cos 45° + \cos 60° \sin 45°$
$= (.866)(.707) + (.5)(.707)$
$= .6124 + .3535 = .9659$

11. $\cos 75° = \cos (45° + 30°)$
$= \cos 45° \cos 30° - \sin 45° \sin 30°$
$= (.707)(.866) - (.707)(.5)$
$= .6124 - .3535 = .2589$

13. $\sin 15° = \sin (60° - 45°)$
$= \sin 60° \cos 45° - \cos 60° \sin 45°$
$= (.866)(.707) - (.5)(.707)$
$= .6124 - .3535 = .2589$

15. $\cos 0° = \cos (30° - 30°)$
$= \cos 30° \cos 30° + \sin 30° \sin 30°$
$= (.866)(.866) + (.5)(.5)$
$= .75 + .25 = 1$

17. $\sin 120° = \sin [2(60°)] = 2 \sin 60° \cos 60°$
$= 2(.866)(.5) = .866$

19. $\cos 270° = \cos [2(135°)]$
$= \cos^2 135° - \sin^2 135°$
$= (-.707)^2 - (.707)^2 = .5 - .5 = 0$

21. $\sin 45° = \sin \left[\frac{1}{2}(90°)\right] = \sqrt{\frac{1 - \cos 90°}{2}}$
$= \sqrt{\frac{1 - 0}{2}} = \sqrt{.5} = .707$

23. $\cos 15° = \cos \left[\frac{1}{2}(30°)\right] = \sqrt{\frac{1 + \cos 30°}{2}}$
$= \sqrt{\frac{1 + .866}{2}} = \sqrt{.933} = .9659$

25. $\sin (A + B) = -\frac{63}{65}$ or $-.9692$ 27. $\sin (2X) = \frac{24}{25}$ or $.96$

Section 10–2

1. $\cos X \cot X - \csc X = \frac{\cos^2 X}{\sin X} - \frac{1}{\sin X}$
$= \frac{\cos^2 X - 1}{\sin X}$
$= \frac{-\sin^2 X}{\sin X} = -\sin X$

3. $\csc^2 A (1 - \cos^2 A) = \csc^2 A \sin^2 A = 1$

5. $\cos (90° + T) = \cos 90° \cos T - \sin 90° \sin T$
$= 0 (\cos T) - 1 (\sin T)$
$= -\sin T$

7. $\sin (A + B) \sin (A - B)$
$= (\sin A \cos B + \cos A \sin B)(\sin A \cos B - \cos A \sin B)$
$= \sin^2 A \cos^2 B - \cos^2 A \sin^2 B$
$= \sin^2 A (1 - \sin^2 B) - (1 - \sin^2 A) \sin^2 B$
$= \sin^2 A - \sin^2 B$

Answers to Odd-Numbered Problems

9. $1 + \cos(2\theta) = 1 + 2\cos^2\theta - 1$
   $= 2\cos^2\theta$

11. $\cos^4 R - \sin^4 R = (\cos^2 R + \sin^2 R)(\cos^2 R - \sin^2 R)$
    $= 1(\cos^2 R - \sin^2 R) = \cos(2R)$

13. $2\sin\left(\dfrac{U}{2}\right)\cos\left(\dfrac{U}{2}\right) = \sin\left[2\left(\dfrac{U}{2}\right)\right]$
    $= \sin U$

15. $2\cos^2\left(\dfrac{V}{2}\right) - \cos V = 2\left[\dfrac{1+\cos V}{2}\right] - \cos V$
    $= 1 + \cos V - \cos V = 1$

17. $\frac{1}{2}[\cos(A-B) - \cos(A+B)]$
    $= \frac{1}{2}[(\cos A \cos B + \sin A \sin B) - (\cos A \cos B - \sin A \sin B)]$
    $= \frac{1}{2}(2\sin A \sin B)$
    $= \sin A \sin B$

19. $\dfrac{\sin 2\theta}{2\sin\theta} + \dfrac{\cos 2\theta}{2\cos\theta} + \dfrac{1}{2\cos\theta} = \dfrac{2\sin\theta\cos\theta}{2\sin\theta} + \dfrac{2\cos^2\theta - 1 + 1}{2\cos\theta}$
    $= \cos\theta + \cos\theta = 2\cos\theta$

21. $\tan(A+B) = \dfrac{\sin(A+B)}{\cos(A+B)}$
    $= \dfrac{\sin A \cos B + \cos A \sin B}{\cos A \cos B - \sin A \sin B}$
    $= \dfrac{\tan A + \tan B}{1 - \tan A \tan B}$

23. $\tan(2A) = \dfrac{\sin(2A)}{\cos(2A)} = \dfrac{2\sin A \cos A}{\cos^2 A - \sin^2 A}$
    $= \dfrac{2\tan A}{1 - \tan^2 A}$

25. $Y = X^2 + 1$

27. $\tan\left(\dfrac{\theta}{2}\right) = \dfrac{\sin\left(\dfrac{\theta}{2}\right)}{\cos\left(\dfrac{\theta}{2}\right)}$
    $= \dfrac{\sqrt{\dfrac{1-\cos\theta}{2}}}{\sqrt{\dfrac{1+\cos\theta}{2}}} \cdot \dfrac{2\sqrt{\dfrac{1+\cos\theta}{2}}}{2\sqrt{\dfrac{1+\cos\theta}{2}}}$
    $= \dfrac{2\sin\left(\dfrac{\theta}{2}\right)\cos\left(\dfrac{\theta}{2}\right)}{2\cos^2\left(\dfrac{\theta}{2}\right)}$

Section 10–3

1. $A = 120°, 240°$
3. $\theta = 150°, 330°$
5. $B = 30°, 90°, 150°, 270°$
7. $A = 120°, 150°, 300°, 330°$
9. $A = 360°, 900°$
11. No solution
13. $A = 0°, 210°, 330°$
15. $X = 45°, 135°, 225°, 315°$
17. $X = 90°, 270°$
19. $B = 0°, 45°, 135°, 180°, 225°, 315°$
21. $t = 0, \dfrac{\pi}{6}, \dfrac{\pi}{2}, \dfrac{5\pi}{6}, \pi, \dfrac{7\pi}{6}, \dfrac{3\pi}{2}, \dfrac{11\pi}{6}$

Section 10–4

1. $X = \frac{1}{2} \arctan Y$
3. $X = \arcsin\left(\frac{Y}{5}\right) - 1$
5. $X = \frac{1}{3} \sin Y$
7. $X = \frac{1}{2} \cos\left(\frac{Y+1}{3}\right)$
9. $Y = 0°$
11. $K = 60°$
13. $\theta = 60°$
15. $R = 90°$
17. $M = 14.4°$
19. $\theta = 210°, 330°$
21. $\theta = 45°, 225°$
23. No solution
25. $\theta = 10°, 170°$
27. $\theta = 240°$
29. $\theta = 21°, 69°$

Section 10–5

1. $\theta = 45°$
3. $\theta = 330°$
5. $\theta = 42°$
7. $\theta = 140°$
9. $\theta = 357°$
11. .9063
13. .3444
15. 1.3054
17. $\dfrac{1}{\sqrt{1-X^2}}$
19. $\dfrac{1}{\sqrt{K^2-1}}$

Section 10–6 Chapter Review

1. $\sin 90° = \sin(45° + 45°)$
   $= \sin 45° \cos 45° + \cos 45° \sin 45°$
   $= (.707)(.707) + (.707)(.707)$
   $= .5 + .5 = 1$
3. $\sin 60° = \sin(90° - 30°)$
   $= \sin 90° \cos 30° - \cos 90° \sin 30°$
   $= 1(.866) - 0(.5) = .866$
5. $\sin 60° = \sin[2(30°)] = 2 \sin 30° \cos 30°$
   $= 2(.5)(.866) = .866$
7. $\sin 15° = \sin\left[\frac{1}{2}(30°)\right] = \sqrt{\dfrac{1 - \cos 30°}{2}}$
   $= \sqrt{\dfrac{1 - .866}{2}} = \sqrt{.067} = .2588$
9. $\tan^2 B - \sec^2 B = \tan^2 B - (1 + \tan^2 B) = -1$
11. $\sin \theta \tan \theta + \cos \theta = \dfrac{\sin^2 \theta}{\cos \theta} + \dfrac{\cos \theta}{1}$
    $= \dfrac{\sin^2 \theta + \cos^2 \theta}{\cos \theta}$
    $= \dfrac{1}{\cos \theta} = \sec \theta$
13. $\cos(2T) + 1 - \cos^2 T = \cos^2 T - \sin^2 T + 1 - \cos^2 T$
    $= 1 - \sin^2 T = \cos^2 T$
15. $\dfrac{\cos X \tan X + \sin X}{\tan X} = \dfrac{\dfrac{\cos X \sin X}{\cos X} + \sin X}{\dfrac{\sin X}{\cos X}}$
    $= 2 \sin X \cdot \dfrac{\cos X}{\sin X}$
    $= 2 \cos X$

Answers to Odd-Numbered Problems

17. $\operatorname{Sin}\left(\dfrac{A}{2}\right)\operatorname{Cos}\left(\dfrac{A}{2}\right) = \sqrt{\dfrac{1 - \operatorname{Cos} A}{2}} \cdot \sqrt{\dfrac{1 + \operatorname{Cos} A}{2}}$

$= \sqrt{\dfrac{1 - \operatorname{Cos}^2 A}{4}}$

$= \sqrt{\dfrac{\operatorname{Sin}^2 A}{4}} = \dfrac{\operatorname{Sin} A}{2}$

19. $\operatorname{Tan}\left(\dfrac{U}{2}\right) = \dfrac{\operatorname{Sin}\left(\dfrac{U}{2}\right)}{\operatorname{Cos}\left(\dfrac{U}{2}\right)} = \dfrac{\sqrt{\dfrac{1 - \operatorname{Cos} U}{2}}}{\sqrt{\dfrac{1 + \operatorname{Cos} U}{2}}} \cdot \dfrac{\sqrt{\dfrac{1 + \operatorname{Cos} U}{2}}}{\sqrt{\dfrac{1 + \operatorname{Cos} U}{2}}}$

$= \dfrac{\sqrt{\dfrac{1 - \operatorname{Cos}^2 U}{4}}}{\dfrac{1 + \operatorname{Cos} U}{2}} = \dfrac{\sqrt{\dfrac{\operatorname{Sin}^2 U}{4}}}{\dfrac{1 + \operatorname{Cos} U}{2}}$

$= \dfrac{\operatorname{Sin} U}{2} \cdot \dfrac{2}{1 + \operatorname{Cos} U} = \dfrac{\operatorname{Sin} U}{1 + \operatorname{Cos} U}$

21. $\theta = 120°, 240°$ **23.** No solution   **25.** $T = 90°, 180°, 270°$
27. $U = 71.6°, 135°, 251.6°, 315°$ **29.** $X = 0°, 180°, 210°, 330°$
31. $180°$ **33.** $35°$ **35.** $.9488$ **37.** $T = \tfrac{1}{2} \arcsin A$
39. $T = \dfrac{1}{5} \arccos\left(\dfrac{U}{.6}\right)$ **41.** $T = \dfrac{1}{3} \operatorname{Sec}\left(\dfrac{Y}{2}\right)$
43. $T = \operatorname{Sin}(M - 3) - 2$ **45.** $2X\sqrt{1 - X^2}$
47. $\dfrac{A}{2} + \dfrac{A}{2}\operatorname{Cos}(2\theta) = \dfrac{A}{2} + \dfrac{A}{2}(2\operatorname{Cos}^2\theta - 1)$

$= \dfrac{A}{2} + A\operatorname{Cos}^2\theta - \dfrac{A}{2} = A\operatorname{Cos}^2\theta$

49. $\dfrac{2v^2}{g}\operatorname{Sin}\theta\operatorname{Cos}\theta = \dfrac{v^2}{g}(2\operatorname{Sin}\theta\operatorname{Cos}\theta)$

$= \dfrac{v^2}{g}\operatorname{Sin}(2\theta)$

51. $\tfrac{1}{2}[\operatorname{Cos}(A - B) - \operatorname{Cos}(A + B)]$
$= \tfrac{1}{2}[(\operatorname{Cos} A \operatorname{Cos} B + \operatorname{Sin} A \operatorname{Sin} B) - (\operatorname{Cos} A \operatorname{Cos} B - \operatorname{Sin} A \operatorname{Sin} B)]$
$= \tfrac{1}{2}(2 \operatorname{Sin} A \operatorname{Sin} B)$
$= \operatorname{Sin} A \operatorname{Sin} B$

53. $t = \dfrac{\arcsin\left(\dfrac{d}{R}\right) - C}{\omega}$ **55.** $A = \arccos\left(\dfrac{a^2 - b^2 - c^2}{-2bc}\right)$

Chapter 11

Section 11–1

1. $3j$ **3.** $-5j$ **5.** $\sqrt{7}j$ **7.** $-\sqrt{5}j$ **9.** $2\sqrt{3}j$
11. $.1j$ **13.** $-2\sqrt{6}j$ **15.** $-.2j$ **17.** j **19.** -1
21. -1 **23.** 1 **25.** 1 **27.** $\dfrac{1}{j}$ or j^{-1}

Section 11–2

1. $7 + 3j$ **3.** $-3 - 4j$ **5.** $32 - 2\sqrt{2}j$ **7.** $7 - j$

9. $-1 - 8j$ **11.** -17 **13.** $A = -4, B = 3$
15. $P = 2, Q = -2$ **17.** $R = -1, S = 2$

Section 11–3

1.

3.

5.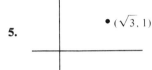

7.

9.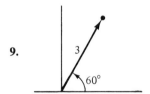

11.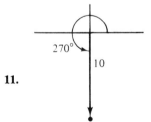

13. $5/126.9°$ **15.** $8.54/249.4°$ **17.** $2.65/49.1°$ **19.** $5.1/101.3°$
21. $9/180°$ **23.** $4/90°$ **25.** $2.3 + 1.9j$ **27.** $3.44 - 4.91j$
29. $-3.99 + .17j$ **31.** $\sqrt{2}$ **33.** $-12j$ **35.** $-2.93 - .62j$

Section 11–4

1. $1 + 10j$ **3.** $2 - 6j$ **5.** $8/60°$ **7.** $3/40°$
9. $11 + 13j$ **11.** $-12 - 24j$ **13.** $60/162°$ **15.** $90/234°$
17. $.5 + 2.5j$ **19.** $-4.5 + 1.5j$ **21.** $2/68°$ **23.** $3/240°$
25. $2 - 11j$ or $11.2/280.2°$ **27.** $-3 - 4j$ or $5/233.1°$
29. $625/80°$ **31.** $4096/138°$

33. 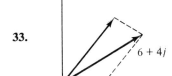   **35.**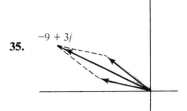

37. $1.4 \times 10^9 \,\underline{/120.5°}$ **39.** $.00177 \,\underline{/199°}$ **41.** $2 \,\underline{/0°}, 2 \,\underline{/90°}, 2 \,\underline{/180°}, 2 \,\underline{/270°}$
43. $\sqrt{2} \,\underline{/9°}, \sqrt{2} \,\underline{/81°}, \sqrt{2} \,\underline{/153°}, \sqrt{2} \,\underline{/225°}, \sqrt{2} \,\underline{/297°}$
45. $1.97 \,\underline{/52.3°}, 1.97 \,\underline{/172.3°}, 1.97 \,\underline{/292.3°}$

Section 11-5

1. $11.5 + 82.4j$ **3.** $105.5 + 113.2j$ **5.** $27.8 + 22.4j$
7. $69.7 + 35.1j$ **9.** $-32.1 + 160.7j$
11. 646.9 mph, $\underline{/36.7°}$ N of W

Section 11-6

1. 11.4 volts   **3.** 1.91 amps   **5.** $Z = 7.2$ ohms $\underline{/-33.7°}$
7. $V_R = 16$ volts, $Z = 9.43$ ohms
     $V_L = 28$ volts, $V_{RLC} = 18.86$ volts
     $V_C = 18$ volts,   Phase angle $= 32°$
9. $Z = 19.7$ ohms $\underline{/-23.97°}$ **11.** $Z = 7$ ohms $\underline{/90°}$
13. $Z = 9.2$ ohms $\underline{/40.6°}$

Section 11-7 Chapter Review

1. $5 - 6j$ **3.** $\sqrt{2} - 2\sqrt{3}\,j$ **5.** $X = 1, Y = -1$
7. $6.4 \,\underline{/321.3°}$ **9.** $10.8 \,\underline{/56.3°}$ **11.** $-3.83 - 3.21j$
13. $2.96 - 6.34j$

15. 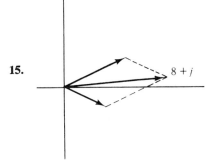   **17.**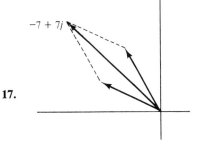

19. $10 - 2j$ **21.** $3j$ **23.** $9 \,\underline{/70°}$ **25.** $3 \,\underline{/300°}$
27. $27 + 6j$ **29.** $32 - 22j$ **31.** $-1.9 + .7j$ **33.** $15 \,\underline{/143°}$

35. $2/\underline{77°}$ **37.** $-9 - 46j$ or $49.9/\underline{258.9°}$ **39.** $81/\underline{288°}$
41. $94.8/\underline{177.2°}$ or $-94.7 + 4.6j$ **43.** $18.1/\underline{115°}$ or $-7.7 + 16.4j$
45. $V_R = 100$ volts, $Z = 23.3$ ohms
$V_L = 90$ volts, $V_{RLC} = 116.5$ volts
$V_C = 150$ volts, Phase angle $= -31°$
47. $V_R = 100$ volts, $Z = 53.6$ ohms
$V_L = 72$ volts, $V_{RLC} = 107.2$ volts
$V_C = 111.2$ volts, Phase angle $= -21.3°$

Chapter 12

Section 12–1

1. $3 = \log_5 125$ **3.** $\frac{2}{3} = \log_{125} 25$ **5.** $-3 = \log_4 \frac{1}{64}$
7. $-3 = \log_2 \frac{1}{8}$ **9.** $2 = \log_{1.1} .01$ **11.** $3^2 = 9$
13. $4^2 = 16$ **15.** $(\frac{1}{3})^{-2} = 9$ **17.** $125^{2/3} = 25$ **19.** $2^5 = 32$
21. $A = 2$ **23.** $b = 4$ **25.** $X = 1$ **27.** $Y = 1$
29. $A = -\frac{1}{2}$

Section 12–2

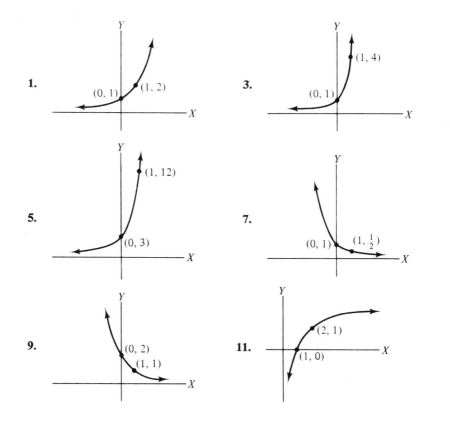

Answers to Odd-Numbered Problems   605

13.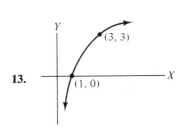

15.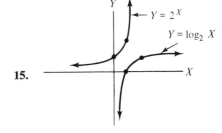

17. $N = 2^t$

Section 12-3

1. 3   3. 2   5. $2 \log_2 5$ 7. $\log_2 .28$ 9. 2
11. $Y = 10$ 13. $Y = \frac{1}{2}$ 15. $Y = 12.5$

Section 12-4

1. 4.6902   3. $7.2529 - 10$ or -2.7471
5. $8.7882 - 10$ or -1.2118 7. $4.9919 - 10$ or -5.0081
9. $7.5985 - 10$ or -2.4015 11. 399   13. .00693
15. .1278   17. 32.5   19. 9.52×10^{-8}

Section 12-5

1. 38893.8   3. .047   5. 7.3×10^6 7. 3.396
9. 41.6   11. .0726   13. 26.07   15. 6.5×10^7
17. .000674   19. 1.923×10^9 21. 3.29 in.

Section 12-6

1. 3.453   3. .336   5. -1.772 7. .157   9. 3.875
11. 9.242   13. 56.895   15. .444   17. 2992.9   19. .999

Section 12-7

1. $A = 1.25$ 3. $X = 14^{10}$ or 2.89×10^{11} 5. $X = 0$
7. $B = 1.87$ 9. $B = 1.47$ 11. $Y = 5.75$ 13. $X = 15.9$
15. $B = 96$ 17. $T = 2$ 19. $X = 3.25$ 21. $E = 2.03$ volts

Section 12-8

1.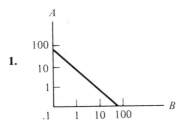

3.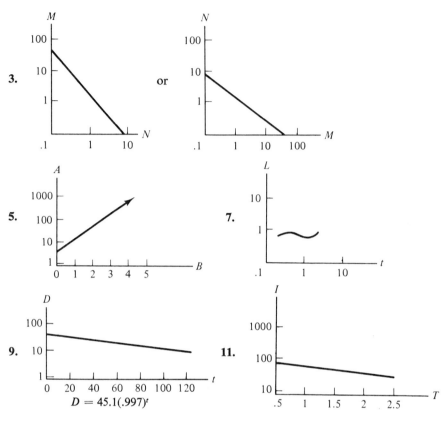

5.

7.

9. $D = 45.1(.997)^t$

11.

Section 12-9 Chapter Review

1. $A = 10$ 3. $b = 512$ 5. $b = .0000508$ 7. No solution

9. $B = 1$ 11.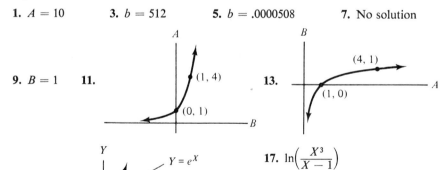

13.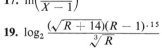

15.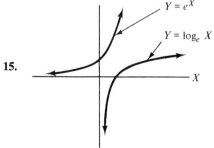

17. $\ln\left(\dfrac{X^3}{X-1}\right)$

19. $\log_2 \dfrac{(\sqrt{R+14})(R-1)^{.15}}{\sqrt[3]{R}}$

21. 1.42
23. $6.8488 - 10$ or -3.1512
25. 1.5913

Answers to Odd-Numbered Problems

27. −.3037	**29.** 20,000	**31.** .00568	**33.** 3.3
35. 1.982	**37.** 73,179.1	**39.** 12,312.5	**41.** .0256
43. 103,804.5	**45.** $X = 2.4$	**47.** $B = 10.3$	**49.** $X = .064$
51. $X = .317$	**53.** $H = 4.95$	**55.** $X = 2.154$	**57.** $X = 2$
59. $B = 13.18$	**61.** 20.12 hr	**63.** 4.81 km	

65. (a)

(b)

67. $P = 1.6L^{.485}$

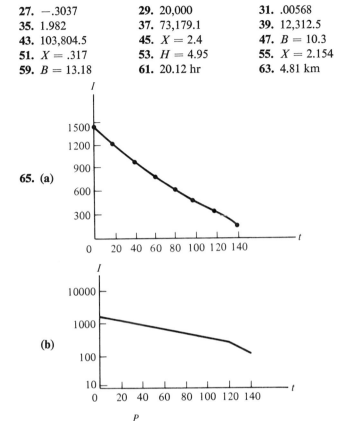

Chapter 13

Section 13–1

1. Yes	**3.** Yes	**5.** Yes	**7.** Yes	**9.** No, No, No
11. Yes, No, No		**13.** No, Yes, No	**15.** Yes	**17.** No
19. No				

Section 13-2

1. $X = 0, 7$
3. $T = \pm 5$
5. $A = -4, 2$
7. $M = -2, 6$
9. $R = -1, -1$
11. $T = -\frac{1}{2}, 3$
13. $Y = -\frac{2}{3}, 4$
15. $S = -\frac{3}{2}, -\frac{3}{2}$
17. $A = -\frac{5}{2}, \frac{4}{3}$
19. $J = -\frac{1}{3}, 4$
21. $I = 50$ amps
23. $-22, -8$

Section 13-3

1. $X = -18, 0$
3. $T = 4 \pm \sqrt{7} = 1.35, 6.65$
5. $B = 6 \pm \sqrt{37} = -.08, 12.08$
7. $V = \frac{4 \pm \sqrt{10}}{2} = .42, 3.58$
9. $Q = \frac{1 \pm \sqrt{43}j}{4}$
11. $S = \frac{2 \pm \sqrt{3}}{2} = .13, 1.87$
13. $R = 4.8$ cm
15. 18 in. by 16 in.

Section 13-4

1. $B = -1, 6$
3. $T = -\frac{1}{5}, 3$
5. $Y = \frac{7 \pm \sqrt{109}}{2} = -1.72, 8.72$
7. $V = \frac{-9 \pm \sqrt{77}}{2} = -8.89, -.12$
9. $L = \frac{-3 \pm \sqrt{35}j}{2}$
11. $P = -\frac{1}{3}, 1$
13. $X = \frac{1 \pm \sqrt{57}}{4} = -1.64, 2.14$
15. $K = 2 \pm \sqrt{2}j$
17. $B = \frac{5 \pm \sqrt{47}j}{4}$
19. $V = \frac{1 \pm \sqrt{193}}{8} = -1.61, 1.86$
21. 4 sec.
23. .29 lb/in., 1.71 lb/in.

Section 13-5

1. Distinct real roots; $X = 0, 3$
3. Distinct complex roots; $W = \pm 2j$
5. Equal real roots; $B = \frac{5}{2}, \frac{5}{2}$
7. Distinct real roots; $V = \frac{3 \pm 3\sqrt{5}}{2} = -1.85, 4.85$
9. Distinct complex roots; $R = \frac{1 \pm \sqrt{139}j}{10}$
11. Sum $= -7$; product $= 0$; $T = 0, -7$
13. Sum $= 0$; product $= -8$; $X = \pm 2\sqrt{2} = \pm 2.83$
15. Sum $= 7$; product $= \frac{49}{4}$; $J = \frac{7}{2}, \frac{7}{2}$
17. Sum $= 3$; product $= \frac{1}{4}$; $Y = \frac{3 \pm 2\sqrt{2}}{2} = .09, 2.91$
19. Sum $= 5$; product $= 7$; $L = \frac{5 \pm \sqrt{3}j}{2}$
21. No, roots are complex

Answers to Odd-Numbered Problems 609

Section 13–6

1.

3.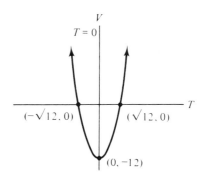

5.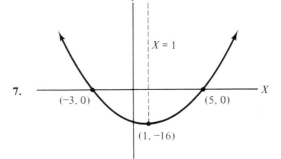

7.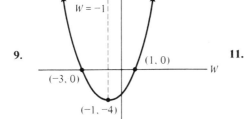

9.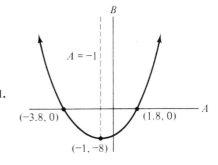

11.

13.

15.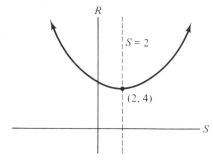

17.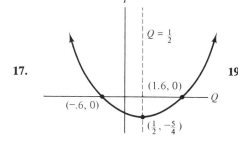

19.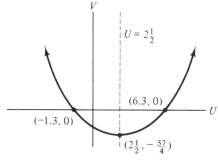

21. (a) $t = 1.3$ sec, 4.7 sec (b) $t = 2$ sec, 4 sec (c) $t = 3$ sec

Section 13-7

1. $T = 0°, 180°$ **3.** No solution **5.** $X = 60°, 109.5°, 250.5°, 300°$
7. $T = 18.4°, 146.3°, 198.4°, 327.3°$ **9.** No solution
11. $V = 11.3°, 78.2°, 191.3°, 258.2°$ **13.** $X = 13.4°, 140.2°, 193.4°, 320.2°$
15. $51.9°, 308.1°$ **17.** No solution **19.** $R = 65.5°, 294.5°$
21. $t = 1.57$ sec, 3.66 sec, 5.76 sec

Section 13-8

1. $T = \pm 2\sqrt{2}, \pm j$ **3.** $K = 81$ **5.** $X = -.45, 1, 3, 4.45$
7. $X = 2$ **9.** $P = 6, 8$ **11.** $R = 5, 21$ **13.** $X = 2.4$
15. $L = 5$ **17.** $S = 4$ **19.** 36 m by 6 m

Section 13-9 Chapter Review

1. $T = -9, 7$ **3.** $R = -\frac{2}{3}, \frac{1}{2}$ **5.** $M = \frac{3}{4}, 2$
7. $X = -4 \pm \sqrt{13} = -7.6, -.4$ **9.** $P = \dfrac{3 \pm \sqrt{15}j}{2}$
11. $N = \dfrac{7 \pm \sqrt{61}}{2} = -.4, 7.4$ **13.** $J = \dfrac{1 \pm \sqrt{11}j}{6}$
15. $X = \dfrac{-11 \pm 3\sqrt{17}}{4} = -5.85, .35$ **17.** $T = 30°, 150°, 210°, 330°$
19. $X = 228.8°, 311.2°$ **21.** $Y = 35.3°, 144.7°, 215.3°, 324.7°$
23. $X = \pm\sqrt{3}, \pm 2$ **25.** $B = 4$ **27.** $X = 12$ **29.** $X = 9$
31. $X = 3.3$ **33.** $A = 5$ **35.** 18 ohms, 12 ohms

Answers to Odd-Numbered Problems

37. $t = 1.25$ sec, 5 sec **39.** $I = 6$ amps
41. 7.88 cm by 7.88 cm by 3 cm

Chapter 14

Section 14–1

1. No   **3.** No   **5.** Yes   **7.** No   **9.** No, No

Section 14–2

1. $X = 0, Y = 2$
$X = -1, Y = 3$

3. $A = -.5, B = 1.9$
$A = .9, B = 1.8$

5. $S = 0, R = 0$
$S = 1, R = 1$
$S = -1, R = -1$

7. $N = -2.0, M = .1$

9. $S = -1, R = -4$
$S = 1.2, R = 3.3$

11. $N = 0, M = 1$
$N = -1.3, M = .3$

13. $X = \frac{\pi}{4} \pm 2K\pi, Y = .7$
$X = \frac{5\pi}{4} \pm 2K\pi, Y = -.7$

15. $R = 2$ **17.** 169.1 m, 7.1 m

Section 14–3

1. $X = \frac{1+\sqrt{7}}{2}, Y = \frac{1-\sqrt{7}}{2}$ or $X = 1.82, Y = -.82$
$X = \frac{1-\sqrt{7}}{2}, Y = \frac{1+\sqrt{7}}{2}$ or $X = -.82, Y = 1.82$

3. $T = \frac{1+\sqrt{41}}{4}, R = \frac{25+\sqrt{41}}{8}$ or $T = 1.85, R = 3.9$
$T = \frac{1-\sqrt{41}}{4}, R = \frac{25-\sqrt{41}}{8}$ or $T = -1.35, R = 2.3$

5. $S = 0, T = \pm\sqrt{2}$

7. $X = 0, Y = 0$
$X = 1, Y = 1$

9. $M = 0, N = 0$
$M = -\sqrt{6}, N = -2\sqrt{6}$
$M = \sqrt{6}, N = 2\sqrt{6}$

11. 28 ft by 20 ft

Section 14–4 Chapter Review

1. $X = 0, Y = -2$
$X = 1.8, Y = .8$
5. $M = \pm 3.1, N = 2.6$
9. $B = 1.3, A = .4$
13. $R = 2, S = 2$
$R = -2, S = -2$
17. $P = \sqrt{2}, Q = \sqrt{2}$
$P = -\sqrt{2}, Q = \sqrt{2}$
$P = -\sqrt{2}, Q = -\sqrt{2}$
$P = \sqrt{2}, Q = -\sqrt{2}$

3. $X = 0, Y = 0$
$X = 1, Y = 1$
7. No solution
11. $X = \pm 1.9, Y = 3.5$
15. $X = 0, Y = 0$
$X = \pm\sqrt{6}, Y = 12$
19. $A = -\frac{10}{3}, B = -\frac{3}{2}$
$A = 1, B = 5$

21. $5

Chapter 15

Section 15–1

1. -22
3. 0
5. 39
7. 32.8125
9. No
11. Yes
13. Yes
15. Yes
17. No
19. No
21. No
23. Yes

Section 15–2

1. $2X^4 + X^3 + 4X^2 + 5X - 1 \quad R(-10)$
3. $Y^3 - 7Y^2 + 20Y - 59 \quad R(77)$
5. $2A^3 + 6A^2 + 17A + 56 \quad R(161)$
7. $2Z^3 + 2Z - 2 \quad R(0)$
9. $K^4 - K^3 + K^2 - K + 1 \quad R(0)$
11. $B^2 - 25 \quad R(0)$
13. Yes
15. No
17. No
19. Yes

Section 15–3

1. $X = -5, -4, 7$
3. $Y = -\frac{1}{2}, -\frac{1}{2}, \pm 4$
5. $S = -2, 1, \dfrac{-1 \pm \sqrt{13}}{2} (-2.3, 2.6)$
7. $B = \pm 4, -3 \pm 2\sqrt{3} \, (-6.46, .46)$
9. $R = -1, -2 \pm j$
11. $Q = -1, -1, \pm 3, \pm 2j$

Section 15–4

1. $X = -3, 1, 1$
3. $S = -3, \frac{1}{2}, 5$
5. $P = \pm 2, \dfrac{-1 \pm \sqrt{7}\,j}{2}$
7. $K = \pm \frac{1}{2}, -\frac{2}{3}, -3$
9. $M = -1, \dfrac{1 \pm \sqrt{3}\,j}{2}$
11. $V = -\frac{1}{2}, \frac{1}{3}, \dfrac{-1 \pm \sqrt{13}}{2} (-2.3, 1.3)$
13. $S = 3$ cm or 6 cm

Section 15–5

1. $X = \pm 1.4, \pm 2.6$
3. $Y = -2.2, -.6, 0, 1.7, 2.2$
5. $M = -2.3, -.4, 1.4, 2.3$
7. $A = 1.53$
9. $Y = 1.38$
11. $S = -.78$
13. $X = 3, \pm 3.16$
15. $N = \pm 1.26$
17. $X = .67$ cm

Section 15–6 Chapter Review

1. -5
3. -19
5. No
7. Yes
9. $5Q^3 + 13Q^2 + 39Q + 125 \quad R(374)$
11. $8V^2 + 8V + 2 \quad R(0)$
13. $X = \pm 2, \pm 2j$
15. $B = -\frac{2}{3}, -\frac{1}{2}, \frac{1}{4}, 3$
17. $Y = 2, 2, 4$
19. $X = -1, -1, \frac{5}{2}$
21. $A = -2, \frac{1}{2}, 3$
23. $X = 3, 3$
25. $Y = -1.3, 2.3$
27. $X = 3.70$
29. $Z = .59$
31. 28 cm by 28 cm by 20 cm
33. 1.55 sec

Answers to Odd-Numbered Problems 613

Chapter 16

Section 16–1

1. $A < -1$
3. $X > 12$
5. $T > -\frac{25}{2}$
7. $A =$ all real numbers
9. $N > 2$
11. All numbers greater than (-1)

Section 16–2

1. $A \geq 12$
3. $N \geq 6$
5. $A \geq \frac{1}{3}$
7. $1 < T < 5$
9. $-12 < T \leq -6$
11. $X =$ all real numbers
13. $-3 \leq A < \frac{13}{2}$
15. $0 \leq B \leq 6\frac{2}{3}$
17. $14° \leq F \leq 59°$

Section 16–3

1. $-2 \leq X \leq 3$
3. $0 < B < 5$
5. $-3 < N < 7$
7. $-\frac{1}{3} < X < \frac{1}{2}$
9. $(1 - \sqrt{5}) \leq T \leq (1 + \sqrt{5})$
11. $X \leq -\frac{1}{2}$ or $0 \leq X \leq 4$
13. $T < -2$ or $0 < T < 2$
15. $Y < -5$ or $-4 \leq Y \leq 2$
17. $A \leq -2$ or $\frac{3}{2} \leq A < 2$
19. $A \leq -\sqrt{2}, A = 0, A \geq \sqrt{2}$
21. $S \neq 9$

Section 16–4

1.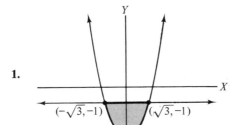

3.

5.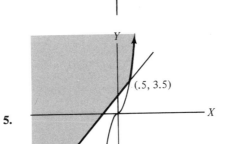

7.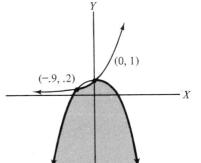

9.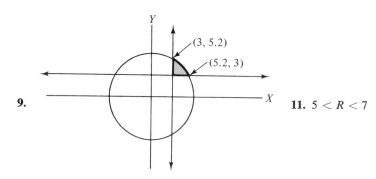

11. $5 < R < 7$

Section 16–5

1. $X < -1$ or $X > 5$
3. $-1 \leq T \leq 9$
5. $-\frac{7}{3} \leq X \leq 1$
7. $-\frac{5}{3} < A < -1$
9. $\frac{1}{4} \leq X \leq \frac{9}{4}$
11. $T =$ all real numbers
13. $|A - 3| > 4$
15. $|N - 5| < 3$
17. $0 < T < 1$ or $2 < T < 3$

Section 16–6 Chapter Review

1. $X \geq 2$
3. $A > 1$
5. $T \leq \frac{7}{2}$
7. $(-1 - \sqrt{5}) \leq A \leq -1 + \sqrt{5}$
9. $-1 \leq A \leq 3$
11. $X < -1$ or $X > 4$
13. $-2 < A < 2$
15. No solution
17. $X < 0$ or $X \geq \frac{1}{2}$
19. $A > 3$
21. $A \leq -3$ or $0 \leq A \leq 2$
23. $0 < R < 2$
25. $A < -1$ or $-\frac{1}{4} < A < 5$
27. $-\frac{9}{2} < A < \frac{5}{2}$
29. $-\frac{9}{2} \leq X \leq \frac{9}{2}$

Answers to Odd-Numbered Problems

31. $-8 \leq A \leq 0$

33. $-2 < A < 4$

35. $5 < B < 7$

37. $X = -\frac{1}{2}$

39. No solution

41. $A < 0$ or $A > 2$

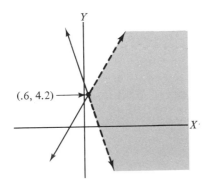

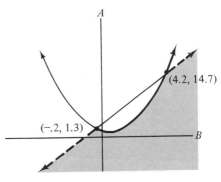

47. $|B - 3| < 4$ **49.** $|4T + 3| > 1$ **51.** $2 \text{ sec} < t < 5 \text{ sec}$
53. $S_1 \geq 65{,}000 \, l$
$S_2 \geq 105{,}000 \, l$

Chapter 17

Section 17–1
No problems

Section 17–2
1. $C = 3d$ **3.** $I = \frac{1}{20}E + 1$ **5.** $d = 88t$ **7.** $L = 2t^2$
9. $T = KV$ **11.** $V = KT$ **13.** 400.9 cm^3 **15.** 16.56 gm
17. 144 ft

Section 17–3
1. $V = \frac{4}{P}$ **3.** $Y = \frac{2}{X}$ or $X = \frac{2}{Y}$ **5.** $I = \frac{10^4}{D}$ or $D = \frac{10^4}{I}$

7. $F = \frac{K}{L}$ **9.** $P = \frac{K}{V}$ **11.** $P = \frac{600}{V}$
13. 61.2 candle power

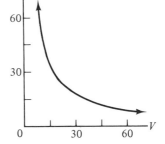

Section 17-4

1. $A = 32$
3. $E = 9$ joules/sec
5. 640 cal
7. No change in Y

Section 17-5 Chapter Review

1. $T = \dfrac{V^2}{100}$ or $V = 10\sqrt{T}$
3. $s = t^2$
5. $s = 1.5t^2$
7. $E = Kv^2$
9. $R = \dfrac{KL}{A}$
11. 400 m
13. 1250 dynes
15. 10 amps
17. 5 sec

Chapter 18

Section 18-1

1. Linear
3. Linear
5. Neither
7. Linear
9. Neither
11. Quadratic
13. Neither
15. Quadratic
17. Neither
19. Neither

Section 18-2

1. $d = \sqrt{61}$ or 7.8, $C = (4.5, 8)$
3. $d = 7$, $(C = (.5, -5)$
5. $d = 12\sqrt{2}$ or 16.97, $C = (-1, -3)$
7. $d = 3\sqrt{10}$ or 9.5, $C = (.5, .5)$
9. $d = \tfrac{1}{2}\sqrt{34}$ or 2.83, $C = (1.75, .25)$
11. $d = \tfrac{1}{3}\sqrt{445}$ or 7.03, $C = (\tfrac{1}{3}, -3.5)$
13. Yes
15. Yes

Section 18-3

1. $2X + Y - 8 = 0$
3. $7X + Y + 10 = 0$
5. $X + Y + 5 = 0$
7. $X - \tfrac{2}{3} = 0$ or $3X - 2 = 0$
9. $2X - Y + 1 = 0$

11. $2X + 5Y + 4 = 0$
13. $Y = 3X + 8$

15. $Y = \tfrac{1}{2}X + \tfrac{7}{4}$

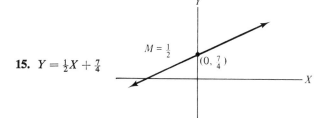

Answers to Odd-Numbered Problems

17. $Y = \frac{5}{4}X - 2$

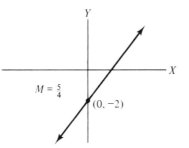

19. Parallel

21. Neither

23. $P = 2L + 40$

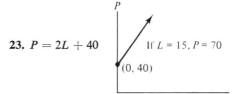

If $L = 15$, $P = 70$

25. $R = .01C + 3.5$

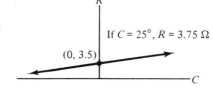

If $C = 25°$, $R = 3.75 \, \Omega$

Section 18–4

1. Vertex $= (0, 0)$
Focus $= (0, 2.5)$
Directrix: $Y = -2.5$

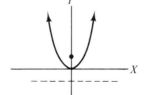

3. Vertex $= (0, 4)$
Focus $= (0, 6.5)$
Directrix: $V = 1.5$

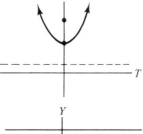

5. Vertex $= (4, -3)$
Focus $= (4, -4.5)$
Directrix: $Y = -1.5$

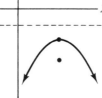

7. Vertex = $(-1, -3)$
   Focus = $(2.5, -3)$
   Directrix: $Q = -4.5$

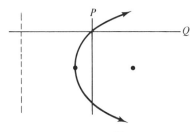

9. Vertex = $(-3, .5)$
   Focus = $(-3, 3.5)$
   Directrix: $Y = -2.5$

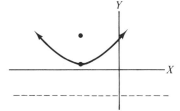

11. $X^2 - 12Y = 0$

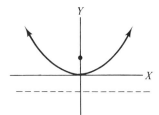

13. $B^2 - 4B + 16A + 4 = 0$

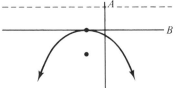

15. $Y^2 - 6Y - 12X - 3 = 0$

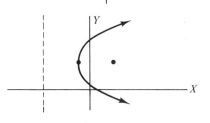

17. $4N^2 - 4N + 64M - 191 = 0$

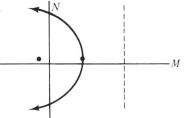

Answers to Odd-Numbered Problems   619

19. $P^2 + 8P - 6Q + 25 = 0$

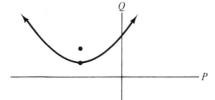

21. $Y - 6 = 0$

23. Vertex = $(0, 0)$
Focus = $\left(0, \dfrac{1}{4\pi}\right)$
Directrix: $A = -\dfrac{1}{4\pi}$

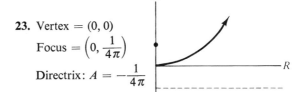

Section 18-5

1.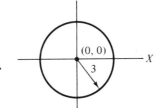

3.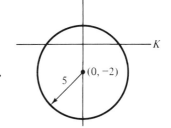

5.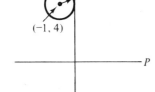

7.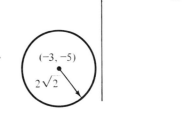

9.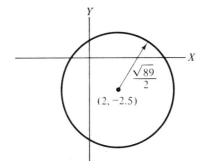

620   *Answers to Odd-Numbered Problems*

11. $X^2 + Y^2 + 8X + 12 = 0$

13. $M^2 + N^2 - 4M + 6N + 2 = 0$

15. $R^2 + S^2 + 2S - 19 = 0$

17. $X^2 + Y^2 + 2X - 6Y - 6 = 0$

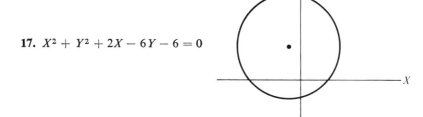

19. $X^2 + Y^2 - 2Y - 19 = 0$

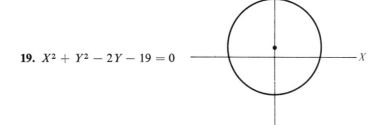

Answers to Odd-Numbered Problems

21.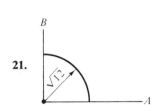

23.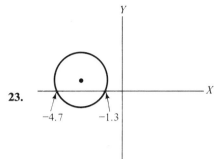

Section 18–6

1. Center = (0, 0)
Vertices = (±3, 0)
Foci = (±√5, 0)
Major Axis = 6
Minor Axis = 4

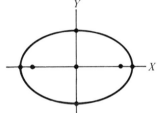

3. Center = (0, 1)
Vertices = (±√10, 1)
Foci = (±2, 1)
Major Axis = 2√10
Minor Axis = 2√6

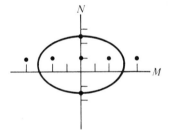

5. Center = (1, −1)
Vertices = (6, −1), and (−4, −1)
Foci = (4, −1), and (−2, −1)
Major Axis = 10
Minor Axis = 8

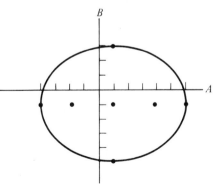

7. Center $= (-2, -4)$
   Vertices $= (-2 \pm \sqrt{7}, -4)$
   Foci $= (-2 \pm \sqrt{5}, -4)$
   Major Axis $= 2\sqrt{7}$
   Minor Axis $= 2\sqrt{2}$

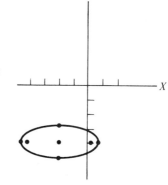

9. Center $= (-3, 5)$
   Vertices $= (-3, 5 \pm \sqrt{15})$
   Foci $= (-3, 5 \pm \sqrt{10})$
   Major Axis $= 2\sqrt{15}$
   Minor Axis $= 2\sqrt{5}$

11. $7X^2 + 16Y^2 - 448 = 0$

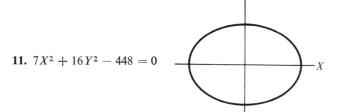

13. $16K^2 + 7L^2 - 32K - 28L - 68 = 0$

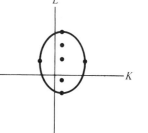

Answers to Odd-Numbered Problems   623

15. $25M^2 + 9N^2 + 100M - 72N - 19 = 0$

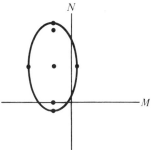

17.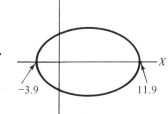

Section 18-7

1. Center $= (0, 0)$
Vertices $= (\pm 3, 0)$
Foci $= (\pm\sqrt{13}, 0)$
Transverse Axis $= 6$
Conjugate Axis $= 4$
Asymptotes: $3Y \pm 2X = 0$

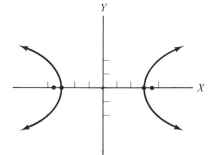

3. Center $= (0, 2)$
Vertices $= (0, 2 \pm \sqrt{2})$
Foci $= (0, 2 \pm \sqrt{7})$
Transverse Axis $= 2\sqrt{2}$
Conjugate Axis $= 2\sqrt{5}$
Asymptotes: $S \pm \dfrac{\sqrt{5}}{\sqrt{2}}R - 2 = 0$

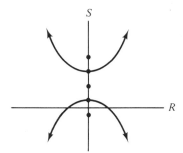

5. Center = (3, 1)
Vertices = $(3 \pm \sqrt{3}, 1)$
Foci = (6, 1) and (0, 1)
Transverse Axis = $2\sqrt{3}$
Conjugate Axis = $2\sqrt{6}$
Asymptotes: $\sqrt{3}\,B \pm \sqrt{6}\,A \mp 3\sqrt{6} - \sqrt{3} = 0$

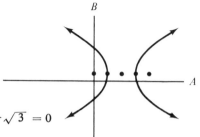

7. Center = (5, −1)
Vertices = (6, −1) and (4, −1)
Foci = $(5 \pm \sqrt{2}, -1)$
Transverse Axis = 2
Conjugate Axis = 2
Asymptotes: $Y - X + 6 = 0$ or $Y + X - 4 = 0$

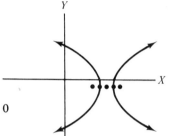

9. Center = (2, −4)
Vertices = $(2, -4 \pm 2\sqrt{2})$
Foci = $(2, -4 \pm \sqrt{11})$
Transverse Axis = $4\sqrt{2}$
Conjugate Axis = $2\sqrt{3}$
Asymptotes: $2\sqrt{2}\,V \pm \sqrt{3}\,T + 8\sqrt{2} \mp 2\sqrt{3} = 0$

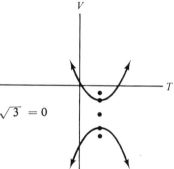

11. $16Y^2 - 9X^2 - 144 = 0$

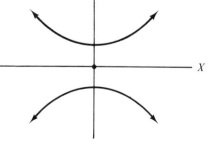

Answers to Odd-Numbered Problems

13. $3N^2 - M^2 + 6N + 4M - 13 = 0$

15. $P^2 - Q^2 + 4P + 4Q - 9 = 0$

17.

19.

21. $LW = 1600$

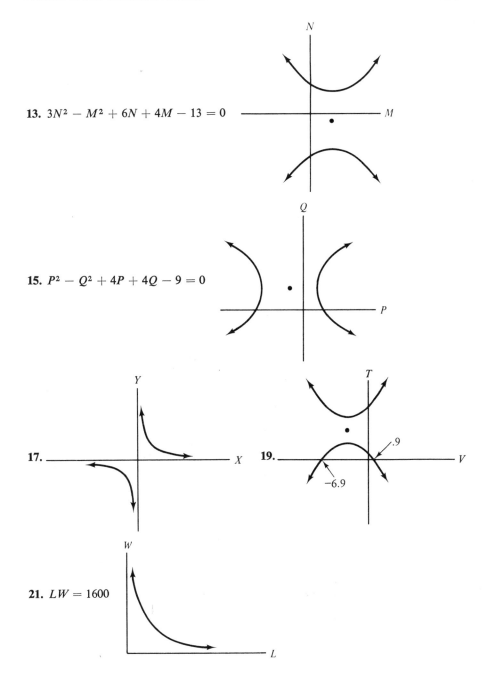

Section 18-8

1. Hyperbola 3. Straight line 5. Straight line
7. Straight line 9. Parabola 11. Parabola 13. Ellipse
15. None 17. Ellipse 19. Ellipse

Section 18-9 Chapter Review

1.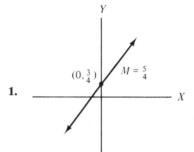

3. Vertex = (2, 4)
Focus = (2, 5.5)
Directrix: $S = 2.5$

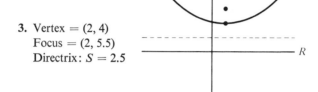

5.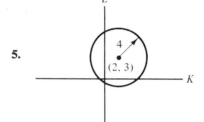

7. Center = (1, 2)
Vertices = $(1 \pm 2\sqrt{2}, 2)$
Foci = (3, 2) and (−1, 2)
Major Axis = $4\sqrt{2}$
Minor Axis = 4

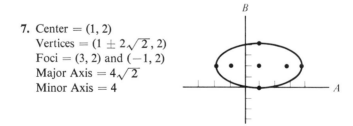

Answers to Odd-Numbered Problems

9. Center $= (0, -1)$
   Vertices $= (\pm\sqrt{2}, -1)$
   Foci $= (\pm\sqrt{5}, -1)$
   Transverse Axis $= 2\sqrt{2}$
   Conjugate Axis $= 2\sqrt{3}$
   Asymptotes: $N \pm \dfrac{\sqrt{3}}{\sqrt{2}} M + 1 = 0$

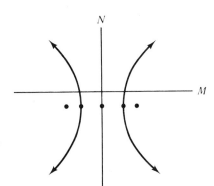

11. $2T - 3V - 11 = 0$

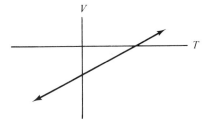

13. $Q^2 - 4Q - 8P - 28 = 0$

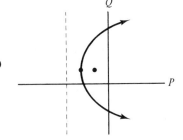

15. $R^2 + S^2 + 6R - 4S + 3 = 0$

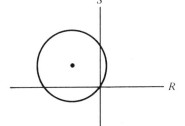

17. $16A^2 + 25B^2 - 100B - 300 = 0$

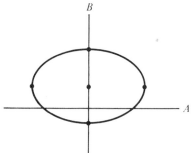

19. $5V^2 - 4T^2 - 10V - 8T - 19 = 0$

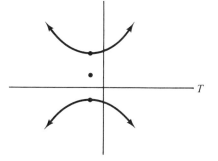

21.

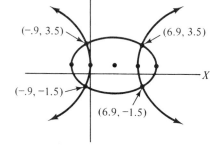

23. $C = 2\pi R$

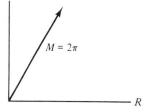

25.

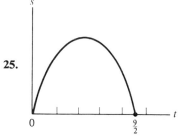

Projectile hits the ground after 4.5 sec.

27.

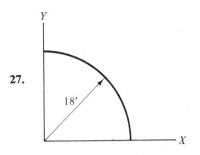

29. Pins should be 6 cm apart. String should be 10 cm long.

Appendix B

Section B–1

1. 2000 and 1 are both exact.
3. Approximate
5. Approximate
7. Approximate
9. 2.2 is approximate, 1 is exact.
11. 2-place precision, 4-digit accuracy
13. 1-place precision, 4-digit accuracy
15. 3-place precision, 6-digit accuracy
17. 41,800 and 42,000
19. 16.4 and 16
21. 2.01 and 2.0
23. 18.921 and 18.92
25. .001 and .00
27. 126.454 and 126.45
29. 102.25
31. 2440
33. 40,000
35. 22.4
37. 28.01
39. 2926
41. 278 m
43. 9 km/hr
45. 18

Section B–2

1. 244.3
3. 104.6
5. 94.2
7. 212.5
9. -16.39
11. -1.2
13. 2690
15. .023
17. 946,000
19. 8.99
21. .0469
23. 2.29
25. .063
27. 1.5
29. .0036
31. -125
33. 44.0
35. 4.38

Section B–3

1. 3.14×10^0
3. 9.46×10^{-3}
5. 4.87×10^7
7. 2.15
9. 3.018×10^{-8}
11. 6.595×10^{15}
13. 6.553×10^5
15. 1.34×10^{-1}
17. 2.923×10^6
19. 6.226×10^3

Section B–4

1. 625
3. 324
5. 161,051
7. 1722.25
9. 393,898.0639
11. 1,860,867
13. .001
15. .0000857338
17. .32694
19. .0005927
21. 3.8729
23. .87177
25. 80.8702
27. .194935
29. 2.8844
31. .742917
33. .89876
35. 38.77539
37. 2.4828
39. 41.266677

Section B–5

1. .5878
3. 1.3866
5. -1.0515
7. $-.9063$
9. $-.0872$
11. .5299
13. $-.1423$
15. .6058

17. .0708 **19.** −.5647 **21.** 66.5°, 293.5° **23.** 1°, 179°
25. 85°, 265° **27.** 155.5°, 335.5° **29.** 143°, 323°
31. 168.1°, 191.9° **33.** .314, 2.828 **35.** .218, 3.359
37. 2.09, 4.19

Section B–6

1. 1.7160 **3.** −.4202 **5.** −1.0410 **7.** .9058
9. −.5031 **11.** 2.1163 **13.** −2.8647 **15.** 9.0768
17. 6.5236 **19.** −4.8358 **21.** 255858.5887 **23.** 123.89
25. 1.18 **27.** .113 **29.** .00027416 **31.** 2.5 **33.** 4.4
35. 4.9998 **37.** 1.076 **39.** .1282

Section B–7

1. 82.6 **3.** 37.861 **5.** 196.108 **7.** 1.2552
9. .67295 **11.** .0052022 **13.** 77.2641 **15.** 45.6035
17. .0000777 **19.** 8.0181

Section B–8

1. 24.698 **3.** 1.1216 **5.** 73.35018 **7.** 155.4252
9. 28.60069 **11.** $X = 2, 3$ **13.** $T = \pm 3$ **15.** $C = -4, \frac{3}{2}$
17. $K = \frac{1}{2}, \frac{4}{3}$ **19.** $S = -1.886, 2.386$ **21.** $L = -6.45, -1.55$

Index

Abscissa, 86
Absolute value, 6, 52, 317, 459
Absolute inequality, 449, 459
Accuracy of a number, 532
Addition of ordinates, 270
Adjacent side, 206
Algebraic expressions, 26
Algebraic operations, 26
Alternating current, 330
Ambiguous case, 242
Amplitude, 260
Angles, 181, 525
  acute, 186, 525
  adjacent, 186, 526
  alternate interior, 187, 526
  central, 219
  complementary, 187, 526
  coterminal, 186
  in standard position, 194
  initial side of, 181
  negative, 181

Angles (cont.),
  obtuse, 186, 525
  of depression, 210
  of elevation, 209
  of inclination, 488
  positive, 181
  quadrantal, 195
  right, 186, 525
  straight, 186, 526
  supplementary, 187, 526
  terminal side of, 181
  vertex of, 181
Antilogarithm, 353
Approximate numbers, 531
Arc length, 219
Area:
  of a sector, 220
  of geometric figures, 525
Argument of a complex number, 317, 328
Associative law, 10

Asymptotes, 91, 510, 513
Averages, 112
Axis:
  conjugate, 510
  coordinate, 86
  major, 504
  minor, 504
  of symmetry, 399, 493
  transverse, 510

Base, 15
Binomial, 27
British system of measurement, 60

Canceling, 40
Capacitance, 331
Center of a circle, 499
Cgs system, 64
Characteristic of a logarithm, 352
Circle, 499, 527
Coefficient, 27
Cofunctions, 207
Common logarithms, 351
Commutative law, 10
Completing the square, 385
Complex numbers, 3, 313
Complex plane, 315
Complex roots for an equation, 434
Components of a vector, 229
Composite numbers, 1
Conditional equation, 45
Conditional inequality, 449
Cone, 529, 530
Conic sections, 480
Conjugate of a complex number, 313, 323, 434
Constant, 6
  of proportionality, 466
Conversions, 67, 71
Coordinates, 86
Counting numbers, 1
Cube, 528
Cycle, 255
Cylinder, 529

Decimal places, 532
Degree:
  as a measure of an angle, 182
  of a polynomial, 27
  of a term, 27
DeMoivre's Theorem, 325
Denominator of a fraction, 36
Dependent system of equations, 150
Dependent variable, 82, 86
Descarte's rule of signs, 438
Determinants, 160, 170
Difference formulas for angles, 285
Dimensions, 60, 61, 64, 69
Dimensional quantity, 69
Directrix, 493
Discriminant, 392
Displacement, 264
Distance formula, 481
Distributive law, 10
Domain, 83
Double angle formulas, 285

e (irrational number), 357
Electric circuits, 330
Elimination, 152, 164
Ellipse, 504
Empirical Data, 97
  graphs of, 97
Empirical equations, 99
Equality, 4
  properties of, 8
Equations:
  basic, 45
  conditional, 45
  empirical, 99
  equivalent, 45
  exponential, 339, 344, 360
  graphical solution of, 103, 441
  higher degree, 424
  linear, 130
  logarithmic, 339, 344, 360
  polynomial, 380, 424
  quadratic, 380, 480
  radical, 50

Equations (cont.):
  roots or solutions of, 45, 131, 392, 435, 441
  systems of, 144, 410
  trigonometric, 294, 401
Exact numbers, 531
Expansion by minors, 175
Exponent, 15
Exponential expression, 15
Exponential form of a complex number, 319
Exponential function, 338
Extraneous roots, 50, 404

Factor, 26
Factoring, 31
Factor theorem, 426
Farad, 331
Foci:
  of a hyperbola, 510
  of an ellipse, 504
Focus of a parabola, 493
Formulas, 51, 551
Fractions, 2, 36
Frequency curve, 125
Frequency distribution, 120
Functions, 78, 79, 84, 97
  exponential, 338, 340
  graphs of, 90
  logarithmic, 338, 342
  notation for, 84
Fundamental theorem of algebra, 431

General equation:
  of a circle, 502
  of a hyperbola, 516
  of an ellipse, 509
  of a parabola, 497
  of a straight line, 488
Geometric figures, 525–530
Geometric formulas, 525–530
Graphical solution of equations, 103, 146, 412, 441

Graphs:
  of conic sections, 493–518
  of empirical data, 97
  of exponential functions, 340
  of functions, 90
  of inequalities, 452–462
  of inverse trigonometric functions, 303
  of linear equations, 130–142
  of logarithmic functions, 340
  of quadratic equations, 396
  of relations, 90
  of straight lines, 130, 486–491
  of trigonometric functions, 254
  on logarithmic and semi-logarithmic paper, 363
Greatest common factor, 37
Grouped data, 120

Half angle formulas, 286
Henry, 331
Higher degree equations, 424
Histogram, 124
Hyperbola, 510
Hypotenuse, 205

Identity, 45
  trigonometric, 289
Identity element, 10
Imaginary numbers, 3, 311
Imaginary unit, 311
Impedance, 332
Inconsistent system of equations, 150
Independent variable, 82, 86
Inductance, 331
Inequalities, 4, 447
  graphs of, 5, 452–462
  properties of, 447
  solutions of, 449–463
  systems of, 457
Integers, 1, 2
Intercepts, 91, 137
Interpolation, 203, 353, 442

Intersecting lines, 146, 525
Interval, 120
Inverse element, 10
Inverse of a function, 300, 303, 338
Inverse of a relation, 299, 300
Inverse trigonometric functions, 303
Inverse trigonometric relations, 299
Irrational numbers, 2
Irrational roots of an equation, 441

j (as the imaginary unit), 311
j-operator, 322

Laws of signed numbers, 8
Linear equations, 130
Linear interpolation, 203, 353, 442
Literal symbols, 5
Logarithmic equations, 339, 344, 360
Logarithmic functions, 338, 342
   graphs of, 340
Logarithms, 338, 547
   common, 351
   natural, 357
   properties of, 347
Lowest common denominator, 38, 49

Magnitude of a vector, 226
Mantissa of a logarithm, 352
Mean, 112
Measurement:
   as a ratio, 42
   systems of, 60
Median, 112
Metric system of measurement, 60, 64
Midpoint:
   formula, 484
   of an interval, 121
   of a straight line segment, 484
Minute:
   as a measure of an angle, 182
   as a measure of time, 61, 64

Mks system, 64
Modal class, 121
Mode, 112
Modulus of a complex number, 317, 328
Monomial, 27
Multinomial, 27

Naperian logarithms, 358
Natural logarithms, 358
Negative exponent, 17
Negative integers, 2
Newton's law, 237
Normal distribution curve, 118, 127
Nth root of a number, 19
Number system, 1
Numerator of a fraction, 36

Ohm, 331
Opposite side, 206
Order of a radical, 19
Ordinate, 86
Origin, 2, 86

Parabola, 396
Parallel lines, 150, 525
Parallelogram, 528
Period of trigonometric functions, 262
Perpendicular lines, 525
Phase angle, 264, 331
Phasor, 331
Point-slope form of the equation of a straight line, 486
Polar form of a complex number, 315
Polynomial, 27
Polynomial equation, 380, 424
Positive integers, 1
Power, 15, 539
Precision of a number, 532
Prime factor, 31
Prime numbers, 1
Principal root of a number, 19

Principal value of inverse trigonometric functions, 303, 304
Projection, 273
Pythagorean theorem, 191, 527

Quadrants, 86
Quadratic equations, 380, 480
Quadratic form, 404
Quadratic formula, 389

Radian, 183
Radical equations, 50
Radicals, 15, 19
Radicand, 19
Radius of a circle, 499
Radius vector, 198
Range, 83, 121
Ratio, 42
Rationalizing a denominator, 323
Rational numbers, 2
Rational roots of an equation, 435
Reactance, 331
Real number line, 3
Real numbers, 2
Reciprocal, 199, 536
Rectangle, 526
Rectangular coordinate system, 86
Rectangular form of a complex number, 313
Rectangular solid, 529
Reduction of fractions, 37
Reductions, 61, 65, 72
Reference angle, 213
Relations, 78, 97
Remainder theorem, 426
Resistance, 331
Resultant vector, 229
Rhombus, 528
Right circular cylinder, 529
Roots:
    of a complex number, 326

Roots (cont.):
    of an equation, 392, 431–443
    of a number, 19, 539

Scalars, 227
Scientific notation, 23, 351, 537
Second:
    as a measure of an angle, 182
    as a measure of time, 61, 64
Semilogarithmic graph paper, 370
Significant digits, 532
Similar terms, 28
Simple harmonic motion, 273
Simultaneous linear equations, 144
Slope:
    formula, 486
    -intercept form of the equation of a straight line, 487
    negative, 488
    of a straight line, 135
    of parallel lines, 489 ·
    of perpendicular lines, 490
    positive, 488
Sphere, 530
Square, 526
Square root of numbers, 19, 539, 557
Squares of numbers, 539, 557
Standard deviation, 115
Standard form for a system of equations, 163, 170
Straight line, 134, 485
Sum formulas for angles, 285
Symmetry, 8
    axis of, 399, 493
Synthetic division, 427
Systems:
    of equations, 144, 410
    of inequalities, 457

Tabulation, 120
Term, 26

Trapezoid, 528
Triangles, 190, 526
  congruent, 191, 527
  equilateral, 190, 526
  isosceles, 190, 527
  oblique, 191, 240, 527
  right, 191, 205, 527
  scalene, 191, 527
  similar, 191, 527
Trigonometric equations, 294, 401
Trigonometric formulas, 283, 567
Trigonometric functions, 198, 544
  graphs of, 254
  of acute angles, 201
  of angles that are not acute, 212
Trigonometric identities, 284, 289
Trinomial, 33

Unit of dimension, 69

Variable, 6
  dependent, 82, 86

Variable (cont.),
  independent, 82, 86
Variation, 465
  direct, 466
  inverse, 470
  joint, 475
Vectors, 226
Velocity:
  angular, 220
  average, 220
  linear, 221
Vertex:
  of an angle, 181
  of a parabola, 493
Vertices:
  of a hyperbola, 510
  of an ellipse, 504
Voltage, 277, 331

Zero, 2
  as an exponent, 16
  operations with, 11
Zeros of a function or relation, 104